MW00838049

TRADES, QUOTES AND PRICES

The widespread availability of high-quality, high-frequency data has revolutionised the study of financial markets. By describing not only asset prices, but also market participants' actions and interactions, this wealth of information offers a new window into the inner workings of the financial ecosystem. In this original text, the authors discuss empirical facts of financial markets and introduce a wide range of models, from the micro-scale mechanics of individual order arrivals to the emergent, macro-scale issues of market stability. Throughout this journey, data is king. All discussions are firmly rooted in the empirical behaviour of real stocks, and all models are calibrated and evaluated using recent data from NASDAQ. By confronting theory with empirical facts, this book for practitioners, researchers and advanced students provides a fresh, new and often surprising perspective on topics as diverse as optimal trading, price impact, the fragile nature of liquidity, and even the reasons why people trade at all.

JEAN-PHILIPPE BOUCHAUD is a pioneer in Econophysics. He co-founded the company Science & Finance in 1994, which merged with Capital Fund Management (CFM) in 2000. In 2007 he was appointed as an adjunct Professor at École Polytechnique, where he teaches a course on complex systems. His work includes the physics of disordered and glassy systems, granular materials, the statistics of price formation, stock market fluctuations, and agent based models for financial markets and for macroeconomics. He was awarded the CNRS Silver Medal in 1995, the Risk Quant of the Year Award in 2017, and is the co-author, along with Marc Potters, of *Theory of Financial Risk and Derivative Pricing* (Cambridge University Press, 2009).

JULIUS BONART is a lecturer at University College London, where his research focuses on market microstructure and market design. Before this, he was a research fellow at CFM and Imperial College London, where he investigated price impact, high-frequency dynamics and the market microstructure in electronic financial markets. Julius obtained his PhD in Statistical Physics from Pierre et Marie Curie University (Paris).

JONATHAN DONIER completed a PhD at University Paris 6 with the support of the Capital Fund Management Research Foundation. He studied price formation in financial markets using tools from physics, economics and financial mathematics. After his PhD, he continued his research career in a music industry start-up that later joined Spotify, where he now serves as a Senior Research Scientist.

MARTIN GOULD currently works in the technology sector. Previously, he was a James S. McDonnell Postdoctoral Fellow in the CFM–Imperial Institute of Quantitative Finance at Imperial College, London. Martin holds a DPhil (PhD) in Mathematics from the University of Oxford, Part III of the Mathematical Tripos from the University of Cambridge, and a BSc (Hons) in Mathematics from the University of Warwick.

TRADES, QUOTES AND PRICES

Financial Markets Under the Microscope

JEAN-PHILIPPE BOUCHAUD

Capital Fund Management, Paris

JULIUS BONART

University College London

JONATHAN DONIER

Capital Fund Management, Paris
& University Paris 6

MARTIN GOULD

CFM–Imperial Institute of Quantitative Finance

CAMBRIDGE
UNIVERSITY PRESS

University Printing House, Cambridge CB2 8BS, United Kingdom

One Liberty Plaza, 20th Floor, New York, NY 10006, USA

477 Williamstown Road, Port Melbourne, VIC 3207, Australia

314-321, 3rd Floor, Plot 3, Splendor Forum, Jasola District Centre, New Delhi - 110025, India

79 Anson Road, #06-04/06, Singapore 079906

Cambridge University Press is part of the University of Cambridge.

It furthers the University's mission by disseminating knowledge in the pursuit of education, learning and research at the highest international levels of excellence.

www.cambridge.org
Information on this title: www.cambridge.org/9781107156050
DOI: 10.1017/9781316659335

First published 2018
3rd printing 2018

A catalogue record for this publication is available from the British Library

Library of Congress Cataloging in Publication data
Names: Bouchaud, Jean-Philippe, 1962– author. | Bonart, Julius, author. | Donier, Jonathan, author. | Gould, Martin, author.
Title: Trades, quotes and prices : financial markets under the microscope / Jean-Philippe Bouchaud, Capital Fund Management, Paris, Julius Bonart, University College London, Jonathan Donier, Spotify, Martin Gould, Spotify.
Description: New York : Cambridge University Press, 2018. | Includes bibliographical references and index.
Identifiers: LCCN 2017049401 | ISBN 9781107156050 (hardback)
Subjects: LCSH: Capital market. | Stocks. | Futures. | Investments.
BISAC: SCIENCE / Mathematical Physics.
Classification: LCC HG4523.B688 2018 | DDC 332.64–dc23
LC record available at https://lccn.loc.gov/2017049401

ISBN 978-1-107-15605-0 Hardback

"Leading physicist and hedge fund manager Jean-Philippe Bouchaud and his co-authors have written an impressive book that no serious student of market microstructure can afford to be without. Simultaneously quantitative and highly readable, *Trades, Quotes and Prices* presents a complete picture of the topic, from classical microstructure models to the latest research, informed by years of practical trading experience."
Jim Gatheral, Baruch College, *CUNY*

"This book describes the dynamics of supply and demand in modern financial markets. It is a beautiful story, full of striking empirical regularities and elegant mathematics, illustrating how the tools of statistical physics can be used to explain financial exchange. This is a tour de force with the square root law of market impact as its climax. It shows how institutions shape human behaviour, leading to a universal law for the relationship between fluctuations in supply and demand and their impact on prices. I highly recommend this to anyone who wants to see how physics has benefited economics, or for that matter, to anyone who wants to see a stellar example of a theory grounded in data."
Doyne Farmer, *University of Oxford*

"This is a masterful overview of the modern and rapidly developing field of market microstructure, from several of its creators. The emphasis is on simple models to explain real and important features of markets, rather than on sophisticated mathematics for its own sake. The style is narrative and illustrative, with extensive references to more detailed work. A unique feature of the book is its focus on high-frequency data to support the models presented. This book will be an essential resource for practitioners, academics, and regulators alike."
Robert Almgren, *New York University and Quantitative Brokers*

Contents

Preface

The Microstructure Age

In recent years, the availability of high-quality, high-frequency data has revolutionised the study of financial markets. By providing a detailed description not only of asset prices, but also of market participants' actions and interactions, this wealth of information offers a new window into the inner workings of the financial ecosystem. As access to such data has become increasingly widespread, study of this field – which is widely known as *market microstructure* – has blossomed, resulting in a growing research community and growing literature on the topic.

Accompanying these research efforts has been an explosion of interest from market practitioners, for whom understanding market microstructure offers many practical benefits, including managing the execution costs of trades, monitoring market impact and deriving optimal trading strategies. Similar questions are of vital importance for regulators, who seek to ensure that financial markets fulfil their core purpose of facilitating fair and orderly trade. The work of regulators has come under increasing scrutiny since the rapid uptake of high-frequency trading, which popular media outlets seem to fear and revere in ever-changing proportions. Only with a detailed knowledge of the intricate workings of financial markets can regulators tackle these challenges in a scientifically rigorous manner.

Compared to economics and mathematical finance, the study of market microstructure is still in its infancy. Indeed, during the early stages of this project, all four authors shared concerns that the field might be too young for us to be able to produce a self-contained manuscript on the topic. To assess the lay of the land, we decided to sketch out some ideas and draft a preliminary outline. In doing so, it quickly became apparent that our concerns were ill-founded, and that although far from complete, the story of market microstructure is already extremely compelling. We hope that the present book does justice both to the main developments in the field and to our strong belief that they represent a truly new era in the understanding of financial markets.

What is Market Microstructure?

Before we embark on our journey, we pause for a moment to ask the question: *What is market microstructure?* As the name suggests, market microstructure is certainly concerned with the details of trading at the micro-scale. However, this definition does little justice to the breadth of issues and themes that this field seeks to address. Most notably, the field also provides insight into the emergence of complex meso- and macro-scale phenomena that have been widely reported but poorly understood for decades. Therefore, we argue that defining market microstructure by limiting its scope to the atomic scale is not the correct approach.

One fundamental decision when embarking on a first-principles analysis of financial markets is whether to start by studying prices, and to dive down into the underlying mechanics, or to start by first analysing the micro-scale actions of individual traders, and to work from the bottom up. We define market microstructure as the field of choosing the latter path. In this way, we regard price formation not as a model input, but rather as a process that emerges from the complex interactions of many smaller parts.

Of course, the story of market microstructure is only just beginning. In the context of standard economic theory, many of the ideas that emerge from this bottom-up approach are currently deeply contentious. Throughout the book, we highlight many cases where the results that we present run contrary to the conventional wisdom about financial markets. We do this not in an attempt to attack other methodologies, but rather in the spirit of scientific discovery. We strongly believe that only by laying bare the hard facts for further debate will true understanding grow and mature.

Aim and Scope of the Book

This book lies somewhere between being theoretical and practical. In all the material that we present, we attempt to maintain a balance between being scientifically rigorous and avoiding overly heavy exposition. For the most part, we adopt a narrative style with informal discussions. This tone may come as something of a surprise to readers more familiar with traditional texts on financial markets, but we believe that a different approach is needed to present this material in an accessible and understandable way.

This is not an economics book, nor is it a mathematical finance book. Most of the material requires only a minimal level of mathematical background. In the few cases where we use more sophisticated techniques, we aim to provide the reader with sufficient details to perform all of the calculations that we present.

We neither propose nor prove theorems regarding how markets should behave. Instead, we focus first and foremost on the data, and place strong emphasis on how the various learnings from the different chapters and sub-narratives fit together to form a cohesive whole.

We divide the book into nine different parts. The early parts deal with more introductory material on the foundations of market microstructure, while the latter parts deal with more advanced material and practical applications. Each part consists of a collection of related chapters with a shared narrative and shared models. Each chapter starts with a brief introduction and concludes with a non-technical summary of the main points, which we call the "Take-Home Messages". We hope that this format will help our readers navigate both familiar and unfamiliar material quickly and easily.

Interspersed with the main text, we also provide details on some advanced topics, which give a more technical foundation for the general topics addressed in the chapter. To separate this more advanced material, we present it in a smaller font and with an indented layout. These advanced sections can be skipped without detracting from the overall narrative of the book.

We have designed this book to be self-contained, but given the huge number of publications on many of the topics that we discuss, we cannot possibly do justice to the full wealth of work in the field. To offer some guidance in navigating the wider literature, we provide many citations throughout the text. We do this in two ways: as footnotes in specific discussions, and as short bibliographies at the end of each chapter. In both cases, the references that we have chosen are the result of a subjective selection, and our choices are certainly far from exhaustive. We therefore strongly encourage readers to venture far and wide among the references contained within the works that we cite.

Data

High-quality data is the cornerstone of market microstructure. Throughout the book, the vast majority of data analysis that we perform is based on the LOBSTER database, which provides a very detailed, event-by-event description of all micro-scale market activity for each stock listed on NASDAQ. We discuss the LOBSTER data in Appendix A.1.

When doing applied science, we firmly believe that there is no substitute for hands-on practical experience. Therefore, we also provide our readers with the opportunity to try all of these calculations first-hand, by downloading a data set from LOBSTER. To do so, simply visit:

https://lobsterdata.com/tradesquotesandprices

The password for the files is “btumohul” (with no quotation marks). Full information about the data can be found on the website. We would like to thank both LOBSTER and NASDAQ for sharing this data with our readers. We hope that it serves to stimulate many future explorations in the field of market microstructure.

Acknowledgements

At the heart of the field of market microstructure lies a rapidly growing community working hard to investigate its many mysteries. As the field matures, so too does the community, which now has a specialised journal for sharing results, regular conferences and workshops, and several academic groups that serve as meeting grounds and melting pots for new ideas. Indeed, all four authors collaborated as part of one such group: the CFM–Imperial Institute at Imperial College, London. We are greatly thankful to the CFM–Imperial Institute, to CFM and to the CFM Research Foundation for unwavering support.

We are also indebted to many of our friends and colleagues, without whose help, input, debate and discussion we could not have written this book. To name but a few, we would like to thank F. Abergel, R. Almgren, E. Bacry, P. Baqué, R. Bénichou, M. Benzaquen, N. Bercot, P. Blanc, G. Bormetti, X. Brokmann, F. Bucci, F. Caccioli, D. Challet, R. Chicheportiche, S. Ciliberti, R. Cont, J. De Lataillade, G. Disdier, Z. Eisler, J. D. Farmer, T. Foucault, V. Filimonov, X. Gabaix, A. Garèche, J. Gatheral, A. Gerig, S. Gualdi, O. Guéant, T. Guhr, S. Hardiman, S. Howison, T. Jaisson, A. Kirman, J. Kockelkoren, A. S. Kyle, L. Laloux, A. Laumonnier, C. Lehalle, Y. Lempérière, F. Lillo, K. Mallick, I. Mastromatteo, A. Menkveld, M. Mézard, J. Muhle-Karbe, F. Patzelt, M. Porter, M. Potters, A. Rej, M. Rosenbaum, P. Seager, E. Sérié, S. Stoikov, N. N. Taleb, D. Taranto, A. Tilloy, B. Tóth, M. Vladkov, H. Waelbroeck, P. Wilmott, M. Wouts, M. Wyart, G. Zérah, Y. C. Zhang and G. Zumbach. We also express our thanks to the many other researchers who contribute to making this new community such a stimulating, exciting and welcoming place to live and work.

PART I

How and Why Do Prices Move?

Introduction

At the heart of all financial markets lies a common ingredient: people. These people act in their best interests, given their environment and the set of rules that govern it. As we know from many other aspects of life, human behaviour is full of complexity and surprises. Therefore, understanding financial markets promises to be a long and colourful journey.

Although the actions of a single person may be difficult to predict, the beauty of statistics is that large-scale systems populated by many different people often exhibit robust regularities that transcend individual behaviours. This property appears in countless manifestations in financial markets. An important first example is the way in which different agents (and their models) interpret "information" – whatever that might be. Predicting the direction of price changes has always proven to be difficult, but one might still aim to grasp some fundamental understanding of other market properties, such as volatility or liquidity, due to the strong regularities that emerge from the diverse actions and reactions of the ensemble.

Of course, there is no good modelling without a proper understanding of a system's rules – and of what motivates agents to play the game to begin with. In this first part of the book, we will address some fundamental questions about trading in financial markets. How do modern markets operate? Who is playing the game? What are their motivations? What challenges do they face? How do they act and interact? And how does all of this affect liquidity and volatility?

In a deep-dive into continuous-time double auctions, we will emphasise the ever-evolving nature of financial markets, and will therefore underline the ever-present need for up-to-date data. We will also touch on some universalities that have transcended time periods and geographical areas – from the subtle nature of the so-called "liquidity game" to the consistent patterns that emerge from it. How does market liquidity look at a zoomed-in and zoomed-out scale? Why is it so thin? And how do market participants' actions affect liquidity, and therefore prices?

Most importantly, we will begin our discussion about what *really* happens when people trade – that is, the *empirical facts* of financial markets. We will introduce the celebrated idea of the Brownian motion, but we will also highlight the limitations of this approach when it comes to modelling prices. For example, we will discuss how price returns exhibit fat tails and are prone to extreme jumps, and how activity is highly clustered in time, while the price is still linearly unpredictable at all scales.

How should we make sense of all these concepts, ideas and facts? Market microstructure is so rich that the scientific community has designed many different approaches and viewpoints, with economists and physicists typically residing at two opposite ends of the spectrum. As we will see, all such approaches have their limitations and puzzles. Developing a coherent understanding of financial markets is a long and complex endeavour – which we embark upon now.

1

The Ecology of Financial Markets

Buyer: How much is it?
Seller: £1.50.
Buyer: OK, I'll take it.
Seller: It's £1.60.
Buyer: What? You just said £1.50.
Seller: That was before I knew you wanted it.
Buyer: You cannot do that!
Seller: It's my stuff.
Buyer: But I need a hundred of those!
Seller: A hundred? It's £1.70 apiece.
Buyer: This is insane!
Seller: It's the law of supply and demand, buddy. You want it or not?

(Translated from "6", by Alexandre Laumonier)

A market is a place where buyers meet sellers to perform trades, and where prices adapt to supply and demand. This time-worn idea is certainly broadly correct, but reality is rather more intricate. At the heart of all markets lies a fundamental tension: buyers want to buy low and sellers want to sell high. Given these opposing objectives, how do market participants ever agree on a price at which to trade?

As the above dialogue illustrates, if a seller was allowed to increase the price whenever a buyer declared an interest to buy, then the price could reach a level so high that the buyer was no longer interested – and vice-versa. If traders always behaved in this way, then conducting even a single transaction would require a long and hard negotiation. Although this might be feasible if trades only occurred very infrequently, modern financial markets involve many thousands of transactions every single day. Therefore, finding a mechanism to conduct this process at scale, such that huge numbers of buyers and sellers can coordinate in real time, is an extremely complex problem.

Centuries of market activity have produced many possible solutions, each with their own benefits and drawbacks. Today, most markets implement an electronic, continuous-time double-auction mechanism based on the essential idea of a *limit order book (LOB)*, which we introduce in Chapter 3. However, as a brief glance at the financial press will confirm, ensuring market stability and "fair and orderly trading" is still elusive, and it remains unclear whether modern electronic markets are any less prone to serious problems than old-fashioned trading pits.

Given the tremendous impact of the digital revolution on society as a whole, why has the advent of computerised trading not solved these age-old problems once and for all? One possible answer is that trading intrinsically leads to instabilities. This viewpoint, which is increasingly supported by a growing body of empirical evidence, will lie at the very heart of our present journey into financial markets.

1.1 The Rules of Trading

We begin our discussion of financial markets by exploring the mechanisms that allow trading to take place, from old-style auctions to modern electronic markets.

1.1.1 The Walrasian Auction

As we emphasised above, markets are attempts to solve the seemingly impossible problem of allowing trades between buyers, who want to buy at an ever-lower price, and sellers, who want to sell at an ever-higher price. One possible way to solve this never-ending back-and-forth problem is to require that whenever a trader specifies a price, he or she makes a firm commitment to trade. This simple idea forms the heart of a classic market organisation called a **Walrasian auction**.[1]

In a Walrasian auction, traders communicate their buying or selling desires to an auctioneer, who collects and records this information. Buyers are invited to post bids that state the maximum price at which they are willing to buy, while sellers are invited to post offers that state the minimum price at which they are willing to sell. When posting bids or offers, each trader enters into a firm commitment to trade if he or she wins the auction. In the language of modern financial markets, this commitment is called **liquidity provision**.

A Walrasian auctioneer gathers these bids and offers into an order book. This order book describes the quantities that are available for purchase or sale at each specified price, as declared by the market participants. In a Walrasian auction, the auctioneer keeps the order book invisible, so that market participants cannot change their minds by observing what others are posting.

[1] The concept of the Walrasian auction first appeared as the design of French mathematical economist Léon Walras (1834–1910), as a gambit to understand how prices can reach their equilibrium such that supply matches demand.

At some instant of time, which can be set by the will of the auctioneer or decided at random, the auctioneer sets a transaction price p^* such that the total volume exchanged at that price is maximised. The transaction price p^* is the only price such that no buyers and no sellers remain unsatisfied after the transaction, in the sense that all remaining buyers have a limit price below p^* and all remaining sellers have a limit price above p^*. For a given price p, let $V_+(p)$ denote the total volume of buy orders at price p and let $V_-(p)$ denote the total volume of sell orders at price p. Formally, the **supply curve** $\mathscr{S}(p)$ is the total volume of sell orders with a price less than or equal to p, and the **demand curve** $\mathscr{D}(p)$ is the total volume of buy orders with a price greater than or equal to p. In a discrete setting, the supply and demand at a given price p can be written as

$$\mathscr{S}(p) = \sum_{p' \le p} V_-(p'); \tag{1.1}$$

$$\mathscr{D}(p) = \sum_{p' \ge p} V_+(p'). \tag{1.2}$$

In words, $\mathscr{S}(p)$ represents the total volume that would be available for purchase by a buyer willing to buy at a price no greater than p. Similarly, $\mathscr{D}(p)$ represents the total volume that would be available for sale by a seller willing to sell at a price no less than p.

For a given price p, the total volume of shares exchanged is given by

$$Q(p) = \min\left[\mathscr{D}(p), \mathscr{S}(p)\right], \tag{1.3}$$

because the volume cannot exceed either the volume for purchase or the volume for sale. As is clear from Equations (1.1) and (1.2), $\mathscr{S}(p)$ is an increasing function of p and $\mathscr{D}(p)$ is a decreasing function of p (see Figure 1.1). Intuitively, for a transaction to occur, the price must be a compromise between the buyers' and sellers' wishes. If the price is too high, buyers will be disinterested; if the price is too low, sellers will be disinterested. Therefore, the auctioneer must find a compromise price p^* such that

$$Q(p^*) = \max_p \min\left[\mathscr{D}(p), \mathscr{S}(p)\right]. \tag{1.4}$$

Note that $Q(p^*) > 0$ if and only if at least one pair of buy and sell orders overlap, in the sense that the highest price offered among all buyers exceeds the lowest price offered among all sellers. Otherwise, $Q(p^*) = 0$, so p^* is ill-defined and no transactions take place.

If $Q(p^*) > 0$, and in the theoretical case where $\mathscr{D}(p)$ and $\mathscr{S}(p)$ are continuous in p, the maximum of $Q(p)$ must occur when supply and demand are equal. The price p^* must therefore satisfy the equality

$$\mathscr{D}(p^*) = \mathscr{S}(p^*). \tag{1.5}$$

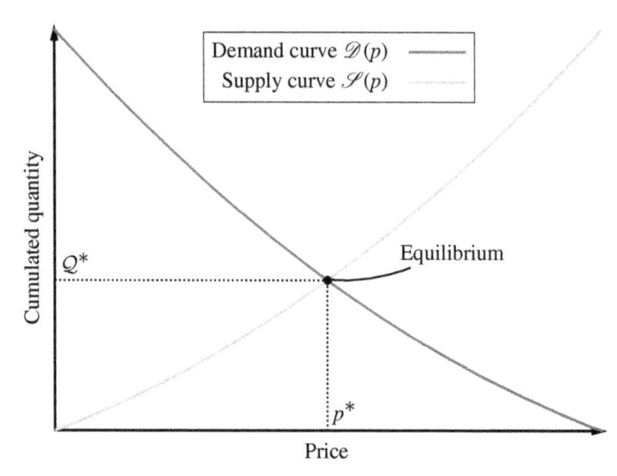

Figure 1.1. Illustration of the (increasing) supply curve $\mathscr{S}(p)$ and (decreasing) demand curve $\mathscr{D}(p)$ as a function of price p. According to Walras' law, the intersection of these curves defines the clearing price p^* and the total exchanged volume Q^*.

When the set of possible prices is restricted to a discrete price grid (as is the case in most markets), it is possible, and indeed common, that Equation (1.5) is not satisfied by the choice of p^* that maximises the total volume of trade. When this happens, some buyers or sellers remain unsatisfied. In this case, determining which of the possible buyers or sellers at a given price are able to trade requires a set of priority rules, which we discuss in detail in Section 3.2.1.

1.1.2 Market-Makers

In practice, Walrasian auctions are unsatisfactory for one major reason: they do not allow for any coordination between buyers and sellers. Therefore, a Walrasian auction might end in one of the following two scenarios:

(i) All buyers and sellers are unreasonably greedy, such that no buy and sell orders overlap and $Q(p^*) = 0$. In this case, there does not exist a price p^* at which any pair of traders is willing to trade, so no transactions can take place.
(ii) One side is unreasonably greedy while the other side is not. For example, the sellers might be unreasonably greedy while the buyers are not, which would result in a sudden increase in p^* and a very small transacted volume $Q(p^*)$. A similar situation can also arise if some external event temporarily causes the number of buyers to far exceed the number of sellers, or vice-versa.

Before the advent of electronic trading, the solution that most markets adopted to remedy this problem was to replace the Walrasian auctioneer (who seeks only to connect buyers and sellers at a reasonable price) with a special category of market participants called **market-makers** (or *specialists*). These special market participants were legally obliged to maintain a fair and orderly market, in exchange for some special privileges. To achieve their goals in the above auction setting, market-makers perform two tasks: quoting and clearing.

- **Quoting**: At all times, a market-maker must provide a *bid-price* b, with a corresponding *bid-volume* V_b, and an *ask-price* a, with a corresponding *ask-volume* V_a. As long as they are not modified by the market-maker, these quotes are binding, in the sense that the market-maker must execute any incoming sell order with volume less than or equal to V_b at the price b, and any incoming buy order with volume less than or equal to V_a at the price a.
- **Clearing**: Once buyers and sellers have submitted orders that specify a price and a volume, the market-maker decides on a price p^* that makes the number of unsatisfied orders as small as possible. Satisfied orders are cleared at price p^*.

A market-maker's quotes play the role of signalling a sensible price to the whole market, and thereby help other market participants to coordinate around a reasonable price. This point is illustrated by the retail foreign exchange (FX) market, in which retailers publicly display their buy and sell prices for a range of different currencies. At any time, if someone is uncertain about the value of the US dollar, then a simple way to gain a reasonably clear picture is to look at the buy and sell prices posted by a currency retailer. In this way, the retailer provides public information about their perception of the price of this asset.

Similarly to a Walrasian auction, buyers and sellers submit orders that specify a price and a volume and that accumulate in a public buy order book $V_+(p)$ and a public sell order book $V_-(p)$. Because market-makers post their quotes publicly, these new orders typically scatter around b (for buys) and a (for sells). At some time chosen by the market-maker, he or she computes the solution (or, in a discrete setting, the closest permissible discrete value) of the equation

$$\sum_{p' \geq p^*} V_+(p') = \sum_{p' \leq p^*} V_-(p'), \tag{1.6}$$

where $V_\pm$ also includes the market-maker's initial quotes at the bid and ask plus any additional buy or sell volume that he or she chooses to add just before the auction (e.g. to manage risk, to manage inventory, or to prevent large excursions of the price if demand temporarily outstrips supply, or vice-versa). If the market-maker is satisfied with p^* and the corresponding volume $Q(p^*)$, then the relevant transactions occur at this price. If $p^* > a$, this signals that the buy pressure was strong and that the market-maker chose not to add more sell volume to compensate. Note that $p^* > a$ can only happen if one or more buyers posted a limit price above the ask, and therefore have a strong urge to buy. In this case, it is likely that the market-maker will increase the quoted prices for the next auction, in an attempt to rebalance supply and demand. By symmetry, similar arguments follow for the case

$p^* < b$. If $b \le p^* \le a$, then the market is relatively balanced, in the sense that every transaction has occurred at a price better than or equal to the posted quotes.

In a market where all trade is facilitated by designated market-makers, participants can be partitioned into one of two categories: the market-makers, who offer trading opportunities to the rest of the market, and the other traders, who have the opportunity to accept them. Traditionally, these categories are named to reflect their members' contribution to the flow of liquidity: market-makers are called **liquidity providers**; the other traders are called **liquidity takers**.

1.1.3 Electronic Markets and Continuous-Time Double Auctions

Over time, markets have evolved away from appointing designated market-makers. Today, most liquid markets – including stocks, futures, and foreign exchange – are electronic, and adopt a **continuous-time double auction** mechanism using a *limit order book* (LOB), in which a transaction occurs whenever a buyer and a seller agree on a price.

LOBs require neither a Walrasian auctioneer nor designated market-makers to facilitate trade.[2] LOBs are updated in real time, and are observable by all traders. They therefore form an important part of the information sets used by traders when deciding how to act. As we discuss at many points throughout the book, the intertwined, interacting sequences of order-flow events generate transactions and price changes in an LOB. Analysing these sequences of events provides a direct route to quantifying and understanding price dynamics "from the bottom up", and therefore lies at the very heart of modern market microstructure.

1.2 The Ecology of Financial Markets

For a trade to take place, a buyer and a seller must agree on a price. Assuming that both parties do so willingly, it seems reasonable to insist that any trader conducting a trade must do so without feeling regrets. However, since assets are quoted every day (and nowadays are even quoted continuously during the day), at least one of the counterparties to a trade will have regrets as soon as the price moves, because he or she could have obtained a better price by waiting. So why do traders trade at all? Answering this puzzling question first requires a detailed understanding of the **ecology** of financial markets.

1.2.1 Trades and Information

Most attempts at explaining why market participants (other than market-makers) trade distinguish between two different types of activity:

[2] Interestingly, however, a category of market participants, who are still called market-makers, perform the useful function of providing some of the liquidity in an LOB. We return to this discussion in Section 1.2.4.

- **Informed trades** are attributed to sophisticated traders with information about the future price of an asset, which these traders buy or sell to eke out a profit.
- **Uninformed trades** are attributed either to unsophisticated traders with no access to (or the inability to correctly process) information, or to liquidity trades (e.g. trades triggered by a need for immediate cash, a need to reduce portfolio risk or a need to offload an inventory imbalance). These trades are often called *noise trades,* because from an outside perspective they seem to occur at random: they do not correlate with future price changes and they are not profitable on average.

In some cases, classifying a trade as informed is relatively straightforward. One clear example is **insider trading**. For example, consider a company with a stock price of \$95 at some time t. If an insider hears that a large corporation seeks to make a takeover bid by offering \$100 per share at some future time $t+T$, and that this information will not become public until time $t+T$, then the insider could buy some shares at the current value of \$95 and realise a near-certain profit of \$5 per share at time $t+T$. Deterministic arbitrage, such as exploiting the mispricing of derivative products, could also fall into the category of informed trading.

In most cases, however, this seemingly intuitive partitioning of trades as informed or uninformed suffers from a problem: information is difficult to measure – and even to define. For example, is an observation of another trade itself information? If so, how much? And how strongly might this impact subsequent market activity? For most large-cap US stocks, about 0.5% of the market capitalisation changes hands every day. Given that insider trading is prohibited by law, can a significant fraction of this vast market activity really be attributed to informed trades of the type described above?

1.2.2 Statistical Information

A less extreme (and perhaps more realistic) view of information can be framed in statistical terms: an informed trade can be defined as a trade whose ex-ante expected profit over some time horizon T is strictly positive, even after including all costs. In other words, information is tantamount to some ability to predict future price changes, *whatever the reasons for these changes.*

When speaking about statistical information, it is customary to decompose trading strategies into two categories:

- **Fundamental analysis** attempts to decide whether an asset is over-priced or under-priced. These strategies seek to use quantitative metrics, such as price-to-earnings ratios, dividend yields, macroeconomic indicators, information on the health and growth of a specific company, and even more qualitative indicators such as the charisma of the CEO.

- **Quantitative analysis** (or technical analysis) attempts to predict price movements by identifying price patterns, some of which are based on solid statistical evidence (such as price trends or mean-reversion), while others are less so (such as chartists' "head and shoulders" distributions, or price "support" and "resistance" levels).

Empirical data suggests that some of these signals are indeed correlated with future price changes, but that this correlation is very weak, in the sense that the dispersion of future price changes is much larger than the mean predicted price change. The time scale of these strategies is also extremely heterogeneous, and spans from months (or even years) for traditional long-only pension funds to just a few minutes (or even seconds) for some intra-day strategies.

In summary, we should expect that informed trades are either very rare and very successful (but unlawful!), like the insider example, or are more common but with weak information content and a low degree of individual success, like statistical arbitrage. Given that very successful trades occur rarely, if many traders really are informed, then the vast majority of such trades must be of the latter type. These trades can be based on information from any combination of a large number of diverse sources, each of which typically provides weak insights into future prices. In practice, this diversity of information signals is reflected in the diversity of market participants, who have different trading strategies, motives and time horizons – from long-term pension funds to day traders, hedgers, and even high-frequency trading (HFT).

1.2.3 To Trade, or Not To Trade?

Despite the prominence of arbitrage strategies, prices are notoriously hard to predict. In fact, as we will illustrate in Chapter 2, prices are close to being martingales, in the sense that the best estimate of the future price is simply the current price. This implies that the real information contained in the signals used by traders or investors is extremely weak. Given that trading entails considerable costs (such as brokerage fees, transaction fees and price impact, which we will discuss in detail throughout the book), the overall nagging feeling is that speculative traders trade too much, probably as a result of overestimating the predictive power of their signals and underestimating the costs of doing so.[3]

Why do traders behave in this way? One possible explanation is that they are blinded by the prospect of large gains and fail to recognise the true costs of their actions. Another is that it is extremely difficult to separate skill from luck in trading performance: even when trading with no information, a lucky trade can lead to a substantial profit. For example, if asset prices followed a simple symmetric random

[3] See, e.g., Odean, T. (1999). Do investors trade too much? *American Economic Review*, 89, 1279–1298.

walk, then in the absence of trading costs, any trade initiated at time t would have a 50% chance of being profitable at time $t+T$, whatever T!

Using numbers from real markets illustrates that evaluating real trading strategies is very difficult. Given a stock with an annual volatility of 15% and a reported annual return of 5%, it would take almost $T = 9$ years to test whether the actual return was statistically significantly different from zero at the one-sigma level.[4] Therefore, it can take a very long time to notice that a seemingly lucrative trading strategy is actually flawed, or vice-versa.

In summary, classifying trades is much less straightforward than the classical "informed-versus-uninformed" dichotomy might suggest. Although truly informed trades (such as insider trades) and truly uninformed trades (such as hedging or portfolio-balancing trades) likely both exist, between these two extremes lies a broad spectrum of other trades based on some sort of information that is difficult to define and even harder to measure. This lack of high-quality information also provides a possible explanation for why the ecology of modern financial markets is so complex, and contains many different types of traders seeking to earn profits on many different time horizons, despite the "no-trade" situation that we described earlier in this section. This important observation will be an overarching theme throughout the book.

1.2.4 Liquidity Providers: The Modern-Day Market-Makers

Most strategies and trading techniques attempt to earn a profit by forecasting the future price of an asset, then buying or selling it accordingly. But much as in the old days, financial markets can only function if some participants commit to providing liquidity. In the ecology of financial markets, liquidity providers offer to *both* purchase and sell an asset, and seek to earn the difference between the buy and sell price. Traders who implement this strategy in modern markets are still called "market-makers". In contrast to those in older markets, however, modern market-makers are not specifically designated market participants with special privileges. Instead, modern markets enable anyone to act as a market-maker by offering liquidity to other market participants.

Market-makers typically aim to keep their net inventory as close to zero as possible, so as not to bear the risk of the asset's price going up or down. Their goal is to earn a profit by buying low (at their bid-price b) and selling high (at their ask-price a), and therefore earning the **bid–ask spread**

$$s := a - b \tag{1.7}$$

for each round-trip trade.

[4] The order of magnitude of the time required is given by the square of the ratio of these two numbers: $(15/5)^2 = 9$ years.

An important consequence of the widespread use of LOBs in modern financial markets is a blurring of the lines between liquidity providers and liquidity takers. For example, if one market-maker noticed another market-maker offering to trade at a very attractive price, it would be illogical not to transact against this price, and to therefore act as a liquidity taker. Indeed, many successful high-frequency market-makers also implement sophisticated short-term prediction tools and exploit profitable high-frequency signals. Similarly, market participants who usually act as liquidity takers might instead choose to provide liquidity if they notice the bid–ask spread to be particularly wide.

Despite this emerging complexity of modern markets, the simple separation of market participants into two classes – speculators (or liquidity takers), who typically trade at medium-to-low frequencies, and market-makers, who typically trade at high frequencies – is a useful first step towards understanding the basic ecology of financial markets. As we will illustrate throughout the book, considering the interactions and tensions between these two groups provides useful insights into the origins of many interesting phenomena.

1.3 The Risks of Market-Making

Throughout this chapter, we have gradually built up a picture of the basic ecology of modern financial markets: market-makers offer opportunities to buy and/or sell, with the aim of profiting from round-trip trades, while speculators buy or sell assets, with the aim of profiting from subsequent price changes.

Based on this simple picture, it seems that market-makers have a much more favourable position than speculators. If speculators make incorrect predictions about future price moves, then they will experience losses, but they will still trade with market-makers, who will therefore still conduct round-trip trades. In this simplistic picture, speculators bear the risk of incorrect predictions, while market-makers seemingly always make a profit from the bid–ask spread. Is it really the case that market-makers can earn risk-free profits?

1.3.1 Adverse Selection

The simple answer to the last question is: no. Market-makers also experience several different types of risk. Perhaps the most important is **adverse selection** (also called the "winner's curse" effect). Adverse selection results from the fact that market-makers must post binding quotes, which can be "picked off" by more informed traders who see an opportunity to buy low or to sell high. This informational asymmetry is a fundamental concern for market-makers and is the final piece of the financial ecosystem that we consider in this chapter.

For a market-maker, the core question (to which we will return in Chapter 16) is how to choose the values of b and a. If the values of b and a remain constant at all times, then the market-maker always earns a profit of s for each round-trip (buy and sell) trade. All else being equal, the larger the value of s, the larger the profit a market-maker earns per trade. However, the larger the value of s, the less attractive a market-maker's buy and sell prices are to liquidity takers.

In situations where several different market-makers are competing, if one market-maker tries to charge too wide a spread, then another market-maker will simply undercut these prices by stepping in and offering better quotes. Most modern financial markets are indeed highly competitive, which prevents the spread from becoming too large. But why does this competition not simply drive s to zero?

Consider a market-maker trading stocks of a given company by offering a buy price of $b = \$54.50$ and a sell price of $a = \$55.50$. If the buy and sell order flow generated by liquidity takers was approximately balanced, then the market-maker would earn $s = \$1.00$ per round-trip trade. If, however, an insider knows that the given company is about to announce a drop in profits, they will revise their private valuations of the stock downwards, say to \$50.00. If the market-maker continues to offer the same quotes, then he or she would experience a huge influx of sell orders from insiders, who regard selling the stock at \$54.50 to be extremely attractive. The market-maker would therefore quickly accumulate a large net buy position by purchasing more and more stocks at the price \$54.50, which will likely be worth much less soon after, generating a huge loss. This is adverse selection.

Although the above example uses a strongly imbalanced order flow to provide a simple illustration, market-makers face the same problem even when facing more moderate imbalances in order flow. If a market-maker holds b and a constant in the face of an imbalanced order flow, then he or she will accumulate a large net position in a short time, with a high probability of being on the wrong side of the trade. As we will discuss at several times throughout the book, market-makers compensate for this potential loss by charging a non-zero spread – even in situations where market-making is fiercely competitive (see Chapter 17).

1.3.2 Price Impact

To mitigate the risk of being adversely selected, market-makers must update their values of b and a to respond to their observations of order-flow imbalance. If a market-maker receives many more buy orders than sell orders, then he or she can attempt to reduce this order-flow imbalance by increasing the ask-price a (to dissuade future buyers), increasing the bid-price b (to encourage future sellers), or both. Similarly, if a market-maker receives many more sell orders than buy orders, then he or she can attempt to reduce this order-flow imbalance by decreasing b, decreasing a, or both. An important consequence of this fact is that trades have **price impact**: on average, the arrival of a buy trade causes prices to rise and the arrival of a sell trade causes prices to fall. This is precisely what the spread s compensates for.

More formally, let

$$m := \frac{a+b}{2} \tag{1.8}$$

denote the **mid-price**, and let $\mathbb{E}[m_\infty]$ denote the expected future value of m. Since a (possibly small) fraction of trades are informed, then if we observe a buy trade, we should expect the future value of m to be greater than the current value of m. Similarly, if we observe a sell trade, then we should expect the future value of m to be less than the current value of m. To express this mathematically, let ε denote the **sign of the trade**, such that $\varepsilon = +1$ for a buy trade and $\varepsilon = -1$ for a sell trade. Let $\mathcal{R}_\infty$ denote the expected long-term impact of a trade,

$$\mathcal{R}_\infty := \mathbb{E}\left[\varepsilon \cdot (m_\infty - m)\right]. \tag{1.9}$$

Because on average the arrival of a buy trade causes m to increase and the arrival of a sell trade causes m to decrease, it follows from the definition of ε that $\mathcal{R}_\infty > 0$.

The concept of price impact is central to the understanding of financial markets, and will be discussed many times throughout this book. The above introduction is highly simplified, but it already provides a quantitative setting in which to consider the problem faced by market-makers. Recall that if b and a remain constant, then a market-maker earns $s/2$ for each of the two legs of a round-trip trade. Since on average the price moves in an adverse direction by an amount $\mathcal{R}_\infty > 0$, a market-maker's net average profit per trade (in this highly simplified framework) is given by $s/2 - \mathcal{R}_\infty$. We provide a much more detailed version of this argument in Chapter 17.

Market-making is thus only profitable if $s/2 > \mathcal{R}_\infty$, i.e. if the mean profit per trade is larger than the associated price impact. The larger the value of $\mathcal{R}_\infty$, the less profitable market-making becomes. The ideal situation for a market-maker is that $\mathcal{R}_\infty = 0$, which occurs when the sign of each trade is uncorrelated with future price moves – i.e. when trades have no price impact.

1.3.3 Skewness

Unfortunately for market-makers, the distribution of price changes after a trade is very broad, with a heavy tail in the direction of the trade. In other words, whereas most trades contain relatively little information and are therefore innocuous for market-makers, some rare trades are triggered by highly informed market participants, such as our insider trader from Section 1.2.1. These traders correctly

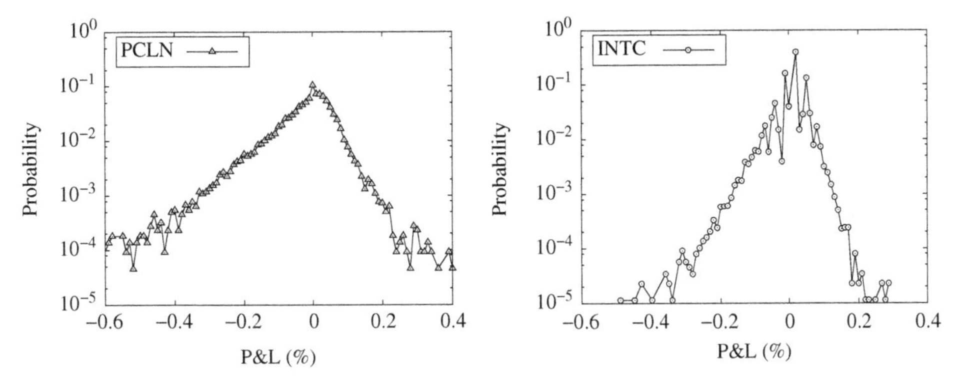

Figure 1.2. Empirical distributions of the P&L of a round-trip trade for (left panel) PCLN and (right panel) INTC. The negative skewness of the distributions is a specific feature of the risk borne by market-makers.

anticipate large price jumps before market-makers are able to update their quotes, and may therefore cause market-makers huge losses (see Figure 1.2).

Let us consider a simple example. Assume that with probability ϕ, an arriving trade is informed, and correctly predicts a price jump of size $\pm J$. Otherwise (with probability $1 - \phi$), the arriving trade is uninformed, and does not predict a future price change. Under the assumption that all trades are independent, it follows that $\mathcal{R}_\infty = \phi J$. To break even on average, a market-maker facing this arriving order flow would need to set $s = 2\phi J$.

The variance σ^2_{MM} of the market-maker's gain per trade is then given by

$$\sigma^2_{\mathrm{MM}} = (1-\phi)\times 0 + \phi \times J^2. \tag{1.10}$$

Similarly, its **skewness** ς_{MM} (which measures the asymmetry of the distribution of the gains) is[5]

$$\varsigma_{\mathrm{MM}} = \frac{(1-\phi)\times 0 - \phi \times J^3}{\sigma^3_{\mathrm{MM}}} = -\phi^{-\frac{1}{2}}. \tag{1.11}$$

Hence, as the probability ϕ that an arriving trade is informed decreases to zero, the relative precision on the average impact decreases as $\sqrt{\phi}$, but the skewness diverges (negatively) as $-1/\sqrt{\phi}$.

This simple calculation illustrates that market-making is akin to selling insurance. Although profitable on average, this strategy may generate enormous losses in the presence of informed trades.

This very simple argument illustrates why liquidity is fragile and can disappear quickly during times of market turbulence: the possible down-side risks of large losses are huge, so liquidity providers are quick to reduce the amount of liquidity that they offer for purchase or sale if they perceive this risk to be too high. Therefore, even in the presence of liquidity providers, organising markets to ensure fair and orderly trading remains a difficult task.

[5] The skewness of a random variable with zero mean is usually defined by the ratio of its third moment to its variance, raised to the power 3/2. Other skewness definitions, less sensitive to outliers, are often used for financial data.

1.4 The Liquidity Game

Understanding the delicate dance between buyers and sellers in financial markets is clearly difficult, but at least one thing is clear: whether informed, uninformed, or even misinformed, all traders want to get the best possible price for what they buy or sell.

An important empirical fact that is crucial to understanding how markets operate is that even "highly liquid" markets are in fact not that liquid after all. Take, for example, a US large-cap stock, such as PCLN (see the right panel of Figure 1.3). Trading for this stock is extremely frequent, with each day containing several thousands of trades that collectively add up to a daily traded volume of roughly 0.5% of the stock's total market capitalisation. At any given time, however, the volume of the stock available for purchase or sale at the best quotes is quite small, and is typically only of the order of about 10^{-4} of the stock's market cap. Liquidity is slightly more plentiful for large-tick stocks, such as CSCO (see the left panel of Figure 1.3), but it is still small compared to the daily traded volume. This phenomenon is also apparent in other markets, such as foreign exchange (FX) and futures, in which trading is even more frantic.

Why is there so little total volume offered for purchase or sale at any point in time? The reason is precisely what we discussed when summarising the difficulties of market-making: adverse selection. Liquidity providers bear the risk of being picked off by an informed trader. To minimise this risk, and perhaps even to bait informed traders and to out-guess their intentions, liquidity providers only offer relatively small quantities for trade. This creates a kind of hide-and-seek game in financial markets: buyers and sellers want to trade, but both avoid showing their hands and revealing their true intentions. As a result, markets operate in a regime of small **revealed liquidity** but large **latent liquidity**. This observation leads to many empirical regularities that we will discuss in this book.

The scarcity of available liquidity has an immediate and important consequence: *large trades must be fragmented.* More precisely, market participants who wish to buy or sell large volumes of a given asset must chop up their orders into smaller pieces, and execute them incrementally over time. For example, it is not uncommon for an investment fund to want to buy 1% (or more) of a company's total market capitalisation. Using the numbers from earlier in this chapter, buying 1% of PCLN would require of the order of 100 or even 1000 individual trades and would correspond to trading twice the typical volume for a whole day. To avoid strongly impacting the market and thus paying a higher price, such a transaction would have to be executed gradually over several days. Therefore, even an inside trader with clear information about the likely future price of an asset cannot use all of this information immediately, lest he or she scares the market and gives away the private information (see Chapter 15). Instead, traders must optimise a trading

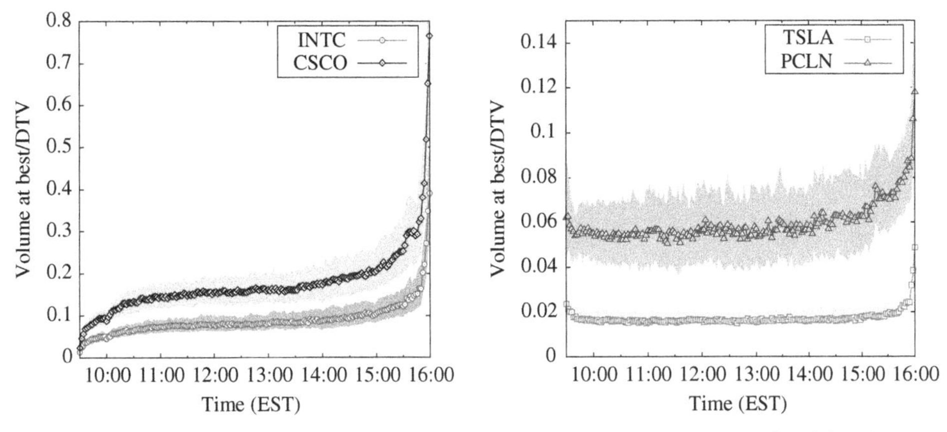

Figure 1.3. Total volume of active orders at the best quotes, normalised by the daily traded volume (during a full trading day), for (left panel) INTC and CSCO, two typical large-tick stocks, and (right panel) TSLA and PCLN, two typical small-tick stocks. The markers show the average over all trading days and the shaded regions show the corresponding lower and upper quartiles.

schedule that seeks to attain the best possible execution price – a topic that we will address in Chapter 21.

From a conceptual viewpoint, the most important conclusion from our discussions in this chapter is that *prices cannot be in equilibrium*, in the traditional sense that supply and demand are matched at some instant in time. Since transactions must be fragmented, the instantaneous traded volume is much smaller than the underlying "true" supply and demand waiting to be matched. Part of the imbalance is necessarily latent, and can only be *slowly* digested by markets. At best, the notion of equilibrium prices can only make sense on longer time scales, but on such time scales, the information set evolves further. In summary, the equilibrium price is an elusive concept – and an ever-moving target.

Take-Home Messages

(i) Organising markets to facilitate fair and orderly trade is a very difficult task.
(ii) Historically, market stability was ensured by designated market-makers, who quoted bid- and ask-prices and cleared matched orders (possibly including their own) in one large auction. Market-making was a risky but potentially lucrative business.
(iii) Most of today's markets implement a continuous-time double-auction mechanism, removing the need for a designated market-maker. In

practice, this mechanism is often implemented using a limit order book (LOB).

(iv) LOBs allow traders to either provide firm trading opportunities to the rest of the market (liquidity provision) or to accept such trading opportunities (liquidity taking).

(v) Today's liquidity providers play a similar role to the market-makers of yesterday. In modern markets, however, any market participant can choose to act as a liquidity provider or a liquidity taker, which makes these activities highly competitive.

(vi) Liquidity providers run the risk of being "picked-off" by better-informed traders who correctly anticipate future price moves. This is known as adverse selection.

(vii) Because of adverse selection, investors typically do not display their full intentions publicly. The revealed liquidity in an LOB is therefore only a small fraction of all trading intentions.

1.5 Further Reading

General

Lyons, R. (2001). *The microstructure approach to foreign exchange rate*. MIT Press.

Biais, B., Glosten, L., & Spatt, C. (2005). Market microstructure: A survey of microfoundations, empirical results, and policy implications. *Journal of Financial Markets*, 8(2), 217–264.

Hasbrouck, J. (2007). *Empirical market microstructure: The institutions, economics, and econometrics of securities trading*. Oxford University Press.

Amihud, Y., Mendelson, H., & Pedersen, L. H. (2012). *Market liquidity: Asset pricing, risk, and crises*. Cambridge University Press. https://www.gov.uk/government/collections/future-of-computer-trading.

Foucault, T., Pagano, M., & Röell, A. (2013). *Market liquidity: Theory, evidence, and policy*. Oxford University Press.

Lehalle, C. A., & Laruelle, S. (2013). *Market microstructure in practice*. World Scientific.

Easley, D., Prado, M. L. D., & O'Hara, M. (2014). *High-frequency trading: New realities for traders, markets and regulators*. Risk Books.

Laumonier, A. "*6*" and "*5*", Zones Sensibles Edt.

Organisation of Markets, Market-Making and Bid-Ask Spreads

Bagehot, W. (1971). The only game in town. *Financial Analysts Journal*, 27(2), 12–14.

Cohen, K. J., Conroy, R. M., & Maier, S. F. (1985). Order flow and the quality of the market. *Market-making and the changing structure of the securities industry* (pp. 93–110). Lexington Books.

Glosten, L. R., & Milgrom, P. R. (1985). Bid, ask and transaction prices in a specialist market with heterogeneously informed traders. *Journal of Financial Economics*, 14(1), 71–100.

Domowitz, I., & Wang, J. (1994). Auctions as algorithms: Computerized trade execution and price discovery. *Journal of Economic Dynamics and Control*, 18(1), 29–60.
Handa, P., & Schwartz, R. A. (1996). Limit order trading. *The Journal of Finance*, 51(5), 1835–1861.
Madhavan, A. (2000). Market microstructure: A survey. *Journal of Financial Markets*, 3(3), 205–258.
Jones, C. M. (2002). A century of stock market liquidity and trading costs. https://ssrn.com/abstract=313681.
See also Section 16.5.

The Ecology of Financial Markets

Handa, P., Schwartz, R. A., & Tiwari, A. (1998). The ecology of an order-driven market. *The Journal of Portfolio Management*, 24(2), 47–55.
Lux, T., & Marchesi, M. (1999). Scaling and criticality in a stochastic multi-agent model of a financial market. *Nature*, 397(6719), 498–500.
Farmer, J. D. (2002). Market force, ecology and evolution. *Industrial and Corporate Change*, 11(5), 895–953.
May, R. M., Levin, S. A., & Sugihara, G. (2008). Complex systems: Ecology for bankers. *Nature*, 451(7181), 893–895.
Bouchaud, J. P., Farmer, J. D., & Lillo, F. (2009). How markets slowly digest changes in supply and demand. In Hens, T. & Schenke-Hoppe, K. R. (Eds.), *Handbook of financial markets: Dynamics and evolution*. North-Holland, Elsevier.
Farmer, J. D., & Skouras, S. (2013). An ecological perspective on the future of computer trading. *Quantitative Finance*, 13(3), 325–346.
Jones, C. M. (2013). What do we know about high-frequency trading? https://ssrn.com/abstract=2236201.
Biais, B., & Foucault, T. (2014). HFT and market quality. *Bankers, Markets & Investors*, 128, 5–19.
Bouchaud, J. P., & Challet, D. (2017). Why have asset price properties changed so little in 200 years. In Abergel, F. et al. (Eds.), *Econophysics and sociophysics: Recent progress and future directions* (pp. 3–17). Springer.
https://en.wikipedia.org/wiki/Trading_strategy.

Noise Trading, Overconfidence and Poor Performance

Black, F. (1986). Noise. *The Journal of Finance*, 41(3), 528–543.
Shleifer, A., & Summers, L. H. (1990). The noise trader approach to finance. *The Journal of Economic Perspectives*, 4(2), 19–33.
Barber, B. M., & Odean, T. (1999). Do investors trade too much? *American Economic Review*, 89(5), 1279–1298.
Barber, B. M., & Odean, T. (2000). Trading is hazardous to your wealth: The common stock investment performance of individual investors. *The Journal of Finance*, 55(2), 773–806.
Barber, B. M., & Odean, T. (2001). Boys will be boys: Gender, overconfidence, and common stock investment. *The Quarterly Journal of Economics*, 116(1), 261–292.
The Case for Index-Fund Investing: https://personal.vanguard.com/pdf/ISGIDX.pdf.
See also Section 20.5.

2

The Statistics of Price Changes: An Informal Primer

If you are going to use probability to model a financial market, then you had better use the right kind of probability. Real markets are wild.

(Benoît B. Mandelbrot)

During the past 40 years, financial engineering has grown tremendously. Today, both the financial industry and its regulators rely heavily on a wide range of models to address many different phenomena on many different scales. These models serve as tools to inform trading decisions and assess risk in a diverse set of applications, including risk management, risk control, portfolio construction, derivative pricing, hedging, and even market design.

Among these models, the most widely used are those that seek to describe changes in an asset's price. Given their prominence, it is important to consider the extent to which these models really reflect empirically observed price series, because models whose assumptions are at odds with real markets are likely to produce poor output. Also, because so much of the modern financial world relies on such models so heavily, widespread application of unsuitable models can create unstable feedback loops and lead to the emergence of system-wide instabilities. For example, the severe market crash in 1987 is often attributed to the prevalence of models that assumed independent Gaussian price returns, and thereby severely underestimated the probability of large price changes. Bizarrely, financial crises can be induced by the very models designed to prevent them.

Market crashes serve as a wake-up call to reject idealistic simplifications and to move towards a more realistic framework that encompasses the real statistical properties of price changes observable in empirical data. Despite considerable recent effort in this direction, this goal remains elusive, due partly to the fact that many of the statistical properties of real price series are highly non-trivial and sometimes counter-intuitive. These statistical properties are called the **stylised facts** of financial price series.

The aim of this chapter is to provide an informal introduction to the most important of these stylised facts. In addition to being interesting in their own right, these stylised facts constitute a set of quantitative criteria against which to evaluate models' outputs. As we discuss throughout the chapter, a model's inability to reproduce one or more stylised facts can be used as an indicator for how it needs to be improved, or even as a reason to rule it out altogether.

Developing a unifying theory that illustrates how these non-trivial statistical properties emerge from some deeper mechanism of individual actions and interactions is a core goal that permeates the field of market microstructure. This general point of view – specifically, a "bottom-up" approach that seeks to explain the emergence of different phenomena on different scales – is natural among physicists. It is also consistent with the work of some economists, such as Richard Lyons, who writes that "microstructure implications may be long-lived" and "are relevant to macroeconomics".[1] If true, these statements are obviously relevant for regulators, who might consider altering the microstructural organisation of financial markets to improve their efficiency and stability (see Chapter 22).

2.1 The Random Walk Model

The first model of price changes dates back to a PhD thesis called *Théorie de la spéculation*, written in 1900 by Louis Bachelier. In his thesis, Bachelier proposed a theory relating Brownian motion to stock markets, five years before Einstein's celebrated paper on Brownian motion.

Bachelier's main arguments were as follows. Since each transaction in a financial market involves both a buyer and a seller, at each instant in time the number of people who think that the price is going to rise balances the number of people who think that the price is going to fall. Therefore, argues Bachelier, future price changes are *de facto* unpredictable. In more technical language, prices are martingales: the best estimate of tomorrow's price is today's price. A simple example, which Bachelier had in mind, is when price changes are independent identically distributed (IID) random variables with zero mean. Bachelier also noted that due to the Central Limit Theorem, aggregate price changes over a given time period (such as a full trading day) should be Gaussian random variables, because they consist of a very large number of small price changes resulting from individual transactions.

These observations led Bachelier to consider price series as **Gaussian random walks** (i.e. random walks whose increments are IID Gaussian random variables). Within this framework, Bachelier was able to derive a large number of interesting results, including statistics describing the first-passage time for a price to reach a

[1] Lyons, R. K. (2001). *The microstructure approach to exchange rates* (Vol. 12). MIT Press.

certain level, and even the price of options contracts – 70 years before Black and Scholes![2]

2.1.1 Bachelier's First Law

The simplest of Bachelier's results states that typical price variations $p_{t+\tau} - p_t$ grow like the square root of the lag τ. More formally, under the assumption that price changes have zero mean (which is a good approximation on short time scales), then the price **variogram**

$$\mathcal{V}(\tau) := \mathbb{E}[(p_{t+\tau} - p_t)^2] \tag{2.1}$$

grows linearly with time lag τ, such that $\mathcal{V}(\tau) = D\tau$.

Subsequent to Bachelier's thesis, many empirical studies noted that the typical size of a given stock's price change tends to be proportional to the stock's price itself. This suggests that price changes should be regarded as multiplicative rather than additive, which in turn suggests the use of *geometric Brownian motion* for price-series modelling. Indeed, geometric Brownian motion has been the standard model in the field of mathematical finance since the 1960s.

Over short time horizons, however, there is empirical evidence that price changes are closer to being additive than multiplicative. One important reason for this is that most markets enforce resolution parameters (such as the tick size – see Section 3.1.5) that dictate the minimum allowable price change. Therefore, at the short time scales that we will focus on in this book, it is usually more appropriate to consider additive models for price changes.

Still, given the prevalence of multiplicative models for price changes on longer time scales, it has become customary to define the **volatility** σ_r in relative terms (even for short time scales), according to the equation

$$D := \sigma_r^2 \bar{p}^2, \tag{2.2}$$

where $\bar{p}$ is either the current price or some medium-term average.

2.1.2 Correlated Returns

How can we extend Bachelier's first law to the case where price changes are not independent? Assume that a price series is described by

$$p_t = p_0 + \bar{p} \sum_{t'=0}^{t-1} r_{t'}, \tag{2.3}$$

[2] Technically, Bachelier missed the hedging strategy and its associated P&L. However, when the drift can be neglected, as assumed by Bachelier, his fair-pricing argument is indeed correct. See, e.g., Bouchaud, J. P., & Potters, M. (2003). *Theory of financial risk and derivative pricing: From statistical physics to risk management.* Cambridge University Press.

where the return series r_t is time-stationary with mean

$$\mathbb{E}[r_t] = 0 \tag{2.4}$$

and variance

$$\mathbb{E}[r_t^2] - \mathbb{E}[r_t]^2 = \sigma_r^2. \tag{2.5}$$

One way to quantify the dependence between two entries $r_{t'}$ and $r_{t''}$ in the return series is via their covariance

$$\text{Cov}\,(r_{t'}, r_{t''}) := \mathbb{E}[r_{t'}r_{t''}] - \mathbb{E}[r_{t'}]\mathbb{E}[r_{t''}] = \mathbb{E}[r_{t'}r_{t''}].$$

Given that the return series is time-stationary, this covariance depends only on the time lag $|t' - t''|$ between the two observations. Therefore, a common way to consider the dependence between $r_{t'}$ and $r_{t''}$ is via the **autocorrelation function** (ACF)

$$C_r(\tau) := \frac{\text{Cov}\,(r_t, r_{t+\tau})}{\sigma_r^2}. \tag{2.6}$$

The case of a random walk with uncorrelated price returns corresponds to $C_r(\tau) = \delta_{\tau,0}$, where $\delta_{\tau,0}$ is the Kronecker delta function

$$\delta_{i,j} = \begin{cases} 1, & \text{if } i = j, \\ 0, & \text{otherwise.} \end{cases} \tag{2.7}$$

For $\tau > 0$, a trending random walk has $C_r(\tau) > 0$ and a mean-reverting random walk has $C_r(\tau) < 0$. How does this affect Bachelier's first law?

One important consideration is that the volatility observed by sampling the price series on a given time scale is itself dependent on that time scale. More precisely, if we sample the p_t series once every τ seconds, then the volatility of our sampled series is given by

$$\sigma^2(\tau) := \frac{\mathcal{V}(\tau)}{\tau \bar{p}^2}, \tag{2.8}$$

where $\mathcal{V}(\tau)$ is the variogram given by Equation (2.1), and by definition $\sigma^2(1) := \sigma_r^2$.

By plugging Equation (2.3) into the definition of $\mathcal{V}(\tau)$ and expanding the square, we can derive a general formula for $\sigma^2(\tau)$ in terms of $C_r(\cdot)$:

$$\sigma^2(\tau) = \sigma_r^2 \left[1 + 2 \sum_{u=1}^{\tau} \left(1 - \frac{u}{\tau} \right) C_r(u) \right]. \tag{2.9}$$

A plot of $\sigma(\tau)$ versus τ is called a volatility **signature plot**. Figure 2.1 shows an example volatility signature plot for the simple case of a return series with

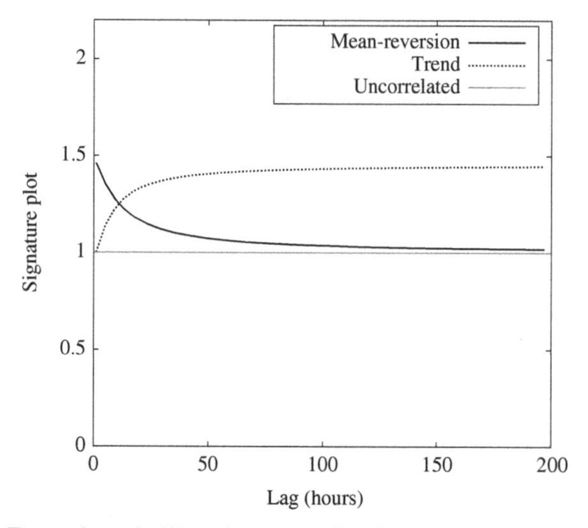

Figure 2.1. Example volatility signature plot for the simple case of a return series with an exponentially decaying ACF $C_r(u) = \rho^u$. An uncorrelated random walk (i.e. with $\rho = 0$) leads to a constant volatility. Positive correlations lead to an increasing $\sigma(\tau)$ whereas negative correlations lead to a decreasing $\sigma(\tau)$. The signature plot is flat only for time series that are neither trending nor mean-reverting on all scales.

an exponentially decaying ACF $C_r(u) = \rho^u$. As the plot illustrates, the case of an uncorrelated random walk (i.e. with $\rho = 0$) leads to a constant volatility. Positive correlations (which correspond to trends) lead to an increase in $\sigma(\tau)$ with increasing τ. Negative correlations (which correspond to mean reversion) lead to a decrease in $\sigma(\tau)$ with increasing τ. In fact, from Equation (2.9), one finds that in the case of an exponentially decaying ACF,

$$\lim_{\tau \to \infty} \frac{\sigma^2(\tau)}{\sigma^2(1)} = \frac{1+\rho}{1-\rho}. \tag{2.10}$$

2.1.3 High-Frequency Noise

Another interesting case occurs when the price p_t is soiled by some high-frequency noise, such as that coming from price discretisation effects, from pricing errors or data problems. In this section, we consider the case where rather than being given by Equation (2.3), p_t is instead assumed to be given by

$$p_t = p_0 + \bar{p} \sum_{t'=0}^{t-1} r_{t'} + \eta_t \tag{2.11}$$

where the noise η_t has mean zero, variance σ_η^2 and is uncorrelated with r_t, but is autocorrelated with an exponential ACF

$$C_\eta(\tau) := \frac{\mathrm{Cov}(\eta_t, \eta_{t+\tau})}{\sigma_\eta^2} = e^{-\tau/\tau_\eta}, \tag{2.12}$$

where τ_η denotes the correlation time of the noise.[3]

One possible interpretation of this model is that the market price can temporarily deviate from the true price (corresponding to $\eta_t = 0$). The mispricing is a random error term that mean-reverts over time scale τ_η (see Section 20.3 for a more detailed discussion). Another common interpretation is that of **microstructure noise**, coming from the fact that there is no unique price p_t but at best two prices, the bid b_t and the ask a_t, from which one has to come up with a proxy of the true price p_t. This necessarily generates some error term, which one can assume to be mean-reverting, leading to what is commonly known as a **bid–ask bounce** effect.

How does this noise on p_t affect the observed volatility? By replacing $C_r(\tau)$ in Equation (2.9) with the ACF $C_\eta(\tau)$, we see that compared to the volatility observed in a price series without noise, the addition of the η_t term in Equation (2.11) serves to increase the lag-τ square volatility as

$$\sigma^2(\tau) \longrightarrow \sigma^2(\tau) + \frac{2\sigma_\eta^2}{\tau}\left(1 - e^{-\tau/\tau_\eta}\right). \tag{2.13}$$

This additional term decays from $2\sigma_\eta^2/\tau_\eta$ for $\tau = 0$, to 0 for $\tau \to \infty$. The effect of this correlated high-frequency noise on a volatility signature plot is thus akin to mean-reversion, in the sense that it creates a higher short-term volatility than long-term volatility.

2.1.4 Volatility Signature Plots for Real Price Series

How well does Bachelier's first law hold for real price series? Quite remarkably, the volatility signature plots of most liquid assets are indeed almost flat for values of τ ranging from a few seconds to a few months (beyond which it becomes dubious whether the statistical assumption of stationarity still holds). To illustrate this point, Figure 2.2 shows the empirical mid-price signature plot from $\tau = 1$ second to $\tau = 10^6$ seconds (which corresponds to about 20 trading days) for the S&P500 E-mini futures contract, which is one of the most liquid contracts in the world. As Figure 2.2 shows, $\sigma(\tau)$ is almost flat over this entire range, and only decreases by about 20%, which indicates a weakly mean-reverting price. The exact form of a volatility signature plot depends on the microstructural details of the

[3] The noise η is often called an Ornstein–Uhlenbeck process, see, e.g., Gardiner, C. W. (1985). *Stochastic methods*. Springer, Berlin-Heidelberg.

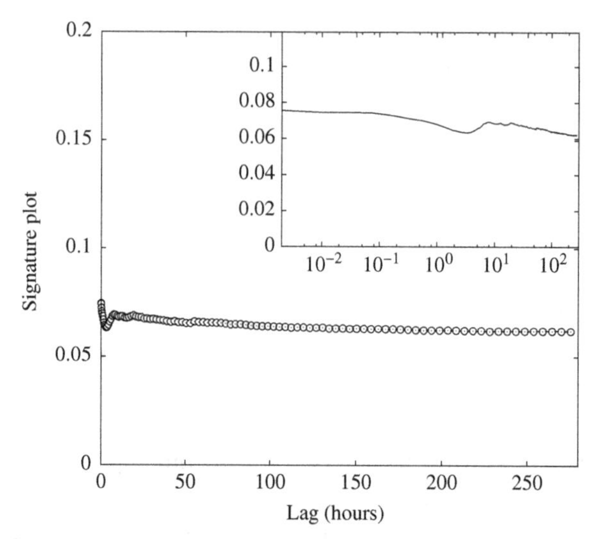

Figure 2.2. Mid-price signature plot for the S&P500 E-mini futures contract, for time lags up to approximately 280 hours (which corresponds to 20 days of trading). The signature plot is almost flat, which implies only weak mean-reversion.

underlying asset, but most liquid contracts have a similar volatility signature plot to that shown in Figure 2.2.

The important conclusion from this empirical result is that long-term volatility is almost identical to the volatility at the shortest time scales, where price formation takes place. Depending on how we view this result, it is either trivial (a simple random walk has this property) or extremely non-intuitive. In fact, as we discuss below, one should expect a rather large fundamental uncertainty about the price of an asset, which would translate into substantially larger high-frequency volatility, as with the η noise described in Section 2.1.3. Although Figure 2.2 shows that high-frequency volatility is slightly larger than low-frequency volatility, the size of this effect is small. This indicates that empirical price series exhibit only weak violations of Bachelier's first law (see Section 2.3.2), and suggests that phenomena happening on short time scales may be relevant for the long-term dynamics of prices.

2.2 Jumps and Intermittency in Financial Markets

2.2.1 Heavy Tails

Many financial models assume that returns follow a **Gaussian distribution**. However, an overwhelming body of empirical evidence from a vast array of financial instruments (including stocks, currencies, interest rates, commodities, and even implied volatility) shows this not to be the case. Instead, the unconditional

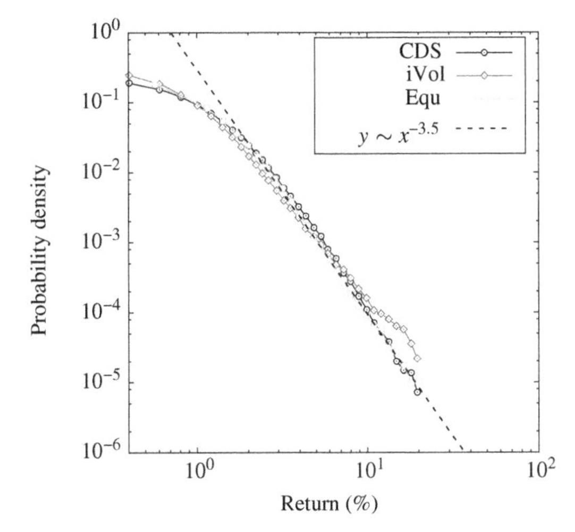

Figure 2.3. Empirical distribution of normalised daily returns of (squares) a family of US stocks, (circles) the spread of the credit default swaps (CDS) on the same stocks, and (diamonds) the average implied volatility of vanilla options again on the same stocks, all during the period 2011–2013. Returns have been normalised by their own volatility before aggregation. The tails of these distributions follow a power-law, with an exponent approximately equal to 3.5.

distribution of returns has **fat tails**, which decay as a power law for large arguments and are much heavier than the corresponding tails of the Gaussian distribution.

On short time scales (between about a minute and a few hours), the empirical density function of returns can be fit reasonably well by a **Student's *t* distribution**, whose probability density function is given by:[4]

$$f_r(r) = \frac{1}{\sqrt{\pi}} \frac{\Gamma[\frac{1+\mu}{2}]}{\Gamma[\frac{\mu}{2}]} \frac{a^\mu}{(r^2+a^2)^{\frac{1+\mu}{2}}} \underset{|r|\gg a}{\longrightarrow} \frac{a^\mu}{|r|^{1+\mu}}, \tag{2.14}$$

where μ is the **tail parameter** and a is related to the variance of the distribution through $\sigma^2 = a^2/(\mu-2)$. Empirically, the tail parameter μ is consistently found to be about 3 for a wide variety of different markets, which suggests some kind of **universality** in the mechanism leading to extreme returns (see Figure 2.3).[5] As we discuss below, this universality hints at the fact that fundamental factors are probably unimportant in determining the amplitude of most price jumps.

If σ^2 is fixed, then the Gaussian distribution is recovered in the limit $\mu \to \infty$. However, the difference between $\mu = 3$ and $\mu \to \infty$ is spectacular for tail events. In a Gaussian world, jumps of size larger than 10σ would only occur with negligible

[4] We assume here and in the following that returns have zero mean, which is appropriate for sufficiently short time scales (see discussion above).

[5] In fact, one can test the hypothesis that the different stocks share the same return distribution. This hypothesis cannot be rejected once volatility clustering effects are taken into account; on this point see: Chicheportiche, R., & Bouchaud, J.-P. (2011). Goodness-of-fit tests with dependent observations. *Journal of Statistical Mechanics: Theory and Experiment*. (09), PO9003.

probability ($\sim 10^{-23}$). For a Student's t distribution with $\mu = 3$, by contrast, this probability is $\sim 4 \times 10^{-4}$, which is several orders of magnitude larger.[6]

2.2.2 Volatility Clustering

Although considering the unconditional distribution of returns is informative, it is also somewhat misleading. Returns are in fact very far from being non-Gaussian IID random variables – although they are indeed nearly uncorrelated, as their flat signature plots demonstrate (see Figure 2.2). Such an IID model would predict that upon time aggregation, the distribution of returns would quickly converge to a Gaussian distribution on longer time scales (provided the second moment is finite, i.e. $\mu > 2$). Empirical data, on the other hand, indicates that returns remain substantially non-Gaussian on time scales up to weeks or even months.

The dynamics of financial markets is highly intermittent, with periods of intense activity intertwined with periods of relative calm. In intuitive terms, the volatility of financial returns is itself a dynamic variable characterised by a very broad distribution of relaxation times. In more formal terms, returns can be decomposed as the product of a time-dependent volatility component σ_t and a directional component ξ_t,

$$r_t := \sigma_t \xi_t. \tag{2.15}$$

In this representation, ξ_t are IID (but not necessarily Gaussian) random variables of unit variance and σ_t are positive random variables that are empirically found to exhibit an interesting statistical property called **long memory**, which we explore in detail in Section 10.1. More precisely, writing $\sigma_t = \sigma_0 e^{\omega_t}$, where σ_0 is a constant that sets the volatility scale, one finds that ω_t is an approximately Gaussian random variable with a variogram (see Section 2.1.1) given by

$$\mathcal{V}_\omega(\tau) = \langle(\omega_t - \omega_{t+\tau})^2\rangle \cong \chi_0^2 \ln[1 + \min(\tau, T)], \tag{2.16}$$

where T is a long cut-off time, estimated to be on the scale of years.[7] The parameter χ_0 sets the scale of the log-volatility fluctuations and is often called the volatility of the volatility (or "vol of vol"). For most assets, its value is found to be $\chi_0^2 \cong 0.05$.

[6] We have used here that for $\mu = 3$, the cumulative distribution of the Student's t distribution is:

$$1 - F_r(r) = \frac{1}{2} - \frac{1}{\pi}\left[\arctan r + \frac{r}{1+r^2}\right] \qquad (\mu = 3, a = 1, \sigma^2 = 1).$$

Another useful formula gives the partial contribution to volatility:

$$\int_0^r \mathrm{d}x\, x^2 f_r(x) = \frac{1}{\pi}\left(\arctan r - \frac{r}{1+r^2}\right) \qquad (\mu = 3, a = 1, \sigma^2 = 1).$$

[7] A recent study suggests rather that "volatility is rough" in the sense that $\mathcal{V}_\omega(\tau) \propto \tau^{2H}$ with $H \cong 0.1$, but note that the two functional forms become identical in the limit $H \to 0$. Gatheral, J., Jaisson, T., & Rosenbaum, M. (2014). Volatility is rough. arXiv:1410.3394.

The variogram in Equation (2.16) is markedly different from the one corresponding to an **Ornstein–Uhlenbeck** log–volatility process, characterised by a single relaxation time τ_ω, which would read

$$\mathcal{V}_\omega(\tau) = \chi_0^2 \left(1 - e^{-\tau/\tau_\omega}\right). \tag{2.17}$$

In other words, volatility fluctuations in financial markets are *multi-time scale*: there are volatility bursts of all sizes, from seconds to years. This remarkable feature of financial time series can also be observed in several other complex physical systems, such as turbulent flows.

It is worth pointing out that volatilities σ and scaled returns ξ are not independent random variables. It is well documented that positive past returns tend to decrease future volatilities and that negative past returns tend to increase future volatilities (i.e. $\langle \xi_t \sigma_{t+\tau} \rangle < 0$ for $\tau > 0$). This is called the **leverage effect**. Importantly, however, past volatilities do not give much information on the sign of future returns (i.e. $\langle \xi_t \sigma_{t+\tau} \rangle \cong 0$ for $\tau < 0$).

2.2.3 Delayed Convergence Towards the Gaussian

Why do returns remain substantially non-Gaussian on time scales up to weeks or even months? Let us consider what happens when, as in Equation (2.3), we sum uncorrelated but dependent random variables such as those described by Equation (2.15). When the number of terms t is large, the Central Limit Theorem (CLT) holds, so the sum converges to a Gaussian random variable. The speed of convergence is a subtle issue, but the simplified picture is that in the case of symmetric IID random variables, the leading correction term (when compared to the Gaussian) scales at large t as t^{-1}.

However, the dependence between the random variables causes the convergence to occur much more slowly than for the IID case. In the presence of long-ranged volatility correlations, such as given by Equation (2.16), one can show that the leading correction to Gaussian behaviour instead decays as $t^{-\zeta}$ with $\zeta = \min(1, 4\chi_0^2)$ (i.e. as $\sim t^{-0.2}$ for the value of χ_0^2 quoted above). Hence, the corrections to the asymptotic Gaussian behaviour are very slow to disappear, and it may take months or even years before the CLT applies.[8]

2.2.4 Activity Clustering

In view of these long-range correlations of the volatility, it is interesting to study the temporal fluctuations of **market activity** itself, defined for example as the frequency of mid-price changes, or as the frequency of market order submission. Even a cursory look at the time series of mid-price changes (see Figure 2.4) suggests a strong degree of **clustering** in the activity as well. What is the relationship between the clustering of market activity and the clustering of volatility?

[8] On this specific point, see the detailed discussion in Bouchaud, J. P., & Potters, M. (2003). *Theory of financial risk and derivative pricing: From statistical physics to risk management*. Cambridge University Press.

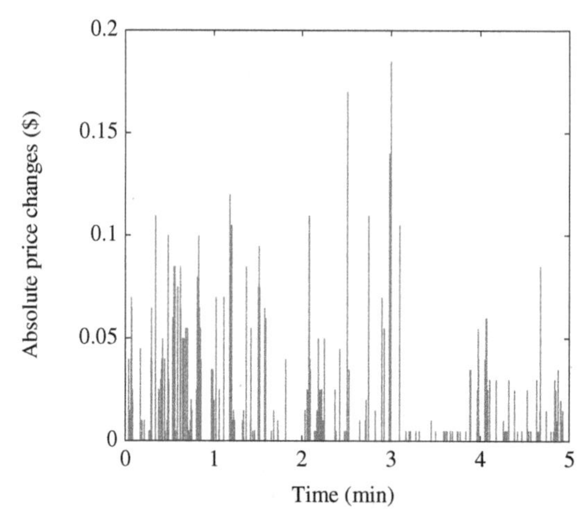

Figure 2.4. Time series of absolute changes in mid-price for TSLA, measured during a typical five-minute time interval. There are several periods with a high concentration of mid-price changes, and several other periods of relative calm.

Consider the following simple model. Assume that the price can only change by one tick ϑ at a time, either up or down, with rate φ per unit time, which quantifies the activity of the market. Assume also that when a price change occurs, up $(+\vartheta)$ and down $(-\vartheta)$ moves occur with equal probability. The price volatility in this model is then simply

$$\sigma^2 = \varphi\,\vartheta^2. \tag{2.18}$$

A more precise way to characterise this clustering property is to choose a time t and a small $\mathrm{d}t$, and count the number $\mathrm{d}N_t$ of price changes that occur during the time interval $[t, t+\mathrm{d}t]$ (i.e. count $\mathrm{d}N_t = 1$ if the mid-price changed or $\mathrm{d}N_t = 0$ if it did not). The empirical average of $\mathrm{d}N_t$ provides a way to define the average market activity $\bar{\varphi}$ as

$$\langle \mathrm{d}N_t \rangle := \bar{\varphi}\,\mathrm{d}t. \tag{2.19}$$

The covariance $\mathrm{Cov}[\frac{\mathrm{d}N_t}{\mathrm{d}t}, \frac{\mathrm{d}N_{t+\tau}}{\mathrm{d}t}]$ describes the temporal structure of the fluctuations in market activity. Figure (2.5) shows that the activity in financial markets is characterised both by long memory (the activity is autocorrelated over very long periods, of 100 days or more) and by an intricate pattern of daily and weekly periodicities.

The relationship between volatility and market activity given in Equation (2.18) however fails to address many other possible ways that prices can change. For example, it does not address the scenario where the price is mostly stable but occasionally makes very large jumps. Prices in financial markets tend to exhibit both types of volatilities: small frequent moves and rare extreme moves.

The broad distribution of price changes discussed in Section 2.2.1 describes exactly this duality between activity and jumps. The relative contribution of jumps

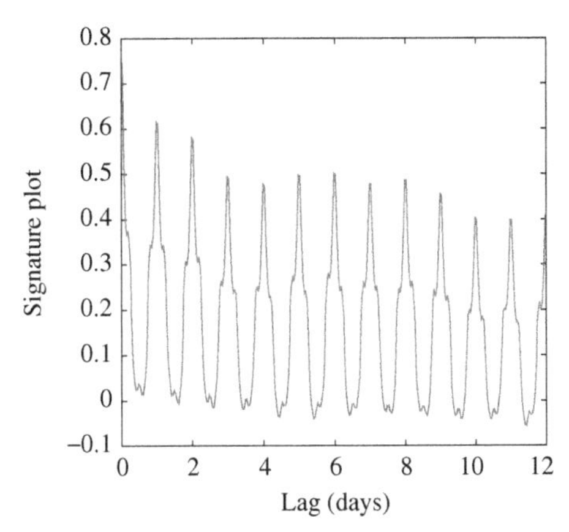

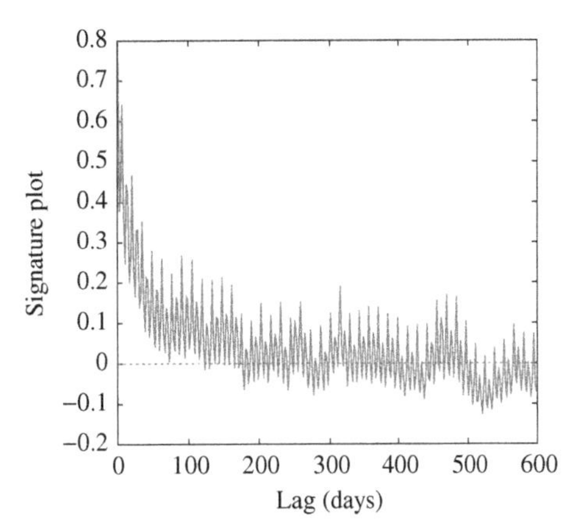

Figure 2.5. Autocorrelation of the trading activity of the S&P500 E-mini futures contract measured as the number of mid-price changes (left panel) per 30-minute interval and (right panel) per day. At the intra-day level, there are clear peaks with daily periodicity associated with the intra-day pattern. At the daily level, there is a clear weekly periodicity and still substantial correlation after 100 days.

to the total volatility depends on the precise definition of a jump, but assuming that returns follow a Student's t distribution, one finds (using the second formula of footnote 6 above) that jumps defined as events of magnitude greater than 4σ contribute about 30% of the total variance. This is quite substantial, especially given that a large fraction of these jumps appear to be unrelated to any clearly identifiable news.

Furthermore, the two types of events are intertwined in a subtle manner: an increased volatility at time t appears to trigger more activity at time $t+\tau$, much like earthquakes are followed by aftershocks. For example, a large jump is usually followed by an increased frequency of smaller price moves. More generally, some kind of **self-excitation** seems to be present in financial markets. This takes place either temporally (some events trigger more events in the future) or across different assets (the activity of one stock spills over to other correlated stocks, or even from one market to another). We present mathematical models to describe these contagion effects in Chapter 9.

2.3 Why Do Prices Move?

2.3.1 The Excess Volatility Puzzle

Why do prices behave like random walks? Often, this question is addressed via the argument that assets have a fundamental value that is known (or computed) by informed traders, who buy or sell the asset according to whether it is under- or over-priced. By the impact of making these profitable trades, the informed traders drive the price back towards its fundamental value. We develop a formal notion of price impact throughout the book, and discuss this topic in detail in Chapter 11.

In this framework, the market price of an asset can only change due to the arrival of unanticipated news (up to short-lived mispricings). The standard picture is as follows: as such news becomes available, fast market participants calculate how it affects the fundamental price, then trade accordingly. After a phase of *tâtonnement*, the price should converge to its new fundamental value, around which it makes random, high-frequency oscillations due to the trades of uninformed market participants, until the next piece of unanticipated news arrives and the whole process repeats.

This idea resides at the very heart of efficient-market theory, but is this picture correct? Can it account for the volatility observed in real markets?

Consider again the case of a typical US large-cap stock, say Apple, which has a daily turnover of around $4–5 billion. Each second, one observes on average six transactions and of the order of 100 events for this stock alone. Compared to the typical time between news arrivals that could potentially affect the price of the company (which are on the scale of one every few days, or perhaps hours), these frequencies are extremely high, suggesting that market activity is not only driven by news.

Perhaps surprisingly, the number of large price jumps is in fact also much higher than the number of relevant news arrivals. In other words, most large price moves seem to be unrelated to news, but rather to arise endogenously from trading activity itself. As emphasised by Cutler, Poterba and Summers, "The evidence that large market moves occur on days without identifiable major news casts doubts on the view that price movements are fully explicable by news."[9]

This is a manifestation of Shiller's **excess-volatility puzzle**: the actual volatility of prices appears to be much higher than the one warranted by fluctuations of the underlying fundamental value.

2.3.2 The Flat Volatility Puzzle

Perhaps even more puzzling is the following: despite the fact that it should take some time for the market to interpret a piece of news and agree on a new price, and despite the fact that liquidity is too thin for supply and demand to be fully and instantaneously revealed, financial time series show very little under-reaction (which would create trend) or over-reaction (which would lead to mean-reversion), leading to almost flat empirical signature plots. How can this be?

[9] Cutler, D. M., Poterba, J. M., & Summers, L. H. (1989). What moves stock prices? *The Journal of Portfolio Management*, 15(3), 4–12.

If price trajectories consisted of a sequence of equilibrium values (around which prices randomly fluctuate) that are interrupted by jumps to another equilibrium price when some unanticipated news becomes available, then signature plots should show a significant decay, as in Figure 2.1. Note that relatively small fluctuations of about 0.1% around the equilibrium value, with a correlation time of 10 minutes, would lead to a high-frequency volatility contribution twice as large as the long-term volatility for the S&P500 future contract. This is clearly not observed empirically.

An alternative picture, motivated by a microstructural viewpoint, is that highly optimised execution algorithms and market-making strategies actively search for any detectable correlations or trends in return series, and implement trading strategies designed to exploit any consequent predictability in asset prices. By doing so, these algorithms and strategies iron out all irregularities in the signature plots. Expressed in terms of the ecology of financial markets (see Chapter 1), higher-frequency strategies feed on the inefficiencies generated by slower strategies, finally resulting in white-noise returns on all time scales. We will return to this important discussion in Chapter 20.

2.3.3 How Relevant is Fundamental Value?

A crucial discussion in financial economics concerns the notion of "efficiency". Are prices **fundamentally efficient** (in the sense that they are always close to some fundamental value) or merely **statistically efficient** (in the sense that all predictable patterns are exploited and removed by technical trading)? After all, purely random trades – think of a market driven by proverbial monkeys (after failing to transcribe the complete works of Shakespeare) – would generate random price changes and a flat signature plot, without reflecting any fundamental information at all. In fact, Black famously argued that prices are correct to within a factor of 2. If this is the case, the anchor to fundamental values can only be felt on a time scale T such that purely random fluctuations $\sigma\sqrt{T}$ become of the order of say 50% of the fundamental price, leading to $T = 6$ years for the stock market with a typical annual volatility of $\sigma = 20\%$.[10]

Such long time scales suggest that the notion of a fundamental price is secondary to understanding the dynamics of prices at the scale of a few seconds to a few days. These are the time scales relevant for the microstructural effects that we study in this book. Such a decoupling allows us to mostly disregard the role of fundamental

[10] This long time scale makes it very difficult to establish statistically whether mean-reversion occurs at all, although several studies have suggested that this is indeed the case, see, e.g., Bondt, W. F., & Thaler, R. (1985). Does the stock market overreact? *The Journal of Finance*, 40(3), 793–805. See also Summers, L. H. (1986). Does the stock market rationally reflect fundamental values? *The Journal of Finance*, 41(3), 591–601.

value in the following discussions, and instead focus on the notion of statistical efficiency (see, however, Chapter 20).

2.4 Summary and Outlook

The main message of this chapter is that price changes are remarkably uncorrelated over a large range of frequencies, with few signs of price adjustments or *tâtonnement* at high frequencies. The long-term volatility is already determined at the high-frequency end of the spectrum. In fact, the frequency of news that would affect the fundamental value of financial assets is much lower than the frequency of price changes. It is as if price changes *themselves* are the main source of news, and induce feedback that creates excess volatility and, most probably, price jumps that occur without any news at all. Interestingly, all quantitative volatility/activity feedback models (such as ARCH-type models,[11] or Hawkes processes, which we discuss in Chapter 9) suggest that at least 80% of the price variance is induced by self-referential effects. This adds credence to the idea that the lion's share of the short- to medium-term activity of financial markets is unrelated to any fundamental information or economic effects.

The decoupling between price and fundamental value opens the exciting prospect of building a theory of price moves that is mostly based on the endogenous, self-exciting dynamics of markets, and not on long-term fundamental effects, which are notoriously hard to model. One particularly important question is to understand the origin and mechanisms leading to price jumps, which seem to have a similar structure on all liquid markets (again indicating that fundamental factors are probably unimportant at short time scales). This is precisely the aim of a microstructural approach to financial markets: reconstructing the dynamics of prices from the bottom up and illustrating how microstructure can indeed be relevant to lower-frequency price dynamics.

Take-Home Messages

(i) Standard Gaussian random-walk models grossly underestimate extreme fluctuations in price returns. In reality, price changes follow fat-tailed, power-law distributions. Extreme events are not as rare as Gaussian models would predict.
(ii) Market activity and volatility are highly intermittent in time, with periods of intense activity intertwined with periods of relative calm.

[11] See, e.g., Bollerslev, T., Engle, R. F., & Nelson, D. B. (1994). ARCH models. In Engle, R. & McFadden, D. (Eds.), *Handbook of econometrics* (Vol. 4, pp. 2959–3028). North-Holland.

(iii) Periods of intense activity/volatility only partially overlap with the arrival of news. In fact, most activity is endogenous, and is triggered by past activity itself.
(iv) A volatility signature plot describes how the volatility of a price series varies with the lag on which it is computed. A decreasing signature plot indicates mean-reversion; an increasing signature plot indicates trending.
(v) Empirical signature plots are remarkably flat over a wide range of time scales, from seconds to months. This is a sign that markets are statistically efficient, in the sense that prices are (linearly) unpredictable.
(vi) Markets exhibit substantial excess volatility: statistical efficiency does not imply fundamental efficiency.

2.5 Further Reading

General

Bachelier, L. (1900). *Théorie de la spéculation*. Gauthier-Villars.
Frisch, U. (1997). *Turbulence: The Kolmogorov legacy*. Cambridge University Press.
Mantegna, R. N., & Stanley, H. E. (1999). *Introduction to econophysics: correlations and complexity in finance*. Cambridge University Press.
Shiller, R. J. (2000). *Irrational exuberance*. Princeton University Press.
Lyons, R. K. (2001). *The microstructure approach to exchange rates* (Vol. 12). MIT Press.
Bouchaud, J. P., & Potters, M. (2003). *Theory of financial risk and derivative pricing: From statistical physics to risk management*. Cambridge University Press.
Andersen, T. G., Davis, R. A., Kreiss, J. P., & Mikosch, T. V. (Eds.). (2009). *Handbook of financial time series*. Springer Science & Business Media.
Slanina, F. (2013). *Essentials of econophysics modelling*. Oxford University Press.

Stylised Facts

Guillaume, D. M., Dacorogna, M. M., Davé, R. R., Muller, U. A., Olsen, R. B., & Pictet, O. V. (1997). From the bird's eye to the microscope: A survey of new stylised facts of the intra-daily foreign exchange markets. *Finance and Stochastics*, 1(2), 95–129.
Gopikrishnan, P., Plerou, V., Amaral, L. A. N., Meyer, M., & Stanley, H. E. (1999). Scaling of the distribution of fluctuations of financial market indices. *Physical Review E*, 60(5), 5305.
Plerou, V., Gopikrishnan, P., Amaral, L. A. N., Meyer, M., & Stanley, H. E. (1999). Scaling of the distribution of price fluctuations of individual companies. *Physical Review E*, 60(6), 6519.
Andersen, T. G. T., Diebold, F. X., & Ebens, H. (2001). The distribution of realised stock return volatility. *Journal of Financial Economics*, 61(1), 43–76.
Cont, R. (2001). Empirical properties of asset returns: Stylised facts and statistical issues. *Quantitative Finance*, 1, 223–236.

Gabaix, X., Gopikrishnan, P., Plerou, V., & Stanley, H. E. (2006). Institutional investors and stock market volatility. *The Quarterly Journal of Economics*, 121(2), 461–504.

Zumbach, G., & Finger, C. (2010). A historical perspective on market risks using the DJIA index over one century. *Wilmott Journal*, 2(4), 193–206.

Diebold, F. X., & Strasser, G. (2013). On the correlation structure of microstructure noise: A financial economic approach. *The Review of Economic Studies*, 80(4), 1304–1337.

Clustering, Intermittency and Power-Laws

Bollerslev, T., Engle, R. F., & Nelson, D. B. (1994). ARCH models. In Engle, R. & McFadden, D. (Eds.), *Handbook of econometrics* (Vol. 4, pp. 2959–3028). North-Holland.

Muzy, J. F., Delour, J., & Bacry, E. (2000). Modelling fluctuations of financial time series: From cascade process to stochastic volatility model. *The European Physical Journal B-Condensed Matter and Complex Systems*, 17(3), 537–548.

Bouchaud, J. P. (2001). Power-laws in economics and finance: Some ideas from physics. *Quantitative Finance*, 1, 105–112.

Sethna, J. P., Dahmen, K. A., & Myers, C. R. (2001). Crackling noise. *Nature*, 410(6825), 242–250.

Cabrera, J. L., & Milton, J. G. (2002). On-off intermittency in a human balancing task. *Physical Review Letters*, 89(15), 158702.

Calvet, L., & Fisher, A. (2002). Multifractality in asset returns: Theory and evidence. *Review of Economics and Statistics*, 84(3), 381–406.

Lux, T. (2008). The Markov-switching multifractal model of asset returns: GMM estimation and linear forecasting of volatility. *Journal of Business and Economic Statistics*, 26, 194–210.

Clauset, A., Shalizi, C. R., & Newman, M. E. (2009). Power-law distributions in empirical data. *SIAM Review*, 51(4), 661–703.

Gabaix, X. (2009). Power laws in economics and finance. *Annual Review of Economics*, 1(1), 255–294.

Patzelt, F., & Pawelzik, K. (2011). Criticality of adaptive control dynamics. *Physical Review Letters*, 107(23), 238103.

Gatheral, J., Jaisson, T., & Rosenbaum, M. (2014). Volatility is rough. arXiv preprint arXiv:1410.3394.

See also Section 9.7.

Probability Theory, Extreme-Value Statistics

Gnedenko, B. V., & Kolmogorov, A. N. (1968). *Limit distributions for sums of independent random variables*, 2nd Edn., Addison-Wesley.

Embrechts, P., Klüppelberg, C., & Mikosch, T. (1997). *Modelling extremal events*. Springer-Verlag.

Malevergne, Y., & Sornette, D. (2006). *Extreme financial risks: From dependence to risk management*. Springer Science & Business Media.

Excess Volatility and Endogeneous Activity

Shiller, R. J. (1980). Do stock prices move too much to be justified by subsequent changes in dividends? *American Economic Review*, 71, 421–436.

Cutler, D. M., Poterba, J. M., & Summers, L. H. (1989). What moves stock prices? *The Journal of Portfolio Management*, 15(3), 4–12.

Fair, R. C. (2002). Events that shook the market. *The Journal of Business*, 75(4), 713–731.

Gillemot, L., Farmer, J. D., & Lillo, F. (2006). There's more to volatility than volume. *Quantitative Finance*, 6(5), 371–384.

Joulin, A., Lefevre, A., Grunberg, D., & Bouchaud, J. P. (2008). Stock price jumps: News and volume play a minor role. *Wilmott Magazine*, September/October, 1–7.

Bouchaud, J. P. (2011). The endogenous dynamics of markets: Price impact, feedback loops and instabilities. In Berd, A. (Ed.), *Lessons from the credit crisis*. Risk Publications.

Cornell, B. (2013). What moves stock prices: Another look. *The Journal of Portfolio Management*, 39(3), 32–38.

Diebold, F. X., & Strasser, G. (2013). On the correlation structure of microstructure noise: A financial economic approach. *The Review of Economic Studies*, 80(4), 1304–1337.

Market Efficiency

Bondt, W. F., & Thaler, R. (1985). Does the stock market overreact? *The Journal of Finance*, 40(3), 793–805.

Black, F. (1986). Noise. *The Journal of Finance*, 41(3), 528–543.

Summers, L. H. (1986). Does the stock market rationally reflect fundamental values? *The Journal of Finance*, 41(3), 591–601.

Shleifer, A., & Summers, L. H. (1990). The noise trader approach to finance. *The Journal of Economic Perspectives*, 4(2), 19–33.

Lyons, R. (2001). *The microstructure approach to foreign exchange rates*. MIT Press.

Schwert, G. W. (2003). Anomalies and market efficiency. In Constantinides, G. M., Harris, M., & Stulz, R. (Eds.), *Handbook of the economies of finance* (Vol. 1, pp. 939–974). Elsevier Science B.V.

Lo, A. W. (2017). *Adaptive markets*. Princeton University Press.

See also Section 20.5.

PART II
Limit Order Books: Introduction

Introduction

Now that the big picture is set, it is time to dive more deeply into the detailed mechanics of trading in modern financial markets. As we will see, practical aspects of market design can substantially affect the microscopic-scale behaviour of individual agents. These actions can then proliferate and cascade to the macroscopic scale.

In this part, we will embark on a detailed discussion of limit order books (LOBs), which are nowadays omnipresent in the organisation of lit markets (i.e. open markets where supply and demand are displayed publicly). We will start by describing the founding principle of LOBs, namely the interaction between limit orders, which offer trading opportunities to the rest of the world, and market orders, which take these opportunities and result in immediate transactions. LOBs are not only a place where buyers meet sellers, but also a place where patient traders (or "liquidity providers") meet impatient traders (or "liquidity takers"). For practical reasons, LOB activity is constrained to a predefined price and volume grid, whose resolution parameters are defined by the *tick size* and the *lot size*, respectively. At any point in time, the state of an LOB is defined by the outstanding limit orders on this grid.

LOBs are an elegant and simple way to channel the trading intentions of market participants into a market price. In an LOB, any buy and sell orders whose prices cross are matched and immediately removed from the LOB. Therefore, an LOB is a collection of lower-priced buy orders and higher-priced sell orders, separated by a bid–ask spread. The persistent aspect of limit orders is key for the stability of the LOB state and the market price.

In the coming chapters, we will discuss several interesting empirical properties that emerge from the interactions of traders in an LOB. We will see that both the distribution of arriving order volumes and the distribution of outstanding volume in an LOB are fat tailed, and we will discuss the consequences of these properties in the context of market stability. We will also discuss the important role of intra-day seasonalities in understanding market dynamics. Last but not least, we will highlight how order flow presents complex conditioning that strongly affects the price-formation process. The understanding of this subtle dance, which is crucial for the modelling of price changes, will be the topic of the next part.

3
Limit Order Books

Though this be madness, yet there is method in't.

(Shakespeare, Hamlet)

Today, most of the world's financial markets use an electronic trading mechanism called a limit order book (LOB) to facilitate trade. The Helsinki, Hong Kong, London, New York, Shenzhen, Swiss, Tokyo, Toronto and Vancouver Stock Exchanges, together with Euronext, the Australian Securities Exchange and NASDAQ, all use LOBs, as do many smaller markets. In this chapter, we provide a detailed introduction to trading via LOBs, and we discuss how price changes in an LOB emerge from the dynamic interplay between liquidity providers and liquidity takers.

3.1 The Mechanics of LOB Trading

In contrast to quote-driven systems (see Chapter 1), in which prices are set by designated market-makers, price formation in an LOB is a self-organised process that is driven by the submissions and cancellations of orders. Each order is a visible declaration of a market participant's desire to buy or sell a specified quantity of an asset at a specified price. Active orders reside in a **queue** until they are either cancelled by their owner or executed against an order of opposite direction. Upon execution, the owners of the relevant orders trade the agreed quantity of the asset at the agreed price.

Whereas in the old days the list of buy and sell orders was only known to the specialist, the queues of outstanding orders in LOBs are now observable in real time by traders from around the world. Because each order within these queues constitutes a firm commitment to trade, analysing the state of the LOB provides a concrete way to quantify the visible liquidity for a given asset.

3.1.1 Orders

An **order** is a commitment, declared at a given submission time, to buy or sell a given volume of an asset at no worse than a given price. An order x is thus described by four attributes:

- its *sign* (or *direction*) $\varepsilon_x = \pm 1$, ($\varepsilon_x = +1$ for buy orders; $\varepsilon_x = -1$ for sell orders),
- its *price* p_x,
- its *volume* $\upsilon_x > 0$, and
- its *submission time* t_x.

We introduce the succinct notation

$$x := (\varepsilon_x, p_x, \upsilon_x, t_x). \tag{3.1}$$

3.1.2 The Trade-Matching Algorithm

Whenever a trader submits a buy (respectively, sell) order x, an LOB's *trade-matching algorithm* checks whether it is possible for x to *match to* existing sell (respectively, buy) orders y such that $p_y \leq p_x$ (respectively, $p_y \geq p_x$). If so, the matching occurs immediately and the relevant traders perform a trade for the agreed amount at the agreed price. Any portion of x that does not match instead becomes *active* at the price p_x, and it remains active until either it matches to an incoming sell (respectively, buy) order or it is *cancelled*. **Cancellation** usually occurs when the owner of an order no longer wishes to offer a trade at the stated price.

3.1.3 Market Orders, Limit Orders and the LOB

Orders that match upon arrival are called **market orders**. Orders that do not match upon arrival are called **limit orders**. A **limit order book** (LOB) is simply a collection of revealed, unsatisfied intentions to buy or sell an asset at a given time. More precisely, an LOB $\mathscr{L}(t)$ is the set of all limit orders for a given asset on a given platform[1] at time t.

The limit orders in $\mathscr{L}(t)$ can be partitioned into the set of buy limit orders $\mathcal{B}(t)$ (for which $\varepsilon_x = +1$) and the set of sell limit orders $\mathcal{A}(t)$ (for which $\varepsilon_x = -1$).

Throughout the book, we will consider the evolution of an LOB $\mathscr{L}(t)$ as a so-called *càglàd process* (*continu à gauche, limite à droite*). Informally, this means that when a new order x is submitted at time t_x, we regard it to be present in the LOB immediately after its arrival, but not immediately upon its arrival. More formally, we introduce the notation

$$\overline{\mathscr{L}}(t) = \lim_{t' \downarrow t} \mathscr{L}(t') \tag{3.2}$$

[1] In modern financial markets, many assets are traded on several different platforms simultaneously. The *consolidated LOB* is the union of all *LOBs* for the asset, across all platforms where it is traded.

to denote the state of $\mathscr{L}(t)$ immediately after time t.[2] Therefore, for a limit order x submitted at time t_x, it holds that

$$x \notin \mathscr{L}(t_x),$$
$$x \in \overline{\mathscr{L}}(t_x).$$

Traders in an LOB are able to choose the submission price p_x, which will classify the order as a limit order (when it does not lead to an immediate transaction) or as a market order (when it is immediately matched to a limit order of opposite sign). Limit orders stand a chance of matching at better prices than do market orders, but they also run the risk of never being matched. Conversely, market orders match at worse prices, but they do not face the inherent uncertainty associated with limit orders. Some trading platforms allow traders to specify that they wish to submit a market order without explicitly specifying a price. Instead, such a trader specifies only a size, and the LOB's trade-matching algorithm sets the price of the order appropriately to initiate the required matching.

The popularity of LOBs is partly due to their ability to allow some traders to demand immediacy (by submitting market orders), while simultaneously allowing others to supply it (by submitting limit orders) and hence to at least partially play the traditional role of market-makers. Importantly, liquidity provision in an LOB is a decentralised and self-organised process, driven by the submission of limit orders from all market participants. In reality, many traders or execution algorithms use a combination of both limit orders and market orders by selecting their actions for each situation based on their individual needs at that time.

The following remark can be helpful in thinking about the relative merits of limit orders and market orders. In an LOB, each limit order can be construed as an option contract written to the whole market, via which the order's owner offers to buy or sell the specified quantity υ_x of the asset at the specified price p_x to any trader wishing to accept. For example, a trader who submits a sell limit order $x = (-1, p_x, \upsilon_x, t_x)$ is offering the entire market a free option to buy υ_x units of the asset at price p_x for as long as the order remains active. Traders offer such options – i.e. submit limit orders – in the hope that they will be able to trade at better prices than if they simply submitted market orders. However, whether or not a limit order will eventually become matched is uncertain, and if it is matched, there is a good chance that the price will continue drifting beyond the limit price, leading to a loss for the issuer of the option. This is an example of adverse selection and its associated skewness, which we already discussed in Section 1.3.2. In a nutshell, as we will show later in the book (see Section 21.3), limit orders typically earn profits when prices mean-revert but suffer losses when prices trend.

[2] The notation $t' \downarrow t$ ($t' \uparrow t$) means that t' approaches t from above (below).

3.1.4 The Bid-, Ask- and Mid-Price

The terms *bid-price*, *ask-price*, *mid-price* and *spread* (all of which we have already encountered in Chapter 1) are common to much of the finance literature. Their definitions can be made specific in the context of an LOB (see Figure 3.1):

- The **bid-price** at time t is the highest stated price among buy limit orders at time t,

$$b(t) := \max_{x \in \mathcal{B}(t)} p_x. \tag{3.3}$$

- The **ask-price** at time t is the lowest stated price among sell limit orders at time t,

$$a(t) := \min_{x \in \mathcal{A}(t)} p_x. \tag{3.4}$$

At any given time, $b(t)$ is therefore the highest price at which it is immediately possible to sell and $a(t)$ is the lowest price at which it is immediately possible to buy. By definition, $a(t) > b(t)$, otherwise some buy and sell limit orders can be immediately matched and removed from the LOB.

- The **mid-price** at time t is

$$m(t) := \frac{1}{2}\left[a(t) + b(t)\right].$$

- The **bid–ask spread** (or simply "spread") at time t is

$$s(t) := a(t) - b(t) > 0.$$

Each of $b(t)$, $a(t)$, $m(t)$ and $s(t)$ are also càglàd processes. We use the same overline notation as in Equation (3.2) to denote the values of these processes immediately after time t. For example,

$$\overline{m}(t) = \lim_{t' \downarrow t} m(t')$$

denotes the mid-price immediately after t.

Figure 3.1 shows a schematic of an LOB at some instant in time, illustrating the above definitions. The horizontal lines within the blocks at each price level denote how the total volume available at that price is composed of different limit orders. As the figure illustrates, an LOB can be regarded as a set of queues, each of which consists of a set of buy or sell limit orders at a specified price. As we will discuss at many points throughout the book, this interpretation of an LOB as a system of queues often provides a useful starting point for building LOB models.

3.1.5 The Lot Size and Tick Size

When submitting an order x, a trader must choose the size υ_x and price p_x according to the relevant lot size and tick size of the given LOB.

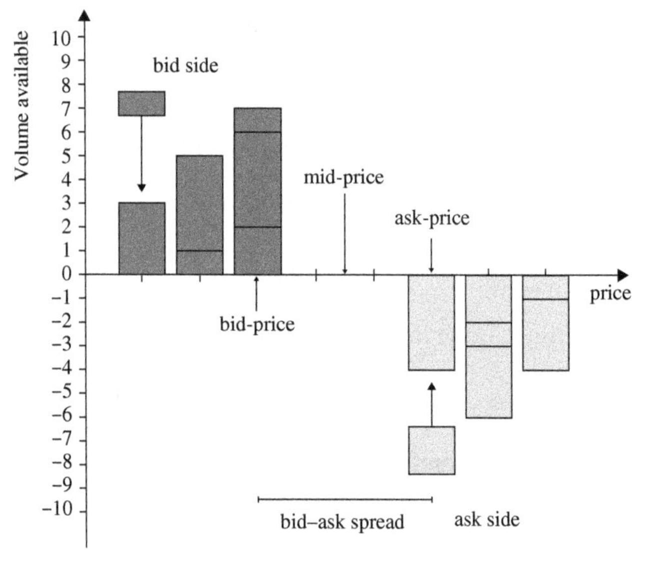

Figure 3.1. A schematic of an LOB, to illustrate the bid-price, the ask-price, the available volumes, the mid-price, and the bid–ask spread.

- The **lot size** υ_0 is the smallest amount of the asset that can be traded within the given LOB. All orders must arrive with a size

$$\upsilon_x \in \{k\upsilon_0 \,|k = 1,2,\ldots\}.$$

- The **tick size** ϑ is the smallest permissible price interval between different orders within a given LOB. All orders must arrive with a price that is a multiple of ϑ.

The lot size and tick size fix the units of order size and order price in a given LOB. For example, if ϑ = \$0.001, then the largest permissible order price that is strictly less than \$1.00 is \$0.999, and all orders must be submitted at a price with exactly three decimal places. The lot size υ_0 and tick size ϑ of an LOB are collectively called its *resolution parameters.*

Values of υ_0 and ϑ vary greatly from between different trading platforms. Expensive stocks are often traded with $\upsilon_0 = 1$ share; cheaper stocks are often traded with $\upsilon_0 \gg 1$ share. In foreign exchange (FX) markets, some trading platforms use values as large as $\upsilon_0 = 1$ million units of the base currency, whereas others use values as small as $\upsilon_0 = 0.01$ units of the base currency.[3] In equity markets, ϑ is often 0.01% of the stock's mid-price $m(t)$, rounded to the nearest power of 10. In US equity markets, the tick size is always ϑ = \$0.01, independently of the stock price. A given asset is sometimes traded with different values of ϑ on different

[3] In FX markets, an XXX/YYY LOB matches exchanges of the *base currency* XXX to the *counter currency* YYY. A price in an XXX/YYY LOB denotes how many units of currency YYY are exchanged for a single unit of currency XXX. For example, a trade at the price \$1.28124 in a GBP/USD market corresponds to 1 pound sterling being exchanged for 1.28124 US dollars.

trading platforms. For example, on the electronic trading platform Hotspot FX, $\vartheta = \$0.00001$ for the GBP/USD LOB and $\vartheta = 0.001$ for the USD/JPY LOB, whereas on the electronic trading platform EBS, $\vartheta = \$0.00005$ for the GBP/USD LOB and $\vartheta = 0.005$ for the USD/JPY LOB.

An LOB's resolution parameters greatly affect trading. The lot size υ_0 dictates the smallest permissible order size, so any trader who wishes to trade in quantities smaller than υ_0 is unable to do so. Furthermore, as we will discuss in Chapter 10, traders who wish to submit large market orders often break them into smaller chunks to minimise their price impact. The size of υ_0 controls the smallest permissible size of these chunks and therefore directly affects traders who implement such a strategy.

The tick size ϑ dictates how much more expensive it is for a trader to gain the priority (see Section 3.2.1) associated with choosing a higher (respectively, lower) price for a buy (respectively, sell) order. In markets where ϑ is extremely small, there is little reason for a trader to submit a buy (respectively, sell) limit order at a price p where there are already other limit orders. Instead, the trader can gain priority over such limit orders very cheaply, by choosing the price $p + \vartheta$ (respectively, $p - \vartheta$) for their limit order. Such a setup leads to very small volumes at any level in the LOB, including the best quotes $b(t)$ and $a(t)$, and therefore leads to extremely frequent changes in $b(t)$ and $a(t)$.

Some market commentators argue that small tick sizes make it difficult for traders to monitor the state of the market in real time. In September 2012, the electronic FX trading platform EBS increased the size of ϑ for most of its currency pairs, to "help thicken top of book price points, increase the cost of top-of-book price discovery, and improve matching execution in terms of percent fill amounts". However, as we will see at many times throughout the book, an asset's tick size can influence order flow in many different ways, some of which are quite surprising. Therefore, understanding how changing the tick size will influence future market activity is far from straightforward.

Even for LOBs with the same or similar resolution parameters υ_0 or ϑ, these can represent vastly different fractions of the typical trade size and price. For example, both the Priceline Group and the Amyris Inc. stocks are traded on NASDAQ with $\vartheta = \$0.01$, yet the typical price for Priceline exceeds \$1000.00 whereas the typical price for Amyris is close to \$1.00. Therefore, ϑ constitutes a much larger fraction of the typical trade size for Amyris than it does for Priceline. For this reason, it is sometimes useful to consider the **relative tick size** ϑ_r, which is equal to the tick size ϑ divided by the mid-price $m(t)$ for the given asset. For example, the price of Priceline Group was on the order of \$1000 in 2014, corresponding to a very small relative tick size ϑ_r of 10^{-5}.

3.1.6 Relative Prices

Because the activity of a single market participant is driven by his/her trading needs, individual actions can appear quite erratic. However, when measured in a suitable coordinate frame that aggregates order flows from many different market participants, robust statistical properties of order flow and LOB state can emerge from the ensemble – just as the ideal gas law emerges from the complicated motion of individual molecules.

Most studies of LOBs perform this aggregation in **same-side quote-relative coordinates**, in which prices are measured relative to the same-side best quote. Specifically, the same-side quote-relative price of an order x at time t is (see Figure 3.1)

$$d(p_x,t) := \begin{cases} b(t) - p_x, & \text{if } x \text{ is a buy limit order,} \\ p_x - a(t), & \text{if } x \text{ is a sell limit order.} \end{cases} \tag{3.5}$$

In some cases (such as when measuring volume profiles, as in Section 4.7), using same-side quote-relative coordinates can cause unwanted artefacts to appear in statistical results. Therefore, instead of measuring prices relative to the same-side best quote, it is sometimes more useful to measure prices in **opposite-side quote-relative coordinates**. The opposite-side quote-relative price of an order x at time t is (see Figure 3.1)

$$d^{\dagger}(p_x,t) := \begin{cases} a(t) - p_x, & \text{if } x \text{ is a buy limit order,} \\ p_x - b(t), & \text{if } x \text{ is a sell limit order.} \end{cases} \tag{3.6}$$

The difference in signs between the definitions for buy and sell orders in Equations (3.5) and (3.6) ensures that an increasing distance from the reference price is always recorded as positive. By definition, all limit orders have a non-negative same-side quote-relative price and a strictly positive opposite-side quote-relative price at all times. Same-side and opposite-side quote-relative prices are related by the bid–ask spread $s(t)$:

$$d^{\dagger}(p_x,t) = d(p_x,t) + s(t).$$

The widespread use of quote-relative coordinates is motivated by the notion that market participants monitor $b(t)$ and $a(t)$ when deciding how to act. There are many reasons why this is the case. For example, $b(t)$ and $a(t)$ are observable to all market participants in real time, are common to all market participants, and define the boundary conditions that dictate whether an incoming order x is a limit order ($p_x < a(t)$ for a buy order or $p_x > b(t)$ for a sell order) or a market order ($p_x \geq a(t)$ for a buy order or $p_x \leq b(t)$ for a sell order). Therefore, the bid and the ask constitute suitable reference points for aggregating order flows across different market participants.

Quote-relative coordinates also provide a useful benchmark for understanding LOB activity. When studying LOBs, it is rarely illuminating to consider the actual price p_x of an order, because this information provides no context for x in relation to the wider activity in $\mathscr{L}(t)$. As we show in Chapter 4, by instead studying the quote-relative price of an order, it is possible to understand the role of the order in relation to the other orders in the market.

3.1.7 The Volume Profile

Most traders assess the state of $\mathscr{L}(t)$ via the **volume profile** (or "depth profile"). The buy-side volume at price p and at time t is

$$V_+(p,t) := \sum_{\{x \in \mathcal{B}(t) | p_x = p\}} \upsilon_x. \tag{3.7}$$

The sell-side volume at price p and at time t, denoted $V_-(p,t)$, is defined similarly using $\mathcal{A}(t)$. The volume profile at time t is the set of volumes at all prices p, $\{V_\pm(p,t)\}$.

Because $b(t)$ and $a(t)$ vary over time, quote-relative coordinates provide a useful approach to studying the volume profile through time, akin to changing reference frame in physics.

3.1.8 Price Changes in an LOB

In an LOB, the rules that govern matching dictate how prices evolve through time. Consider a buy (respectively, sell) order $x = (\pm 1, p_x, \upsilon_x, t_x)$ that arrives at time t.

- If $p_x \leq b(t)$ (respectively, $p_x \geq a(t)$), then x is a limit order. It does not cause $b(t)$ or $a(t)$ to change.
- If $b(t) < p_x < a(t)$, then x is also a limit order. It causes $b(t)$ to increase (respectively, $a(t)$ to decrease) to p_x at time t_x.
- If $p_x \geq a(t)$ (respectively, $p_x \leq b(t)$), then x is a market order that matches to one or more buy (respectively, sell) limit orders upon arrival. Whenever such a matching occurs, it does so at the price of the limit orders, which can lead to "price improvement" (compared to the price p_x of the incoming market order). Whether or not such a matching causes $b(t)$ (respectively, $a(t)$) to change at time t_x depends on the volume available at the bid $V_+(b(t_x),t)$ (respectively, at the ask $V_-(a(t_x),t)$) compared to υ_x. In particular, the new bid-price $\overline{b}(t_x)$ immediately after the arrival of a sell market order x is set as follows. One first computes the *largest* price b^* such that such that all the volume between b^* and $b(t_x)$ is greater

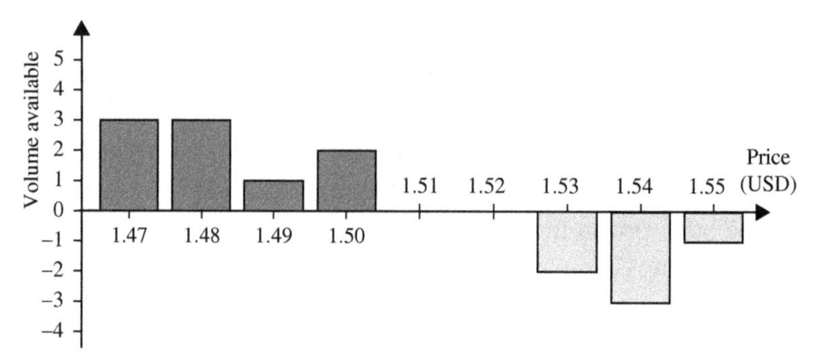

Figure 3.2. A specific example of an LOB, as presented in Table 3.1.

than or equal to υ_x:

$$\sum_{p=b^*}^{b(t_x)} V_+(p,t) \geq \upsilon_x,$$

$$\sum_{p=b^*+\vartheta}^{b(t_x)} V_+(p,t) < \upsilon_x.$$

Then $\overline{b}(t_x)$ is equal to b^* whenever $p_x \leq b^*$ and to the limit price p_x otherwise. In the latter case, the market order is not fully executed.

Similarly, the new ask-price $\overline{a}(t_x)$ immediately after the arrival of a buy market order x is the smallest of a^* or p_x, where:

$$\sum_{p=a(t_x)}^{a^*} V_-(p,t) \geq \upsilon_x,$$

$$\sum_{p=a(t_x)}^{a^*-\vartheta} V_-(p,t) < \upsilon_x.$$

To illustrate the above expressions, Table 3.1 lists several possible market events that could occur to the LOB displayed in Figure 3.2 and the resulting values of $\overline{b}(t_x), \overline{a}(t_x), \overline{m}(t_x)$ and $\overline{s}(t_x)$ after their arrival. Figure 3.3 shows how the arrival of an order $(+1, \$1.55, 3, t_x)$, as described by the third line of Table 3.1, would impact the LOB shown in Figure 3.2.

3.2 Practical Considerations

Many practical details of trading vary considerably across LOB platforms. In this section, we highlight some of these practical differences and discuss how they can influence both the temporal evolution of an LOB and the actions of traders within it.

Table 3.1. *Example to illustrate how specified order arrivals would affect prices in the LOB in the top panel of Figure 3.3. The tick size is $\vartheta = 0.01$ and the lot size is $\upsilon_0 = 1$ (see Section 3.1.5).*

Arriving order x	Values before arrival (USD)				Values after arrival (USD)			
$(\varepsilon_x, p_x, \upsilon_x, t_x)$	$b(t_x)$	$a(t_x)$	$m(t_x)$	$s(t_x)$	$\overline{b}(t_x)$	$\overline{a}(t_x)$	$\overline{m}(t_x)$	$\overline{s}(t_x)$
$(+1, \$1.48, 3, t_x)$	1.50	1.53	1.515	0.03	1.50	1.53	1.515	0.03
$(+1, \$1.51, 3, t_x)$	1.50	1.53	1.515	0.03	1.51	1.53	1.52	0.02
$(+1, \$1.55, 3, t_x)$	1.50	1.53	1.515	0.03	1.50	1.54	1.52	0.04
$(+1, \$1.55, 5, t_x)$	1.50	1.53	1.515	0.03	1.50	1.55	1.525	0.05
$(-1, \$1.54, 4, t_x)$	1.50	1.53	1.515	0.03	1.50	1.53	1.515	0.03
$(-1, \$1.52, 4, t_x)$	1.50	1.53	1.515	0.03	1.50	1.52	1.51	0.02
$(-1, \$1.47, 4, t_x)$	1.50	1.53	1.515	0.03	1.48	1.53	1.505	0.05
$(-1, \$1.50, 4, t_x)$	1.50	1.53	1.515	0.03	1.49	1.50	1.495	0.01

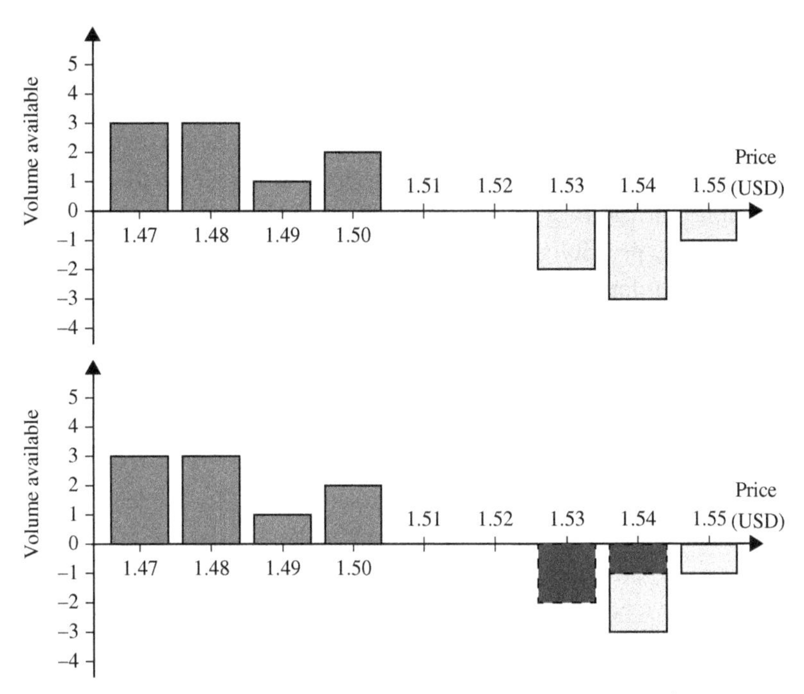

Figure 3.3. An illustration of how the arrival of an order $(+1, \$1.55, 3, t_x)$, as described by the third line of Table 3.1, would impact the LOB shown in Figure 3.2. The dashed boxes denote the limit orders that match to the incoming market order, and that are therefore removed from the LOB. The ask-price immediately after the market order arrival is $\overline{a}(t_x) = \$1.54$.

3.2.1 Priority Rules

As shown in Figure 3.1, several different limit orders can reside at the same price at the same time. Much like priority is given to limit orders with the best (i.e. highest buy or lowest sell) price, LOBs also employ a *priority system* for limit orders within each individual price level.

By far the most common rule currently used is **price-time priority**. That is, for buy (respectively, sell) limit orders, priority is given to the limit orders with the highest (respectively, lowest) price, and ties are broken by selecting the limit order with the earliest submission time t_x.

Another priority mechanism, commonly used in short-rate futures markets, is **pro-rata priority**. Under this mechanism, when a tie occurs at a given price, each relevant limit order receives a share of the matching equal to its fraction of the available volume at that price. Traders in pro-rata priority LOBs are faced with the substantial difficulty of optimally selecting limit order sizes: posting limit orders with larger sizes than the quantity that is really desired for trade becomes a viable strategy to gain priority.

Different priority mechanisms encourage traders to behave in different ways. Price-time priority encourages traders to submit limit orders early (even at prices away from the best quotes) and increase the available liquidity. Indeed, without a priority mechanism based on time, there is no incentive for traders to show their hand by submitting limit orders earlier than is absolutely necessary. Pro-rata priority rewards traders for placing larger limit orders and thus for providing greater liquidity at the best quotes. Because they directly influence the ways in which traders act, priority mechanisms play an important role, both for models and for market regulation.

3.2.2 Order Types

The actions of traders in an LOB can be expressed solely in terms of the submission or cancellation of orders of elementary size υ_0. For example, a trader who sends a sell market order of $4\upsilon_0$ units of the traded asset in the LOB displayed in Figure 3.2 can be regarded as submitting two sell orders of size υ_0 at the price \$1.50, one sell order of size υ_0 at the price \$1.49, and one sell order of size υ_0 at the price \$1.48. Similarly, a trader who posts a sell limit order of size $4\upsilon_0$ at the price \$1.55 can be regarded as submitting four sell orders of size υ_0 at a price of \$1.55 each.

Because all traders' actions can be decomposed in this way, it is customary to study LOBs in terms of these simple building blocks. In practice, many platforms offer traders the ability to submit a wide assortment of order types, each with complicated rules regarding its behaviour. Although such orders are rarely discussed in the literature on LOBs (because it is possible to decompose the resulting order flow into elementary limit and/or market orders), in practice their use is relatively widespread. We therefore provide a brief description of the main order types.[4]

- **Stop orders**: A stop order is a buy (respectively sell) market order that is sent as soon as the price of a stock reaches a specified level above (respectively below) the current price. This type of order is used by investors who want to avoid big losses or protect profits without having to monitor the stock performance. Stop orders can also be limit orders.
- **Iceberg orders**: Investors who wish to submit a large limit order without being detected can submit an iceberg order. For this order type, only a fraction of the order size, called *peak volume*, is publicly disclosed. The remaining part is not visible to other traders, but usually has lower time priority, i.e. regular limit orders at the same price that are submitted later will be executed first. When the publicly disclosed volume is filled and a hidden volume is still available, a new peak volume enters the book. On exchanges where iceberg orders are allowed, their use can be quite frequent. For example, according to studies of iceberg orders on Euronext, 30% of the book volume is hidden.
- **Immediate-or-cancel orders**: If a market participant wants to profit from a trading opportunity that s/he expects will last only a short period of time, s/he can send an "immediate-or-cancel" (IOC) order. Any volume that is not matched is cancelled, such that it never enters the LOB, and leaves no visible trace unless it is executed. In particular, if zero volume is matched, everything is as if the IOC order had never been sent. This order type is similar to the all-or-nothing (or fill-or-kill) order. Orders of the latter are either executed completely or are not executed at all. In this way, the investor avoids revealing his/her intention to trade if the entire quantity was not available.
- **Market-on-close orders**: A market-on-close (MOC) order is a market order that is submitted to execute as close to the closing price as possible. If a double-sided closing auction takes place at the end of the day (such as on the New York Stock Exchange (NYSE) or the Tokyo Stock Exchange), the order will participate in the auction. Investors might want to trade at the close because they expect liquidity to be high.

3.2.3 Opening and Closing Auctions

Many exchanges are closed overnight and suspend standard LOB trading at the beginning and end of the trading day. During these two periods, volumes are so large that LOB trading would be prone to instabilities, so exchanges prefer to use an auction system to match orders. For example, the LSE's flagship order book SETS has three distinct trading phases in each trading day:

[4] From Kockelkoren, J. (2010). *Encyclopedia of quantitative finance*. Wiley.

- a ten-minute **opening auction** between 07:50 and 08:00;
- a **continuous trading** period between 08:00 and 16:30 (during which the standard LOB mechanism is used); and
- a five-minute **closing auction** between 16:30 and 16:35.

During the opening and closing auctions, traders can place orders as usual, but no orders are matched. Due to the absence of matching, the highest price among buy orders can exceed the lowest price among sell orders. All orders are stored until the auction ends. At this time, for each price p at which there is non-zero volume available, the trade-matching algorithm calculates the total volume $Q(p)$ of trades that could occur by matching buy orders with a price greater than or equal to p to sell orders with a price less than or equal to p (this is precisely the mechanism described by Equation (1.4)). It then calculates the **auction price**:

$$p^* = \arg\max_p Q(p). \tag{3.8}$$

In contrast to standard LOB trading, all trades take place at the same price p^*. Given p^*, if there is a smaller volume available for sale than there is for purchase (or vice-versa), ties are broken using time priority.

Throughout the opening and closing auctions, all traders can see what the value of p^* would be if the auction were to end at that moment. This allows all traders to observe the evolution of the price without any matching taking place until the process is complete, and to revise their orders if needed. Such a price-monitoring process is common to many markets.

Take-Home Messages

(i) Most modern markets use limit order books (LOBs) to facilitate trade. LOBs allow liquidity takers to conduct transactions with the liquidity posted by liquidity providers.

(ii) In an LOB, traders interact via orders. An order consists of a direction (buy/sell), a price, a volume, and a submission time. The price and volume must be multiples of the tick and lot sizes, respectively.

(iii) Patient orders (i.e. orders that do not trigger an immediate transaction, and that are therefore added to the LOB) are called limit orders. Impatient orders (i.e. orders that trigger an immediate transaction against an existing limit order) are called market orders.

(iv) At any given time, the state of an LOB is simply the set of all the limit orders. The price of the best buy limit order is called the bid-price b, and the price of the best sell limit order is called the ask-price a.

(v) The quantity $m := (b+a)/2$ is called the mid-price. The quantity $s := a-b$ is called the bid–ask spread.

(vi) The mid-price and bid–ask spread both change whenever the bid-price or ask-price change. Such changes can be caused by the disappearance of the corresponding limit orders, or by the arrival of a new limit order inside the spread.

(vii) When a market order arrives, limit orders are executed according to priority rules. Limit orders with better prices (i.e. a higher buy price or lower sell price) always have priority over limit orders with worse prices. In case of price equality, other rules (such as time- or size-priority) are used to break the tie. In this way, an LOB can be regarded as a queuing system, with limit orders residing in queues at different prices.

3.3 Further Reading

Stoll, H. R. (1978). The pricing of security dealer services: An empirical study of NASDAQ stocks. *The Journal of Finance*, 33(4), 1153–1172.

Cohen, K. J., Conroy, R. M., & Maier, S. F. (1985). Order flow and the quality of the market. In *Market-making and the changing structure of the securities industry* (pp. 93–110). Lexington Books.

Glosten, L. R. (1994). Is the electronic open limit order book inevitable? *The Journal of Finance*, 49(4), 1127–1161.

Handa, P., & Schwartz, R. A. (1996). Limit order trading. *The Journal of Finance*, 51(5), 1835–1861.

Harris, L. (2003). *Trading and exchanges: Market microstructure for practitioners*. Oxford University Press.

Biais, B., Glosten, L., & Spatt, C. (2005). Market microstructure: A survey of microfoundations, empirical results, and policy implications. *Journal of Financial Markets*, 8(2), 217–264.

Hasbrouck, J. (2007). *Empirical market microstructure: The institutions, economics, and econometrics of securities trading*. Oxford University Press.

Parlour, C. A., & Seppi, D. J. (2008). Limit order markets: A survey. In Thakor, A. V. and Boot, A. (Eds.), *Handbook of financial intermediation and banking* (Vol. 5, pp. 63–95). North-Holland.

Foucault, T., Pagano, M., & Röell, A. (2013). *Market liquidity: Theory, evidence, and policy*. Oxford University Press.

Lehalle, C. A., & Laruelle, S. (2013). *Market microstructure in practice*. World Scientific.

4

Empirical Properties of Limit Order Books

When my information changes, I alter my conclusions. What do you do, sir?
(John Maynard Keynes)

Many LOBs record comprehensive digital transcriptions of their participants' submissions and cancellations of orders. These event-by-event records describe the temporal evolution of visible liquidity at the microscopic level of detail. During the past two decades, a vast number of empirical studies have analysed LOB data to address a wide array of questions regarding the high-frequency activity and price formation in financial markets. This work has served to illustrate many important aspects of the complex interplay between order flow, transactions and price changes. In recent years, several survey articles and books (see the list of references at the end of this chapter) have established new microstructural stylised facts and have highlighted both similarities and differences between different assets and different market organisations.

Although the breadth of such empirical work is substantial, an important consideration when reviewing previous studies of LOBs is that the high-frequency actors in financial markets have evolved rapidly in recent years. Market-making strategies, trading algorithms, and even the rules governing trade have changed over time, so old empirical observations may not accurately describe current LOB activity. Therefore, maintaining a detailed and up-to-date understanding of modern financial markets requires empirical analysis of recent, high-quality LOB data.

In this chapter, we present some up-to-date statistical results regarding order flow and LOB state for a collection of stocks that we study throughout the book:[1] PCLN (Priceline Group Inc.), TSLA (Tesla Motors Inc.), CSCO (Cisco Systems Inc.) and INTC (Intel Corp.). The stocks CSCO and INTC are large-tick stocks (i.e. with a spread close to the minimum value, $s \approx \vartheta$) and the stocks PCLN and TSLA are small-tick stocks (i.e. with a spread much larger than the tick, $s \gg \vartheta$)

[1] For a detailed discussion of our data and sample, see Appendix A.1.

(see Section 4.8 for more precise statements). As we will see at several points throughout this chapter, the statistical properties of large-tick stocks and small-tick stocks can be radically different.

4.1 Summary Statistics

Table 4.1 lists a range of summary statistics that describe the four stocks' aggregate activity and spread. A few notable features, common to all stocks, should be highlighted:

(i) The daily turnover (as a fraction of market capitalisation) is of the order of 0.5%. This result in fact holds across a wide selection of international stocks. This number has roughly doubled from 1995 to 2015.
(ii) The total volume displayed in the LOB within 1% of the mid-price (roughly half the daily volatility) is between 1% and 3% of the daily traded volume.
(iii) The activity at the best quotes takes place at a sub-second time scale (sometimes milliseconds), with much more activity for large-tick stocks than for small-tick stocks. This reflects the importance of queue position for large-tick stocks (see Section 21.4).
(iv) The number of trade-through market orders (i.e. orders that match at several different prices and therefore walk up the order book) is on the order of a few percent for small-tick stocks, and a few per thousand for large-tick stocks.

Point (ii) shows that the total outstanding volume in the LOB at any instant of time is only a small fraction of the total daily activity. This is a fundamental observation that has important consequences for trading, as we will discuss in Section 10.5.

4.2 Intra-day Patterns

Many properties of order flow and LOB state follow strong **intra-day patterns**, which we illustrate in this section. It is important to bear these patterns in mind when considering other order-flow and LOB statistics. For example, the distribution of the total volume at the best quotes will look very different when pooling together different times of day than when restricting observations to a specific time window. For the statistics that we calculate throughout the remainder of this chapter, we restrict our attention to market activity between 10:30 and 15:00 local time, where most quantities have an approximately flat average profile.

Table 4.1. *Summary statistics of aggregate activity for PCLN, TSLA, CSCO and INTC between 10:30 and 15:00 on all trading days during 2015. To remove effects stemming from the opening and closing auctions, we disregard the first and last hour of the regular trading time, except for variables marked with a dagger†. Cancellations include total order deletions and partial cancellations. We calculate the total market capitalisation, the daily traded volume on NASDAQ and the NASDAQ market share by using data from Compustat for the full trading days.*

		PCLN	TSLA	CSCO	INTC
Average share price (dollars)		1219.0	229.9	27.86	32.15
Average quoted spread† (dollars)		1.366	0.180	0.0105	0.0103
Average spread before transactions (dollars)		0.946	0.130	0.0109	0.0108
Fraction of MO that match at worse prices than the best quote		0.038	0.021	0.0017	0.0026
Fraction of MO that match a hidden order inside the spread		0.050	0.038	0.0311	0.0337
Mean total volume at best quotes (dollars $\times 10^3$)		129.6	38.35	248.4	163.6
Mean daily traded volume on NASDAQ† (dollars $\times 10^6$)		209.8	204.0	142.1	181.4
Mean fraction of daily traded volume that is hidden		0.340	0.241	0.086	0.074
Mean total volume of active orders within 1% of mid-price (dollars $\times 10^6$)		2.701	1.690	3.885	3.342
Average stock-specific market share of NASDAQ†		50.2%	36.0%	39.2%	39.2%
Average total market capitalisation† (dollars $\times 10^9$)		61.15	30.50	143.2	149.6
Average daily number (Best Quote Orders $\times 10^3$)	Limit Orders	8.959	15.83	84.90	116.5
	Market Orders	1.342	3.932	3.123	4.394
	Cancellations	5.873	10.44	75.584	101.2
Mean Inter-Arrival Time of Best Quote Orders (seconds)	Limit Orders	2.393	1.208	0.220	0.158
	Market Orders	14.264	4.831	5.895	4.230
	Cancellations	3.620	1.812	0.247	0.183
Median Inter-Arrival Time of Best Quote Orders (seconds)	Limit Orders	0.136	0.063	0.00013	0.00005
	Market Orders	0.081	0.094	0.22063	0.15983
	Cancellations	0.110	0.051	0.00038	0.00018
Mean Size of Best Quote Orders (dollars $\times 10^3$)	Limit Orders	82.12	23.63	15.23	10.44
	Market Orders	75.88	25.95	22.23	20.37
	Cancellations	80.26	21.54	14.18	9.769

4.2.1 Market Activity

In Figure 4.1, we plot the mean total volume of executed market orders (MO) and submitted limit orders (LO) during two-minute intervals throughout the trading day. In both cases, the activity exhibits a **U-shape profile**, or better a "J-shape profile" with asymmetric peaks at the beginning and end of the day and a minimum at around midday. Activity during the busiest periods is about four times greater than activity during the quietest periods.

The J-shaped profile is similar across all stocks, independently of the tick size and market capitalisation. Why should this be so? One possible explanation is that the intense spike shortly after the market opens is caused by company news revealed during the previous overnight period, when markets are closed. The spike at the close is possibly due to traders who previously hoped to get a better deal having to speed up to finish their trades for the day. Another possible explanation is that individual traders implement non-uniform execution patterns to minimise their price impact (see Section 21.2).

Figure 4.2 shows how the volume at the best quotes varies according to the time of day. Here, we see a striking difference between large-tick and small-tick stocks: whereas small-tick stocks exhibit a J-shaped profile similar to those in Figure 4.1, large-tick stocks show a lack of liquidity in the early minutes of trading, followed by a slow increase throughout the day, then a final, steep rise shortly before the end of the trading day. The volumes available just before close are about 20 times larger than those just after open.

In each of Figures 4.1 and 4.2, the statistics are (within statistical errors) symmetric between buy orders and sell orders, so we only present the average of the two. Of course, this does not imply that this symmetry holds for a given time on a given day, where order imbalance can be locally strong, but rather that the symmetry emerges when averaging market activity over long times.

4.2.2 Bid-Ask Spread

Figure 4.2 also shows the intra-day average values of the bid–ask spread $s(t) = a(t) - b(t)$. The spread narrows throughout the trading day, quite markedly for small-tick stocks and much more mildly for large-tick stocks. As we will discuss in Chapter 16, the narrowing of the bid–ask spread is often interpreted as a reduction of the adverse selection faced by liquidity providers, as overnight news gets progressively digested by market participants.

In summary, Figure 4.2 paints an interesting picture that suggests that liquidity is much more scarce at the open than it is at the close. Compared to submitting a limit order, the average cost of immediacy (i.e. of submitting a market order) is much larger in the morning than it is later in the day: not only is the spread wider,

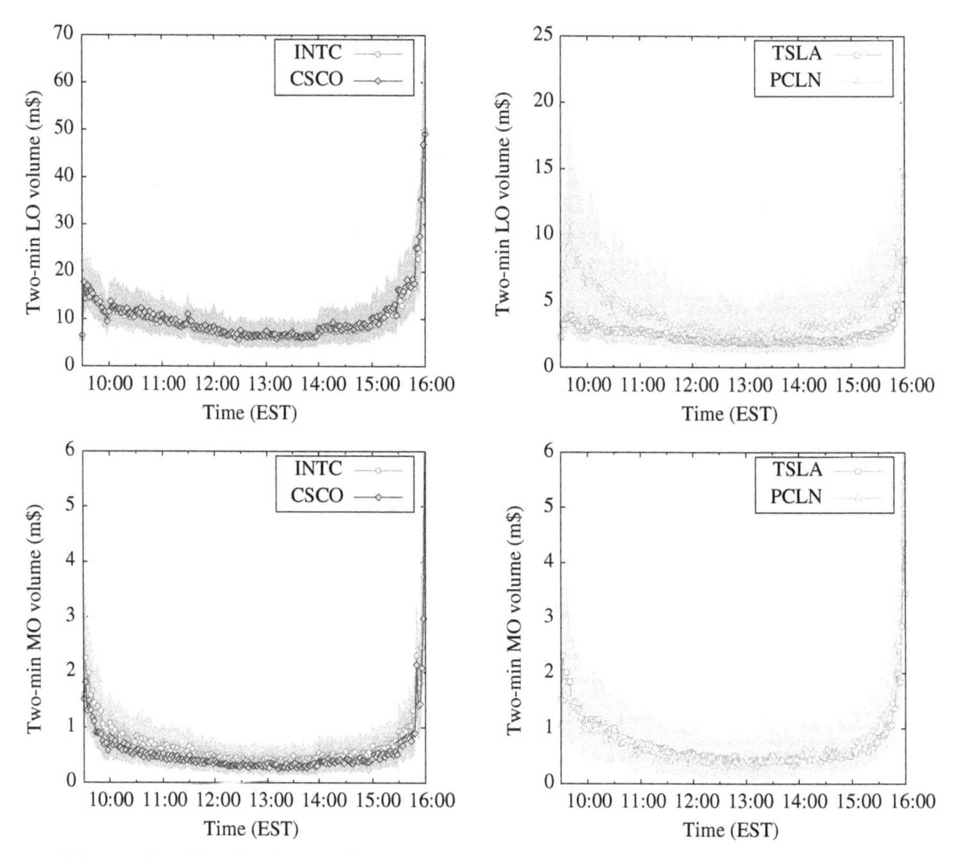

Figure 4.1. Total volume of (top panels) limit order and (bottom panels) market order arrivals during each two-minute interval throughout the trading day, for (left panels) the large-tick stocks INTC and CSCO, and (right panels) the small-tick stocks TSLA and PCLN. The markers show the average over all trading days and the shaded regions show the corresponding lower and upper quartiles.

but market orders need to penetrate deeper into the LOB to find their requested volume.

4.3 The Spread Distribution

Figure 4.3 shows the **spread distribution** for a selection of different stocks. As the figure illustrates, the distributions are very different for small-tick and large-tick stocks. For large-tick stocks, the distribution is sharply peaked at the minimum spread (which is equal to one tick, ϑ), with rare moments when the spread opens to 2ϑ or, in extreme cases, 3ϑ (or larger). For small-tick stocks, by contrast, the distribution is much wider, and the value of $s(t)$ can vary between one tick and several tens of ticks. The upper tail of the distribution decays approximately like an exponential.

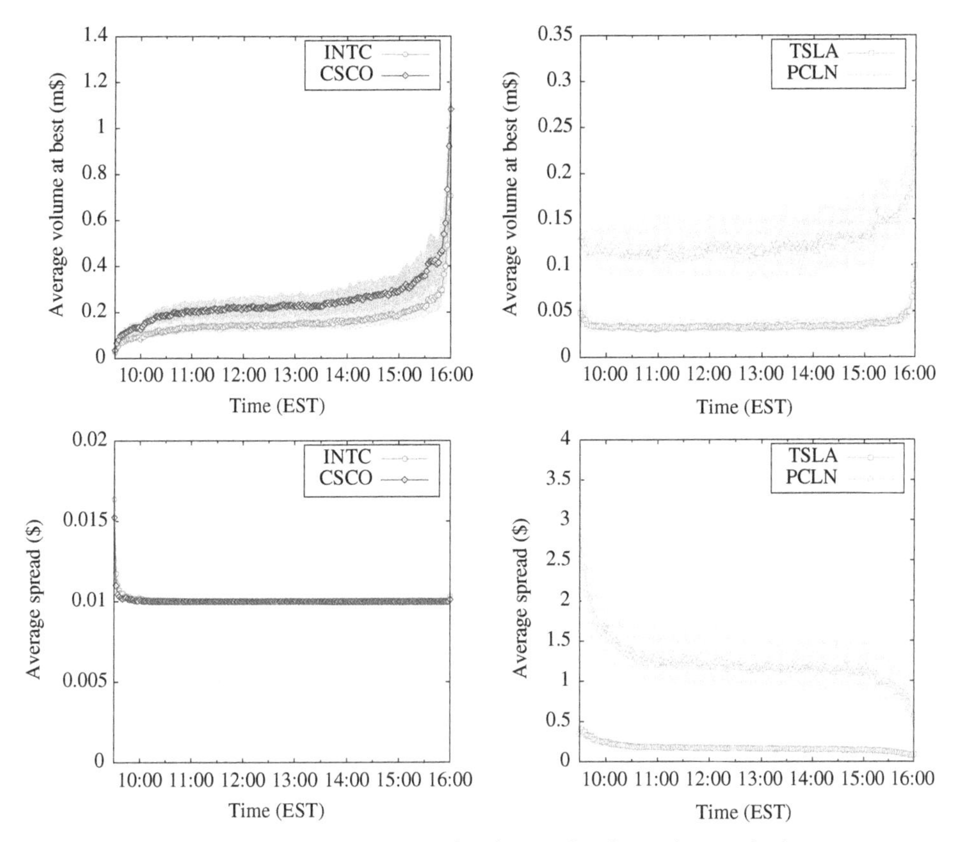

Figure 4.2. (Top panels) Average total volume of active orders at the best quotes and (bottom panels) average bid–ask spread, during each two-minute interval throughout the trading day for (left panels) the large-tick stocks INTC and CSCO, and (right panels) the small-tick stocks TSLA and PCLN. The markers show the average over all trading days and the shaded regions show the corresponding lower and upper quartiles.

Note that the bid–ask spread distribution can be measured in several different ways, such as at random instants in calendar time, at random instants in event-time, immediately before a transaction, etc. These different ways of measuring the spread do not yield identical results (see Figure 4.4). In a nutshell, the spread distribution is narrower when measured before transactions than at random instants. This makes sense, since liquidity takers carefully select the submission times of their market orders to benefit from relatively tight spreads.

4.4 Order Arrivals and Cancellations

Provided that their sizes do not exceed the volume available at the best quote, market orders can only be executed at the best bid or at the best ask. Limit orders,

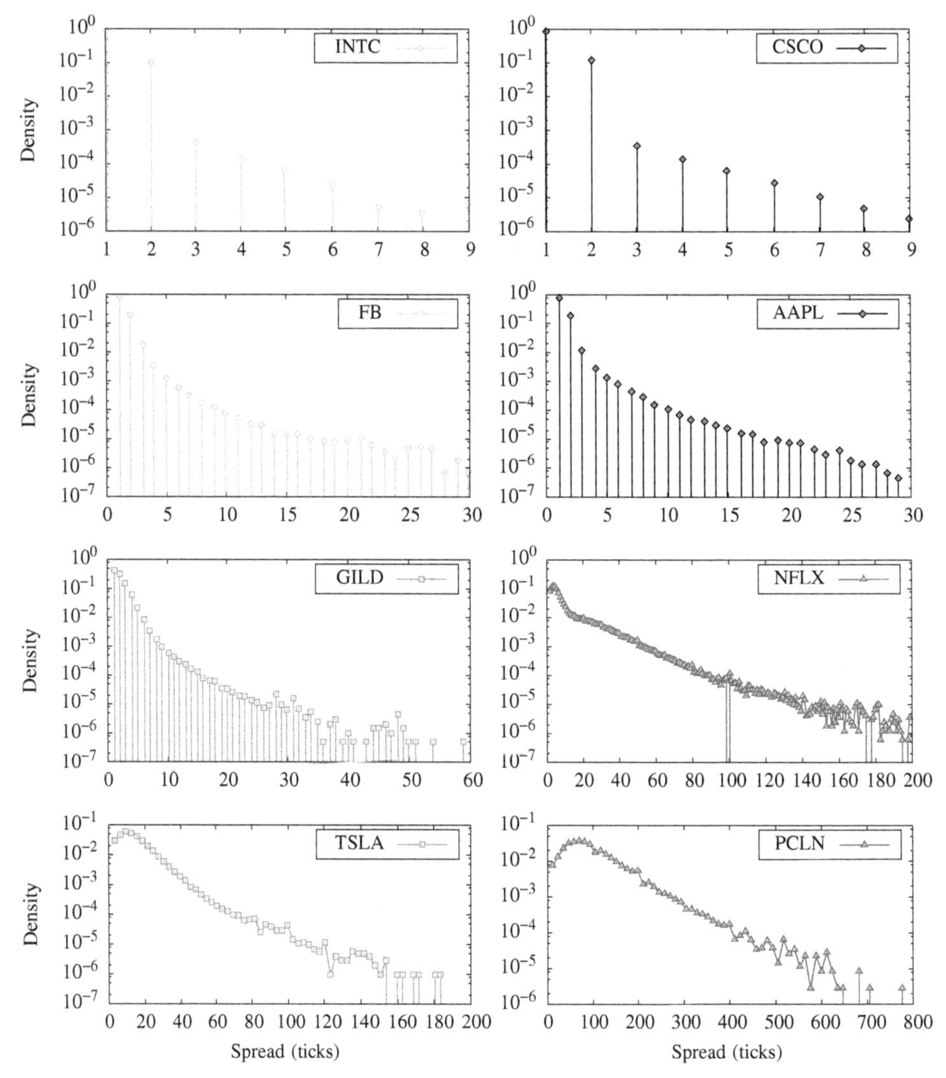

Figure 4.3. Empirical distributions of the bid–ask spread $s(t) = a(t) - b(t)$, measured at the times of events at the best quotes, for the stocks INTC, CSCO, FB, AAPL, GILD, NFLX, TSLA and PCLN. The stocks are ordered (from top left to bottom right) by their relative tick sizes.

on the other hand, can arrive and be cancelled at any of a wide range of prices. It is thus interesting to estimate the distributions of relative prices for arriving and cancelled limit orders, because (together with the arrivals of market orders) they play an important role in determining the temporal evolution of the state $\mathscr{L}(t)$ of the LOB (see for example Chapter 8).

Figure 4.5 shows the distribution of d, the same-side quote-relative prices for arriving limit orders, again for our selection of large-tick and small-tick

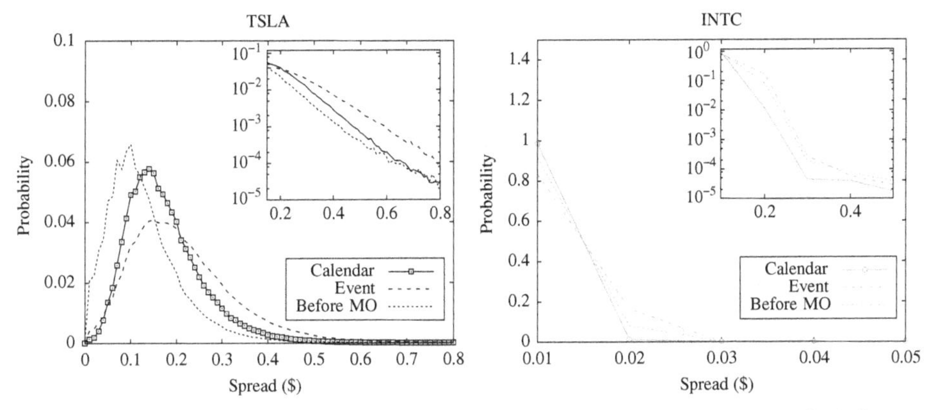

Figure 4.4. Empirical distribution of the bid–ask spread $s(t) = a(t) - b(t)$ for (left panel) TSLA and (right panel) INTC. The solid curves with markers show the results when measuring the spread at a random calendar time, the dashed curves show the results when measuring the spread at the times of events at the best quotes, and the dotted curves show the results when measuring the spread immediately before transactions.

stocks. In both cases, the **deposition probability** peaks at the best quotes, but there is also significant activity deeper in the LOB. Indeed, when plotting the distributions in semi-logarithmic coordinates, it becomes clear that the deposition probabilities decay rather slowly[2] with increasing relative price d. For large-tick stocks, the decay of the distribution is approximately monotonic. For such stocks, the probability of observing a spread $s(t) > \vartheta$ is in fact very small (see Figure 4.3), and when this happens, the overwhelming probability is that the next limit order will arrive within the spread. For small-tick stocks, by contrast, the LOB is usually sparse, with many empty price levels. There is also a secondary peak in the distribution at a distance comparable to the bid–ask spread. This suggests that it is the value of the spread itself, rather than the tick size, that sets the typical scale of the gaps between non-empty price levels in the LOB. Note that there is also substantial intra-spread activity for small-tick stocks (i.e. for $d < 0$), which makes sense because there are typically many empty price levels within the spread.

One can also study the corresponding distributions for limit order cancellations. For all non-negative relative prices, these plots (not shown) are remarkably similar to the distributions of limit order arrivals in Figure 4.5. This suggests that, to a first approximation, the cancellation rate is simply proportional to the arrival rate. One possible explanation for this result is that much LOB activity is associated with many limit orders being placed then rapidly cancelled.

[2] Several older empirical studies have reported this decay to follow a power law.

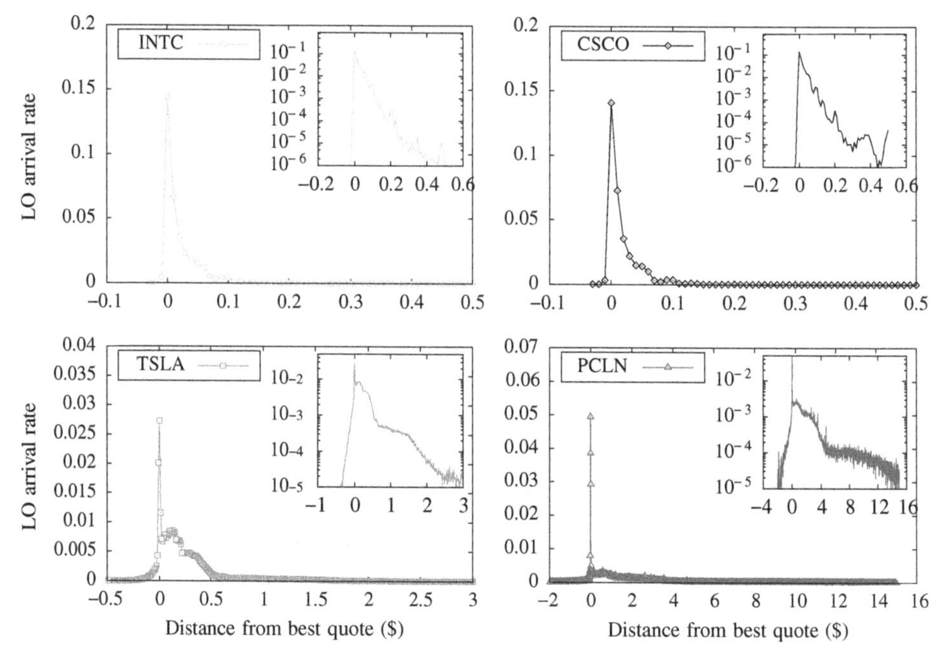

Figure 4.5. Empirical distributions of same-side relative prices d for arriving limit orders for (top panels) the large-tick stocks INTC and CSCO, and (bottom panels) the small-tick stocks TSLA and PCLN. The main plots show the body of the distributions and the inset plots show the upper tails in semi-logarithmic coordinates.

4.5 Order Size Distributions

Figure 4.6 shows the empirical cumulative density functions (ECDFs) for the sizes of limit orders and market orders, expressed in both lots and US dollars. The plot is in log-log coordinates, to emphasise that these distributions are extremely broad. For both limit and market orders, the most common order sizes are relatively small (a few thousands of dollars), but some orders are much larger. The upper tail of the ECDF appears to decay approximately according to a power law, with an exponent scattered around $-5/2$ (as represented by the dotted line). Although this power-law behaviour has been reported by empirical studies of many different markets, the value of the tail exponent varies quite significantly across these studies. Despite their quantitative differences, the same key message applies: order sizes are not clustered around some average size, but rather are distributed over a very broad range of values.

4.6 Volume at the Best Quotes

Figure 4.7 shows the ECDFs for the total volumes V_b and V_a available at the bid- and ask-prices, respectively. For both small-tick and large-tick stocks, the

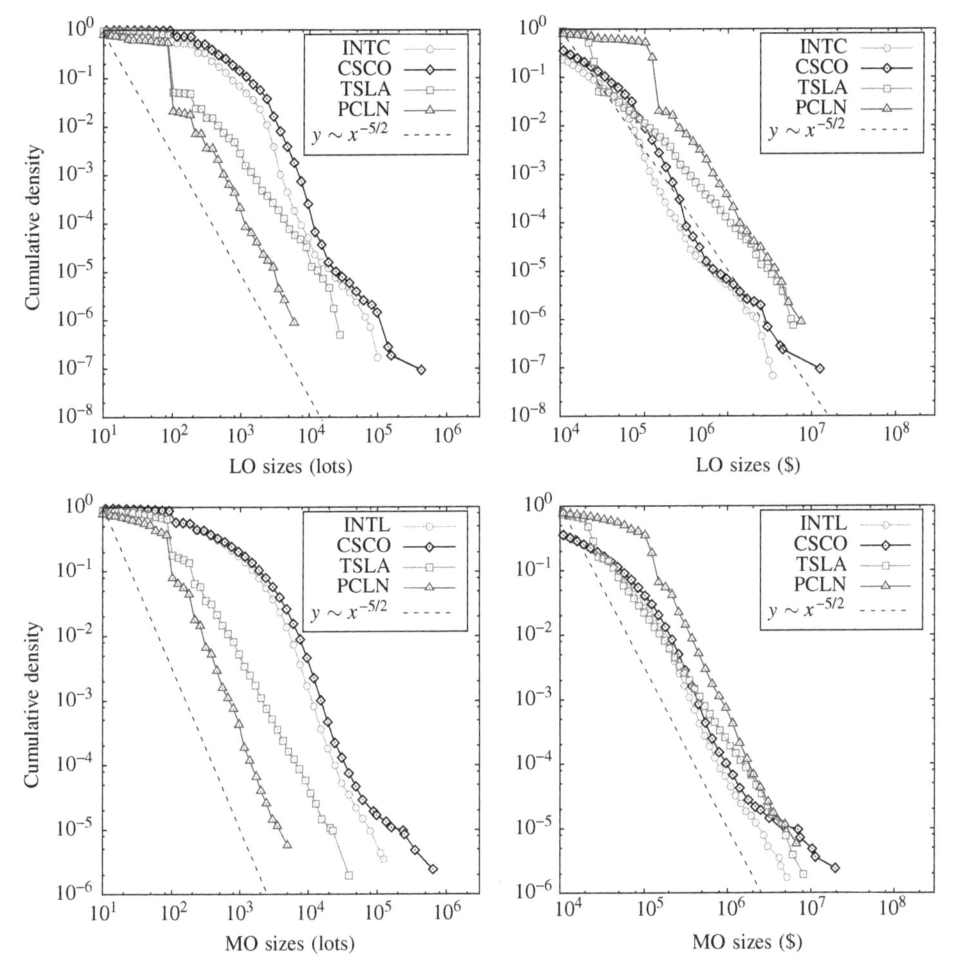

Figure 4.6. Empirical cumulative density functions (ECDFs) for the sizes of (top panels) market orders and (bottom panels) limit orders, expressed in (left column) number of shares and (right column) US dollars, for INTC, CSCO, TSLA and PCLN. The plots are in doubly logarithmic coordinates, such that power-law distributions appear as straight lines. The thin dotted line has slope $-5/2$.

distributions are approximately symmetric for buy and sell orders, and the tails of the distributions are similar to those of the distributions of order sizes in Figure 4.6. This makes sense because a single, large limit order arriving at $b(t)$ (respectively, $a(t)$) contributes a large volume to V_b (respectively, V_a). For small volumes, however, the distributions are quite different from those in Figure 4.6, because while small limit order sizes are common, observing a small value of V_b or V_a is much rarer. This is because the total volume available at $b(t)$ and $a(t)$ is typically comprised of several different limit orders.

As for the spread distribution, the distribution of queue volumes can be measured in calendar-time, in event-time or immediately before a market order.

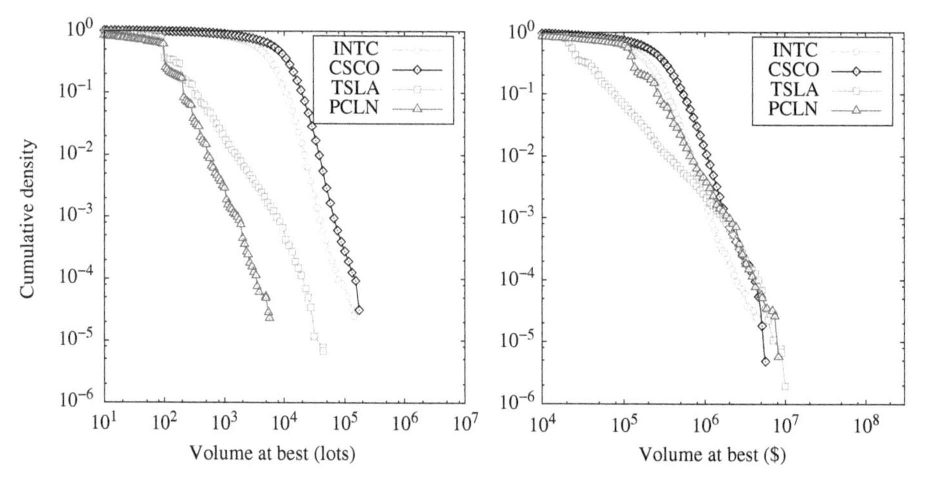

Figure 4.7. ECDFs for the total volumes available at the best quotes, for INTC, CSCO, TSLA and PCLN, expressed in (left panel) number of shares and (right panel) US dollars.

We again observe conditioning effects: for small-tick stocks, the volume at the best quote is higher immediately before being hit by a market order, indicating that liquidity takers choose to submit their orders when the opposite volume is relatively high. However, for large-tick stocks, the volume of the queue is *smaller* before transactions. In this case, as we discuss in Section 7.2, the queue volume itself conveys significant information about the direction of future price changes, because when liquidity is small at the ask (bid) and large at the bid (ask), the price typically moves up (down, respectively). When the queue is small, liquidity takers rush to take the remaining volume before it disappears.

4.7 Volume Profiles

Figure 4.8 shows the mean relative volume profiles, measured as a function of the distance to the opposite-side quote $d^{\dagger}$ (see Section 3.1.6). For this figure, and throughout this section, we use opposite-side distances because (by definition) there is always a non-zero volume available at the same-side best quotes. This causes a spurious sharp peak in the mean relative volume profile when measured as a function of the same-side distance d, which disappears for opposite-side distances.

The mean relative volume profiles first increase for small distances, then reach a maximum before decreasing very slowly for large distances. This shows that there is significant liquidity deep in the book even at very large distances from the best quotes. We also observe strong round-number effects: liquidity providers seem to prefer round distances (from the opposite best) for their limit orders, such as multiples of half dollars.

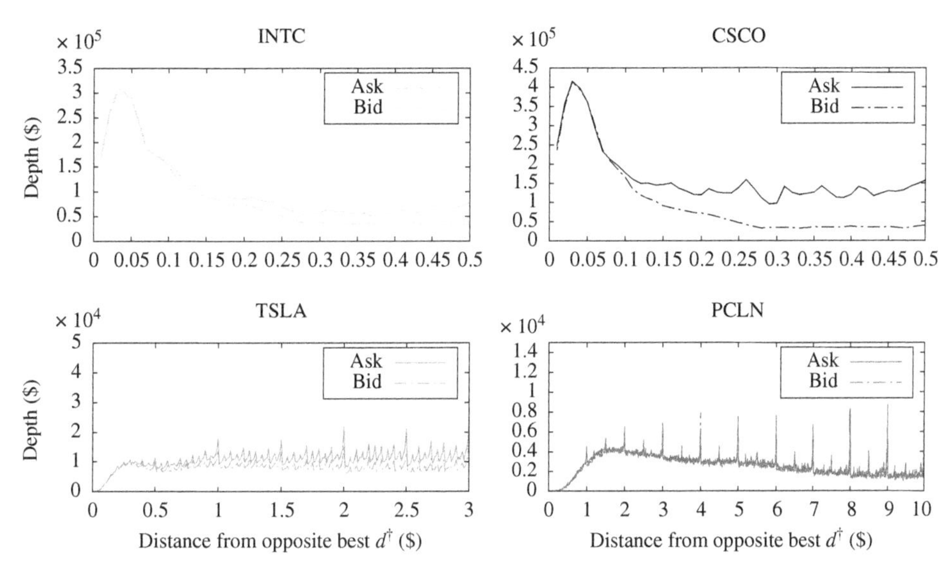

Figure 4.8. Mean relative volume profiles (measured as a function of the distance $d^{\dagger}$ to the opposite-side quote) of the (dash-dotted curves) bid side and the (solid curves) ask side of the LOBs of (top panels) the large-tick stocks INTC and CSCO, and (bottom panels) the small-tick stocks TSLA and PCLN.

To what extent are these average relative volume profiles representative of a typical volume profile snapshot? To address this question, we also plot snapshots of the volume profiles, taken at 10:30:00 on 3 August 2015, in Figure 4.9. In contrast to the mean relative volume profiles, for which liquidity is (on average) available over a wide range of consecutive prices, LOB snapshots tend to be sparse, in the sense that they contain many prices with no limit orders. This effect is particularly apparent for small-tick stocks, but less so for large-tick stocks (see Figure 4.5).

4.8 Tick-Size Effects

As we discussed in Section 3.1.5, the relative tick size ϑ_r varies considerably across different assets. In this section, we illustrate several ways in which LOBs with different relative tick sizes behave very differently. To illustrate our findings on a wide range of different stocks, throughout this section we consider a sample of 120 different US stocks traded on NASDAQ, for which the (non-relative) tick size is fixed to $\vartheta = \$0.01$ but for which the relative tick size ϑ_r varies considerably, because prices themselves can vary between a few cents (so called "penny stocks") and thousands of dollars (see, e.g., PCLN in Table 4.1).

4.8.1 Tick and Spread

First, we address how ϑ_r impacts the mean bid–ask spread $\langle s \rangle$ (see Figure 4.10). As the figure illustrates, when ϑ_r is small (i.e. when the stock price is large), the

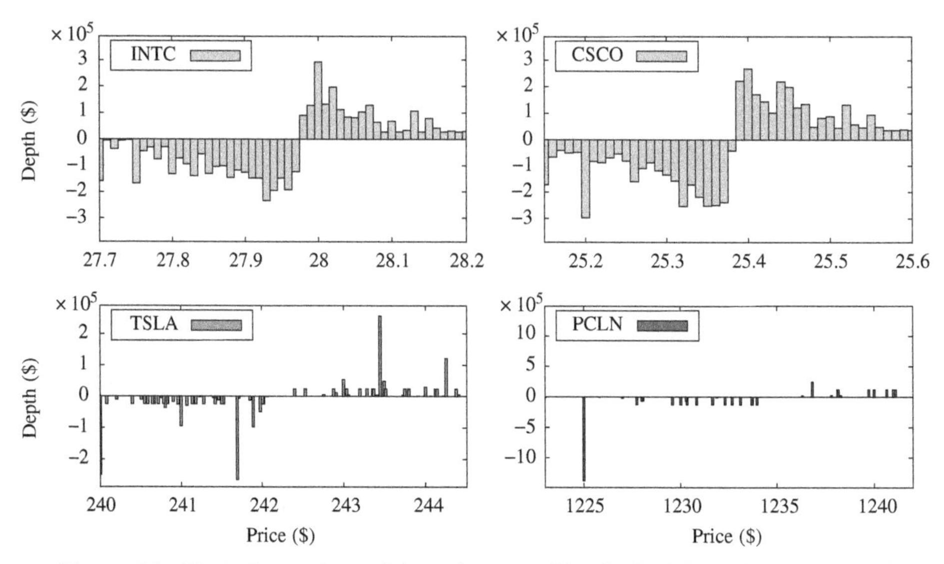

Figure 4.9. Typical snapshots of the volume profiles for INTC, CSCO, TSLA and PCLN, taken at 10:30:00 (EST) on 3 August 2015.

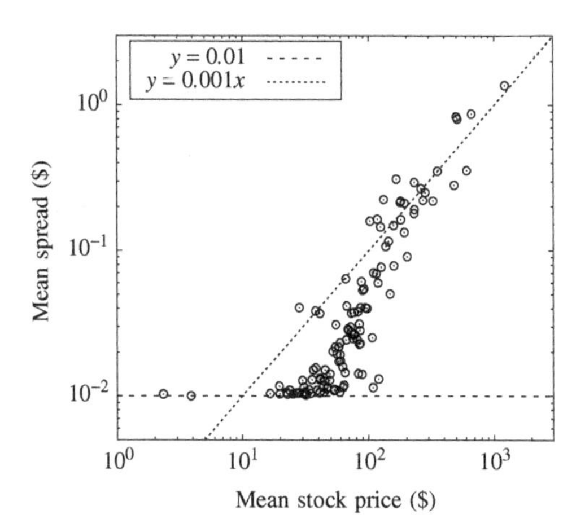

Figure 4.10. Mean bid–ask spread versus the mean share price, for a sample of 120 different US stocks traded on NASDAQ. The tick size on NASDAQ constrains the spread from below (dashed black line).

average spread is roughly proportional to the price of the stock itself. Empirically, the relationship $\langle s \rangle \cong 0.001 \times m(t)$ appears to hold approximately. When the price is smaller than \$10, this proportionality breaks down, because the spread cannot be smaller than $\vartheta = \$0.01$. The mean relative spread is thus much larger for large-tick stocks.

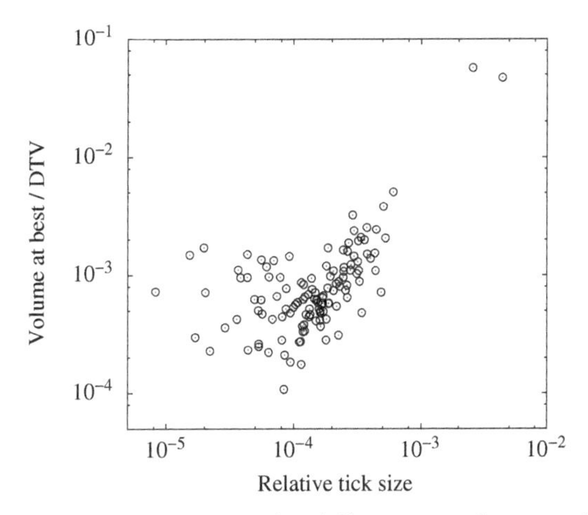

Figure 4.11. Average ratio between the daily mean volume at the best quotes and the daily traded volume versus the relative tick size ϑ_r, for a sample of 120 different US stocks traded on NASDAQ.

4.8.2 *Tick and Volume at Best*

Second, we address how ϑ_r impacts the mean depths at the best quotes $\langle V_b \rangle$ and $\langle V_a \rangle$, expressed as a fraction of the daily traded volume (see Figure 4.11). For small-tick stocks, $\langle V_b \rangle$ and $\langle V_a \rangle$ are about $\cong 1\%$ of the daily traded volume. For very large-tick stocks, by contrast, $\langle V_b \rangle$ and $\langle V_a \rangle$ can reach $\cong 10\%$ of the daily traded volume. Note that very small relative tick sizes correspond to very large stock prices, since the absolute tick is fixed to $\vartheta = \$0.01$. This can explain why the ratio in Figure 4.11 appears to saturate for the very small-tick stocks, as a single limit order containing 100 shares of PCLN (our smallest-tick stock) already corresponds to a volume of $120,000!

4.8.3 *Tick and Volume of Trades*

Finally, we consider how ϑ_r impacts the ratio of the average volume of a market order to the average volume at the best quotes (see Figure 4.12). For large-tick stocks, market orders are typically small ($< 10\%$) compared to the mean volume at the best quotes. For small-tick stocks, by contrast, arriving market orders typically consume more than half of the outstanding volume (which is itself quite small in this case).

4.9 Conclusion

In this chapter, we have presented a selection of empirical properties of order flow and LOB state. Some of these statistical properties vary considerably across

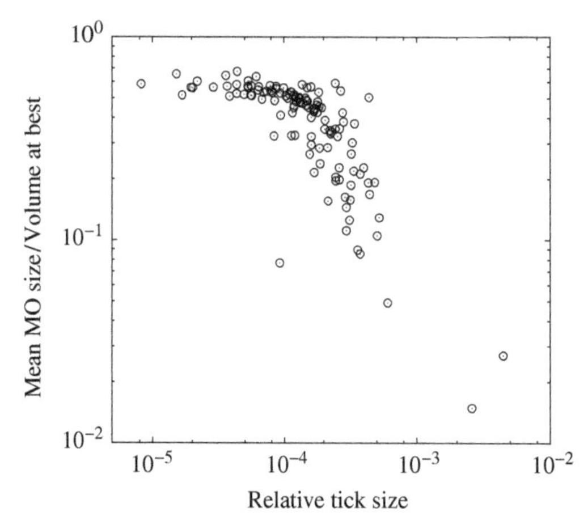

Figure 4.12. Average ratio between the daily mean market order size and the daily mean volume at the best quotes versus the relative tick size ϑ_r, for a sample of 120 different US stocks traded on NASDAQ.

different stocks, whereas others appear to be more universal. The most important feature distinguishing different stocks is the relative tick size ϑ_r, which itself varies considerably across different stocks – even among stocks with the same (absolute) tick size ϑ.

Although the statistics in this chapter provide an interesting glimpse into the behaviour of real LOBs, they are just the first small step towards a more comprehensive understanding of real market activity. Developing such an understanding requires addressing not only specific LOB properties in isolation, but also the interactions between them. In the subsequent chapters, we will consider some of the most relevant questions to help quantify both these interactions and the complex phenomena that emerge from them.

Take-Home Messages

(i) At any instant of time, the total volume in an LOB is a small fraction of the corresponding daily traded volume.

(ii) When measured over a suitably long time horizon, all market statistics are approximately symmetric between buys and sells.

(iii) Many market quantities present a strong average daily profile: volumes and activity exhibit a J-shaped pattern, with most activity happening close to the open and close of the market, whereas the spread undergoes a steep decrease after the open, then decreases more gradually throughout the remainder of the day.

(iv) For large-tick stocks, the spread is almost always equal to one tick. Small-tick stocks show a broader distribution of spread values.
(v) Most activity takes place close to or inside the spread. This is partly due to the fact that traders have little incentive to display publicly their trading intentions long in advance (except to gain queue priority in the case of large-tick stocks).
(vi) The size distributions of limit orders, market orders and best queue volumes have heavy tails.

4.10 Further Reading

General

Madhavan, A. (2000). Market microstructure: A survey. *Journal of Financial Markets*, 3(3), 205–258.
Biais, B., Glosten, L., & Spatt, C. (2005). Market microstructure: A survey of microfoundations, empirical results, and policy implications. *Journal of Financial Markets*, 8(2), 217–264.
Hasbrouck, J. (2007). *Empirical market microstructure: The institutions, economics, and econometrics of securities trading*. Oxford University Press.
Chakraborti, A., Toke, I. M., Patriarca, M., & Abergel, F. (2011). Econophysics review: II. Agent-based models. *Quantitative Finance*, 11(7), 1013–1041.
Cont, R. (2011). Statistical modeling of high-frequency financial data. *IEEE Signal Processing Magazine*, 28(5), 16–25.
Gould, M. D., Porter, M. A., Williams, S., McDonald, M., Fenn, D. J., & Howison, S. D. (2013). Limit order books. *Quantitative Finance*, 13(11), 1709–1742.
Abergel, F., Chakraborti, A., Anane, M., Jedidi, A., & Toke, I. M. (2016). *Limit order books*. Cambridge University Press.

Empirical Studies of LOBs

Biais, B., Hillion, P., & Spatt, C. (1995). An empirical analysis of the limit order book and the order flow in the Paris Bourse. *The Journal of Finance*, 50(5), 1655–1689.
Bollerslev, T., Domowitz, I., & Wang, J. (1997). Order flow and the bid–ask spread: An empirical probability model of screen-based trading. *Journal of Economic Dynamics and Control*, 21(8), 1471–1491.
Gourieroux, C., Jasiak, J., & Le Fol, G. (1999). Intra-day market activity. *Journal of Financial Markets*, 2(3), 193–226.
Kempf, A., & Korn, O. (1999). Market depth and order size. *Journal of Financial Markets*, 2(1), 29–48.
Challet, D., & Stinchcombe, R. (2001). Analyzing and modeling 1+ 1d markets. *Physica A: Statistical Mechanics and Its Applications*, 300(1), 285–299.
Bouchaud, J. P., Mézard, M., & Potters, M. (2002). Statistical properties of stock order books: Empirical results and models. *Quantitative Finance*, 2(4), 251–256.
Zovko, I., & Farmer, J. D. (2002). The power of patience: A behavioural regularity in limit-order placement. *Quantitative Finance*, 2(5), 387–392.
Farmer, J. D., Gillemot, L., Lillo, F., Mike, S., & Sen, A. (2004). What really causes large price changes? *Quantitative Finance*, 4(4), 383–397.

Hollifield, B., Miller, R. A., & Sandas, P. (2004). Empirical analysis of limit order markets. *The Review of Economic Studies*, 71(4), 1027–1063.

Huang, W., Lehalle, C. A., & Rosenbaum, M. (2015). Simulating and analysing order book data: The queue-reactive model. *Journal of the American Statistical Association*, 110(509), 107–122.

PART III
Limit Order Books: Models

Introduction

In this part, we will start our deep-dive into the mechanisms of price formation at the most microscopic scale. In the coming chapters, we will zoom in – in both space and time – to consider how the interactions between single orders contribute to the price-formation process in an LOB.

In line with our overall effort to start with elementary models before adding any layers of complexity, we will initially focus on purely stochastic, rather than strategic, behaviours. In short, we will assume that agents' actions are governed by simple rules that can be summarised by stochastic processes with rate parameters that depend only on the current state of the world – or, more precisely, on the current state of the LOB.

We will start with the study of the volume dynamics of a single queue of orders at a given price. Such a queue grows due to the arrival of new limit orders, or shrinks due to limit orders being cancelled or executed against incoming market orders. We will investigate the behaviour of the average volume of a queue, its stationary distribution and its time to depletion when starting from a given size. We will repeat each of these analyses with different assumptions regarding the behaviour of the order flows, to illustrate how these quantities depend on the specific details of the modelling framework.

From there, extending the model to account for the joint behaviour of the bid and ask queues will be a natural next step. We will introduce the "race to the bottom" between the best bid and ask queues. This race dictates whether the next price move will be upwards (if the ask depletes first) or downwards (if the bid depletes first). We will then fit these order flows directly to empirical data, in a model-free attempt to analyse the interactions between order flow and LOB state.

We will end this modelling effort by introducing a stochastic model for the whole LOB. This approach, which is often called the *Santa Fe* model, will allow us to make some predictions concerning the LOB behaviour within and beyond the bid–ask spread. Comparison with real-world data will evidence that such a "zero-intelligence" approach of a purely stochastic order flow succeeds at explaining some market variables, but fails at capturing others.

This part teaches us an important lesson: although useful and intuitive, simple, Markovian models cannot on their own fully account for the more complex phenomena that occur in real LOBs. More realistic approaches will require us to introduce additional ingredients – like *strategies* and *long-range dependences* – as we will see in the subsequent parts.

5

Single-Queue Dynamics: Simple Models

Nothing is more practical than a good theory.

(L. Boltzmann)

Modelling the full dynamics of an LOB is a complicated task. As we discussed in Chapter 3, limit orders can be submitted or cancelled at a wide range of different prices, and can also be matched to incoming market orders. Limit orders of many different sizes often reside at the same price level, where they queue according to a specified priority system (see Section 3.2.1). The arrival and cancellation rates of these orders also depend on the state of the LOB, which induces a feedback loop between order flow and liquidity and thereby further complicates the problem. Due to the large number of traders active in some markets, and given that each such trader can own many different limit orders at many different prices, even keeping track of an LOB's temporal evolution is certainly a challenge.

Despite these difficulties, there are many clear benefits to developing and studying LOB models. For example, analysing the interactions between different types of orders can help to provide insight into how best to act in given market situations, how to design optimal execution strategies, and even how to address questions about market stability. Therefore, LOB modelling attracts a great deal of attention from practitioners, academics and regulators.

Throughout the next four chapters, we introduce and develop a framework for LOB modelling. In the present chapter, we begin by considering the core building block of our approach: the temporal evolution of a single queue of limit orders, using highly simplified models. In Chapter 6, we extend our analysis to incorporate several important empirical facts into our theoretical description of single queues. In Chapter 7, we consider the joint dynamics of the best bid- and ask-queues together, from both a theoretical and an empirical point of view. Finally, in Chapter 8, we discuss how to extend these models to describe the dynamics of a full LOB. In all of these chapters, we aim to derive several exact results within the framework of simplified stochastic models, and approximate results for more realistic models calibrated to market data.

5.1 The Case for Stochastic Models

An immediate difficulty with modelling LOBs concerns the reasons that order flows exist at all. At one extreme lies the approach in which orders are assumed to be submitted by *perfectly rational* traders who attempt to maximise their utility by making trades in markets driven by exogenous information. This approach has been the starting point for many microstructural models within the economics community. However, this extreme assumption has come under scrutiny, both because such models are (despite their mathematical complexity) unable to reproduce many salient empirical stylised facts and because perfect rationality is difficult to reconcile with direct observations of the behaviour of individuals, who are known to be prone to a variety of behavioural and cognitive biases. Therefore, motivating order submissions in a framework of perfect rationality is at best difficult and at worst misleading.

At the other extreme lies an alternative approach, in which aggregated order flows are simply assumed to be governed by stochastic processes. Models that adopt this approach ignore the strategies employed by individual market participants, and instead regard order flow as random. Due to its exclusion of explicit strategic considerations, this approach is often called **zero-intelligence** (ZI) modelling. In a zero-intelligence LOB model, order flow can be regarded as a consequence of traders blindly following a set of stochastic rules without strategic considerations.

The purpose of ZI models is not to claim that real market participants act without intelligent decision making. Instead, these models serve to illustrate that in some cases, the influence of this intelligence may be secondary to the influence of the simple rules governing trade.

In many situations, ZI frameworks are too simplistic to reproduce the complex dynamics observed in real markets. In some cases, however, ZI models serve to illustrate how some seemingly non-trivial LOB properties or behaviours can actually emerge from the interactions of simple, stochastic order flows.

Much like perfect rationality, the assumptions inherent in ZI models are extreme simplifications that are inconsistent with some empirical observations. However, the ZI approach has the appeal of leading to quantifiable models without the need for auxiliary assumptions. Therefore, we adopt this approach for the simple benchmark LOB models that we consider in the following chapters. ZI models can often be extended to incorporate "boundedly rational" considerations, which serve to improve their predictive power by incorporating simple constraints that ensure the behaviour of traders acting within them is not obviously irrational, or to remove obvious arbitrage opportunities.

5.2 Modelling an Order Queue

Given that an LOB consists of many different order queues at many different prices, a sensible starting point for developing a stochastic model of LOB dynamics is to first concentrate on just one such queue. This enables us to ignore the interactions (including any possible correlations in order flow) between the activity at different price levels. This is the approach that we take throughout this chapter. Specifically, we consider the temporal evolution of a single queue of orders that is subject to limit order arrivals, market order arrivals and cancellations.

Let $V(t)$ denote the total volume of the orders in the queue at time t. Because we consider only a single order queue, the models studied in this chapter can be used for both the bid-queue and for the ask-queue. Assume that:

- all orders are of unit size $\upsilon_0 = 1$;
- limit orders arrive at the queue as a Poisson process with rate λ;
- market orders arrive at the queue as a Poisson process with rate μ;
- limit orders in the queue are cancelled as a Poisson process with some state-dependent rate (which we will specify in the following sections).

Despite the apparent simplicity of this modelling framework, we will see throughout the subsequent sections that even understanding the single-queue dynamics created by these three interacting order flows is far from trivial, and leads to useful insights about the volume dynamics at the best quotes.

5.3 The Simplest Model: Constant Cancellation Rate

To gain some intuition of the basic dynamics of the model in Section 5.2, we first consider the case in which the probability per unit time that exactly one order in a queue is cancelled is constant ν, independent of instantaneous queue length $V(t)$.

For $V \geq 1$, the model corresponds to the following stochastic evolution rules between times t and $t + \mathrm{d}t$:

$$\begin{cases} V \to \quad V+1 & \text{with rate} \quad \lambda \mathrm{d}t \quad \text{(deposition)}, \\ V \to \quad V-1 & \text{with rate} \quad (\mu+\nu)\mathrm{d}t \quad \text{(execution + cancellation)}. \end{cases} \tag{5.1}$$

Intuitively, our model regards the queue volume as a random walker with "position" V, with moves that occur as a Poisson process with rate $(\lambda + \mu + \nu)$. When a move occurs, the probability that it is an upward move is $\lambda/(\lambda+\mu+\nu)$ and the probability that it is a downward move is $(\mu+\nu)/(\lambda+\mu+\nu)$.

The case where the queue is empty requires further specification, because executions and cancellations are impossible when $V = 0$. To keep our calculations as simple as possible, we assume that as soon as the queue size depletes to zero, the queue is immediately replenished with some volume $V \geq 1$ of limit orders, chosen from a certain distribution $\varrho(V)$. We return to this discussion at many points throughout the chapter.

5.3.1 The Master Equation

To analyse our model in detail, we first write the master equation, which describes the temporal evolution of the probability $P(V,t)$ of observing a queue of length V at time t. Counting the different Poisson events with their respective probabilities, one readily obtains, for $V \geq 1$:

$$\frac{\partial P(V,t)}{\partial t} = -(\lambda + \mu + \nu)P(V,t) + \lambda P(V-1,t) + (\mu + \nu)P(V+1,t) + J(t)\varrho(V), \quad (5.2)$$

where the assumption that the queue replenishes as soon as it reaches $V = 0$ implies that $P(V = 0, t) = 0$. The probability per unit time of such depletion events is $J(t)$, which justifies the presence of the **reinjection current** (equal here to the **exit flux**) $J(t)\varrho(V)$, which represents replenished queues with initial volume V.

Given a dynamic equation such as in Equation (5.2), it is often desirable to seek a **stationary solution** $P_{\text{st.}}(V)$, for which

$$\frac{\partial P(V,t)}{\partial t} = 0.$$

In the stationary solution, the probability of observing a queue with a given length V does not evolve over time. In this case, it must also hold that $J(t)$ does not depend on time.

When $\lambda > \mu + \nu$, there cannot be any stationary state since the limit order arrival rate is larger than the departure rate (see Equation (5.6) below). In other words, there is a finite probability that the queue size grows unboundedly, without ever depleting to 0. This behaviour is clearly unrealistic for bid- and ask-queues, so we restrict our attention to the case where $\lambda \leq \mu + \nu$.

Let us first assume that the reinjection process is such that $\varrho(V) = 1$ for $V = V_0$ and $\varrho(V) = 0$ otherwise – in other words, that all newborn queues have the same initial volume V_0. To find the stationary solution in this case, we must solve Equation (5.2) when the left-hand side is equal to 0 and when $J(t)$ is equal to some constant J_0 that matches the exit flow. In this case, observe that the ansatz $f(V) = k + a^V$ is a solution, provided that

$$(\mu + \nu)a^2 - (\lambda + \mu + \nu)a + \lambda = 0. \quad (5.3)$$

Equation (5.3) has two roots,

$$a_+ = 1; \qquad a_- = \frac{\lambda}{\mu+\nu} \leq 1.$$

The solution that tends to zero at large V reads

$$P_{\text{st.}}(V \leq V_0) = A + Ba_-^V,$$
$$P_{\text{st.}}(V > V_0) = Ca_-^V,$$

with constants A, B, and C such that the following conditions hold:

(i) By considering $V = 1$, it must hold that

$$-(\lambda+\mu+\nu)P_{\text{st.}}(1) + (\mu+\nu)P_{\text{st.}}(2) = 0,$$

or

$$A\lambda = Ba_-[(\mu+\nu)a_- - (\lambda+\mu+\nu)],$$

which can be simplified further to $A = -B$.

(ii) By considering $V = V_0$, it must hold that

$$J_0 + (\mu+\nu)[(C-B)a_-^{V_0+1} - A] = 0.$$

(iii) By considering the exit flux, it must hold that

$$J_0 = (\mu+\nu)P_{\text{st.}}(1) = (\mu+\nu)(A + Ba_-).$$

(iv) By considering the normalisation of probabilities, it must hold that

$$\sum_{V>0} P_{\text{st.}}(V) = 1.$$

Rearranging these equations yields

$$C = A(a_-^{-V_0} - 1); \qquad J_0 = A(\mu+\nu-\lambda). \tag{5.4}$$

Finally, the normalisation condition simplifies to

$$AV_0 = 1. \tag{5.5}$$

The stationary solution $P_{\text{st.}}(V)$ then reads:

$$P_{\text{st.}}(V \leq V_0) = \frac{1}{V_0}\left(1 - a_-^V\right),$$
$$P_{\text{st.}}(V > V_0) = \frac{1}{V_0}\left(a_-^{-V_0} - 1\right)a_-^V.$$

The average volume $\bar{V} = \sum_{V=1}^{\infty} V P_{\text{st.}}(V)$ can be computed exactly for any V_0, but its expression is messy. It simplifies in the limit $a_- \to 1$, where it becomes:

$$\bar{V} \approx (1 - a_-)^{-1} = \frac{\mu + \nu}{\mu + \nu - \lambda}, \tag{5.6}$$

independently of V_0. This result clearly shows that the average volume diverges as $\lambda \uparrow (\mu + \nu)$, beyond which the problem admits no stationary state.

Beyond being a good exercise in dealing with master equations, the above calculations also provide insight into how we might approach the more general case, in which reinjection occurs not with some constant volume V_0, but instead with some arbitrary reinjection probability $\varrho(V_0)$. By linearity of the equations, we can solve this case as a linear superposition of the above solutions for different values of V_0, each with weight $\varrho(V_0)$. In this case, the normalisation in Equation (5.5) now reads:

$$A \sum_{V_0} V_0 \varrho(V_0) = 1. \tag{5.7}$$

How might we choose the function ϱ? One possible choice is simply to self-consistently use the stationary distribution, $P_{\text{st.}}(V)$. This corresponds to the simplifying assumption that whenever a queue depletes to 0, it is replaced by a new queue (behind it) drawn from the same stationary distribution. We will discuss this further in Section 5.3.6.

5.3.2 First-Hitting Times

Another quantity of interest is the length of time that elapses before a queue with a given length V first reaches $V = 0$. We call this time the **first-hitting time** $T_1(V)$. Given that we assume the queue dynamics to be stochastic, it follows that $T_1(V)$ is a random variable. In this section, we use the master equation (5.2) to derive several interesting results about its behaviour.

Let $\mathbb{E}[T_1|V]$ denote the mean first-hitting time from state V. We first turn our attention to the case where $V = V_0$. Recall that we can consider $V(t)$ as the state of a random walker with dynamics specified by Equation (5.1), and that the exit flux J_0 denotes the probability per unit time that the queue hits zero (and is reinjected at V_0). The quantities $\mathbb{E}[T_1|V_0]$ and J_0 are then related by the equation

$$J_0 = \frac{1}{\mathbb{E}[T_1|V_0]}. \tag{5.8}$$

To see why this is the case, imagine observing the system for a very long time T. By definition of the exit flux J_0, the expected number of hits during time T is equal to $J_0 T$. If the average length of time between two successive hits is $\mathbb{E}[T_1]$, then the average number of hits during T is also $T/\mathbb{E}[T_1]$. We thus obtain a simple formula for the mean first-hitting time:

$$\mathbb{E}[T_1|V_0] = \frac{1}{J_0} = \frac{V_0}{\mu + \nu - \lambda}. \tag{5.9}$$

Intuitively, this result makes sense because $\mu + \nu - \lambda$ is the velocity with which the $V(t)$ process moves towards zero, and V_0 is the initial position.

We next turn our attention to deriving the full distribution of first-hitting times T_1. More precisely, we consider the question: given that the total volume in the queue is currently V, what is the probability $\Phi(\tau, V)$ that the first-hitting time is $T_1 = \tau$?

Let τ_1 denote the time at which the queue first changes volume (up or down). For the queue length to have remained constant between 0 and τ_1 requires that no event occurs during that time interval. Because of the Poissonian nature of volume changes, this happens with probability $e^{-(\nu+\mu+\lambda)\tau_1}$. Then, between times τ_1 and $\tau_1 + \mathrm{d}\tau_1$, the queue length will either increase with probability $\lambda \mathrm{d}\tau_1$ or decrease with probability $(\nu + \mu)\mathrm{d}\tau_1$. The first-hitting time problem then restarts afresh, but now from a volume $V + 1$ or $V - 1$ and first-hitting time $\tau - \tau_1$. Summing over these two different possibilities, one finds, for $V \geq 1$:

$$\Phi(\tau, V) = \int_0^\tau \mathrm{d}\tau_1 \lambda e^{-(\nu+\mu+\lambda)\tau_1} \Phi(\tau - \tau_1, V + 1) \tag{5.10}$$

$$+ \int_0^\tau \mathrm{d}\tau_1 (\mu + \nu) e^{-(\nu+\mu+\lambda)\tau_1} \Phi(\tau - \tau_1, V - 1), \tag{5.11}$$

with, trivially, $\Phi(\tau, V = 0) = \delta(\tau)$.

By introducing the **Laplace transform** of $\Phi(\tau, V)$,

$$\widehat{\Phi}(z, V) := \int_0^\infty \mathrm{d}\tau \Phi(\tau, V) e^{-z\tau}, \tag{5.12}$$

and using standard manipulations (see Appendix A.2), one can transform this equation for $\Phi(\tau, V)$ to arrive at:

$$\widehat{\Phi}(z, V) = \frac{\lambda}{\lambda + \mu + \nu + z} \widehat{\Phi}(z, V + 1) + \frac{\mu + \nu}{\lambda + \mu + \nu + z} \widehat{\Phi}(z, V - 1). \tag{5.13}$$

One can check directly that $\widehat{\Phi}(z = 0, V) = 1$ solves the above equation for $z = 0$. Recalling Equation (5.12), this just means that

$$\int_0^\infty \mathrm{d}\tau \Phi(\tau, V) = 1, \qquad \text{for all } V,$$

which implies that the probability that the queue will eventually deplete at some time in the future is equal to one (when $\lambda \leq \mu + \nu$).

By expanding Equation (5.12) for small z, we can also derive expressions for the moments of the distribution of first-hitting times:

$$\widehat{\Phi}(z,V) = 1 - z\mathbb{E}[T_1|V] + \frac{z^2}{2}\mathbb{E}[T_1^2|V] + O(z^3), \qquad (z \to 0). \tag{5.14}$$

Expanding Equation (5.13) to first-order in z therefore leads to:

$$-z\mathbb{E}[T_1|V] = -z\frac{\lambda}{\lambda+\mu+\nu}\mathbb{E}[T_1|V+1] - z\frac{\mu+\nu}{\lambda+\mu+\nu}\mathbb{E}[T_1|V-1] - z\frac{1}{\lambda+\mu+\nu}, \tag{5.15}$$

which implies that:

$$\mathbb{E}[T_1|V] = \frac{\lambda}{\lambda+\mu+\nu}\mathbb{E}[T_1|V+1] + \frac{\mu+\nu}{\lambda+\mu+\nu}\mathbb{E}[T_1|V-1] + \frac{1}{\lambda+\mu+\nu}. \tag{5.16}$$

This has a transparent interpretation: the average time for the queue to reach length 0 from an initial length of V is equal to the average time to make a move up or down, plus the average time to reach 0 from $V+1$ times the probability of making an upward move, plus the average time to reach 0 from $V-1$ times the probability of making a downward move.

In the case where $\lambda < \mu + \nu$, we can interpret the difference between $(\mu + \nu)$ and λ as the mean drift of the queue length towards 0. Since this drift does not depend on V, this suggests that when seeking solutions to $\mathbb{E}[T_1|V]$, we should consider expressions that take the form

$$\mathbb{E}[T_1|V] = A'V, \tag{5.17}$$

where, using Equation (5.16),

$$A'(\lambda - \mu - \nu) + 1 = 0. \tag{5.18}$$

Substituting Equation (5.18) into Equation (5.17), we finally recover

$$\mathbb{E}[T_1|V] = \frac{V}{\mu+\nu-\lambda},$$

which is exactly what we derived in Equation (5.9) by a different method.

An interesting point is that Equation (5.13) can be solved for any z, and that $\widehat{\Phi}(z,V) = a(z)^V$ is a solution, provided that

$$\lambda a^2(z) - (\lambda+\mu+\nu+z)a(z) + \mu + \nu = 0, \tag{5.19}$$

which can be rearranged to

$$a_{\pm}(z) = \frac{1}{2\lambda}\left[(\lambda+\mu+\nu+z) \pm \sqrt{(\lambda+\mu+\nu+z)^2 - 4\lambda(\mu+\nu)}\right]. \tag{5.20}$$

Note that $a_+(0) = (\mu + \nu)/\lambda > 1$ and $a_-(0) = 1$. Hence, the constraint $\widehat{\Phi}(0, V) \equiv 1$ eliminates the exponentially growing contribution $a_+(z)^V$, which finally leads to a very simple result:

$$\ln \widehat{\Phi}(z, V) = V \ln a_-(z). \tag{5.21}$$

Expanding to second order in z directly allows one to derive the variance of the first-hitting time (see Appendix A.2):

$$\mathbb{V}[T_1|V] = V \frac{\mu + \nu + \lambda}{(\mu + \nu - \lambda)^3}. \tag{5.22}$$

If an explicit solution is required, Equation (5.21) can be inverted to yield the exact hitting time distribution for any V. However, the analytical form of this solution is messy and not particularly instructive. More interestingly, Equation (5.21) shows that all cumulants of the hitting time distribution behave linearly in V. This situation is identical to the well-known **Central Limit Theorem**, where all cumulants of the sum of V IID random variables behave linearly in V (see Appendix A.2).[1] Therefore, Equation (5.21) can be used to show that for large V, the distribution of first-hitting times approaches a Gaussian distribution with mean $V/(\mu + \nu - \lambda)$ and variance $V(\mu + \nu + \lambda)/(\mu + \nu - \lambda)^3$.

5.3.3 Long Lifetimes: The Critical Case

So far in this section, we have considered the case where $\lambda < \mu + \nu$. The case $\lambda = \mu + \nu$, where the flow is exactly balanced, is quite different. By inspection, one sees that the expression for the mean first-hitting time in Equation (5.9) diverges when $\lambda = \mu + \nu$. Therefore, the expected first-hitting time in this case is infinite. However, this does not imply that the queue never empties. In fact, the queue empties with probability 1, although this can take a very long time.

In this special case, expanding Equations (5.20) and (5.21) in the limit $z \to 0$ leads to the singular expansion:

$$\ln \widehat{\Phi}(z, V) = V \ln a_-(z) \underset{z \to 0}{\approx} -V \sqrt{\frac{z}{\lambda}} + O(z). \tag{5.23}$$

The $\sqrt{z}$ behaviour for small z implies that for large τ, the distribution of the hitting time $\Phi(\tau, V)$ decays as $(V/\sqrt{\lambda})\tau^{-3/2}$ (see Appendix A.2). This distribution is very broad; in fact, it has an infinite mean, which is characteristic of unbiased one-dimensional random walks. This behaviour can be traced back to the very long upward excursions of the queue length, as the force driving to make it smaller is exactly zero in the critical case. When $\lambda \uparrow (\mu + \nu)$, the $\tau^{-3/2}$ tail is truncated beyond a time scale $\propto (\mu + \nu - \lambda)^{-2}$, much larger than the average exit time $\mathbb{E}[T_1|V]$.

[1] In fact, the first-hitting time from V is equal to the first-hitting time of $V - 1$ from V plus the first-hitting time from $V - 1$, which recursively demonstrates that $T_1(V)$ is indeed the sum of V independent random variables.

In fact, the only way that the above (highly simplified) random walk model can be used to represent the dynamics of long queues is when the parameters are chosen to be close to the critical case – otherwise, order queues would never become large – see Equation (5.6). However, as we will see in the subsequent sections of this chapter, this fine-tuning to a critical case is not necessary when we consider more complex models in which the aggregate cancellation rate increases with the total volume of limit orders in the queue.

5.3.4 The Continuum Limit and the Fokker–Planck Equation

Throughout this section, we have studied a model in which the state space of V is discrete, and where V can only change by one unit at a time. We now consider the limit where the volume in the queue becomes large, such that $V \gg 1$ most of the time. In this situation, volume changes can be considered as infinitesimal. In this **continuum limit** (where V is now considered as a continuous variable), the master equation from Section 5.3.1 instead becomes a **Fokker–Planck equation**.

Consider again the dynamics described by Equation (5.1), and assume that $P(V,t)$ is sufficiently smooth to allow the following Taylor expansion:

$$P(V \pm 1, t) \approx P(V,t) \pm \partial_V P(V,t) + \frac{1}{2}\partial^2_{VV} P(V,t). \tag{5.24}$$

Substituting this expansion into the master equation (5.2) with $J(t) = 0$:

$$\begin{aligned}\frac{\partial P(V,t)}{\partial t} &\approx -(\lambda+\mu+\nu)P(V,t) + \lambda\left(P(V,t) - \partial_V P(V,t) + \frac{1}{2}\partial^2_{VV}P(V,t)\right)\\ &\quad + (\mu+\nu)\left(P(V,t) + \partial_V P(V,t) + \frac{1}{2}\partial^2_{VV}P(V,t)\right),\\ &\approx -(\lambda-\mu-\nu)\partial_V P(V,t) + (\lambda+\mu+\nu)\frac{1}{2}\partial^2_{VV}P(V,t).\end{aligned}$$

Introducing the **drift constant** $F = \lambda - \mu - \nu$ and the **diffusion constant** $D = (\lambda + \mu + \nu)/2$ yields

$$\frac{\partial P(V,t)}{\partial t} \approx -F\partial_V P(V,t) + D\partial^2_{VV}P(V,t). \tag{5.25}$$

Equation (5.25) is the simplest case of a Fokker–Planck equation, called the **drift–diffusion equation**. Working directly with this equation provides an alternative (and often simpler) route to calculating some of the quantities and distributions that we calculated for the case of strictly positive order sizes earlier in this section. For example, let us consider the stationary distribution $P_{\text{st.}}(V)$, which we calculate in Section 5.4.1 by solving the discrete master equation. We now consider Equation (5.25) with the left-hand side set to 0. We add an extra current contribution $J_0\delta(V - V_0)$ to the right-hand side, to account for the reinjection of

V_0 when the queue depletes to 0. For $F < 0$, the solution such that $P_{\text{st.}}(V = 0) = 0$ (because of the absorbing condition there) and $P_{\text{st.}}(V \to \infty) = 0$ takes the form:

$$P_{\text{st.}}(V \le V_0) = A(1 - e^{-|F|V/D}),$$
$$P_{\text{st.}}(V \ge V_0) = A(e^{|F|V_0/D} - 1)e^{-|F|V/D}.$$

The discontinuity of the slope at $V = V_0$ is fixed by the extra current contribution $J_0\delta(V - V_0)$:

$$A|F|e^{-|F|V_0/D} - J_0 = -A|F|(1 - e^{-|F|V_0/D}),$$

which yields

$$A = J_0/|F|.$$

Finally, the normalisation of $P_{\text{st.}}(V)$ fixes A such as $AV_0 = 1$, leading to $J_0 = |F|/V_0$. The mean first-hitting time is thus given by $\mathbb{E}[T_1] = V_0/|F|$.

5.3.5 Revisiting the Laplace Transform

We can also consider the continuum limit within the Laplace transform in Equation (5.12), to find:

$$s\widehat{\Phi}(z, V) = F\partial_V\widehat{\Phi}(z, V) + D\partial^2_{VV}\widehat{\Phi}(z, V). \tag{5.26}$$

Looking for solutions of the form $\widehat{\Phi}(z, V) \propto e^{a(z)V}$, it holds that $a(z)$ must obey

$$Da^2(z) + Fa(z) - z = 0,$$

so

$$a_\pm(z) = \frac{1}{2D}\left[-F \pm \sqrt{F^2 + 4Dz}\right].$$

Note that the problem is only well defined when $F < 0$ (i.e. when the drift is towards $V = 0$). The constraint that $\widehat{\Phi}(z = 0, V) = 1$ for all V eliminates the positive $a_+(z)$ solution. Hence

$$\ln \widehat{\Phi}(z, V) = a_-(z)V, \tag{5.27}$$

which is very similar to Equation (5.21). Note that $a_-(0) = 0$, such that the condition $\ln \widehat{\Phi}(0, V) = 0$ is indeed satisfied.

5.3.6 A Self-Consistent Solution

We now discuss the *self-consistent* case, in which reinjection occurs according to the stationary distribution $P_{\text{st.}}(V)$, such that the probability of observing a reinjection of a given size is precisely equal to the stationary probability of observing a queue of that size. This corresponds to the idea that when a queue empties, it is replaced by the queue just behind, which we assume to have the same stationary distribution. This is of course only an approximation, since the second-best queue is shielded from market orders, and might not be described by the same deposition and cancellation rates as the best queue.

The Fokker–Planck equation for $P_{\text{st.}}(V)$ reads

$$\frac{\partial P_{\text{st.}}(V)}{\partial t} = -F\partial_V P_{\text{st.}}(V) + D\partial^2_{VV} P_{\text{st.}}(V) + J_0 P_{\text{st.}}(V),$$

and by stationarity of $P_{\text{st.}}(V)$ it holds that

$$\frac{\partial P_{\text{st.}}(V)}{\partial t} = 0,$$

so we seek to solve

$$-F\partial_V P_{\text{st.}}(V) + D\partial^2_{VV} P_{\text{st.}}(V) + J_0 P_{\text{st.}}(V) = 0. \tag{5.28}$$

To find the solution that vanishes at $V = 0$, we proceed exactly as in the previous section, but replacing $\widehat{\Phi}(z, V)$ in Equation (5.26) with $P_{\text{st.}}$, replacing z with $-J_0$, and replacing F with $-F$. Since now both solutions $a_+(-J_0)$ and $a_-(-J_0)$ are negative for $F < 0$, we have:

$$P_{\text{st.}}(V) = A(-e^{a_+(-J_0)V} + e^{a_-(-J_0)V}).$$

The exit flux at $V = 0$ is given by $J_0 = -D\partial_V P_{\text{st.}}$, leading to:

$$J_0 = DA(a_+(-J_0) - a_-(-J_0)) = A\sqrt{F^2 - 4DJ_0}.$$

The normalisation condition is

$$1 = -A\left(\frac{1}{a_+(-J_0)} - \frac{1}{a_-(-J_0)}\right),$$

which turns out to be satisfied for any A. We can thus arbitrarily choose A, which amounts to choosing an arbitrary exit flux $J_0 \leq F^2/4D$. The corresponding stationary solution reads:

$$P_{\text{st.}}(V) = \frac{2J_0}{\sqrt{F^2 - 4DJ_0}} e^{-|F|V/2D} \sinh\left(\frac{\sqrt{F^2 - 4DJ_0}\, V}{2D}\right). \tag{5.29}$$

Interestingly, this model can self-consistently produce both long queues and short queues in the limit $J_0 \to 0$.

5.4 A More Complex Model: Linear Cancellation Rate

Throughout Section 5.3, we assumed that the total rate of cancellations was constant. Although this framework led to convenient mathematics, it is not a realistic representation of cancellations in real markets. In particular, the model requires fine-tuning of its parameters to produce long queues, such as those often observed in empirical data for large-tick assets.

How might we change the model to address this problem? One way is to assume that each individual order in the queue has cancellation rate ν independent of the queue length V, such that the total cancellation rate grows linearly as νV (instead of being constant, as in the previous model). This makes intuitive sense: the more orders in the queue, the larger the probability that one of them gets cancelled. Specifically, consider the model

$$\begin{cases} V \to & V+1 \quad \text{with rate} \quad \lambda \mathrm{d}t \quad \text{(deposition)}, \\ V \to & V-1 \quad \text{with rate} \quad (\mu + V\nu)\mathrm{d}t \quad \text{(execution + cancellation)}. \end{cases} \tag{5.30}$$

To lighten exposition in the subsequent sections, we introduce the notation $W_{\pm}(V)$ as the **transition rate** from $V \to V \pm 1$. For the model described by Equation (5.30),

$$W_{+}(V) = \lambda, \tag{5.31}$$

$$W_{-}(V) = \mu + V\nu. \tag{5.32}$$

Observe that the only difference between this model and the model specified by Equation (5.1) is the inclusion of the $V\nu$ term, instead of the ν term, to account for cancellations.

Importantly, the inclusion of this $V\nu$ term ensures that queue sizes self-stabilise for any choices of $\lambda, \mu, \nu > 0$. In other words, there exists some value

$$V^* = \frac{\lambda - \mu}{\nu} \tag{5.33}$$

of the queue length such that the total out-flow rate of orders $W_{-}(V^*)$ matches the in-flow rate of orders $W_{+}(V^*)$. If $V < V^*$, then $W_{+}(V) > W_{-}(V)$ and the queue length tends to grow on average; if $V > V^*$, then $W_{+}(V) < W_{-}(V)$ and the queue length tends to shrink on average. Those self-regulating mechanisms lead to queues that fluctuate around their **equilibrium size** V^*. Given a large value of λ, it is therefore possible to observe long queues without requiring a fine-tuned relation between λ, μ, and ν to ensure stability (as was the case in Section 5.3).

5.4.1 Quasi-Equilibrium

The master equation corresponding to the dynamics specified by Equation (5.30) is now given by

$$\frac{\partial P(V,t)}{\partial t} = -(\lambda + \mu + \nu V)P(V,t) + \lambda P(V-1,t) + (\mu + \nu(V+1))P(V+1,t); \quad V > 1. \tag{5.34}$$

For the moment, we fix the boundary condition such that whenever $V = 0$, the particle is immediately reinjected at state $V = 1$. Therefore, the exit flux is 0, there is no particle at $V = 0$ and the missing equation for $V = 1$ reads

$$\frac{\partial P(1,t)}{\partial t} = -\lambda P(1,t) + (\mu + 2\nu)P(2,t).$$

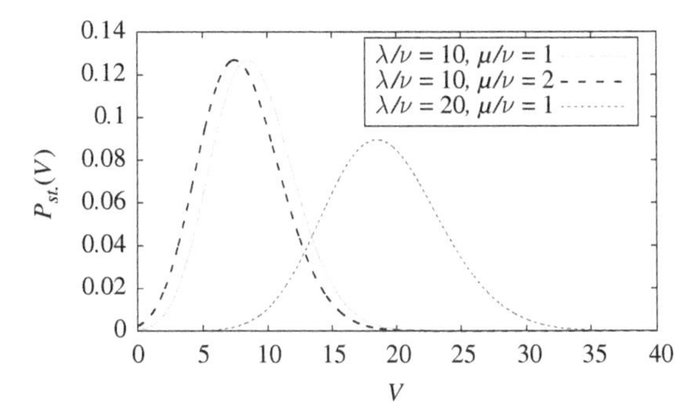

Figure 5.1. $P_{\text{st.}}(V)$ from Equation (5.35) for different values of λ/ν and μ/ν.

In this case, one can readily check that for $V \geq 2$, the ansatz

$$P_{\text{st.}}(V) = A \frac{\left(\frac{\lambda}{\nu}\right)^V}{\Gamma\left[V + 1 + \frac{\mu}{\nu}\right]} \tag{5.35}$$

is an exact stationary state, in the sense that it causes the right-hand side of Equation (5.34) to equal 0 for all $V > 1$. The value of $P_{\text{st.}}(1)$ is simply given by $(\mu + 2\nu)P_{\text{st.}}(2)$ and the constant A is fixed by the normalisation condition $\sum_{V=1}^{\infty} P_{\text{st.}}(V) = 1$.

To make sense of this solution when depletion and reinjection are present, let us assume that we are in a situation where queues are typically very large, corresponding to the limit $\lambda \gg \mu, \nu$, such that $V^* \gg 1$. Figure 5.1 shows the shape of $P_{\text{st.}}(V)$ for this case. As the figure illustrates, the stationary distribution is peaked, with a very low probability of small queues.

More precisely, consider the relative probability of observing the queue in state $V = 1$ rather than in state $V = V^*$. This is given by

$$\frac{P_{\text{st.}}(1)}{P_{\text{st.}}(V^*)} \propto V^{*\frac{1}{2}+\frac{\mu}{\nu}} e^{-V^*}. \tag{5.36}$$

For increasingly large choices of V^*, this ratio becomes very small, very quickly. This suggests that even if we apply a boundary condition with non-zero exit flux, the contribution of this exit flux (which is proportional to $P_{\text{st.}}(V = 1)$) to the overall dynamics will be very small, so the stationary state will still be given by Equation (5.35) up to exponentially small corrections $\propto e^{-V^*}$.

As we discuss in Section 5.4.5, the order of magnitude of the time required to reach the stationary distribution is ν^{-1}. Equation (5.36) suggests that compared to ν^{-1}, the first-hitting time $T_1(V)$ is very large – indeed, it is exponentially larger in V^*, as we will confirm in Equation (5.42) below. Therefore, even in the presence of a non-zero exit flux, the system reaches a **quasi-equilibrium** (where the distribution is approximately given by $P_{st.}(V)$) much faster than the time needed for the queue to empty.

5.4.2 First-Hitting Times

We now study the first-hitting time for the model with a constant cancellation rate per order. Using the notation $W_+(V)$ and $W_-(V)$ from Equations (5.31) and (5.32), we can implement the Laplace transform from Equation (5.12) to arrive at

$$\widehat{\Phi}(z,V) = \frac{W_+(V)}{W_+(V)+W_-(V)+z}\widehat{\Phi}(z,V+1) + \frac{W_-(V)}{W_+(V)+W_-(V)+z}\widehat{\Phi}(z,V-1). \quad (5.37)$$

As we did for the constant-cancellation model in Section 5.3.2, we can expand Equation (5.37) to first order in z to derive an expression for the mean first-hitting time $\mathbb{E}[T_1|V]$:

$$(W_+(V)+W_-(V))\mathbb{E}[T_1|V] - W_+(V)\mathbb{E}[T_1|V+1] - W_-(V)\mathbb{E}[T_1|V-1] = 1. \quad (5.38)$$

Setting

$$U(V) := \nu[\mathbb{E}[T_1|V+1] - \mathbb{E}[T_1|V]]$$

yields the recursion

$$W_+(V)U(V) - W_-(V)U(V-1) = -\nu. \quad (5.39)$$

Note that from the definition of $U(V)$, the mean first-hitting time of state V when starting from state $V+1$ is $U(V)/\nu$.

The recursion in Equation (5.39) can be solved by using the generating function

$$\widetilde{U}(y) = \sum_{V=0}^{\infty} \frac{y^V}{V!} U(V), \quad (5.40)$$

which obeys a simple ordinary linear differential equation.[2] The problem then resides in finding appropriate boundary conditions to impose on the solution. When $y \to 0$, the solution should be 0, but when $y \to \infty$, the solution should not grow exponentially in y.

In general, finding a solution that satisfies these boundary conditions is difficult. However, when $\mu/\nu \ll 1$, the solution is simple. As can be verified by inserting it into Equation (5.39), the solution reads:[3]

$$U(V) = \frac{V!}{V^{*V+1}} \sum_{K=V+1}^{\infty} \frac{V^{*K}}{K!}. \quad (5.41)$$

Equation (5.41) contains all the information we need to infer how the mean first-hitting time depends on the volume of the queue. When $V^* \gg 1$ (i.e. when the mean queue length is large), the function $U(V)$ has two regimes:

$$U(V < V^*) \approx \frac{V!}{V^{*V+1}} e^{V^*},$$

$$U(V > V^*) \approx \frac{1}{V}.$$

[2] Another possible method for solving Equation (5.39) is to use continued fractions (see, e.g., Cont, R., Stoikov, S., & Talreja, R. (2010). A stochastic model for order book dynamics. *Operations Research*, 58(3), 549–563). For a related calculation, see also Godrèche, C., Bouchaud, J. P., & Mézard, M. (1995). Entropy barriers and slow relaxation in some random walk models. *Journal of Physics A*, 28, L603–L611.

[3] For arbitrary μ/ν, the solution is obtained by replacing $V!$ by $\Gamma[V+1+\mu/\nu]$ and $K!$ by $\Gamma[K+1+\mu/\nu]$. This does not change the qualitative behaviour of $\mathbb{E}[T_1|V]$ when $V^* \gg 1$.

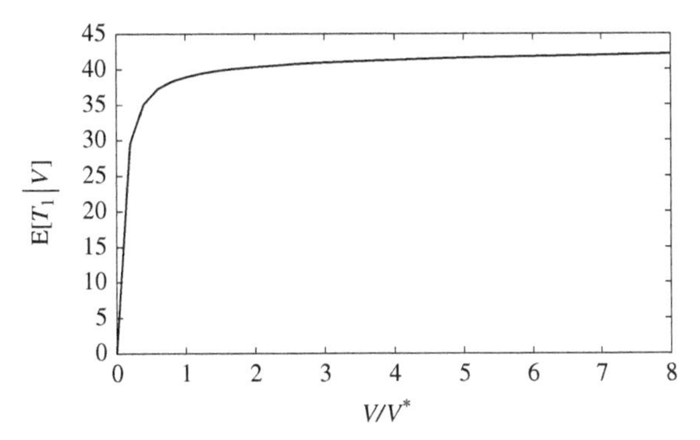

Figure 5.2. The average first-hitting time $\mathbb{E}[T_1|V]$ versus V/V^* for $V^* = 5$. A quasi-plateau value $\approx e^{V^*}/\nu V^*$ is reached quickly as the initial volume V increases.

Observing that

$$\nu\mathbb{E}[T_1|V] = U(0) + U(1) + \cdots + U(V-1)$$

allows us to infer the behaviour of the mean first-hitting time as a function of V. As Figure 5.2 illustrates, $\mathbb{E}[T_1|V]$ grows rapidly towards a quasi-plateau that extends over a very wide region around V^*, before finally growing as $\ln V$ when $V \to \infty$. We provide an intuitive explanation for this logarithmic behaviour in Section 5.4.4, where we discuss the continuous limit of the model.

The plateau value defines the mean first-hitting time when starting from a queue size $V_0 \approx V^*$, and is given by

$$\mathbb{E}[T_1|V^*] \approx \frac{e^{V^*}}{\nu V^*} \times \left[1 + \frac{1!}{V^*} + \frac{2!}{(V^*)^2} + \cdots\right]. \tag{5.42}$$

The mean first-hitting time thus grows exponentially in V^*, as anticipated by the arguments of Section 5.4.1.

5.4.3 Poissonian Queue Depletions

We now show that in the limit when $\mathbb{E}[T_1|V^*]$ is large, the distribution of first-hitting times is exponential, such that the counting process of queue depletions is Poissonian. Consider again Equation (5.37) for the Laplace transform of the distribution of first-hitting times. Expanding $\widehat{\Phi}(z,V)$ in powers of z generates the moments of this distribution (see Appendix A.2):

$$\widehat{\Phi}(z,V) = 1 + \sum_{n=1}^{\infty} \frac{(-z)^n}{n!}\mathbb{E}[T_1^n|V]. \tag{5.43}$$

Identifying terms of order z^n in Equation (5.37) leads to the following exact recursion equation:

$$(W_+(V) + W_-(V))\mathbb{E}[T_1^n|V] - W_+(V)\mathbb{E}[T_1^n|V+1] - W_-(V)\mathbb{E}[T_1^n|V-1] =$$

$$n!\sum_{k=1}^{n} \frac{(W_+(V)+W_-(V))^{-k}}{(n-k)!}\left[W_+(V)\mathbb{E}[T_1^{n-k}|V+1] + W_-(V)\mathbb{E}[T_1^{n-k}|V-1]\right]. \tag{5.44}$$

When $n = 1$, we recover Equation (5.38). When $n = 2$, and for V in the wide plateau region (where the average hitting time is nearly constant) defined in Section 5.4.2, the right-hand side of Equation (5.44) contains two types of terms:

- $k = 1$: terms of order $\mathbb{E}[T_1|V^*]$, which grow exponentially in V^*;
- $k = 2$: terms of order $\mathbb{E}[T_1^0|V]$, which are of order unity.

Neglecting the latter contribution, we find

$$(W_+(V) + W_-(V))\mathbb{E}[T_1^2|V] - W_+(V)\mathbb{E}[T_1^2|V+1] - W_-(V)\mathbb{E}[T_1^2|V-1] \approx 2\mathbb{E}[T_1|V^*].$$

Therefore the equation for $\mathbb{E}[T_1^2|V]$ is exactly the same as for $\mathbb{E}[T_1|V]$, except that the right-hand side is now equal to $2\mathbb{E}[T_1|V^*]$ instead of 1. Since the equation is linear, we can conclude that

$$\mathbb{E}[T_1^2|V] \approx 2\mathbb{E}[T_1|V^*]\mathbb{E}[T_1|V],$$

which again does not depend heavily on V in a wide region around V^*. We can then extend this argument by recursion to arbitrary n. Up to the leading exponential contribution, the dominant term is always $k = 1$, hence

$$\mathbb{E}[T_1^n|V] \approx n!\mathbb{E}[T_1|V]^n,$$

which are the moments of an exponential distribution with mean $\mathbb{E}[T_1|V]$. Consequently, the distribution of hitting times is

$$\Phi(\tau, V) \approx \frac{1}{\mathbb{E}[T_1|V]} \exp -\left[\frac{\tau}{\mathbb{E}[T_1|V]}\right].$$

Consistently with the physical picture that we develop in Section 5.4.5, this result illustrates that the sequences of events that cause the queue to empty occur so rarely that we can regard them as independent, leading to a Poisson depletion process. This rather remarkable property will enable us to discuss the race between two queues in a very simple manner in Chapter 7.

5.4.4 The Continuum Limit: The CIR Process

We now consider the continuum limit for the model with a constant cancellation rate per order. Similarly to our approach in Section 5.3.4, we assume that $P(V,t)$ is sufficiently smooth to allow the same Taylor expansion as in Equation (5.24). The Fokker–Planck equation describing the linear problem is then:

$$\frac{\partial P(V,t)}{\partial t} \approx -\partial_V [F(V)P(V,t)] + \partial_{VV}^2 [D(V)P(V,t)], \tag{5.45}$$

with drift

$$F(V) = W_+(V) - W_-(V) = \lambda - \mu - \nu V$$

and diffusion

$$D(V) = \frac{W_+(V) + W_-(V)}{2} = \frac{\lambda + \mu + \nu V}{2}.$$

The equilibrium size,

$$V^* = (\lambda - \mu)/\nu, \tag{5.46}$$

corresponds to the point where the drift $F(V)$ vanishes.

This model is known as a **modified CIR process**.[4] Given that we are considering the specific context of a queue, we will call the model the Q-CIR process (where

[4] The model is also sometimes called a Heston process or a mean-reverting Bessel process. The associated stochastic PDE, defining the dynamics of V, is $\mathrm{d}V = (\lambda - \mu - \nu V)\mathrm{d}t + \sqrt{\lambda + \mu + \nu V}\mathrm{d}W_t$, where W_t is a Brownian motion. The standard CIR process is obtained upon defining $X := \lambda + \mu + \nu V$.

Q stands for queuing). Note that the model holds quite generally; for example, it is not necessary to assume that all orders have the same size to obtain such a Q-CIR process in the continuum limit.

If we impose an absorbing boundary condition at $V = 0$, then, in the absence of reinjection, Equation (5.45) does not have any non-trivial stationary solutions, because for all $V > 0$, it follows that $\lim_{t\to\infty} P(V,t) = 0$.

If we instead impose a reflecting boundary condition at $V = 0$, such that queues are not allowed to empty and instead bounce back to a strictly positive value, then the exit flux at $V = 0$ is zero, so the stationary solution of Equation (5.45) obeys

$$-F(V)P_{\text{st.}}(V) + \partial_V \left[D(V)P_{\text{st.}}(V)\right] = 0.$$

Solving this expression, we arrive at

$$P_{\text{st.}}(V) = A\left(\frac{\lambda+\mu}{\nu} + V\right)^{\frac{4\lambda}{\nu}-1} e^{-2V},$$

where A is fixed by the normalisation of $P_{\text{st.}}(V)$.

In the case where the deposition rate λ becomes very large, such that $\mu \ll \lambda$ and $V^* \approx \lambda/\nu \gg 1$, the stationary state can be written as

$$P_{\text{st.}}(V) \approx A\,(V^* + V)^{4V^*} e^{-2V},$$

which behaves very similarly to its discrete counterpart in Equation (5.35) (see also Figure 5.1). In this case, $P_{\text{st.}}(V)$ first grows exponentially (as e^{2V} for $V \ll V^*$), before reaching a maximum at $V = V^*$ and decaying back to zero as $V^{4V^*}e^{-2V}$ for $V \gg V^*$. Similarly to in Equation (5.36), we can calculate the following ratio of probabilities

$$\frac{P_{\text{st.}}(0)}{P_{\text{st.}}(V^*)} \approx e^{-(4\ln 2-2)V^*}.$$

As in the discrete model, this ratio behaves like e^{V^*}, so for increasingly large choices of V^*, this ratio becomes very small, very quickly. However, note that the continuum limit predicts a coefficient of $4\ln 2 - 2 \approx 0.76$ in the exponential, whereas the discrete case in Equation (5.36) predicts a coefficient of 1 in the exponential. This highlights that adopting the assumptions required to study the model in the continuum limit can induce subtle quantitative differences that may affect the evaluation of rare events.

Despite these quantitative differences, the continuum limit allows us to gain useful intuition about the dynamics of the process. The reason for the exponentially small probability of observing small queues is quite simple: since the drift $F(V)$ is strongly positive when $V \ll V^*$, it tends to drive the queue back to its equilibrium size V^*. In the same way, when $V \gg V^*$, the drift $F(V)$ is negative, and tends to drive the queue back to V^*. Correspondingly, it takes a very long time to overcome this drift and reach $V = 0$.

In the limit $\lambda \gg \mu, \nu$, one can in fact show that

$$\mathbb{E}[T_1|V] \approx \nu^{-1}\sqrt{\frac{\pi}{2V^*}}\, e^{(4\ln 2-2)V^*}, \tag{5.47}$$

which again shows very little dependence on V in a wide region around V^*.

Similarly to our findings in Section 5.4.1, we recover the result that in the presence of a non-zero exit flux, the system reaches a quasi-equilibrium (where the distribution is approximately given by $P_{\text{st.}}(V)$) exponentially faster than the time needed for the queue to empty. In fact, this result still holds when re-initialising the queue size to almost any value V_0 after the queue depletes to $V = 0$, because the queue size re-equilibrates to $P_{\text{st.}}(V)$ so quickly (in a time whose magnitude is of the order ν^{-1}). This result is important because it illustrates that it is not necessary to specify the exact reinjection mechanism to obtain a precise prediction for the shape of the stationary distribution $P_{\text{st.}}(V)$. In practice, the distribution of initial values V_0 should be the size distribution of either the second-best queue, or of an incipient new queue (depending on the cases; see Section 6.1 below). But in both cases, $P_{\text{st.}}(V)$ will quickly settle in, with very little dependence on the re-initialisation mechanism.

5.4.5 Effective Potential Barriers

How can one estimate the mean first-hitting time using the Fokker–Planck formalism? In the continuum limit, the analogues of Equations (5.37) and (5.38) read

$$z\widehat{\Phi}(z,V) = F(V)\partial_V\widehat{\Phi}(z,V) + D(V)\partial^2_{VV}\widehat{\Phi}(z,V) \tag{5.48}$$

and

$$F(V)\partial_V\mathbb{E}[T_1|V] + D(V)\partial^2_{VV}\mathbb{E}[T_1|V] = -1, \tag{5.49}$$

respectively, with boundary conditions

$$\mathbb{E}[T_1|V=0] = 0; \qquad \lim_{V\to\infty}\partial_V\mathbb{E}[T_1|V] = 0.$$

This second-order, inhomogeneous ODE can be solved by standard techniques, to produce the result shown in Equation (5.47).

As in the discrete case, the mean first-hitting time is only weakly dependent on the starting size V_0 (provided V_0 is neither too large or too small) and grows exponentially in the equilibrium size V^*. To understand these (perhaps surprising) results, we provide the following physical interpretation of the system. Consider a Brownian particle (with a V-dependent diffusion constant) in a parabolic-like potential well that reaches its minimum at V^* with curvature $\mathcal{W}''(V^*) = 1/V^*$ (see

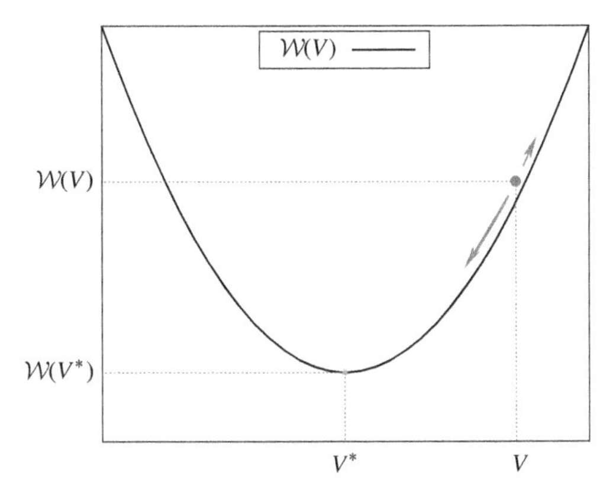

Figure 5.3. A potential well $\mathcal{W}(V)$, within which the effective "particle" (which represents the size of the queue) oscillates randomly before overcoming the barrier.

Figure 5.3). Consider the particle's **effective potential energy** $\mathcal{W}(V)$, defined such that $F(V) := -D(V)\mathcal{W}'$, and with $\mathcal{W}(0) = 0$. In the present linear case, and again in the limit $\lambda \gg \mu$, it follows that

$$\mathcal{W}(V) = 2V - 4V^* \ln\left(1 + \frac{V}{V^*}\right). \tag{5.50}$$

Intuitively, the particle's typical trajectories are such that starting from almost any V_0, the particle first rolls down the potential well, to reach the minimum after a time whose magnitude is of order ν^{-1}. Then, driven by the Brownian noise, the particle fluctuates around this minimum. It takes an extremely rare fluctuation to allow the particle to climb all the way uphill (from V^* to $V = 0$) and find the exit. As we saw in Section 5.4.3, these fluctuations in fact lead to a Poissonian exit process. The mean escape time of the particle is given by

$$\mathbb{E}[T_1|V] \approx \frac{1}{D(V^*)|\mathcal{W}'(0)|}\sqrt{\frac{2\pi}{\mathcal{W}''(V^*)}}e^{-\mathcal{W}(V^*)}.$$

In the Q-CIR case (see Section 5.4.4), where $\mathcal{W}$ is given by Equation (5.50), one can check that this expression recovers exactly Equation (5.47).

Finally, our physical picture can also shed light on the logarithmic behaviour of $\mathbb{E}[T_1|V]$ for $V \to \infty$. In this case, the Brownian particle must first scurry from this large V down to V^*, from which it then starts attempting to escape the potential well. The initial roll-down phase is governed by a quasi-deterministic evolution

$$\left.\frac{dV}{dt}\right|_{V\to\infty} \approx F(V) \approx -\nu V.$$

In this regime, we can neglect the Brownian noise contribution, because it is so small compared to the drift. Integrating this equation of motion reveals why it takes a time whose magnitude is of order $\ln V$ to reach the bottom of the potential well, hence explaining the logarithmic growth of $\mathbb{E}[T_1|V]$ for large V.

5.5 Conclusion

In this chapter, we have considered many useful ideas about single-queue systems. Many of these results hold more generally, independently of the details of the model used to describe the dynamics of long queues. To summarise the most important intuitions that we have garnered:

- When the total cancellation rate grows with the queue size V, queues self-stabilise around a well-defined equilibrium size V^*.
- When V^* is sufficiently large (i.e. when the order submission process is more intense than either cancellations or market order execution), the queue dynamics can be neatly decomposed into a fast part, where a quasi-equilibrium state quickly sets in, and a slow part, where a Poisson process leads to an emptying of the queue.
- Although we have only considered the case where the cancellation rate grows linearly with the queue size, the resulting separation of time scales is in fact generic and holds whenever the total cancellation rate grows with the queue size, say as V^ζ with $\zeta > 0$.
- The case $\zeta = 0$ corresponds to the first model that we considered in this chapter. It is special in the sense that the only way to obtain long queues is to fine-tune the parameters of the model such that it sits at its "critical point", beyond which no stationary solution exists.

Empirical data suggests that the total cancellation rate grows *sublinearly* with the queue size (i.e. $0 < \zeta < 1$; see Section 6.4). A simple argument to understand this is time priority, which is extremely valuable in large-tick markets, because being at the front of a long queue increases the probability of execution, while minimising adverse selection (see Section 21.4 for a detailed discussion of this point). This means that cancelling an order at the front of a long queue does not make much sense economically. The only orders likely to be cancelled are those at the back of the queue. Therefore, we expect that only a small fraction of the orders in a given queue have an appreciable probability of being cancelled.

In the next chapter, we will consider more general specifications of single-queue models, and discuss how to calibrate such models directly to empirical data, without making many assumptions on the microscopic order flow. The above conclusions will hold for a large family of such models provided the effective potential barrier introduced in Section 5.4.5 is sufficiently high.

Take-Home Messages

(i) Stochastic models can be used to model markets with a "zero-intelligence" framework. These models seek to make predictions based on the statistical properties of order flow, rather than trying to understand what drives it in the first place.
(ii) The volume available at each level of the price grid can be modelled as a queuing system. These queues grow due to the arrival of new limit orders and shrink due to the cancellation of existing limit orders. At the best quotes, arriving market orders also cause the queues to shrink.
(iii) When all are described by mutually independent, homogeneous Poisson processes whose rate parameters are independent of the queue size, many quantities of interest can be derived analytically. Only when the total growth rate is nearly equal to the total shrink rate does the queue exhibit large excursions, as observed empirically. In this case, the distribution of times between two queue depletions is extremely broad.
(iv) When assuming a constant cancellation rate *per limit order*, the queue stabilises around a stationary length. If this stationary length is large, the excursions around it can be described by a CIR process, and depletions occur as a Poisson process.
(v) In the limit of large volumes and/or small increments, the queue dynamics can be described in a continuous setting using a Fokker–Planck equation, which can be solved analytically using standard PDE tools.

5.6 Further Reading

General

Feller, W. (1971). *An introduction to probability theory and its applications* (Vol. 2). John Wiley and Sons.
Van Kampen, N. G. (1983). *Stochastic processes in physics and chemistry*. North-Holland.
Risken, H. (1984). *The Fokker–Planck equation*. Springer, Berlin-Heidelberg.
Chen, H., & Yao, D. D. (2001). *Fundamentals of queueing networks: Performance, asymptotics, and optimisation* (Vol. 46). Springer Science & Business Media.
Gardiner, C. W. (1985). *Stochastic methods*. Springer, Berlin-Heidelberg.
Redner, S. (2001). *A guide to first-passage processes*. Cambridge University Press.
Whitt, W. (2002). *Stochastic process limits*. Springer-Verlag.
Oksendal, B. (2003). *Stochastic differential equations: An introduction with applications*. Springer.
Abergel, F., Chakraborti, A., Anane, M., Jedidi, A., & Toke, I. M. (2016). *Limit order books*. Cambridge University Press.

Queue Models for LOB Dynamics

Cont, R., Stoikov, S., & Talreja, R. (2010). A stochastic model for order book dynamics. *Operations Research*, 58(3), 549–563.

Avellaneda, M., Reed, J., & Stoikov, S. (2011). Forecasting prices from Level-I quotes in the presence of hidden liquidity. *Algorithmic Finance*, 1(1), 35–43.

Cont, R. (2011). Statistical modelling of high-frequency financial data. *IEEE Signal Processing Magazine*, 28(5), 16–25.

Cont, R., & De Larrard, A. (2012). Order book dynamics in liquid markets: Limit theorems and diffusion approximations. https://ssrn.com/abstract=1757861.

Cont, R., & De Larrard, A. (2013). Price dynamics in a Markovian limit order market. *SIAM Journal on Financial Mathematics*, 4(1), 1–25.

Garèche, A., Disdier, G., Kockelkoren, J., & Bouchaud, J. P. (2013). Fokker-Planck description for the queue dynamics of large-tick stocks. *Physical Review E*, 88(3), 032809.

Guo, X., Ruan, Z., & Zhu, L. (2015). Dynamics of order positions and related queues in a limit order book. https://ssrn.com/abstract=2607702.

Huang, W., Lehalle, C. A., & Rosenbaum, M. (2015). Simulating and analyzing order book data: The queue-reactive model. *Journal of the American Statistical Association*, 110(509), 107–122.

Huang, W., & Rosenbaum, M. (2015). Ergodicity and diffusivity of Markovian order book models: A general framework. arXiv preprint arXiv:1505.04936.

Muni Toke, I. (2015). The order book as a queueing system: Average depth and influence of the size of limit orders. *Quantitative Finance*, 15(5), 795–808.

Muni Toke, I. (2015). Stationary distribution of the volume at the best quote in a Poisson order book model. arXiv preprint arXiv:1502.03871.

Lakner, P., Reed, J., & Stoikov, S. (2016). High frequency asymptotics for the limit order book. *Market Microstructure and Liquidity*, 2(01), 1650004.

Queuing Theory

Harrison, J. M. (1978). The diffusion approximation for tandem queues in heavy traffic. *Advances in Applied Probability*, 10(04), 886–905.

Abate, J., & Whitt, W. (1995). Numerical inversion of Laplace transforms of probability distributions. *ORSA Journal on Computing*, 7(1), 36–43.

Abate, J., & Whitt, W. (1999). Computing Laplace transforms for numerical inversion via continued fractions. *INFORMS Journal on Computing*, 11(4), 394–405.

CIR Processes

Cox, J. C., Ingersoll Jr, J. E., & Ross, S. A. (1985). A theory of the term structure of interest rates. *Econometrica: Journal of the Econometric Society*, 53, 385–407.

Heston, S. L. (1993). A closed-form solution for options with stochastic volatility with applications to bond and currency options. *Review of Financial Studies*, 6(2), 327–343.

Göing-Jaeschke, A., & Yor, M. (2003). A survey and some generalisations of Bessel processes. *Bernoulli*, 9(2), 313–349.

Fokker–Planck Equation and First Passage Times

Hänggi, P., Talkner, P., & Borkovec, M. (1990). Reaction-rate theory: Fifty years after Kramers. *Reviews of Modern Physics*, 62(2), 251.

6

Single-Queue Dynamics for Large-Tick Stocks

The calculus of probability can doubtless never be applied to market activity, and the dynamics of the Exchange will never be an exact science. But it is possible to study mathematically the state of the market at a given instant.

(Louis Bachelier)

In Chapter 5, we studied two simple models of a single order queue. Despite the apparent simplicity of our modelling assumptions, we saw that deriving expressions about the queue's behaviour required some relatively sophisticated machinery. In this chapter, we take the next step towards developing a stochastic model of an LOB by extending the models of the previous chapter towards more realistic situations. We focus on large-tick stocks, for which two major simplifications of the resulting queue dynamics occur.[1]

First, the bid–ask spread $s(t)$ is constrained heavily by the tick size ϑ, such that $s(t) = \vartheta$ for the vast majority of the time. Therefore, traders are unable to submit limit orders inside the spread, so price changes only occur when either the bid-queue or the ask-queue depletes to $V = 0$. Second, considerable volumes of limit orders typically accumulate at the bid-price and at the ask-price. In fact, V_b and V_a alone can correspond to a non-negligible fraction of the daily traded volume (see Table 4.1). For small-tick stocks, by contrast, V_b and V_a typically correspond to about 10^{-2} of the daily traded volume, it is common for limit orders to arrive inside the spread, and some arriving market orders cause price changes of several or even several tens of ticks. These empirical facts make modelling of LOBs much more difficult for small-tick stocks.

Because of the large volumes at the best quotes for large-tick stocks, and because price changes correspond to the times when either the bid-queue or the ask-queue deplete to 0, such events occur relatively infrequently, so that the typical daily

This chapter is not essential to the main story of the book.

[1] Recall from Chapter 4 that a large-tick stock is a stock whose tick size ϑ is a relatively large fraction of its price (typically tens of basis points or even some percentage points).

price change for a large-tick stock is usually between a few ticks and a few tens of ticks. For example, consider a stock with relative tick size $\vartheta_r \approx 0.002$ and a daily volatility of 2%. The number N of (random) daily mid-price changes satisfies

$$0.02 \approx 0.002\sqrt{N},$$

so $N \approx 100$. If this stock's trading day lasts eight hours, then on average the mid-price only changes value (usually by a single tick) about once every five minutes, during which a very large number of events take place at the best quotes.

Given that their depletion constitutes a relatively rare event, the dynamics of the bid- and ask-queues for a large-tick stock becomes an interesting modelling topic in its own right. As we will see in Section 6.2, we only require a few, relatively small extensions of our single-queue model from Chapter 5 to produce a useful model for large-tick stocks.

6.1 Price-Changing Events

We start our exploration of more realistic models precisely where we left off in Chapter 5: with a Fokker–Planck description of a single queue, in the continuum limit. As we will see in this chapter, this description is much more general than the specific cases that we have considered previously. The power of this approach is well known in many different disciplines, including queueing theory, where it is called the *heavy traffic limit* (see Queue Models for LOB Dynamics in Section 5.6) and statistical physics.

In the present context, the core of this approach is as follows. Provided that the change in queue size is the result of many small events that are weakly correlated in time, then the long-time dynamics of the queue are given by the general Fokker–Planck equation

$$\frac{\partial P(V,t)}{\partial t} = -\partial_V [F(V)P(V,t)] + \partial^2_{VV} [D(V)P(V,t)], \tag{6.1}$$

where $F(V)$ is the drift and $D(V)$ is the diffusion. In the Q-CIR model of Section 5.4.4, we assumed that $F(V)$ and $D(V)$ depended linearly on V, but in general this dependence can take any linear or non-linear form. In this chapter, we derive several results within this framework, then use empirical data to calibrate the drift and diffusion terms, and compare the corresponding model predictions to real queue dynamics.

Equation (6.1) models the stochastic evolution of a single queue. Assume now that this queue is either the best bid- or ask-queue in the LOB for a large-tick stock. In addition to the changes in queue volume, the dynamics of an LOB are also influenced by events that change the price of the best bid or ask. Therefore, to extend our single-queue analysis from Chapter 5 into the context of an LOB, we must also supplement our approach to account for these price-changing events.

Suppose, for concreteness, that we focus on the bid-queue, such that $V(t) = V_b(t)$ denotes the volume at the best bid at time t. Because we study large-tick stocks, throughout this chapter we assume that the spread is equal to its minimum size of one tick. Therefore, no limit orders can arrive inside the spread, so price changes can only occur when a queue depletes to $V = 0$. Whenever a queue depletes in this way, we assume that some new volume must arrive in the gap immediately (otherwise, the spread would be greater than one tick). This new volume can be either a buy limit order, in which case the bid- and ask-prices both remain the same as they were before the depletion, or a sell limit order, in which case both the bid- and ask-prices decrease by one tick from the previous values before the depletion.

More formally, we assume that:

(i) If the bid-queue volume reaches $V = 0$, then one of the following two possibilities must occur:
  - With some probability ϕ_0, a new buy limit order arrives, such that the bid-queue replenishes and the bid- and ask-prices do not change. The size $\delta V > 0$ of the new buy limit order is distributed according to some distribution $\varrho(\delta V|V = 0)$.
  - With probability $1 - \phi_0$, a new sell limit order arrives (with a volume distribution $\varrho(\delta V|V = 0)$) and the bid- and ask-prices both decrease by one tick. The new bid-queue is then the queue that was previously the second-best bid-queue, whose volume we assume is distributed according to some distribution $P_{2\text{-best}}(V)$.

(ii) Similarly, if the ask-queue volume reaches $V = 0$, then one of the following two possibilities must occur:
  - With some probability ϕ_0, a new sell limit order arrives, such that the ask-queue replenishes and the bid- and ask-prices do not change. The size $\delta V > 0$ of the new sell limit order is distributed according to some distribution $\varrho(\delta V|V = 0)$.
  - With probability $1 - \phi_0$, a new buy limit order arrives (with a volume distribution $\varrho(\delta V|V = 0)$) and the bid- and ask-prices both increase by one tick. The new ask-queue is then the queue that was previously the second-best ask-queue, whose volume we assume is distributed according to some distribution $P_{2\text{-best}}(V)$.

How can we implement these modelling assumptions into our Fokker–Planck framework? In Chapter 5, we introduced the concept of exit flux $J_0(t)$ to account for queue depletions. In that chapter, we did not consider the corresponding price changes that occur after a bid- or ask-queue depletion, but in this chapter we seek to address this effect directly. Therefore, we extend the concept of a single exit flux $J_0(t)$ to also encompass possible asymmetries in the bid- and ask-queues. Specifically, we let $J_b(t)$ denote the probability per unit time that the bid-queue empties and $J_a(t)$ denote the probability per unit time that the ask-queue empties. The mechanisms above then add some reinjection terms on the right-hand side of Equation (6.1) to govern the distribution of volume at the bid:

$$\mathcal{J}(V,t) = J_b(t)\left(\phi_0 \varrho(V|0) + (1-\phi_0)P_{\text{2-best}}(V)\right) + J_a(t)(1-\phi_0)\left(\varrho(V|0) - P(V,t)\right). \tag{6.2}$$

The first two terms in Equation (6.2) correspond to the bid-queue depleting to 0, and the last two terms correspond to the ask-queue depleting to 0. Finally, the absorbing condition fixes $P(0,t) = 0$, and the probability of the bid-queue hitting zero is given by:

$$J_b(t) = \partial_V \left[D(V)P(V,t)\right]\Big|_{V=0}.$$

By symmetry of the system, a similar result holds for J_a. One can directly check that $\partial_t \int \mathrm{d}V P(V,t) = 0$, which shows that we have correctly accounted for all outgoing and incoming order flow. This fixes the complete description of the queue dynamics, which allows us to derive several results, for example concerning the stationary distribution of the bid- or the ask-queue.

Throughout this chapter, we restrict our attention to the symmetric, stationary case where $J_a(t) = J_b(t) = J_0$. Using this symmetry, we arrive at

$$\partial_V \left[F(V)P_{\text{st.}}(V)\right] - \partial^2_{VV}\left[D(V)P_{\text{st.}}(V)\right] = J_0\left[\varrho(V|0) + (1-\phi_0)\left(P_{\text{2-best}}(V) - P_{\text{st.}}(V)\right)\right], \tag{6.3}$$

where the exit flux J_0 must satisfy

$$J_0 = \partial_V \left[D(V)P_{\text{st.}}(V)\right]\Big|_{V=0}. \tag{6.4}$$

6.2 The Fokker–Planck Equation

Throughout this chapter, the **Fokker–Planck equation** (6.3) is our central focus. Given its importance, we first take a moment to consider where it comes from and why it holds.

Assume that there exists some time $\tau_c \geq 0$ such that on time scales beyond τ_c, order flow can be assumed to be uncorrelated. Let $\varrho(\delta V|V)$ denote the probability that the change in queue size in the next time interval τ_c is δV, given that the current queue length is V. Values $\delta V > 0$ correspond to limit order arrivals, and values $\delta V < 0$ correspond to limit order cancellations and market order arrivals. We assume that all of these events occur with a probability that depends on the queue volume at time t. The general master equation for the size distribution then reads

$$P(V, t+\tau_c) = \sum_{\delta V} P(V-\delta V, t)\varrho(\delta V | V-\delta V). \tag{6.5}$$

Assume for now that there is no change of the bid- or ask-prices between t and $t+\tau_c$. Assume further that within a time interval of length τ_c, the typical change of queue size is relatively small, such that $\delta V \ll V$. In this framework, one can expand the master equation (6.5) in powers of δV. The general expansion is called the **Kramers–Moyal expansion**, which when truncated to second order reads

$$\begin{aligned} P(V, t+\tau_c) \approx \sum_{\delta V} \Bigg[& \left(P(V,t) - \delta V \partial_V P(V,t) + \frac{1}{2}(\delta V)^2 \partial^2_{VV} P(V,t) \right) \\ & \times \left(\varrho(\delta V|V) - \delta V \partial_V \varrho(\delta V|V) + \frac{1}{2}(\delta V)^2 \partial^2_{VV} \varrho(\delta V|V) \right) \Bigg]. \end{aligned} \tag{6.6}$$

Regrouping the terms by order:

$$P(V, t+\tau_c) \approx P(V,t) - \partial_V \left[\sum_{\delta V} \delta V \varrho(\delta V|V) P(V,t) \right] + \frac{1}{2} \partial^2_{VV} \left[\sum_{\delta V} (\delta V)^2 \varrho(\delta V|V) P(V,t) \right]. \tag{6.7}$$

Finally, substituting the approximation $P(V, t+\tau_c) \approx P(V,t) + \tau_c \partial_t P(V,t)$ into Equation (6.7) produces the Fokker–Planck equation (6.1), where

$$F(V) = \frac{1}{\tau_c} \sum_{\delta V} \delta V \varrho(\delta V|V); \qquad D(V) = \frac{1}{2\tau_c} \sum_{\delta V} (\delta V)^2 \varrho(\delta V|V) \tag{6.8}$$

are the first two moments of the conditional distribution of size changes, $\varrho(\delta V|V)$. The special Q-CIR model in Section 5.4.4 corresponds to the case

$$\begin{aligned} \tau_c &= \mathrm{d}t, \\ \varrho(\delta V = +1|V) &= \lambda \mathrm{d}t, \\ \varrho(\delta V = -1|V) &= (\mu + \nu V)\mathrm{d}t, \\ \varrho(\delta V = 0|V) &= 1 - (\lambda + \mu + \nu V)\mathrm{d}t, \end{aligned}$$

from which it again follows that

$$\begin{aligned} F(V) &= \lambda - \mu - \nu V, \\ D(V) &= \frac{\lambda + \mu + \nu V}{2}. \end{aligned}$$

The Fokker–Planck formalism allows for a much more general specification. In particular, one can remove the assumption that all events are of unit volume, and one can accommodate any dependence of the market order arrival and limit order arrival/cancellation rates on the volume V.

The general conditions under which the Fokker–Planck approximation is valid are discussed in many books.[2] Intuitively, $P(V,t)$ must vary slowly, both in V (on the scale of typical values of δV) and in time (on the correlation time scale τ_c). It is important to check on a case-by-case basis that these conditions hold.

6.2.1 The Boltzmann–Gibbs Measure

Because Equation (6.3) is a linear second-order ODE, one can in principle write down its explicit solution in full generality. However, as we saw in Chapter 5, this solution can be quite complicated, and is therefore not very useful for developing intuition about the system. A better approach is to follow a similar path to that in Section 5.4.5, and to look for cases where the effective potential

$$\mathcal{W}(V) = -\int_0^V \mathrm{d}V' \frac{F(V')}{D(V')} \tag{6.9}$$

has a minimum at $V = V^*$ sufficiently far away from $V = 0$ and sufficiently deep to keep the queue size in the vicinity of V^* for a substantial amount of time.

Analysis of this system is further complicated by the fact that the reinjection term in Equation (6.3) is non-zero. This makes finding general solutions particularly difficult. However, there are two situations in which we can avoid this difficulty. The first is where $\mathcal{W}(V^*)$ is large, so J_0 is very small (recall that J_0 decays exponentially in $\mathcal{W}(V^*)$). In this case, we can simply approximate the reinjection term in Equation (6.3) as 0. The second is where the best and second-best queues have similar stationary profiles (such that $P_{2\text{-best}}(V) \approx P_{\text{st.}}(V)$) and where the size of refilled queues is equal to a small elementary volume υ_0 (such that $\varrho(V|0) \approx \delta(V - \upsilon_0)$). In these cases, the equation for the stationary distribution when $V > 0$ takes the simpler form

$$F(V)P_{\text{st.}}(V) - \partial_V [D(V)P_{\text{st.}}(V)] = 0. \tag{6.10}$$

As can be verified by inspection, the expression

$$P_{\text{st.}}(V) = \frac{A}{D(V)} e^{-\mathcal{W}(V)} \tag{6.11}$$

is an explicit solution for Equation (6.10). The value of A is fixed by the normalisation of $P_{\text{st.}}(V)$. Equation (6.11) is called the **Boltzmann–Gibbs measure**.

By comparing $P_{\text{st.}}(0)$ to $P_{\text{st.}}(V^*)$, we can again guess that the exit flux J_0 is indeed of the order of $e^{\mathcal{W}(V^*)}$, because J_0 is itself proportional to $P_{\text{st.}}(0)$. Therefore, J_0 is very small as soon as the potential depth $|\mathcal{W}(V^*)|$ is larger than about 5. This corresponds to the regime of large, long-lived queues. In this regime, the equilibrium in the effective potential $\mathcal{W}(V)$ is reached much faster than the time needed for the queue to empty.

[2] See for example: Gardiner, C. W. (1985). *Stochastic methods*. Springer, Berlin-Heidelberg and Section 6.6.

The equilibration time (i.e. the time required to reach the stationary state) can be approximated by considering the first-order expansion of $F(V)$ in the vicinity of V^*,

$$F(V \approx V^*) \approx 0 + \kappa(V^* - V) + O((V^* - V)^2),$$

with $\kappa > 0$. Locally, the process is thus an **Ornstein–Uhlenbeck** process[3] with relaxation time κ^{-1}. In the Q-CIR model of Chapter 5, κ is simply equal to ν.

In the regime $J_0\kappa^{-1} \ll 1$ (i.e. where the relaxation is so fast that the exit flux can be neglected), the stationary distribution $P_{\text{st.}}(V)$ is nearly independent of the reinjection mechanism, as we found in Chapter 5. In particular, deriving $P_{\text{st.}}(V)$ does not require knowledge of either $P_{\text{2-best}}(V)$ or $\varrho(V|0)$. This remarkable property, which we discussed qualitatively in Section 5.4.4, will enable us to test empirically the relevance of the Fokker–Planck framework for describing the dynamics of bid–ask-queues. We present these empirical results in Section 6.4.

6.2.2 First-Hitting Times

Recall from Section 5.4.5 that the mean first-hitting time $\mathbb{E}[T_1(V)]$ obeys Equation (5.49). As can be verified by inspection, the general expression

$$\mathbb{E}[T_1(V)] = \int_0^V \mathrm{d}V' \int_{V'}^{\infty} \mathrm{d}V'' \frac{1}{D(V'')} e^{\mathcal{W}(V')-\mathcal{W}(V'')} \tag{6.12}$$

is a solution for Equation (5.49).

In the limit $|\mathcal{W}(V^*)| \gg 1$, one can carry out an approximate Laplace estimation of this integral to conclude that for a large region of initial conditions, the mean first-hitting time is to a very good approximation independent of the starting point V, and is given by:

$$\mathbb{E}[T_1(V)] \approx \frac{1}{D(V^*)|\mathcal{W}'(0)|} \sqrt{\frac{2\pi}{\mathcal{W}''(V^*)}} e^{|\mathcal{W}(V^*)|}.$$

Therefore, $\mathbb{E}[T_1(V)]$ grows exponentially in the barrier height. Recalling that

$$J_0 \approx \frac{1}{\mathbb{E}[T_1(V^*)]},$$

we verify that J_0 decays exponentially in $\mathcal{W}(V^*)$ and can therefore indeed be neglected in Equation (6.3).

Note also that using the quasi-independence of $\mathbb{E}[T_1(V)]$ on V (see Section 5.4.3) and following the calculation presented for the discrete case in Section 5.4.3, one can establish that the hitting process is also Poissonian for the general Fokker–Planck equation (6.3) in the limit of large barrier heights.

6.3 Sweeping Market Orders

As we discussed in Section 6.2, the Fokker–Planck approach requires individual changes in V to be small. For large-tick assets, this is indeed the case for a large

[3] An Ornstein–Uhlenbeck process is described by a linear stochastic differential equation: $\mathrm{d}V = -\kappa(V - V^*)\mathrm{d}t + \mathrm{d}W_t$, where W_t is a Brownian motion. See, e.g., Gardiner (1985).

fraction of events. Sometimes, however, very large market orders consume all the remaining volume in the best bid- or ask-queue. Due to their large size, such market orders fall outside the scope of the standard Fokker–Planck formalism for computing the first-hitting time. As we illustrate in Section 6.4, such large market orders are not particularly frequent, especially when queues are large, but are still important since they lead to immediate depletion. In this section, we introduce a way to account for these events.

6.3.1 An Extended Fokker–Planck Description

Let $\Pi(V)$ denote the probability per unit time of the arrival of a market order that consumes the entire queue under consideration, and let $\Pi^{\dagger}(V)$ denote the probability per unit time of the arrival of a market order that consumes the entire opposite-side best queue. Using this notation, we can extend Equation (6.3) to account for these events by re-writing $P(V,t)$ as

$$\begin{aligned}\frac{\partial P(V,t)}{\partial t} &= -\partial_V [F(V)P(V,t)] + \partial^2_{VV} [D(V)P(V,t)] + \mathcal{J}(V,t) \\ &\quad - \Pi(V)P(V,t) + \bar{\Pi}(t)\,(\phi_0 \varrho(V|0) + (1-\phi_0)P_{\text{2-best}}(V)) \\ &\quad - (1-\phi_0)\Pi^{\dagger}(V)P(V,t) + (1-\phi_0)\bar{\Pi}^{\dagger}(t)\varrho(V|0), \end{aligned} \qquad (6.13)$$

where $\mathcal{J}$ is given by Equation (6.2) and

$$\bar{\Pi}(t) = \int \mathrm{d}V \Pi(V)P(V,t),$$

and similarly for $\bar{\Pi}^{\dagger}(t)$. The second line in Equation (6.13) says that if the queue is instantaneously depleted to zero (which occurs with probability $\bar{\Pi}(t)$), then it either bounces back (with probability ϕ_0) or is replaced by the second-best queue (with probability $1-\phi_0$). The third line describes what happens when the opposite queue is instantaneously depleted to zero (which occurs with probability $\bar{\Pi}^{\dagger}(t)$). With probability $1-\phi_0$, the queue we are looking at becomes the second-best queue and is replaced by a new queue with volume taken from the distribution $\varrho(V|0)$; otherwise (with probability ϕ_0), it bounces back so nothing changes.

As we saw in Chapter 5, diffusion processes typically produce an extremely small exit flux. Depending on the frequency with which they occur, large market order arrivals can be a much more efficient mechanism than standard diffusions for emptying the queue. In other words, the $\Pi(V)$ and $\Pi^{\dagger}(V)$ terms in Equation (6.13) can cause a considerable impact on queue dynamics.

Let us define

$$\Pi^* := \max\left[\Pi(V^*), \Pi^{\dagger}(V^*)\right]. \qquad (6.14)$$

Three situations can occur:

(i) $\Pi^* \ll J_0 \ll \kappa$. In this case, the probability of large market orders is so small that their effect can be neglected.
(ii) $J_0 \ll \Pi^* \ll \kappa$. In this case, large market order arrivals are the primary mechanism driving queue depletions. However, since the typical time between these large market order arrivals (which is of order $1/\Pi^*$) is still very large compared to the equilibration time $1/\kappa$, the stationary distribution $P_{\text{st.}}(V)$ is still approximately equal to the Boltzmann–Gibbs measure in Equation (6.11).
(iii) $\kappa \lesssim \Pi^*$. In this case, large market orders arrive so frequently that the queue does not have time to equilibrate between their arrivals. Therefore, the whole Fokker–Planck approach breaks down. Assets with small relative tick size (see Section 3.1.5) typically fall into this case, because the size of a typical queue for such assets is similar to the size of a typical market order.

Empirically, one finds Π^* to be smaller than but comparable to J_0 (see Table 6.1), which leaves us in an intermediate regime between cases (i) and (ii).

In the limit of small Π and $\Pi^\dagger$, and if the term $\mathcal{J}(V,t)$ is still negligible, one can still find an explicit form for the stationary solution. Assuming for simplicity that

$$P_{\text{2-best}}(V) \approx P_{\text{st.}}(V)$$

and that the incipient queues are small, i.e.

$$\varrho(V|0) \approx \delta(V - \upsilon_0),$$

the stationary equation for $V > 0$ now reads:

$$0 = \partial_V [F(V)P_{\text{st.}}(V)] - \partial^2_{VV} [D(V)P_{\text{st.}}(V)] +$$
$$+ \epsilon\left[\Pi(V) + (1 - \phi_0)(\Pi^\dagger(V) - \bar{\Pi}_{\text{st.}})\right] P_{\text{st.}}(V),$$

where $\bar{\Pi}_{\text{st.}}$ is defined as:

$$\bar{\Pi}_{\text{st.}} := \int \mathrm{d}V \Pi(V) P_{\text{st.}}(V).$$

We also introduced $\epsilon \ll 1$ as a device to highlight that all of $\Pi(V)$, $\Pi^\dagger(V)$ and $\bar{\Pi}_{\text{st.}}$ are small. This device is convenient for keeping track of the order to which we expand the solution. Looking for a stationary solution of the form

$$P_{\text{st.}}(V) = \frac{A(V)}{D(V)} e^{-\mathcal{W}(V)}, \tag{6.15}$$

one finds that $B(V) = A'(V)/A(V)$ obeys the so-called Ricatti equation

$$B^2 + B' - \mathcal{W}'B - \epsilon\widehat{\Pi} = 0; \qquad \widehat{\Pi}(V) := \frac{\Pi(V) + (1 - \phi_0)(\Pi^\dagger(V) - \bar{\Pi}_{\text{st.}})}{D(V)}.$$

When the probability of large market orders is sufficiently small, one can look for a solution $B \propto \epsilon$ and neglect the B^2 term, which is of order ϵ^2. The resulting linear equation leads to the following result, valid to order ϵ:

$$\ln A(V) = -\int_0^V \mathrm{d}V' \int_{V'}^{\infty} \mathrm{d}V'' \widehat{\Pi}(V'') e^{\mathcal{W}(V') - \mathcal{W}(V'')}. \tag{6.16}$$

If needed, this solution can be plugged back into the Ricatti equation to find the correction to order ϵ^2.

6.3.2 When Large Market Orders Dominate

If $F(V)$ is always positive, then in the absence of large market order arrivals, the queue volume tends to infinity and the Fokker–Planck equation has no stationary state. However, the arrivals of large market orders can keep the queue length finite in this case. To build intuition, let's consider the simplest case where

$$F(V) = F_0 > 0,$$
$$D(V) = D_0 > 0,$$
$$\Pi(V) = \Pi^{\dagger}(V) = \Pi_0 > 0$$

are all strictly positive constants. We also assume that $P_{\text{2-best}}(V) = P_{\text{st.}}(V)$. In this case, the diffusive flux $\mathcal{J}(V,t)$ is very small. Therefore, in the stationary limit, Equation (6.13) becomes, for $V > 0$,

$$-F_0 \partial_V P_{\text{st.}}(V) + D_0 \partial^2_{VV} P_{\text{st.}}(V) - \Pi_0 P_{\text{st.}}(V) = 0, \qquad (6.17)$$

where $\varrho(V|0)$ is again assumed to be localised around $V = 0$.

Because Equation (6.17) is a linear, second-order ODE, we look for a solution of the form $P_{\text{st.}}(V) \propto e^{aV}$, which leads to the following second-degree equation for a:

$$D_0 a^2 - F_0 a - \Pi_0 = 0.$$

This equation has solutions

$$a_{\pm} = (F_0 \pm \sqrt{F_0^2 + 4\Pi_0 D_0})/2D_0.$$

For $P_{\text{st.}}(V)$ to be normalisable to 1, we require the solution in which $P_{\text{st.}}(V)$ tends to 0 for large V. Therefore, we require $a < 0$, so we choose $a = a_-$, and

$$P_{\text{st.}}(V) \approx |a_-| e^{-|a_-|V},$$

which decays exponentially in V. In summary, the arrival of sweeping market orders makes the probability of large queues exponentially small, even when the drift $F(V)$ is positive.

6.4 Analysing Empirical Data

6.4.1 Calibrating the Fokker–Planck Equation

We now turn our attention to fitting the Fokker–Planck model to empirical data, to see how well the results we have derived throughout the chapter are consistent with the behaviour observed in real markets.

As we discussed in Section 4.2, order flows in financial markets undergo strong intra-day patterns. Attempting to fit the model without acknowledging these intra-day seasonalities is likely to produce poor estimates of the input parameters and poor outputs. To avoid this problem, we take the simple approach of only studying a period of time within which order flow is approximately stationary. As in the previous chapters, we simply discard the first and last hour of trading activity each day. This removes the strongest element of the intra-day pattern. In the following, we work in rescaled units $u := V/\bar{V}$, where $\bar{V}$ is the mean queue length, allowing the comparison between different stocks (and different time periods if volumes evolve substantially between these periods). Indeed, a remarkable property of the dynamics of queues is approximate **scale invariance**, i.e. once rescaled by the local average volume, all statistical properties of queue sizes are similar. Throughout this section, we will call the quantity u the **rescaled volume**.

Then, one can measure the drift $F(u)$ and diffusion $D(u)$ as conditional empirical averages of volume changes in a unit event-time interval $\delta n = 1$, using:[4]

$$F(u) = \frac{1}{\delta n} \langle u(n+\delta n) - u(n) \rangle |_{u(n)=u},$$

$$D(u) = \frac{1}{2\delta n} \left\langle (u(n+\delta n) - u(n))^2 \right\rangle \Big|_{u(n)=u}.$$

For simplicity, we restrict our analysis to the most natural choice $\delta n = 1$. Because we consider single-queue dynamics, event-time in this section refers to events only pertaining to the queue under consideration. Thus, each event necessarily changes the queue volume. (Note that we exclude all market order events that only execute hidden volume, which are irrelevant for the dynamics of the visible volume in the queue.)

Table 6.1 lists several summary statistics for the queue dynamics of CSCO and INTC, both of which are large-tick stocks (see Table 4.1). There are some noteworthy features:

- The rescaled volume changes per event are small ($\langle |\delta u| \rangle \cong 6\%$), which vindicates the use of a Fokker–Planck approach for the dynamics of large-tick queues.
- The initial rescaled volume of incipient queues (i.e. queues born when the spread momentarily opens) is around 10–15% on average, meaning that the choice $\varrho(V|0) = \delta(V - 0^+)$ is only a rough first approximation.
- The average number of events before depletion is large (~ 100), which suggests that queues are typically in the stationary regime of the Fokker–Planck dynamics.
- The average time to depletion is smaller for incipient queues than when second-best queues become best queues. This makes sense since the initial volumes are smaller in the former case.
- More than 20% of the depletion events lead to immediate refill, with no price change.

[4] Choosing $\delta n > 1$ is also possible, although some ambiguity is introduced in this case, because the bid- or ask-prices may have changed between n and $n + \delta n$. In particular, if the price changes and the queue is replaced by a new queue with different volume, conditioning to $u(n) = u$ makes little sense, as the volume might be very different for the last events. However, one can adopt reasonable conventions to give a meaning to u, leading in the end to very similar results for $\delta n = 10$ and $\delta n = 1$.

Table 6.1. *Summary statistics of queue-depleting events for INTC and CSCO. The initial volume is typically small for incipient queues (i.e. for queues born in an open spread), and large when the second-best queue becomes the best queue. The empirical values of the average time to depletion imply that the exit flux J_0 is of the order of* 0.01. *This is (see Figure 6.2) roughly ten times the probability of a large sweeping* MO *hitting an average queue:* $\Pi^* \cong 0.001$.

	INTC	CSCO
Number of events per queue in 5 min.	4117	3031
Average queue volume [shares]	5112	9047
Average rescaled volume change per event $\langle\lvert\delta u\rvert\rangle$	0.065	0.060
Average rescaled volume of incipient queues	0.140	0.114
Average rescaled volume of former 2nd-best queues	1.45	1.46
Probability that next event is LO	0.506	0.505
Probability that next event is CA	0.472	0.474
Probability that next event is MO	0.022	0.021
Probability that next event is a sweeping MO $\bar{\Pi}_{\text{st.}}$	0.0059	0.0046
Probability that next event depletes the queue	0.0067	0.0052
Refill probability after depletion ϕ_0	0.219	0.234
Average event-time to depletion (from incipient)	63	78
Average event-time to depletion (from 2nd-best)	180	248

- The total probability of sweeping market orders is $\cong 5 \times 10^{-3}$ per event. Most such market orders are executed when the queue is already small (see Figure 6.2).

We now turn to the shape of the drift $F(u)$ and diffusion $D(u)$ curves, shown in Figure 6.1. Interestingly, these shapes are quite similar for the two stocks considered here, and for a variety of other stocks as well.[5] One sees that $F(u)$ is positive for small queues, vanishes for some rescaled volume $u_c \cong 1.5$, and becomes negative for $u > u_c$. As one would expect, small queues grow on average and long queues shrink on average. The shape of $F(u)$ is however more complex than those assumed in the models considered in Chapter 5, where $F(u)$ was either *independent* of u and slightly negative (Section 5.3.4), or a linear function of u with a negative slope (Section 5.4.4). Empirically, $F(u)$ shows some convexity, which can be roughly fitted as $F(u) = F_0 - F_1 u^{\zeta}$ with F_0, F_1 constants and ζ an exponent between 0 and 1. This shape for $F(u)$ corresponds to a model intermediate between the diffusion model and the Q-CIR model, where the cancellation rate grows sublinearly with the size of the queue. Note the scale of the y-axis, which reveals

[5] On this point, see: Garèche, A., Disdier, G., Kockelkoren, J., & Bouchaud, J. P. (2013). Fokker-Planck description for the queue dynamics of large-tick stocks. *Physical Review E*, 88(3), 032809.

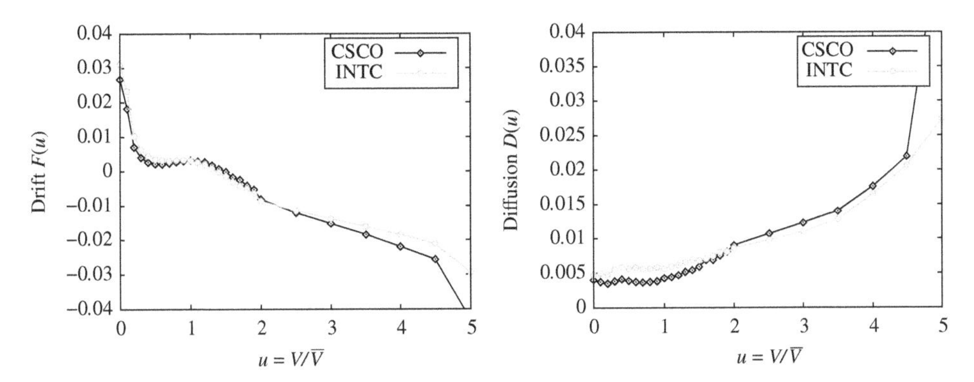

Figure 6.1. (Left panel) Drift $F(u)$ and (right panel) diffusion $D(u)$ as a function of the normalised queue volume $u = V/\bar{V}$, for INTC and CSCO, measured with $\delta n = 1$. Note that $F(u)$ changes sign at $u = u_c \cong 1.5$.

the small magnitude of the drift: a few percent of volume change on average, even for large queues with $u \sim 5$.

The diffusion coefficient $D(u)$ grows with volume, meaning that typical volume changes themselves grow with the size of the queue. This reflects the fact that the size of incoming market orders and incoming limit orders are both typically larger when the size of the queue is larger. In the simple models of Chapter 5, $D(u)$ is either flat or linear in u (see Section 5.4.4), whereas the empirically determined $D(u)$ grows faster with u.

Figure 6.2 shows the probability of sweeping market orders $\Pi(u)$ and $\Pi^{\dagger}(u)$. As expected, the probability of a queue disappearing by being fully consumed by a single market order is a strongly decreasing function of its volume (note that the y-axis is in log-scale). The probability that a queue is hit by a sweeping order is an increasing function of the opposite-side volume. This suggests that a large volume imbalance between the two queues tends to trigger a sweeping market order that consumes the smaller of the two volumes. We provide a detailed discussion of this phenomenon in Section 7.2.

6.4.2 Predicting the Stationary Distribution of Queue Sizes

We can now evaluate the accuracy of the Fokker–Planck formalism by comparing the predicted stationary queue size distribution (in event-time) with the corresponding empirical distribution. From the knowledge of $F(u)$, $D(u)$, $\Pi(u)$ and $\Pi^{\dagger}(u)$, we can use Equation (6.16) to estimate the stationary distribution of rescaled queue sizes, $P_{\text{st.}}(u)$. Figure 6.3 compares the empirically determined distribution with the theoretical Fokker–Planck prediction, Equation (6.16). The overall shape of $P_{\text{st.}}(u)$ is rather well reproduced by the Fokker–Planck prediction, in particular in the right tail ($u > u_c$) where the negative drift $F(u)$ prevents the queues from

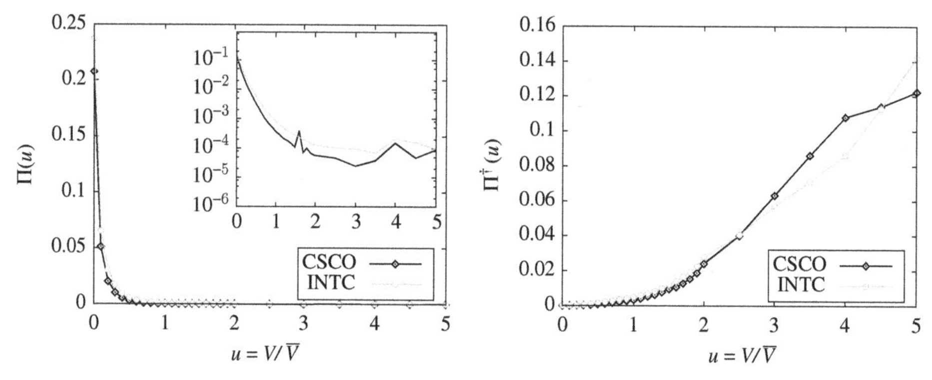

Figure 6.2. Probability of a sweeping market order, conditioned on (left panel) the same-side volume $\Pi(u)$ and (right panel) the previous opposite-side volume $\Pi^{\dagger}(u)$, as a function of the normalised queue volume $u = V/\bar{V}$, for INTC and CSCO.

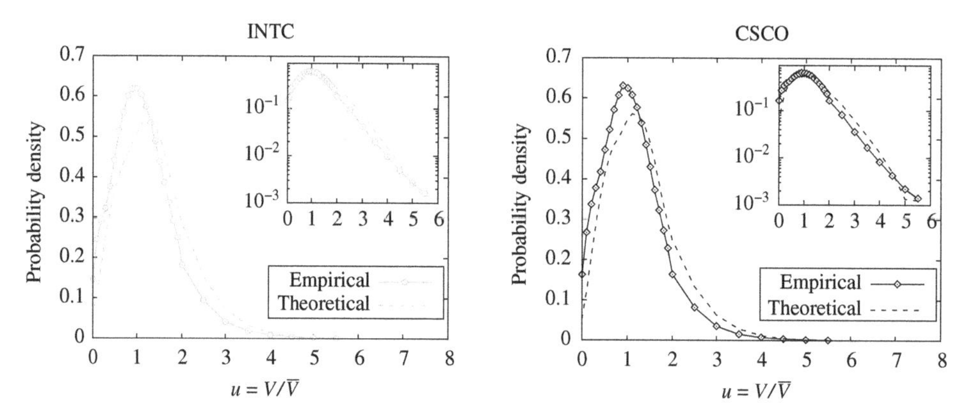

Figure 6.3. (Solid curves with markers) Empirical and (dashed curves) theoretical distribution of rescaled queue sizes as a function of $u = V/\bar{V}$, for (left panel) INTC and (right panel) CSCO.

becoming very large. This result again vindicates the use of the Fokker–Planck formalism to describe the dynamics of queues for large-tick assets.[6]

Notice that $P_{\text{st.}}(u)$ is hump-shaped, but not sharply peaked (in the sense that the ratio $P_{\text{st.}}(u^*)/P_{\text{st.}}(0)$ is not very large for these stocks). This means that the "barrier height" for depletion is not high, which is contrary to the simplifying assumption that we made in Section 6.2.2. In other words, the potential

[6] Recall that Equation (6.16) further assumes that $\varrho(u|0) = \delta(u - 0^+)$ and that $P_{\text{2-best}}(u) \approx P_{\text{st.}}(u)$, two approximations that are quite rough and that could be easily improved.

well $\mathcal{W}(u^*) = -\int^{u^*} \mathrm{d}u\, F(u)/D(u)$ is too shallow to block diffusion-induced depletion.

6.5 Conclusion

In this chapter, we have introduced an intuitive and flexible description of the queue dynamics for large-tick assets, based on the Fokker–Planck formalism. Interestingly, this description includes some dependence on the current state of the queue, through the explicit dependence of the drift and diffusion constant on the queue size. However, this approach required no restrictive assumptions on the dependence of the arrival or cancellation rates on queue size, nor on the size of queue changes. The Fokker–Planck approach can also accommodate jump events, which correspond to sudden changes of the bid or ask prices.

All of the quantities involved in this approach can be calibrated using event-resolved data that describes the order flow at the best quotes. One important observation was that the dynamical process is approximately scale invariant, in the sense that the only relevant variable is the ratio of the volume of the queue V to its average value $\bar{V}$. While the latter shows intra-day seasonalities and strong variability across stocks and time periods, the dynamics of the rescaled volumes is, to first approximation, universal.

The validity of the Fokker–Planck approximation relies on elementary volume changes δV being small. Although this assumption fails for small-tick assets, it is well justified for large-tick assets (see Table 6.1). Correspondingly, when solved to find the stationary distribution of rescaled volumes in a queue, the Fokker–Planck equation calibrated on dynamical data fares quite well at reproducing the empirical distribution.

For the two large-tick stocks that we investigated in this chapter (INTC and CSCO), the distribution of queue sizes is not strongly peaked around its maximum V^*, as the Q-CIR model would predict. This observation can be traced back to the fact that the ratio of the drift to the diffusion is *not* very large, which implies that the depletion time is large but not extremely large. The Q-CIR model, on the other hand, would predict that queues empty exceedingly rarely (see also Chapter 8).

As we will discuss in the next chapter, the fact that the depletion time is moderate allows one to predict which of the bid-queue or the ask-queue is most likely to empty first. As we will conclude in Section 7.3.3, reality seems to be best described by a model intermediate between the diffusive model (with a constant cancellation rate per queue) and the Q-CIR model (with a constant cancellation rate per order).

Take-Home Messages

(i) Under the assumptions that the best bid- and ask-queues evolve independently and via small increments, it is possible to write 1-D Fokker–Planck equations for the resulting queue dynamics.
(ii) It is straightforward to calibrate the Fokker–Planck equation on real data. The theoretical stationary queue distribution can be found using the Boltzmann–Gibbs measure, and is close to the empirical distribution.
(iii) Taking large (sweeping) market orders into account significantly improves the fit to the data. This suggests that the small-increments hypothesis central to the Fokker–Planck approach is too restrictive for modelling real order flow.

6.6 Further Reading

See also Section 5.6.

General

Feller, W. (1971). *An introduction to probability theory and its applications* (Vol. 2). John Wiley and Sons.
Van Kampen, N. G. (1983). *Stochastic processes in physics and chemistry.* North-Holland.
Risken, H. (1984). *The Fokker–Planck equation.* Springer, Berlin-Heidelberg.
Gardiner, C. W. (1985). *Stochastic methods.* Springer, Berlin-Heidelberg.
Hänggi, P., Talkner, P., & Borkovec, M. (1990). Reaction-rate theory: Fifty years after Kramers. *Reviews of Modern Physics*, 62(2), 251.
Redner, S. (2001). *A guide to first-passage processes.* Cambridge University Press.

Fokker–Planck Equations for Modelling LOBs

Cont, R., & De Larrard, A. (2012). Order book dynamics in liquid markets: Limit theorems and diffusion approximations. https://ssrn.com/abstract=1757861.
Garèche, A., Disdier, G., Kockelkoren, J., & Bouchaud, J. P. (2013). Fokker-Planck description for the queue dynamics of large-tick stocks. *Physical Review E*, 88(3), 032809.
Yang, T. W., & Zhu, L. (2016). A reduced-form model for level-1 limit order books. *Market Microstructure and Liquidity*, 2(02), 1650008.

7

Joint-Queue Dynamics for Large-Tick Stocks

What is the logic of all this? Why is the smoke of this pipe going to the right rather than to the left? Mad, completely mad, he who calculates his bets, who puts reason on his side!

(Alfred de Musset)

In Chapter 6, we considered how to model either the best bid-queue or the best ask-queue in an LOB for a large-tick stock. Although this approach enabled us to derive many interesting and useful results about single-queue dynamics, it is clear that the dynamics of the bid-queue and the dynamics of the ask-queue mutually influence each other. In this chapter, we therefore extend our framework to consider the joint dynamics of the best bid- and ask-queues together.

We begin the chapter by discussing the bid- and ask-queues' "race to the bottom", which is the core mechanism that drives price changes in LOBs for large-tick stocks. We present some empirical results to illustrate how such races play out in reality, then turn our attention to modelling such races. We first consider a race in which the bid- and ask-queues are assumed to evolve independently, and derive several results from the corresponding model. We next turn our attention to a more general case, in which we allow the dynamics of the bid- and ask-queues to be correlated. We end the chapter by introducing an even more versatile framework, by extending the Fokker–Planck framework from the previous two chapters to also encompass the coupled dynamics of the two queues. We illustrate that this approach provides a consistent description of the joint dynamics observed in real LOBs for large-tick stocks.

This chapter is not essential to the main story of the book.

7.1 The Race to the Bottom

Recall from Chapter 3 that in an LOB for a large-tick stock, the bid–ask spread is very often equal to its minimum possible size, $s(t) = \vartheta$. This has an important consequence for LOB dynamics, because it removes the possibility that a new limit order will arrive inside the spread. Therefore, whenever $s(t) = \vartheta$, the only way for either the bid-price or the ask-price to change is for one of the order queues at the best quotes to deplete to $V = 0$.

If the bid-queue depletes before the ask-queue, then the bid-price moves down (but might bounce back up immediately – see below). If the ask-queue depletes before the bid-queue, then the ask-price moves up (but might bounce back down immediately). Therefore, price changes in this situation can be regarded as the consequence of a simple race between the bid- and ask-queues, in which the winner (i.e. the queue that depletes first) dictates the direction of the next price change.

By considering some initial volumes V_b^0 and V_a^0, then considering the subsequent market order arrivals, limit order arrivals, and cancellations that occur at $b(t)$ and $a(t)$, we can derive models for the two-dimensional trajectory $(V_b(t), V_a(t))$ (see Figure 7.1), in which $V_b(t)$ and $V_a(t)$ may be regarded as racing to zero, until at some time t_1 one queue empties and the spread momentarily widens. After t_1, a limit order will very likely rapidly fill the (now widened) spread. What happens in that case will be discussed in depth in Section 7.5.

Many interesting questions about the dynamics of the bid- and ask-queues spring to mind. At some randomly chosen time t:

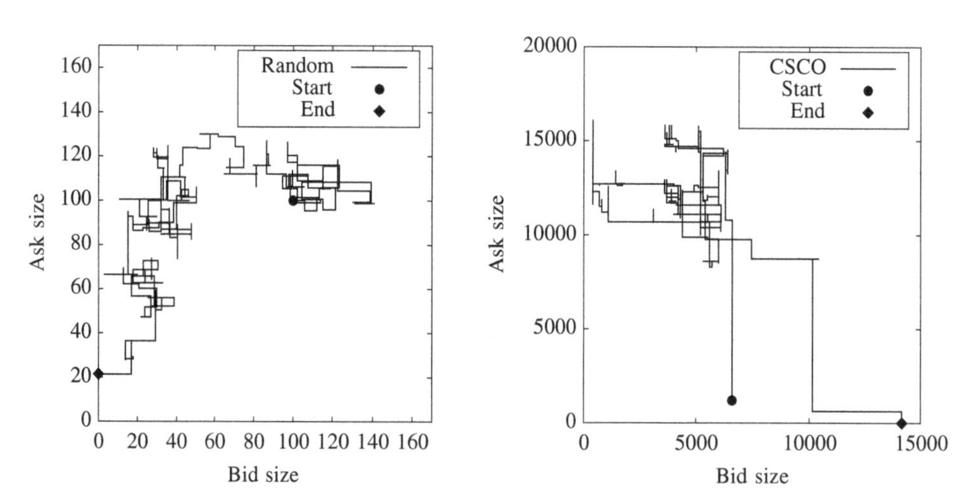

Figure 7.1. Typical trajectories in the plane (V_b, V_a), starting (circle) at (V_b^0, V_a^0) and ending (diamond) by the depletion of one of the two queues. The left panel shows simulation data for a pair of uncoupled random walks; the right panel shows the real data for CSCO.

- What is the joint probability $P(V_b, V_a)$ that the ask volume is $V_a(t) = V_a$ and the bid volume is $V_b(t) = V_b$?
- Having observed the two volumes $(V_b(t), V_a(t))$, what is the probability $p_\pm$ that the next mid-price move is upwards?
- Having observed the two volumes $(V_b(t), V_a(t))$, what is the rate $\Phi(\tau)$ for the next mid-price move (up or down) to take place at time $t + \tau$?
- Having observed the two volumes $(V_b(t), V_a(t))$, what is the probability that an order with a given priority at the bid or at the ask will be executed before the mid-price changes?

Throughout the remainder of this chapter, we address the first two topics and some other questions about joint-queue dynamics for large-tick stocks. The last two questions, as well as many others, can be studied using similar tools.

7.2 Empirical Results

We first present some empirical results to illustrate joint-queue dynamics for large-tick stocks in real markets. Given that the state space we consider is the two-dimensional space (V_b, V_a), one possible approach could be to estimate empirically the fraction of times that a race ended by an upward price movement (rather than a downward price movement) for a given initial choice of (V_b, V_a). However, this approach suffers the considerable drawback of needing to estimate and interpret a two-dimensional surface, for which many choices of initial conditions have relatively few data points because they occur relatively infrequently in real markets. Furthermore, the **scale invariance** property of the queue dynamics, noted in the previous chapter and approximately valid when queues are large, suggests to investigate this problem not by considering the two-dimensional input quantity (V_b, V_a), but instead by considering the one-dimensional **queue imbalance** ratio

$$I := \frac{V_b}{V_b + V_a}. \tag{7.1}$$

The quantity I measures the (normalised) imbalance between V_b and V_a, and thereby provides a quantitative assessment of the relative strengths of buying and selling pressure in an LOB. Observe that I is invariant upon rescaling both volumes by the same factor α (i.e. $V_a \to \alpha V_a$ and $V_b \to \alpha V_b$), and takes values in the open interval $(0, 1)$. A value $I \approx 0$ corresponds to a situation where the ask-queue is much larger than the bid-queue, which suggests that there is a net positive selling pressure in the LOB that will most likely push the price downwards. A value $I \approx 1$ corresponds to a situation where the bid-queue is much larger than the ask-queue, which suggests that there is a net positive buying pressure in the LOB that will most likely push the price upwards. A value $I \approx \frac{1}{2}$ corresponds to a situation in

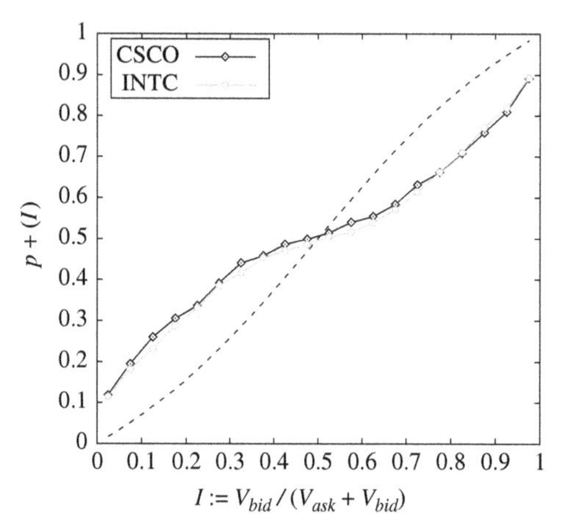

Figure 7.2. Probability that the ask-queue depletes before the bid-queue, as a function of the queue imbalance I for INTC and CSCO. We also show (dashed curves) the result of a simple model where the volume of the bid- and ask-queues undergo independent random walks, given by Equation (7.6).

which the bid- and ask-queues have approximately equal lengths, which suggests that the buying and selling pressures in the LOB are approximately balanced.

We now turn our attention to an empirical analysis of how the volume imbalance I impacts the outcome of a given race. To do so, we consider the value of I at some times chosen uniformly at random, then fast-forward to see which of the bid- or ask-queues depleted to zero first. We can then estimate the probability $p_+(I)$ as the fraction of races for which the subsequent price movement was upwards. We plot these empirical results in Figure 7.2. The graph reveals a strong, monotonic correlation between the queue imbalance and the direction of the next price move.[1]

Throughout the remainder of this chapter, we will introduce several different models for the joint-queue dynamics of large-tick stocks. For each such model, we will estimate the relevant parameters from data, then plot the model's predicted output for the probability that a race with a given input imbalance will result in an upward price movement. We will then compare each model's outputs with our empirical results in Figure 7.2 to assess how well (or otherwise) each approach is able to reproduce realistic market dynamics.

7.3 Independent Queues

We first consider a race in which the bid- and ask-queues evolve independently, according to the dynamics that we first introduced in Chapter 5 (in which limit order arrivals, market order arrivals, and cancellations are each assumed to occur

[1] This has been called the "worst kept secret of high-frequency traders", see, e.g., http://market-microstructure.institutlouisbachelier.org/uploads/91_3%20STOIKOV%20Microstructure_talk.pdf

according to independent Poisson processes). In Section 5.3, we studied a model in which cancellations occur as a Poisson process with fixed rate for the whole queue together, and in Section 5.4, we studied a model in which each individual limit order is cancelled with some rate (such that at time t, the total cancellation rate for the whole bid-queue is proportional to $V_b(t)$ and the total cancellation rate for the whole ask-queue is proportional to $V_a(t)$). For describing the dynamics of either the bid-queue or ask-queue in isolation, these approaches can be written:

$$\begin{cases} V \to & V+1 \qquad \text{with rate} \quad \lambda \mathrm{d}t \\ V \to & V-1 \qquad \text{with rate} \quad (\mu + \nu V^{\zeta})\mathrm{d}t, \end{cases} \tag{7.2}$$

where $\zeta = 0$ for the first model and $\zeta = 1$ for the second model.

To describe the joint dynamics of the bid- and ask-queues together, we replace the parameters λ, μ, and ν with the parameters λ_b, λ_a, μ_b, μ_a, ν_b and ν_a, where the subscript denotes the relevant queue. To begin with, let us suppose that the initial bid- and ask-queue volumes are (V_b, V_a), that their subsequent evolution is governed by (7.2), with no coupling between the two queues, and that the rate parameters obey

$$\lambda_b = \lambda_a,$$

$$\mu_b = \mu_a,$$

$$\nu_b = \nu_a.$$

When queues are independent, there are several different approaches for obtaining the probability p_+ that the ask-queue empties first. One such approach is to consider an integral that counts all the possible outcomes where the first-hitting time of the bid-queue is larger than that of the ask-queue,

$$p_+(V_b, V_a) = \int_0^{\infty} \mathrm{d}\tau \Phi(\tau, V_a) \int_{\tau}^{\infty} \mathrm{d}\tau' \Phi(\tau', V_b), \tag{7.3}$$

where $\Phi(\tau, V)$ is the distribution of first-hitting times considered in Section 5.3.2.

An equivalent approach is to follow a similar reasoning to the one that we employed in Equation (5.16):

$$\begin{aligned} p_+(V_b, V_a) = & \frac{W_+(V_b)}{Z(V_b, V_a)} p_+(V_b + 1, V_a) + \frac{W_+(V_a)}{Z(V_b, V_a)} p_+(V_b, V_a + 1) \\ & + \frac{W_-(V_b)}{Z(V_b, V_a)} p_+(V_b - 1, V_a) + \frac{W_-(V_a)}{Z(V_b, V_a)} p_+(V_b, V_a - 1), \end{aligned} \tag{7.4}$$

where $Z(V_b, V_a) = W_+(V_b) + W_-(V_b) + W_+(V_a) + W_-(V_a)$, and with the boundary conditions $p_+(V_b > 0, V_a = 0) = 1$ and $p_+(V_b = 0, V_a > 0) = 0$. We will exploit this general framework in simple cases in the next sections.

7.3.1 Races between Diffusive Queues

We first consider the diffusive case, where cancellations occur as a Poisson process with fixed rate for the whole queue together. This corresponds to the case $\zeta = 0$ in Equation (7.2) and the case $W_+(V) = \lambda$ and $W_-(V) = \mu + \nu$ in Equation (7.4), for both the bid- and the ask-queues.

As we discussed in Chapter 5, the non-critical case (where $\lambda < \mu + \nu$) is not very interesting, because it cannot produce long queues. Recall from Section 5.3.2 that the mean first-hitting time of a single queue is a Gaussian random variable with mean $V/(\mu + \nu - \lambda)$ and variance proportional to V. If one neglects fluctuations (of order $\sqrt{V}$), it is clear that the shortest queue empties first. There is significant uncertainty only if the width of the first-hitting time distribution becomes larger than the difference $|V_a - V_b|/(\mu + \nu - \lambda)$. For large queues, this can only happen when $V_a \sim V_b$, with $|V_a - V_b|$ of the order of $\sqrt{V_a}$ ($\sim \sqrt{V_b}$). In other words, the probability that the ask-queue empties first is close to 1 when $V_a \ll V_b$ and close to 0 when $V_a \gg V_b$, with a sharp, step-like transition of width $\sqrt{V_{a,b}}$ around $V_a = V_b$.

The more interesting case is the critical case (where $\lambda = \mu + \nu$). We study this case in the continuum limit, (see Section 5.3.4).[2] In this case, $W_+(V) = W_-(V) = \lambda$ and Equation (7.4) becomes **Laplace's equation** in the first quadrant ($V_b > 0, V_a > 0$):

$$\lambda\left(\partial^2_{V_b V_b} + \partial^2_{V_a V_a}\right) p_+(V_b, V_a) = 0; \qquad p_+(V_b, 0) = 1; \qquad p_+(0, V_a) = 0. \tag{7.5}$$

One approach to solving this equation is to introduce polar coordinates (R, θ), with

$$R = \sqrt{V_b^2 + V_a^2},$$
$$\theta = \arctan(V_b / V_a).$$

In polar coordinates, Laplace's equation becomes

$$\left(\partial^2_{RR} + R^{-2}\partial^2_{\theta\theta}\right) p_+(R, \theta) = 0.$$

One can check that the only acceptable solution to this equation is of the form $p_+ = A\theta$ (i.e. independent of R). The constant A is fixed by the boundary conditions, so that the solution finally reads:

$$p_+(I) := p_+(V_b, V_a) = \frac{2}{\pi}\arctan\left(\frac{I}{1 - I}\right), \tag{7.6}$$

with

$$\frac{I}{1 - I} = \frac{V_b}{V_a}. \tag{7.7}$$

Therefore, we find that p_+ is scale invariant, i.e. it depends only on the ratio V_b / V_a (or equivalently on the volume imbalance I), but not on V_a and V_b separately.

[2] For a detailed discussion of the discrete case, see Cont, R. & De Larrard, A. (2013). Price dynamics in a Markovian limit order market. *SIAM Journal on Financial Mathematics*, 4(1), 1–25.

As a sanity check, any solution must obey $p_+(I) = p_-(1 - I) = 1 - p_+(1 - I)$, by the symmetry of the system. We can verify that Equation (7.6) indeed satisfies these conditions by direct substitution. Figure 7.2 compares Equation (7.6) with empirical results. Clearly, the independent, diffusive queue model is unable to predict the observed concavity of $p_+(I)$.

This framework can be extended to the case where $\lambda_b \neq \lambda_a$, where both queues are still critical but one queue is more active than the other. The Laplace equation then becomes:

$$\begin{aligned}\left(\lambda_b \partial^2_{V_b V_b} + \lambda_a \partial^2_{V_a V_a}\right) p_+(V_b, V_a) &= 0;\\ p_+(V_b, 0) &= 1;\\ p_+(0, V_a) &= 0.\end{aligned}$$

By setting

$$\begin{aligned}V_a &= \sqrt{\lambda_a} u_a,\\ V_b &= \sqrt{\lambda_b} u_b,\end{aligned}$$

we arrive back at the same system as in Equation (7.5), leading to:

$$p_+(V_b, V_a) = \frac{2}{\pi} \arctan\left(\sqrt{\frac{\lambda_a}{\lambda_b}} \frac{V_b}{V_a}\right).$$

This result provides a simple interpretation of the dynamics in the critical case ($\lambda = \mu + \nu$): for queues of the same size, the fastest-changing queue is more likely to empty first. For example, if $V_a = V_b$ but $\lambda_a > \lambda_b$, then $p_+ > 1/2$ (see Figure 7.3).

7.3.2 Diffusive Queues with Correlated Noise

As we saw in Section 7.3.1, p_+ is an increasing function of I, even when the order flow is completely random – meaning that short queues tend to disappear first. This empirical behaviour is well known by traders, which causes it to induce a feedback loop between traders' observations and actions. For example, when the bid-queue increases in size, traders might think that the price is more likely to go up, and therefore some sellers with limit orders in the ask-queue might cancel, while some buyers might rush to buy quickly (before the anticipated price move). Both of these actions lead to a decreased volume of the ask-queue.

Through this feedback mechanism, the dynamics of the bid- and ask-queue become negatively correlated (see Section 7.4.2). How might we extend our models of joint-queue dynamics to reflect this empirical behaviour? The simplest approach is to consider a diffusive framework with correlated noise acting on the

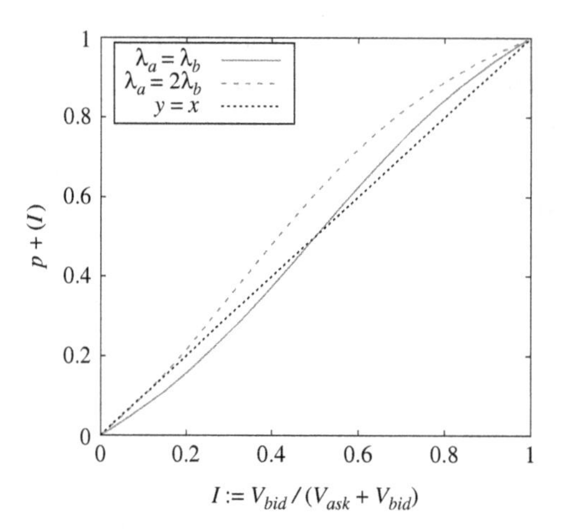

Figure 7.3. Probability that the ask-queue depletes before the bid-queue, as a function of the queue imbalance I, for diffusive queues with (solid line) $\lambda_a = \lambda_b$ and (dashed line) $\lambda_a = 2\lambda_b$. For similar volumes, the fastest changing queue has a higher probability of depleting first. The dotted line represents the diagonal.

two queues. More formally, this amounts to saying that the random increments can be written as

$$\delta V_b = \xi\sqrt{1+\rho} + \widehat{\xi}\sqrt{1-\rho},$$
$$\delta V_a = \xi\sqrt{1+\rho} - \widehat{\xi}\sqrt{1-\rho},$$

where ξ and $\widehat{\xi}$ are independent noise terms with zero mean and arbitrary but equal variances, and ρ is the correlation coefficient between the increments at the bid- and ask-queues.

From this decomposition, one sees that the sum

$$S = V_a + V_b$$

and the difference

$$\Delta = V_b - V_a$$

evolve independently, but with noises of different variances (respectively, $1+\rho$ for S and $1-\rho$ for Δ). In the (S,Δ) plane (with $S \geq 0$ and $-S \leq \Delta \leq S$), the Laplace equation from Section 7.3.1 becomes

$$\left((1+\rho)\partial^2_{SS} + (1-\rho)\partial^2_{\Delta\Delta}\right) p_+(S,\Delta) = 0, \tag{7.8}$$

with boundary conditions

$$p_+(S,S) = 1,$$
$$p_+(S,-S) = 0.$$

Defining

$$S' = \frac{S}{\sqrt{1+\rho}},$$
$$\Delta' = \frac{\Delta}{\sqrt{1-\rho}},$$

we recover exactly the independent queue problem of Section 7.3.1. In these new coordinates, the boundaries $\Delta = \pm S$ now have slopes $\pm\sqrt{(1+\rho)/(1-\rho)}$, so the wedge has an opening half-angle

$$\psi = \arctan\left(\sqrt{\frac{1+\rho}{1-\rho}}\right).$$

In terms of the new coordinates, the initial conditions read

$$S' = \frac{V_a + V_b}{\sqrt{1+\rho}},$$
$$\Delta' = \frac{V_b - V_a}{\sqrt{1-\rho}}.$$

Therefore, the angle corresponding to the initial condition (V_b, V_a) is given by

$$\theta_0 = \arctan\left(\sqrt{\frac{1+\rho}{1-\rho}}\,\frac{V_b - V_a}{V_b + V_a}\right). \tag{7.9}$$

The solution $p_+ = A\theta$ easily generalises for a wedge of arbitrary half-angle ψ, as:

$$p_+ = \frac{1}{2} + \frac{\theta_0}{2\psi}. \tag{7.10}$$

In the limit of zero correlations, $\rho = 0$, so $\psi = \pi/4$ and one indeed recovers (after simple trigonometric manipulations)

$$p_+ = \frac{2}{\pi}\arctan(V_b/V_a),$$

which is the same result as we found in Equation (7.6).

In the limit of perfectly anti-correlated queues, $\rho = -1$, so the wedge becomes extremely sharp ($\theta_0, \psi \to 0^+$) and the problem becomes the famous one-dimensional **gambler's ruin**, for which the behaviour is linear in the imbalance:

$$p_+ = I.$$

Figure 7.4 shows p_+ as a function of I, for the three values $\psi = \pi/4$, $\psi = 3\pi/8$ and $\psi = 0^+$, which correspond to $\rho = 0$, $\rho = 0.71$ and $\rho = -1$, respectively. For all

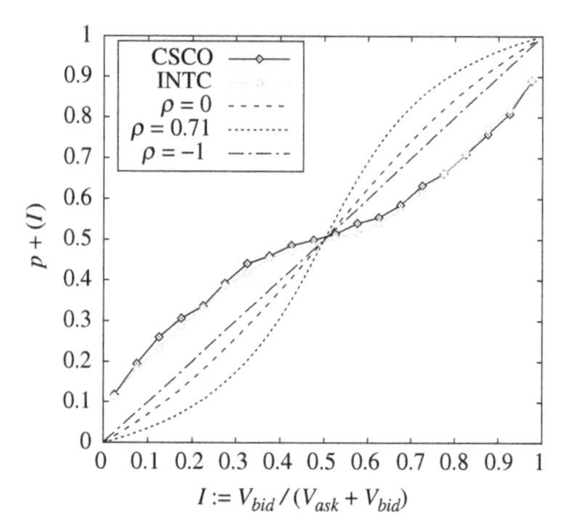

Figure 7.4. Probability that the ask-queue depletes before the bid-queue, as a function of the queue imbalance I. The solid curves denote empirical estimates for INTC and CSCO; the other curves denote theoretical predictions for (dotted curve) $\rho = 0$, (dashed curve) $\rho = 0.71$ and (dash-dotted curve) $\rho = -1$.

values of correlations, the dependence of p_+ on I has an inverted concavity when compared with empirical data. We now turn to a model that correctly captures this concavity.

7.3.3 Races between Long, Self-Stabilising Queues

We now consider the Q-CIR case (see Section 5.4.4), with $W_+(V) = \lambda$ and $W_-(V) = \mu + \nu V$, in the limit of long equilibrium sizes $V^* = (\lambda - \mu)/\nu \gg 1$. In the notation of Section 7.2, this corresponds to the case $\zeta = 1$.[3] Using the fact that the distribution of first-hitting times is Poissonian (see Section 5.4.3), Equation (7.3) becomes very simple:

$$p_+(V_b, V_a) \approx \frac{\mathbb{E}_b[T_1]}{\mathbb{E}_a[T_1] + \mathbb{E}_b[T_1]},$$

where $\mathbb{E}_b[T_1]$ and $\mathbb{E}_a[T_1]$ denote the mean first-hitting times of the bid- and ask-queues, respectively.

This result illustrates that the queue with the smaller mean first-hitting time has a larger probability of emptying first. However, because the mean first-hitting time is nearly independent of the initial queue lengths when V^* is large, we discover that in the present framework, the probability that a given queue empties first only depends very weakly on the queue imbalance I, except when at least one of the queues become small.

Figure 7.5 shows how p_+ depends on I, with an initial condition chosen with the stationary measure. Interestingly, the plateau region for $I \approx 1/2$ is less pronounced

[3] In fact, the results that we present here actually hold for any $\zeta > 0$.

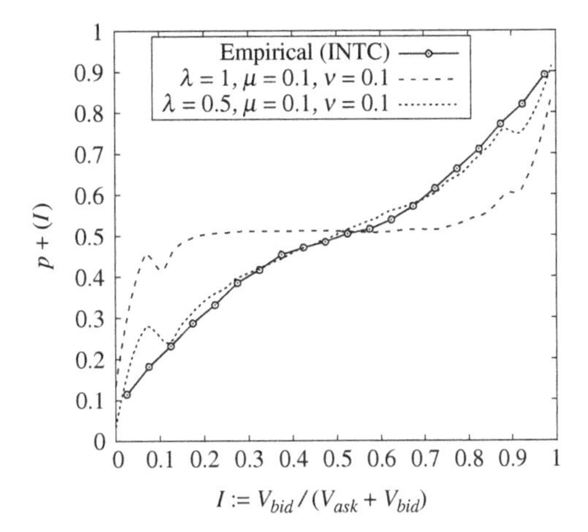

Figure 7.5. Probability that the ask-queue depletes before the bid-queue, as a function of the queue imbalance I within the stationary Q-CIR model with (dashed line) $\lambda = 1$ and $\mu = \nu = 0.1$ (corresponding to $V^* = 10$), and (dotted line) $\lambda = 0.5$ and $\mu = \nu = 0.1$ (corresponding to $V^* = 5$). We also show the same empirical data as in Figure 7.2 for INTC; its concavity is well captured by the Q-CIR model with a moderate value of V^*.

when V^* is not too large, and the curves start resembling quite closely those observed empirically. In particular, the Q-CIR model (i.e. with $\zeta = 1$) provides qualitative insights into the concavity of $p_+(I)$ that are not correctly predicted by the diffusive model (i.e. with $\zeta = 0$).

We thus recover the same conclusion as in Section 6.5: the queue dynamics of large-tick stocks is in a regime where drift effects are comparable to diffusion effects – in other words, that the effective potential defined in Sections 5.4.5 and 6.2.1 is shallow. Reality is probably best described by a value of ζ somewhere between 0 and 1 (see also Figure 6.1).

7.4 The Coupled Dynamics of the Best Queues

In the previous section, we postulated some simple dynamics for the volumes of the queues. We now consider a more flexible approach to modelling the joint-queue dynamics, which more clearly elicits the nature of the coupling between the two queues, and can be calibrated on empirical data.

7.4.1 A Two-Dimensional Fokker–Planck Equation

Generalising the arguments from Section 6.2, one finds (neglecting all price-changing events) that the bivariate distribution of the bid- and ask-queue sizes obeys a **two-dimensional Fokker–Planck equation**:

$$\begin{aligned}\frac{\partial P(V_b, V_a, t)}{\partial t} &= -\partial_{V_b}\left[F_b(V_b, V_a)P(V_b, V_a, t)\right] \\ &\quad - \partial_{V_a}\left[F_a(V_b, V_a)P(V_b, V_a, t)\right] \\ &\quad + \partial^2_{V_b V_b}\left[D_{bb}(V_b, V_a)P(V_b, V_a, t)\right] \\ &\quad + \partial^2_{V_a V_a}\left[D_{aa}(V_b, V_a)P(V_b, V_a, t)\right] \\ &\quad + 2\partial^2_{V_b V_a}\left[D_{ba}(V_b, V_a)P(V_b, V_a, t)\right], \end{aligned} \tag{7.11}$$

where $\vec{F} := (F_b, F_a)$ is a two-dimensional vector field called the **flow field** and $\mathbf{D} = (D_{aa}, D_{ab}; D_{ba}, D_{bb})$ is a space-dependent, two-dimensional **diffusion matrix**. Similarly to Equation (6.8), these objects are defined as:

$$\begin{aligned} F_i(V_b, V_a) &:= \frac{1}{\tau_c}\sum_{\delta V_i} \delta V_i\, \varrho(\delta V_i | V_b, V_a), \qquad i \in \{b, a\}; \\ D_{ij}(V_b, V_a) &:= \frac{1}{2\tau_c}\sum_{\delta V_i, \delta V_j} \delta V_i \delta V_j \varrho(\delta V_i, \delta V_j | V_b, V_a), \qquad i, j \in \{b, a\}, \end{aligned} \tag{7.12}$$

where ϱ is the distribution of elementary volume changes, conditioned to the current size of the queues.

Note that D_{ba} and D_{ab} are the only terms that require explicit knowledge of the joint distribution of simultaneous size changes in both queues. (By symmetry, these terms are equal). These are the only terms that are sensitive to correlation between the order flow at the best bid- and ask-queues. However, even if these terms were absent, the dynamics of the two queues are coupled by the explicit dependence of the drift field $\vec{F}$ and the diagonal elements of $\mathbf{D}$ on the sizes of both the bid- and ask-queues. Such a drift-induced coupling is completely absent from the models in Sections 7.3 and 7.3.2.

7.4.2 Empirical Calibration

It is interesting to consider how the flow field $\vec{F} = (F_b, F_a)$ behaves empirically. We calculate the statistics of the flow using $\delta n = 10$ events (i.e. we measure the queue lengths once every 10 events, where each event corresponds to whenever one of the two queues changes size). Figure 7.6 shows $\vec{F}$ for INTC and CSCO, as a function of rescaled volumes $u_b = V_b/\bar{V}_b$, $u_a = V_a/\bar{V}_a$. As the figure illustrates, the flow field is very similar in both cases. As we expect from **buy/sell symmetry**, the flow field obeys

$$F_b(V_b, V_a) \cong F_a(V_a, V_b).$$

These plots reveal that when both queues are small, their volumes tend to grow on average, and when both queues are large, their volumes tend to decrease on

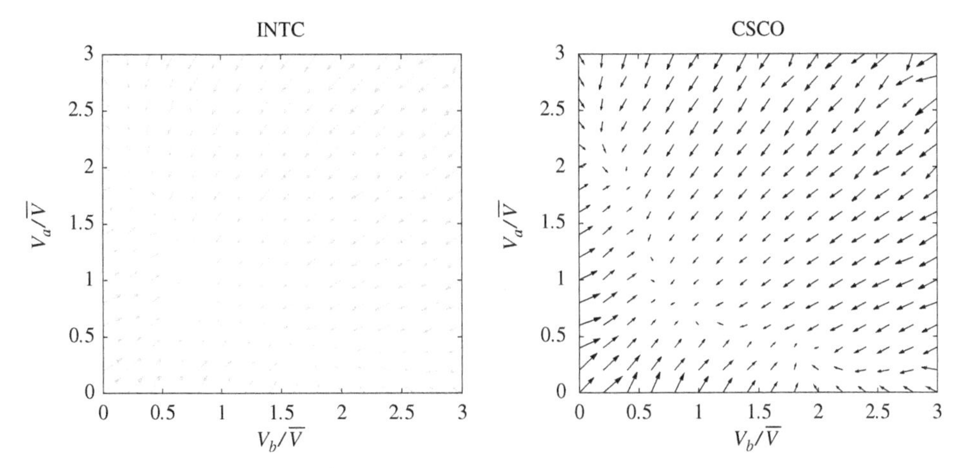

Figure 7.6. Flow field $\vec{F}$ of volume changes at the best bid and ask for (left panel) INTC and (right panel) CSCO. To improve readability, the displayed vector norms correspond to the square-root of the original vector lengths, rescaled by a factor 10.

average. The flow vanishes (i.e. $|\vec{F}| = 0$) at only one point $u_a^* = u_b^* \cong 1$. The flow lines are quite regular and appear to be nearly **curl-free**, which implies that $\vec{F}$ can be approximately written as a pure gradient $\vec{F} = -\vec{\nabla} U$, where U has a funnel-like shape pointing towards (u_b^*, u_a^*).

Of course, a significant dispersion of the dynamics around the average flow pattern occurs through the diffusion matrix. By symmetry, one expects

$$D_{bb}(V_b, V_a) = D_{aa}(V_a, V_b),$$

which is indeed well-obeyed by the data.

Empirically, the variance of volume changes is found to depend mostly on the same queue and very little on the opposite queue, so

$$D_{bb}(V_b, V_a) \cong D(V_b),$$
$$D_{aa}(V_b, V_a) \cong D(V_a),$$

where the function $D(V)$ is as in Section 6.4. The cross-diffusion terms $D_{ba} = D_{ab}$ are found to be negative and of amplitude $\cong 20\%$ of the diagonal components. This negative correlation was expected from the arguments given in Section 7.3.2, but is rather weak.

7.4.3 Stationary State

In Section 6.2.1, we saw that it is always possible to write the stationary state of the one-dimensional Fokker–Planck equation as a Boltzmann–Gibbs measure, in terms of an effective potential $\mathcal{W}(V)$. Unfortunately, this is not the case in higher dimensions (including the present two-dimensional case), except in special cases. For example, when

$$D_{aa}(V_a, V_b) = D_{bb}(V_a, V_b) = D(V_a, V_b),$$
$$D_{ab} = D_{ba} = 0,$$

and provided the flow can be written as a gradient,

$$F_{a,b}(V_a, V_b) := -D(V_a, V_b)\partial_{a,b}\mathcal{W}(V_a, V_b),$$

then the stationary state (again neglecting any reinjection term) can be written as a Boltzmann–Gibbs measure

$$P_{\text{st.}}(V_a, V_b) = \frac{A}{D(V_a, V_b)} e^{-\mathcal{W}(V_a, V_b)}. \tag{7.13}$$

This result cannot be exactly applied to data, because D_{aa} and D_{bb} are not equal to each other for all V_a, V_b. Still, because the flow field is nearly curl-free (as noted above), Equation (7.13) is an acceptable approximation in the regime where V_a and V_b are not too large, such that D is approximately constant. Such an approximation immediately leads to the conclusion that $P_{\text{st.}}(V_a, V_b)$ is peaked in the vicinity of the point where the flow field $\vec{F}$ vanishes. This agrees well with empirical observations.

7.5 What Happens After a Race Ends?

To understand the full price dynamics of large-tick stocks, one has to discuss what happens at the end of the race between the two best quotes. Imagine that the bid-queue has emptied first. Two things can happen at the now vacant price level (see Section 6.1):

- With probability ϕ_0, a buy limit order immediately refills the old bid position. In this case, the mid-price reverts to its previous position (after having briefly moved down by half a tick).
- With probability $1-\phi_0$, a sell limit order immediately refills the old bid position. In this case the mid-price has moved down by one tick; the new ask is the old bid and the new bid is the old second-best bid.

In the second case, the price moves down, but the incipient ask-queue now faces the old second-best bid-queue. The new ask-queue is therefore typically much smaller than the bid-queue. The volume imbalance I introduced in Section 7.2 would then suggest that the new ask is more likely to disappear first. This would lead to strong mean-reversion effects where the mid-price moves back up with a large probability. In the purely stochastic models considered in the previous chapters and in the next chapter, this is indeed what happens.

Therefore at least one extra ingredient must be present to counteract this mean-reversion mechanism. As the flow lines of Figure 7.6 indicate, the new ask-queue will in fact grow quickly at first, until the system equilibrates around $I \approx 1/2$. This is the spirit of the assumption made in Cont and De Larrard

(2013), where the volume of both the new ask-queue and the new bid-queue are independently chosen from the stationary distribution $P_{\text{st.}}(V)$. Such an assumption allows one to erase all memory of the past queue configuration, and automatically generates a diffusive mid-price.

7.6 Conclusion

This ends a series of three chapters on the dynamics of long queues. Here, we have mostly focused on the joint dynamics of the bid- and ask-queues, for which the Fokker–Planck formalism offers an intuitive and flexible modelling framework. As emphasised in the last chapter, the Fokker–Planck formalism can accommodate any empirical dependence of the market order/limit order/cancellations rates simultaneously on both bid- and ask-queue volumes.

Of particular interest is the outcome of the "race to the bottom" of the bid and ask, which determines the probability of the direction of the next price change. This probability is empirically found to be correlated with volume imbalance I, with a non-trivial dependence on I. One of the most interesting conclusions of the chapter is that purely diffusive models fail to describe the quantitative shape of this function, even in the presence of noise correlations. In particular, the concavity of the empirical relation is not reproduced by the theoretical formulas.

This discrepancy can be mitigated by the introduction of a drift component in the queue dynamics, whereby small queues tend to grow on average and large queues tend to shrink on average. The dynamics of real queues are well described by a (two-dimensional) Fokker–Planck equation where the drift component is moderate compared to diffusion, preventing depletion times from becoming exceedingly large. Many dynamical properties of queues can be inferred from the interesting structure of the flow field (see Figure 7.6) and the volume dependence of the diffusion tensor (Equation (7.12)), which are found to be mostly independent of the chosen large-tick stock, provided volumes are rescaled by their average values.

Take-Home Messages

(i) Because the spread is almost always locked to one tick, the price dynamics of large-tick stocks boils down to describing the dynamics of the best bid- and ask-queues. Price changes occur when a queue empties ("wins the race").

(ii) When a queue empties, either the volume bounces back and the race continues without a change in price, or an order of opposite type fills the hole and a new race begins around the new mid-price, with the previous second-best queue on the opposite side becoming the new best queue.

(iii) In an LOB, the imbalance between the lengths of the bid- and ask-queues is a strong predictor of the direction of the next price change. The probability that the next price change is in the direction of the shorter queue is a convex function of this queue imbalance.

(iv) By assuming that queues are diffusive, one can compute the theoretical probability of a price change in a given direction as a function of the queue imbalance. However, this approach leads to predicting that this probability is a concave function of imbalance (or a linear function if the bid and ask queue lengths are perfectly anti-correlated).

(v) In the case of long, self-stabilising and independent queues, the probability of a price change in a given direction no longer depends on the queue imbalance. Instead, the price moves up or down with some fixed Poisson rate.

(vi) A more flexible modelling approach is to consider the joint dynamics of the best queues in a Fokker–Planck framework that takes into account the parameter dependence on each queue length.

7.7 Further Reading

General

Feller, W. (1971). *An introduction to probability theory and its applications* (Vol. 2). John Wiley and Sons.

Van Kampen, N. G. (1983). *Stochastic processes in physics and chemistry*. North-Holland.

Risken, H. (1984). *The Fokker–Planck equation*. Springer, Berlin-Heidelberg.

Gardiner, C. W. (1985). *Stochastic methods*. Springer, Berlin-Heidelberg.

Hänggi, P., Talkner, P., & Borkovec, M. (1990). Reaction-rate theory: Fifty years after Kramers. *Reviews of Modern Physics*, 62(2), 251.

Redner, S. (2001). *A guide to first-passage processes*. Cambridge University Press.

Races and Volume Imbalances

Avellaneda, M., Reed, J., & Stoikov, S. (2011). Forecasting prices from Level-I quotes in the presence of hidden liquidity. *Algorithmic Finance*, 1(1), 35–43.

Cartea, A., Donnelly, R. F., & Jaimungal, S. (2015). Enhanced trading strategies with order book signals. https://ssrn.com/abstract=2668277.

Huang, W., Lehalle, C. A., & Rosenbaum, M. (2015). Simulating and analysing order book data: The queue-reactive model. *Journal of the American Statistical Association*, 110(509), 107–122.

Gould, M. D., & Bonart, J. (2016). Queue imbalance as a one-tick-ahead price predictor in a limit order book. *Market Microstructure and Liquidity*, 2(02), 1650006.

Lachapelle, A., Lasry, J. M., Lehalle, C. A., & Lions, P. L. (2016). Efficiency of the price formation process in presence of high frequency participants: A mean field game analysis. *Mathematics and Financial Economics*, 10(3), 223–262.

Fokker–Planck Equation for LOBs

Cont, R., & De Larrard, A. (2012). Order book dynamics in liquid markets: Limit theorems and diffusion approximations. https://ssrn.com/abstract=1757861.

Garèche, A., Disdier, G., Kockelkoren, J., & Bouchaud, J. P. (2013). Fokker-Planck description for the queue dynamics of large-tick stocks. *Physical Review E*, 88(3), 032809.

Yang, T. W., & Zhu, L. (2016). A reduced-form model for level-1 limit order books. *Market Microstructure and Liquidity*, 2(02), 1650008.

8

The Santa Fe Model for Limit Order Books

Done properly, computer simulation represents a kind of "telescope for the mind", multiplying human powers of analysis and insight just as a telescope does our powers of vision. With simulations, we can discover relationships that the unaided human mind, or even the human mind aided with the best mathematical analysis, would never grasp.

(Mark Buchanan)

In this chapter, we generalise the single-queue model that we first encountered in Chapter 5 to now account for the dynamics of all queues in an LOB. As we will discuss, the resulting model is simple to formulate but quickly leads to difficult mathematics. By implementing a wide range of mathematical techniques, and also some simplifications, we illustrate how the model is able to account for several important qualitative properties of real LOBs, such as the distribution of the bid–ask spread and the shape of the volume profiles.

We also discuss how some of the model's assumptions fail to account for important empirical properties of real LOBs, such as the long-range correlations in order flow that we discuss in Chapter 10, and we note that the model leads to conditions that allow highly profitable market-making strategies, which would easily be spotted, exploited and eliminated in real markets. We therefore argue that more elaborate assumptions motivated by empirical observations must be included in the description at a later stage. Still, considering the model in the simple form that we study throughout this chapter has important benefits, including understanding the mathematical frameworks necessary to address the many interacting order flows and eliciting the degree of complexity of the corresponding dynamics.

8.1 The Challenges of Modelling LOBs

Despite the apparent simplicity of the rules that govern LOBs, building models that are both tractable and useful has proven to be an extremely difficult task. We begin this chapter by highlighting some of the challenges that make LOB modelling so difficult.

8.1.1 State-Space Complexity

One key difficulty with modelling LOBs is that their set of possible states is so large. Measured in units of the lot size υ_0, the volume at each price level in an LOB can take any integer value (positive for buys, negative for sells). Therefore, if a given LOB offers N_P different choices for price, then the state space of the volume profile is $(N_P+1)\mathbb{N}^{N_P}$, since there are N_P+1 different possible positions for the mid-price separating buys from sells. In addition, the constraint that buy and sell orders should not overlap introduces a non-linear coupling between the two types of orders, thus forbidding the use of standard linear methods (see Chapter 19 for a model that explicitly accounts for this constraint).

This huge state space is an example of the so-called *curse of dimensionality*, which affects the modelling and analysis of high-dimensional systems in many different fields. The curse of dimensionality creates important difficulties for LOB modelling: even given huge quantities of LOB data, the number of independent data points that correspond to a specified LOB state is often very small. Therefore, performing robust calibration of an LOB model from empirical data can be extremely difficult unless the dependence on the state of the LOB is specified in a parsimonious way. A key objective common to many LOB models is to find a way to simplify the evolving, high-dimensional state space, while retaining an LOB's most important features.

8.1.2 Feedback and Control Loops

The state of an LOB clearly depends on the order flow generated by the market participants trading within it. This order flow is not static but instead fluctuates considerably through time. One key ingredient that drives the order flow generated by the market participants – especially liquidity providers – is the state of the LOB itself. Therefore, the temporal evolution of an LOB is governed by complex, dynamic feedback and control loops between order flow and liquidity.

One example of such a dependency is illustrated by the flow lines in Figure 7.6, which captures the influence of the relative volume at the best quotes on their subsequent average evolution. Writing expressions that capture this feedback in a general LOB model is a difficult task.

Another key difficulty with many stylised LOB models is that they can end up in a state in which the full LOB is completely depleted. By contrast, the feedback and control loops in real markets tend to cause the LOB to replenish as the total number of limit orders becomes small. This makes the probability of observing a real LOB with no limit orders extremely small. However, this situation is not impossible, because the stabilising feedback loops sometimes break down. Such situations can lead to price jumps and flash crashes. This is a very important theme that we will discuss in the very last chapter of this book, Chapter 22.

8.2 The Santa Fe Model

We now describe the LOB model that we will consider in this chapter, and that will serve as the foundation for many of our subsequent discussions. In a nutshell, the model generalises the single-queue model that we considered in Chapter 5 by allowing orders to reside at any price on a discrete grid. We call this model the **Santa Fe model** because it was initially proposed and developed by a group of scientists then working at the Santa Fe Institute.[1]

Consider the continuous-time temporal evolution of a set of particles on a doubly infinite, one-dimensional lattice with mesh size equal to one tick ϑ. Each location on the lattice corresponds to a specified price level in the LOB. Each particle is either of type A, which corresponds to a sell order, or of type B, which corresponds to a buy order. Each particle corresponds to an order of a fixed size υ_0, which we can arbitrarily set to 1 (see Figure 8.1). Whenever two particles of opposite type occupy the same point on the pricing grid, an annihilation $A + B \rightarrow \emptyset$ occurs, to represent the matching of a buy order and a sell order in $\mathscr{L}(t)$. Particles can also evaporate, to represent the cancellation of an order by its owner. As above, $a(t)$ is the ask-price, defined by the position of the leftmost A particle, and $b(t)$ is the bid-price, defined by the position of the rightmost B particle. As always, the mid-price is $m(t) = (b(t) + a(t))/2$.

In this **zero-intelligence** model, order flows are assumed to be governed by the following stochastic processes,[2] where all orders have size $\upsilon_0 = 1$:

- At each price level $p \leq m(t)$ (respectively $p \geq m(t)$), buy (respectively sell) limit orders arrive as a Poisson process with rate λ, independently of p.
- Buy and sell market orders arrive as Poisson processes, each with rate μ.
- Each outstanding limit order is cancelled according to a Poisson process with rate ν.
- All event types are mutually independent.

[1] This group of scientists, led by J. D. Farmer, published several papers on their model throughout the early 2000s; see Section 8.9 for detailed references.

[2] In this chapter, we will only consider the symmetric case where buy and sell orders have the same rates. The model can be extended by allowing different rate parameters on the buy and sell sides of the LOB.

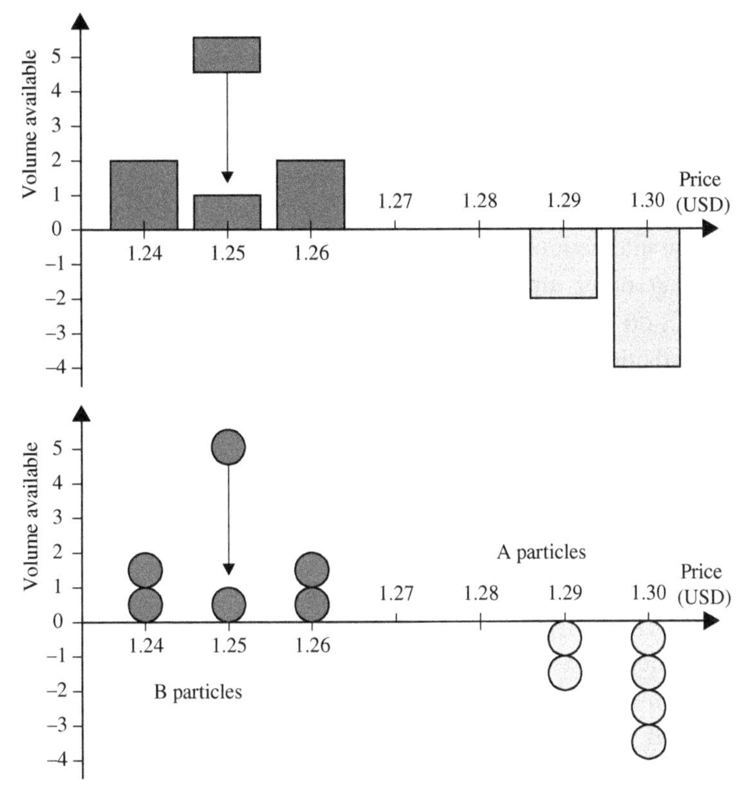

Figure 8.1. (Top) An LOB $\mathscr{L}(t)$ at some moment in time, and (bottom) its corresponding representation as a system of particles on a one-dimensional pricing lattice.

The limit order rule means that the mid-price $m(t)$ is the reference price around which the order flow organises. Whenever a buy (respectively, sell) market order x arrives, it annihilates a sell limit order at the price $a(t_x)$ (respectively, buy limit order at the price $b(t_x)$), and thereby causes a transaction. The interacting flows of market order arrivals, limit order arrivals, and limit order cancellations together fully specify the temporal evolution of $\mathscr{L}(t)$.

When restricted to the queues at the bid- and ask-prices, this model is exactly equivalent to the linear model described by Equation 7.2, with a total cancellation rate that is proportional to V^{ζ} with $\zeta = 1$. The model is also the same at all other prices, but without the possibility of a market order arrival (so $\mu = 0$ at all prices except for the best quotes).

8.3 Basic Intuitions

Before investigating the model in detail, we first appeal to intuition to discuss three of its more straightforward properties. First, each of the parameters λ, μ and

ν are rate parameters, with units of inverse time. Therefore, in order for the units to cancel out correctly, any observable quantity related to the equilibrium distribution of volumes, spreads or gaps between filled prices can only depend on ratios of λ, μ and ν. This is indeed the case for the formulae that we describe in the remainder of this section.

Second, the approximate distributions of queue sizes can be derived by considering the interactions of the different types of order flows at different prices. Because market order arrivals only influence activity at the best quotes, and because queues do not interact with one another, it follows that very deep into the LOB, the distribution of queue sizes reaches a stationary state that is independent of the distance from $m(t)$. For $V \in \mathbb{N}$ (where V is in units of the lot size υ_0), this distribution is given by Equation (5.35) with $\mu = 0$:

$$P_{\text{st.}}(V) = e^{-V^*} \frac{V^{*V}}{V!}, \qquad V^* = \frac{\lambda}{\nu}. \tag{8.1}$$

Two extreme cases are possible:

- A **sparse LOB**, corresponding to $V^* \ll 1$, where most price levels are empty while the others are only weakly populated. This case corresponds to the behaviour of the LOB for very small-tick assets.
- A **dense LOB**, corresponding to $V^* \gg 1$, where all price levels are populated with a large number of orders. This corresponds to the behaviour of the LOB for large-tick assets (at least close enough to the mid-price so that the assumption that λ is constant is reasonable).

In real LOBs, λ decreases quite steeply with increasing distance from $m(t)$ (see Figure 4.5). Therefore, even in the case of large-tick stocks, we expect a crossover between a densely populated LOB close to the best quotes and a sparse LOB far away from them.[3]

The distribution in Equation (8.1) does not hold for prices close to $m(t)$. If d denotes the distance between a given price and $m(t)$, then for smaller values of d, it becomes increasingly likely that a given price was actually the best price in the recent past. Correspondingly, limit orders are depleted not only due to cancellations but also due to market orders that have hit that queue. Heuristically, one expects that for $\lambda > \mu$, the average size of the queues at a distance d from the mid-price is given by

$$V^* \approx \frac{\lambda - \mu\phi_{\text{eff}}(d)}{\nu},$$

where $\phi_{\text{eff}}(d)$ is the fraction of time during which the corresponding price level was the best quote in the recent past (of duration ν^{-1}, beyond which all memory is lost).

[3] This does not, however, imply that there are no buyers or sellers who wish to trade at prices far away from the current mid-price. As we will argue in Chapter 18, most of the corresponding liquidity is *latent*, and only becomes revealed as $m(t)$ changes.

This formula interpolates between Equation (5.46) for $d = 0$ and Equation (8.1) for d large and says that queues tend to be smaller on average in the immediate vicinity of the mid-price, simply because the market order flow plays a greater role in removing outstanding limit orders. One therefore expects that the average depth profile is an increasing function of d, at least close to $d = 0$ (where the limit order arrival rate can be considered as a constant). This is indeed what we observed in empirical data of LOB volume profiles (see Section 4.7). We discuss a simple formula that describes the approximate shape of an LOB's volume profile in Section 8.7, which confirms the above intuition.

Next, we consider the size of the spread $s(t)$. For large-tick stocks, the bid- and ask-queues will both typically be long and $s(t)$ will spend most of the time equal to its smallest possible value of one tick, $s(t) = \vartheta$. For small-tick stocks, for which $V^* \lesssim 1$, the spread may become larger. Introducing the notation $\widehat{s}$ for the spread (measured in ticks), the probability per unit time that a new limit order (either buy or sell) arrives inside the spread can be estimated as $(\widehat{s} - 1)\lambda$. The probability per unit time that an order at one of the best quotes is removed by cancellation or by an incoming market order is given by $2(\mu + \nu)$, since the most probable volume at the bid or at the ask is $V = 1$. The *equilibrium spread* size is such that these two effects compensate,

$$s_{\text{eq.}} \approx \vartheta\left[1 + 2\frac{\mu + \nu}{\lambda}\right].$$

Although hand-waving, this argument gives a good first approximation of the average spread. Indeed, the simple result that a large rate μ of market orders opens up the spread sounds reasonable. We will return to a more precise discussion of spread sizes in Section 8.6.3.

8.4 Parameter Estimation

We now turn to the task of fitting the Sante Fe model to empirical data. At first sight, it appears that estimating most of the model's parameters should be extremely straightforward. However, one of the model's simplifying assumptions causes a considerable difficulty for model fitting. Specifically, the model assumes that the values of λ and ν do not vary with increasing distance from the best quotes. Estimating the (assumed constant) limit order arrival rate in the model requires dividing the total size of all limit order arrivals by the width of the price interval studied. However, this assumption is at odds with empirical data (see Section 4.4), because the rate of limit order arrivals tends to decrease with increasing d. Therefore, if this simplistic method is used to estimate λ from real data, then the resulting estimate of λ decreases with increasing width of the price interval studied.

One simple way to address this problem is to estimate λ only from the limit orders that arrive reasonably close to the current mid-price. In our implementations of the Santa Fe model, we choose to restrict our estimation of λ to the set $\mathcal{X}_{LO}$ of limit orders that arrive either at the best quotes or within the spread. We similarly write $\mathcal{X}_C$ to denote the set of order cancellations that occur at the best quotes. We write $\mathcal{X}_{MO}$ to denote the set of market orders (which, by definition, occur at the best quotes). We write N_{LO}, N_C and N_{MO} to denote, respectively, the number of limit order arrivals at the best quotes, cancellations at the best quotes, and market order arrivals, within a certain time window. We perform all counts independently of the corresponding order signs (recall that we assume symmetry between buy and sell activities).

We estimate the model's parameters as follows:

- To estimate the order size υ_0, we calculate the mean size of arriving limit orders

$$\upsilon_0 = \frac{1}{N_{LO}} \sum_{x \in \mathcal{X}_{LO}} \upsilon_x, \tag{8.2}$$

  where υ_x is the size of order x. Using the mean size of arriving market orders produces qualitatively similar results.
- To estimate the total (buy + sell) market order arrival rate per event, $2\tilde{\mu}$, we calculate the total size of arriving market orders (buy or sell), expressed in units of υ_0 and excluding hidden-order execution volume, then divide by the total number of events:

$$2\tilde{\mu} = \frac{1}{N_{MO} + N_{LO} + N_C} \sum_{x \in \mathcal{X}_{MO}} (\upsilon_x / \upsilon_0). \tag{8.3}$$

- To estimate the total limit order arrival rate per event, $2\tilde{\lambda}_{\text{all}}$, we simply divide the total number of limit orders by the total number of events:

$$2\tilde{\lambda}_{\text{all}} = \frac{N_{LO}}{N_{MO} + N_{LO} + N_C} \equiv \frac{1}{N_{MO} + N_{LO} + N_C} \sum_{x \in \mathcal{X}_{LO}} (\upsilon_x / \upsilon_0). \tag{8.4}$$

  The limit order arrival rate in the Santa Fe model is a rate per unit price, so to estimate $\tilde{\lambda}$, we divide $\tilde{\lambda}_{\text{all}}$ by the mean number n of available price levels inside the spread and at the best quotes, measured only at the times of limit order arrivals:

$$\tilde{\lambda} = \frac{\tilde{\lambda}_{\text{all}}}{n}, \tag{8.5}$$

  with

$$n := 2\left(1 + \left\langle \lfloor \frac{\widehat{s}}{2} \rfloor \,|\, \text{event=LO} \right\rangle\right),$$

Table 8.1. *Estimated values of υ_0, $\tilde{\mu}$, $\tilde{\nu}$ and $\tilde{\lambda}$ for each of the ten stocks in our sample. We also give the average volume at the best quotes, and the spread $\langle s \rangle$ for each stock, sampled uniformly in time.*

	$\tilde{\lambda}$	$\tilde{\nu}$	$\tilde{\mu}$	υ_0 [shares]	$\bar{V}/\upsilon_0$	$\langle s \rangle$ [ticks]
SIRI	0.236	0.0041	0.013	2387	114.9	1.08
INTC	0.222	0.012	0.019	328	36.4	1.17
CSCO	0.229	0.012	0.014	545	38.1	1.14
MSFT	0.220	0.013	0.022	238	36.1	1.18
EBAY	0.208	0.022	0.029	168	23.4	1.21
FB	0.169	0.041	0.031	140	13.0	1.48
TSLA	0.023	0.109	0.062	103	3.1	21.4
AMZN	0.018	0.107	0.055	92	3.1	32.6
GOOG	0.014	0.118	0.049	84	3.0	39.2
PCLN	0.0037	0.132	0.033	68	2.8	156.7

where $\lfloor x \rfloor$ means the integer part of x. This value of n takes into account the fact that both buy and sell limit orders can fill the mid-price level when the spread (in ticks) is an even number.

- To estimate the total cancellation rate *per unit volume* and per event, $2\tilde{\nu}$, we proceed similarly and write:

$$2\tilde{\nu} = \frac{1}{N_{MO} + N_{LO} + N_C} \sum_{x \in \mathcal{X}_C} \frac{\upsilon_x}{\bar{V}}; \qquad \bar{V} := \frac{\bar{V}_a + \bar{V}_b}{2}. \tag{8.6}$$

Note that the above estimation procedures all lead to "rates per event" rather than "rates per unit time". In other words, we have not determined the continuous-time values of μ, ν, and λ, but rather the ratios $\mu/(\mu+\nu+\lambda)$, $\nu/(\mu+\nu+\lambda)$ and $\lambda/(\mu+\nu+\lambda)$. However, for the statistical properties of the LOB, these ratios are the important quantities. The only effect of considering their overall scale is to "play the movie" of order arrivals and cancellations at different speeds.

Table 8.1 lists the estimates of υ_0, $\tilde{\mu}$, $\tilde{\nu}$ and $\tilde{\lambda}$ for each stock in our sample.

8.5 Model Simulations

As we have already emphasised, deriving analytical results about the behaviour of the Santa Fe model is deceptively difficult. By contrast, simulating the model is relatively straightforward. Before we turn our attention to an analytical study of the model, we therefore perform a simulation study. Specifically, we estimate the model's input parameters from empirical data, then measure several of the model's outputs and compare them to the corresponding properties of the same data.

The Santa Fe model is extremely rich, so many output observables can be studied in this way. We consider the following four topics:

(i) the mean and distribution of the bid–ask spread;
(ii) the ratio between the mean first gap behind the best quote (i.e. the price difference between the best and second-best quotes) and the mean spread;
(iii) the volatility and signature plot of the mid-price;
(iv) the mean impact of a market order and the mean profit of market-making.

When simulating the Santa Fe model there is a trade-off between computation time and unwanted finite-size effects, induced by an artificially truncated LOB beyond some distance from the mid-price. A suitable choice is a system of size at least ten times larger than the average spread, which allows the system to equilibrate relatively quickly. We choose the initial state of the LOB such that each price level is occupied by exactly one limit order. Because small-tick LOBs exhibit gaps between occupied price levels, and because each price level in large-tick LOBs is typically occupied by multiple limit orders, this initial condition in fact corresponds to a rare out-of-equilibrium state whose evolution allows one to track the equilibration process. Once the system is in equilibrium, the standard event-time averages of the quantities listed at the start of this section can be computed from the synthetic Santa Fe time series, and then compared to empirical data.[4]

We will see in the following sections that the model does a good job of capturing some of these properties, but a less good job at capturing others, due to the many simplifying assumptions that it makes. For example, the model makes good predictions of the mean bid–ask spreads, but it predicts volatility to be too small for large-tick stocks and too large for small-tick stocks. Moreover, the model creates profitable opportunities for market-making strategies that do not exist in real markets. These weaknesses of the model provide insight into how it might be improved by including either some additional effects such as the long-range correlations of order flow (which we discuss in Chapter 10) or some simple strategic behaviours from market participants.

8.5.1 The Bid–Ask Spread

Figure 8.2 shows the empirical mean bid–ask spread versus the mean bid–ask spread generated by simulating the Santa Fe model with its parameters estimated from the same data, for each of the ten stocks in our sample. As the plot illustrates, the model fares quite well, but slightly underestimates the mean bid–ask spread for all of the stocks. This could be a simple consequence of our specific parameter estimation method (see Section 8.4), which only considers the arrivals

[4] Recall that throughout this chapter event-time refers to all events occurring at the best quotes.

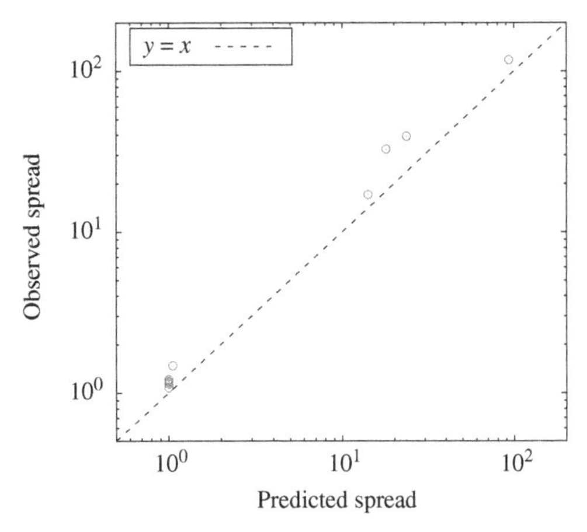

Figure 8.2. Empirical observations versus Sante Fe model predictions of the mean bid–ask spread, measured in units of the tick size, using the rate parameters in Table 8.1. The dashed line indicates the diagonal. The model does a good job at predicting the mean spread for all stocks in our sample.

and cancellations of limit orders that occur at the best quotes or inside the spread, and therefore likely causes us to overestimate the value of the limit order arrival rate.[5]

In any case, considering only the mean does not provide insight into whether the model does a good job at capturing the full distribution of the bid–ask spread. This full distribution is particularly interesting for small-tick stocks, for which the bid–ask spread can (and often does) take any of a wide range of different values. Figure 8.3 shows the distribution of the bid–ask spread for three small-tick stocks in our sample (i.e. PCLN, AMZN, TSLA), together with the corresponding results from the Santa Fe model, rescaled in such a way that the mean spread coincides in all cases. Once rescaled properly, the model does in fact reproduce quite well the bid–ask spread distribution.

8.5.2 The Gap-to-Spread Ratio

We now consider the mean gap between the second-best quote and the best quote, compared to the mean spread itself. We call this ratio the **gap-to-spread ratio**. The gap-to-spread ratio is interesting since it is to a large extent insensitive to the problem of calibrating λ correctly and of reproducing the mean bid–ask spread exactly. It is also important because this gap determines the impact of large market orders (see Equation (8.8) below).

Figure 8.4 shows the empirical gap-to-spread ratio versus the gap-to-spread ratio generated by simulating the Santa Fe model, for the same ten stocks. For large-tick

[5] Recall that the Santa Fe model also assumes that limit orders and market orders all have the same size.

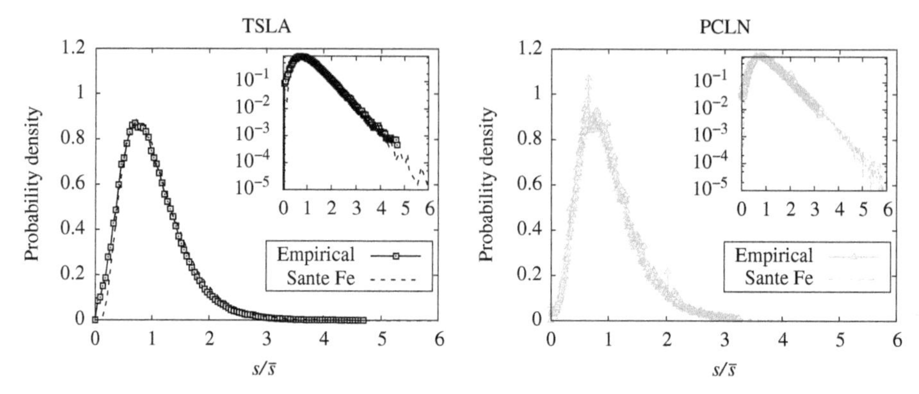

Figure 8.3. Distribution of the rescaled bid–ask spread for (left panel) TSLA and (right panel) PCLN. The markers denote the empirical values and the dashed lines denote the predictions from the Santa Fe model. The model's predictions are remarkably accurate.

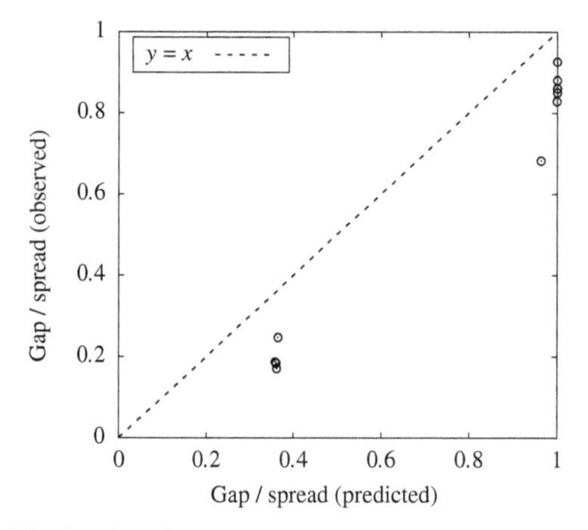

Figure 8.4. Empirical ratio of the gap between the best and second-best quote to the spread, versus the corresponding values predicted from the Santa Fe model, using the rate parameters in Table 8.1. The dashed line denotes the diagonal.

stocks, the model predicts that both the spread and the gap between the second-best and best quotes are nearly always equal to one tick, so the ratio is simply equal to 1. In reality, however, the situation is more subtle. The bid–ask spread actually widens more frequently than the Santa Fe model predicts, such that the empirical gap-to-spread ratio is in fact in the range 0.8–0.95. This discrepancy becomes even more pronounced for small-tick stocks, for which the empirical gap-to-spread ratio typically takes values around 0.2, but for which the Santa Fe model predicts a gap-to-spread ratio as high as 0.4. In other words, when the spread is large, the second-best price tends to be much closer to the best quote in reality than it is in the model.

This analysis of the gap-to-spread ratio reveals that the Santa Fe model is missing an important ingredient. When the spread is large, real liquidity providers tend to place new limit orders inside the spread, but still close to the best quotes, and thereby typically only improve the quote price by one tick at a time. This leads to gaps between the best and second-best quotes that are much smaller than those predicted by the assumption of uniform arrivals of limit orders at all prices beyond the mid-price. One could indeed modify the Santa Fe specification to account for this empirically observed phenomenon of smaller gaps for limit orders that arrive inside the spread, but doing so comes at the expense of adding extra parameters.[6]

8.5.3 Volatility

Simulating the Santa Fe model reveals that the signature plot $\sigma(\tau)/\sqrt{\tau}$ (see Section 2.1.4) exhibits a typical **mean-reversion** pattern, in particular when both the cancellation rate ν and the average size of the best queues $V^* \approx (\lambda - \mu)/\nu$ are small (see Figure 8.5).

Clearly, when ν is large, the memory of the LOB is completely erased after a short time ν^{-1}, beyond which the model has Markovian dynamics and the price is diffusive. Suppose now that ν is small and V^* is large, and that both queues are initially in equilibrium, with comparable volumes of order V^*. The first-hitting time T_1 after which one of the best queues depletes (and the mid-price changes) then grows exponentially with V^* (see Equation (5.47)). If the bid (say) manages to deplete, the empty level just created can refill with either a buy limit order or a sell limit order with equal probability. However, these two situations are not exactly symmetrical: if the new limit order is a buy, then the position of the old bid is restored, facing again the same ask-queue as before, with volume of order V^*. If however the new limit order is a sell, then the incipient ask-queue faces the previous second-best bid, which was previously shielded from the flow of market orders and is therefore typically longer than the best queues.

In other words, the initial volume imbalance I is more favourable to the old bid than to the new ask, leading to some mean reversion. The relative difference in volume imbalance is of order μ/λ, because of the shielding effect. Therefore, the effect is small unless μ is comparable to or greater than λ, in which case the LOB is sparse and V^* small (see Section 8.6). In this case, mean-reversion effects can be substantial (see Figure 8.5). In the other limit when ν is small and V^* large, the volatility of the mid-price is very small (because the depletion time T_1 grows exponentially with V^*, leading to very infrequent price changes) and the signature plot is nearly flat.

We can now turn to a comparison of the Santa Fe model's prediction about volatility with the empirical volatility of our pool of stocks. First, from the results

[6] On this point, see Mike, S., & Farmer, J. D. (2008). An empirical behavioral model of liquidity and volatility. *Journal of Economic Dynamics and Control*, 32, 200–234.

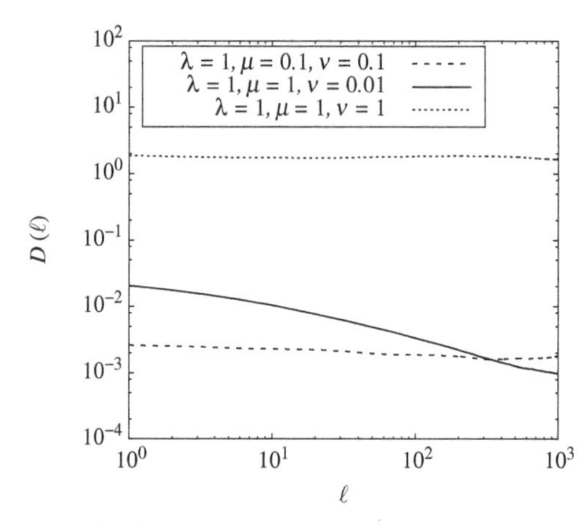

Figure 8.5. Signature plot for the Santa Fe model, with parameters (dashed curve) $\lambda = 1$ and $\nu = \mu = 0.1$, (dotted curve) $\lambda = 1$ and $\nu = \mu = 1$, and (solid curve) $\lambda = 1$, $\nu = 0.01$ and $\mu = 1$. The model exhibits mean-reversion effects only in the regime where $\nu \ll \lambda$ and $\nu \ll \mu$.

in Table 8.1, we can see that large-tick stocks are in a regime where V^* is much larger than 1. From the discussion above, we understand that the volatility predicted by the Santa Fe model is very small, and grossly underestimates the empirical volatility (see Figure 8.6). In reality, these extremely long depletion times are tempered, partly because the assumption of a constant cancellation rate per order is inadequate, as discussed in Sections 6.5 and 7.3.3.

For small-tick stocks, the model predicts values of volatility higher than those observed empirically (see Figure 8.6). The main reason for this weakness is the absence of a mechanism that accurately describes how order flows adapt when prices change. In the model, once the best quote has disappeared, the order flow immediately adapts around the new mid-price, irrespective of whether the price change was caused by a cancellation, a market order arrival or a limit order arrival inside the spread. In the language of Chapter 11, the permanent impact of all of these events is identical, whereas in reality the permanent impact of a market order arrival is much larger than that of cancellations.[7] This causes volatility in the Santa Fe model to be higher than the volatility observed empirically for small-tick stocks.

8.5.4 Impact of Market Orders and Market-Making Profitability

We now analyse in more detail the **lag-dependent impact** of market orders in the Santa Fe model. We define this lag-τ impact as

$$\mathcal{R}(\tau) := \langle \varepsilon_t \cdot (m_{t+\tau} - m_t) \rangle_t, \tag{8.7}$$

[7] On this point, see Eisler, Z., Bouchaud, J. P., & Kockelkoren, J. (2012). The price impact of order book events: Market orders, limit orders and cancellations. *Quantitative Finance*, 12(9), 1395–1419.

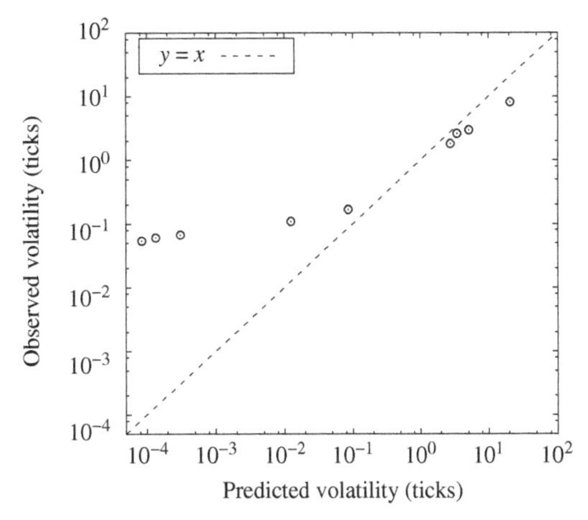

Figure 8.6. Empirically observed volatility and corresponding predictions from the Santa Fe model, for the ten stocks listed in Table 8.1. For the large-tick stocks, we estimate the volatility using the theoretical first-hitting time formula, $\sigma = \sqrt{2/T_1}$, with $T_1 = \exp(V^*)/(\nu V^*)$ and $V^* = (\lambda - \mu)/\nu$. SIRI has a predicted volatility of 10^{-12} and is not plotted here. The model overestimates the observed volatility of (rightmost points) small-tick stocks and grossly underestimates the observed volatility of (leftmost points) large-tick stocks. The dashed line denotes the diagonal.

where ε_t denotes the sign of the market order at event-time t, where t increases by one unit for each market order arrival. As we will discuss in Chapter 11, in real markets the impact function $\mathcal{R}(\tau)$ is positive and grows with τ before saturating for large τ. In the Santa Fe model, however, $\mathcal{R}(\tau)$ is constant, independent of τ, when mean-reversion effects are small (as discussed in the previous section). The lag-τ impact is then well approximated by:

$$\mathcal{R}(\tau) \approx \mathbb{P}(V_{\text{best}} = 1) \times \frac{1}{2} \langle \text{first gap} \rangle, \tag{8.8}$$

where $\mathbb{P}(V_{\text{best}} = 1)$ is the probability that the best queue is of length 1. This approximation holds because a market order of size 1 impacts the price if and only if it completely consumes the volume at the opposite-side best quote, and if it does so, it moves that quote by the size of the first gap, and thus moves the mid-price by half this amount. Since order flow in the Santa Fe model is uncorrelated and is always centred around the current mid-price, the impact of a market order is instantaneous and permanent. Therefore, the model predicts that $\mathcal{R}(\tau)$ is constant in τ for small-tick stocks. In real markets, by contrast, the signs of market orders show strong positive autocorrelation (see Chapter 10), which causes $\mathcal{R}(\tau)$ to increase with τ.

This simple observation has an important consequence: the Santa Fe model specification leads to **profitable market-making strategies**, even when the

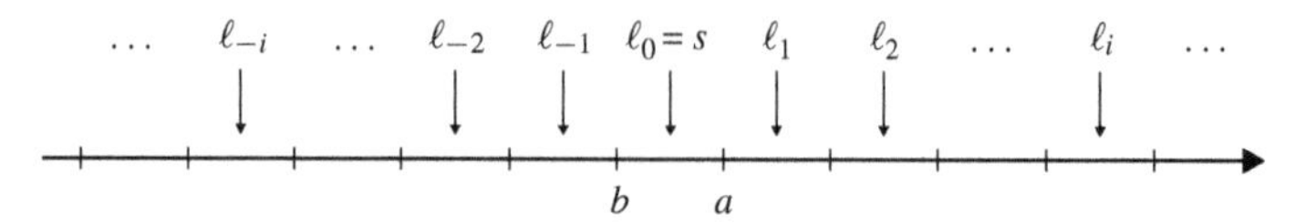

Figure 8.7. Illustration of the kinematics of the model.

signature plot is flat (i.e. when prices are diffusive). In Section 1.3.2, we argued that market-making is profitable on average if the mean bid–ask spread is larger than twice the long-term impact $\mathcal{R}_\infty$. In the Santa Fe model, the mean first gap is always smaller than the bid–ask spread. Therefore, from Equation (8.8), one necessarily has $\mathcal{R}_\infty < \langle s \rangle/2$, so market-making is easy within this framework. If we want to avoid the existence of such opportunities, which are absent in real markets, we need to find a way to extend the Santa Fe model. One possible route for doing so is incorporating some strategic behaviour into the model, such as introducing agents that specifically seek out and capitalise on any simple market-making opportunities that arise. Another route is to modify the model's assumptions regarding order flow to better reflect the empirical properties observed in real markets. For example, introducing the empirically observed autocorrelation of market order signs would increase the long-term impact $\mathcal{R}_\infty$ and thereby reduce the profitability of market-making.

8.6 Some Analytical Results

In this section, we turn our attention to deriving some analytical results from the Santa Fe model. Despite the simplicity of its modelling assumptions, analytical treatment of the full model (as specified in Section 8.2) is extremely difficult. However, as we will see throughout this section, if we adopt the assumption that the sizes of gaps in the LOB are independent, then it is possible to derive several interesting results from the model, such as the distribution of the bid–ask spread. The methods that we use to obtain these results are also interesting in their own right, and are useful in many other contexts.

Throughout this section, we use the index i to label the different intervals (i.e. gaps between successive prices at which limit orders reside) in the LOB. We use positive indices to label these gaps on the sell-side, negative indices to label these gaps on the buy-side, and the index $i = 0$ to label the bid–ask spread (see Figure 8.7).

We write ℓ_i to denote the size of gap i. Due to the finite tick size in an LOB (see Section 3.1.5), the ℓ_i must always take values

$$\ell_i \in \{0, \vartheta, 2\vartheta, \dots\}, \qquad i = \dots, -2, -1, 0, 1, 2, \dots.$$

We write $P_i(\ell)$ to denote the probability distribution of ℓ_i. The value $\ell_i = 0$ corresponds to the case where the distance between the ith and $(i+1)^{\text{th}}$ occupied prices is exactly one tick (i.e. there is no gap in the LOB). In this notation, we can express the bid–ask spread as $s = \ell_0 + \vartheta$ and the ith gap as $\ell_i + \vartheta$. In the following, we will set $\vartheta = 1$, such that all gap sizes are measured in tick units.

As we mentioned at the start of this section, we will assume that all of the ℓ_i are *independent* random variables. Under this assumption, it is possible to write an exact recursion equation for the P_i, by the following argument. Due to the nature of order flow in the model, each queue in the LOB disappears (i.e. depletes to 0) with a certain

Poisson rate (see Section 5.4.3). Let γ_i denote this Poisson rate of disappearing for the ith level. We also introduce a corresponding variable λ_i to denote the rate per unit time and per unit length that a new limit order falls in the ith interval. In the Santa Fe model, λ_i is simply a constant equal to λ for all i. In this section, we consider a slight extension, where λ_i can be arbitrary, for example $\lambda_i = \lambda$ for all $i \neq 0$, but with λ_0 possibly different, to reflect that the arrival rate of limit orders inside the spread can differ considerably from the arrival rate of limit orders at other prices in the LOB.

We now consider the temporal evolution of the gap sizes. For concreteness, we discuss an interval $i \geq 1$ (i.e. on the sell-side of the LOB); similar arguments hold by symmetry for the buy-side of the LOB (the case $i = 0$ is treated separately). Several different events can contribute in changing the gap size ℓ_i between t and $t + \mathrm{d}t$:

(i) With probability $\lambda_i \mathrm{d}t \times \ell_i$, a new sell limit order arrives inside the ith gap and cuts it into two intervals of size ℓ'_i and ℓ'_{i+1}, such that $\ell'_i + \ell'_{i+1} + 1 = \ell_i$ (one tick is now occupied with the new limit order). The size of the ith gap is ℓ'_i. In the Santa Fe model, limit order arrivals occur at a uniform rate at all prices in the gap, so ℓ'_i is uniformly distributed between 0 and $\ell_i - 1$.
(ii) With probability $\lambda_{i-1} \mathrm{d}t \times \ell_{i-1}$, a new sell limit order arrives inside the $i - 1^{th}$ gap and cuts it into two intervals of size ℓ'_{i-1} and ℓ'_i, such that $\ell'_{i-1} + \ell'_i + 1 = \ell_{i-1}$. The new size of the ith gap is ℓ'_i, which is again uniformly distributed between 0 and $\ell_{i-1} - 1$.
(iii) With probability $\mathrm{d}t \times \sum_{j=0}^{i-2} \lambda_j \ell_j$, a new sell limit order arrives inside any interval to the left of $i - 1$, and thereby increases the label of i by one, such that the new ℓ_i is equal to the old ℓ_{i-1}.
(iv) With probability $\gamma_{i+1} \mathrm{d}t$, the right boundary level of the ith gap disappears, which causes ℓ_i to increase to $1 + \ell_{i+1}$.
(v) With probability $\sum_{j=1}^{i} \gamma_j \mathrm{d}t$, one of the queues to the left of interval i disappears, which causes the old $i + 1^{th}$ interval to become the new i^{th} interval, such that the new ℓ_i is equal to the old ℓ_{i+1}.

8.6.1 The Master Equation

Assuming that all ℓ_i are independent, the above enumeration of all possible events enables us to write an evolution equation for the marginal distributions $P_i(\ell)$. Let

$$\bar{\ell}_i = \sum_{\ell=0}^{\infty} \ell P_i(\ell) \tag{8.9}$$

denote the mean of ℓ_i. Removing the explicit t-dependence of the P_i (to simplify the notation), one has:

$$\begin{aligned} \frac{\partial P_i(\ell)}{\partial t} &= -\lambda_i \ell_i P_i(\ell) + \lambda_i \sum_{\ell'=\ell+1}^{\infty} P_i(\ell') + \lambda_{i-1} \sum_{\ell'=\ell+1}^{\infty} P_{i-1}(\ell') \\ &+ \gamma_{i+1} \sum_{\ell'=0}^{\ell-1} P_i(\ell') P_{i+1}(\ell - \ell' - 1) \\ &- \left(\sum_{j=1}^{i+1} \gamma_j + \sum_{j=0}^{i-1} \lambda_j \bar{\ell}_i \right) P_i(\ell) + \sum_{j=1}^{i} \gamma_j P_{i+1}(\ell) + \sum_{j=0}^{i-2} \lambda_j \bar{\ell}_j P_{i-1}(\ell) \end{aligned} \tag{8.10}$$

for all $i \geq 1$.

This equation looks extremely complicated, but is actually relatively friendly in Laplace space. Introducing

$$\widehat{P}_i(z) := \sum_{\ell=0}^{\infty} e^{-z\ell} P_i(\ell),$$

we arrive (after some simple manipulations) at

$$\frac{\partial \widehat{P}_i(z)}{\partial t} = \lambda_i \widehat{P}'_i(z)) + \frac{1}{1-e^{-z}}\left(\lambda_i(1-\widehat{P}_i(z)) + \lambda_{i-1}(1-\widehat{P}_{i-1}(z))\right) + \gamma_{i+1}e^{-z}\widehat{P}_i(z)\widehat{P}_{i+1}(z)$$
$$-\left(\sum_{j=1}^{i+1}\gamma_j + \sum_{j=0}^{i-1}\lambda_j\bar{\ell}_j\right)\widehat{P}_i(z) + \sum_{j=1}^{i}\gamma_j\widehat{P}_{i+1}(z) + \sum_{j=0}^{i-2}\lambda_j\bar{\ell}_j\widehat{P}_{i-1}(z), \tag{8.11}$$

where $\widehat{P}'(z)$ is the derivative of $\widehat{P}(z)$ with respect to z.

If we consider the case $i = 0$, we can write the equation for the distribution of the bid–ask spread. Recalling that in the Santa Fe model, order flow is symmetric on the buy and sell-sides of the LOB, and noting that both buy and sell limit and market orders can influence the bid–ask spread, we see that:

$$\frac{\partial \widehat{P}_0(z)}{\partial t} = \lambda_0 \widehat{P}'_0(z) + \frac{\lambda_0}{1-e^{-z}}\left(1-\widehat{P}_0(z)\right) + 2\gamma_1 e^{-z}\widehat{P}_0(z)\widehat{P}_1(z) - 2\gamma_1\widehat{P}_0(z). \tag{8.12}$$

In the following we will consider the case where $\lambda_{|i|\geq 1} := \lambda$, as assumed in the original version of the model. We will determine the stationary solutions such that the left-hand side of Equations (8.11) and (8.12) are zero, both deep in the book and for the bid–ask spread.

8.6.2 The "Deep" Solution

Deep inside the LOB (i.e. when $i \gg 1$), one expects that $\gamma_i \to \gamma^*$ and that $\widehat{P}_i(z) \to Q^*(z)$. Plugging this into Equation (8.11), one finds:

$$0 = \lambda Q^{*\prime}(z) + \frac{2\lambda}{1-e^{-z}}(1-Q^*(z)) + \gamma e^{-z}Q^{*2}(z) - (\gamma^* + \lambda\bar{\ell}^*)Q^*(z), \tag{8.13}$$

where $\bar{\ell}^*$ is the mean gap length in the limit $i \to \infty$. This non-linear, ordinary differential equation admits the following solution:

$$Q^*(z) = \frac{1}{1+\bar{\ell}^*(1-e^{-z})}, \quad \text{with} \quad \bar{\ell}^* = \frac{\gamma^*}{\lambda}. \tag{8.14}$$

This corresponds to a geometric distribution

$$P^*(\ell) = \frac{1}{\bar{\ell}^*}\left(1-\frac{1}{\bar{\ell}^*}\right)^{\ell} \tag{8.15}$$

for the gap size ℓ, with a mean interval length equal to γ/λ.

The value of γ can be calculated from the single-queue problem that we considered in Chapter 5. At queues deep in the LOB, one can neglect the influence of the market orders and set $\mu = 0$. Equation (5.35) then leads to the stationary distribution of the volume at each occupied price:

$$P_{\text{st.}}(V) = \frac{1}{e^{\lambda/\nu}-1}\frac{\left(\frac{\lambda}{\nu}\right)^V}{V!}, \text{ for } V \geq 0.$$

The rate γ^* at which an occupied price becomes unoccupied (i.e. the total volume at the given price depletes to 0) is given by

$$\gamma^* = \nu P_{\text{st.}}(V=1),$$
$$= \frac{\lambda}{e^{\lambda/\nu}-1},$$

which finally leads to

$$\bar{\ell}^* = \frac{\gamma^*}{\lambda} = \frac{1}{e^{\lambda/\nu} - 1}.$$

In the limit $\nu \gg \lambda$ (i.e. where the cancellation rate is much higher than the deposition rate, so the LOB is sparse, as typically occurs for small-tick stocks) the mean gap size is very large, of order ν/λ. In the limit $\nu \ll \lambda$ (i.e. where the deposition rate is much higher than the cancellation rate, so the LOB is densely populated, as typically occurs for large-tick stocks), the mean gap size is very small, because the gaps are equal to zero most of the time, and are rarely equal to one tick (and even more rarely equal to two ticks, etc.). Although these two regimes lead to very different LOB states, the gap distribution is always a geometric distribution.

8.6.3 The Bid–Ask Spread

We now consider Equation (8.12) in the stationary limit, for which

$$\lambda_0 Q_0'(z) + \frac{\lambda_0}{1 - e^{-z}}\left(1 - \widehat{P}_0(z)\right) + 2\gamma_1 e^{-z}\widehat{P}_0(z)\widehat{P}_1(z) - 2\gamma_1 \widehat{P}_0(z) = 0. \tag{8.16}$$

Let us first assume that the tick size is very large, such that all gaps with $|i| \geq 1$ have a very small mean length $\bar{\ell}_i \ll 1$. As a first approximation, we can assume $\widehat{P}_1(z) \approx 1$, and thus simplify Equation (8.16) to yield

$$\widehat{P}_0(z) \approx 1 + (e^z - 1)\int_z^{\infty} \mathrm{d}z' \frac{e^{-d_0(z' - z + e^{z'} - e^z)} - 1}{4\sinh^2(z'/2)}; \qquad d_0 = \frac{2\gamma_1}{\lambda_0}.$$

Taking the limit $z \to 0$ of the above integral leads to a general expression for the mean bid–ask spread. Two limiting cases are interesting:

- If $d_0 \ll 1$ (which corresponds to an intense in-flow of limit orders), then the mean spread is close to one tick, and the mean gap size is given by

$$\bar{\ell}_0 \approx d_0 \ll 1.$$

  In this case, the probability that the spread is equal to exactly one tick is given by $(1 + d_0)^{-1}$.
- If $d_0 \gg 1$ (which corresponds to the case where the market order rate μ is large compared to the flow of limit orders inside the spread), then

$$\bar{\ell}_0 \approx 1.6449\ldots d_0.$$

  This is a strange (but possible) case, since we still assume that the first gap behind the best quote is nearly always equal to zero, while the spread itself is very large.

In the limit where all gaps are large (as typically occurs for small-tick stocks), the relevant range of z values is $z \ll 1$, which corresponds to the continuum limit of the model. In this case, Equation (8.16) can be solved to produce

$$\widehat{P}_0(z) = 1 - d_0 z \int_z^{\infty} \mathrm{d}z' \frac{1 - \widehat{P}_1(z')}{z'} e^{-d_0 \int_z^{z'} \mathrm{d}z''(1 - \widehat{P}_1(z''))}. \tag{8.17}$$

Writing $\widehat{P}_0(z \to 0) \approx 1 - z\bar{\ell}_0$, one finds

$$\bar{\ell}_0 = d_0 \int_0^{\infty} \mathrm{d}z' \frac{1 - \widehat{P}_1(z')}{z'} e^{-d_0 \int_0^{z'} \mathrm{d}z''(1 - \widehat{P}_1(z''))}.$$

In the limit $d_0 \gg 1$, this leads to an interesting, universal result that is independent of the precise shape of $\widehat{P}_1(z)$. Due to the rapid decay of the exponential term, it follows that z' must remain small, so that the result reads

$$\bar{\ell}_0 \approx d_0 \bar{\ell}_1 \int_0^\infty \mathrm{d}z' e^{-\frac{d_0}{2} z'^2 \bar{\ell}_1} = \sqrt{\frac{\pi}{2}} \times \sqrt{d_0 \bar{\ell}_1}. \tag{8.18}$$

By introducing a scaled variable $u = s\bar{\ell}_0$, one can in fact analyse the full equation for $\widehat{P}_0(z)$ in the regime $d_0 \gg 1$. In this case, Equation (8.17) leads to the result

$$\widehat{P}_0(u) = 1 - u e^{u^2/2} \int_u^\infty \mathrm{d}u' e^{-u'^2/2}.$$

The Laplace transform can be inverted and leads to the following distribution for $x = \ell_0/\bar{\ell}_0$:

$$P_0(x) = x e^{-x^2/2}.$$

The case $\mu \gg \nu \gg \lambda$ is also very interesting, because one can characterise the distribution of gap sizes for all i to find an extended power-law regime:

$$\bar{\ell}_i \approx \frac{1}{\lambda}\sqrt{\frac{\mu\nu}{|i|}}; \qquad 1 \ll |i| \ll \frac{\mu}{\nu}.$$

In particular, $\bar{\ell}_1 \sim \sqrt{\mu\nu}/\lambda$. Therefore, using $\bar{\ell}_1$, Equation (8.18) and the approximation $d_0 \approx \mu/\lambda$, we finally arrive at the scaling relation

$$\bar{\ell}_0 \sim \frac{\mu^{3/4}\nu^{1/4}}{\lambda}$$

for the mean bid–ask spread in this regime. Note that $\bar{\ell}_i \sim |i|^{-1/2}$ corresponds to a linear volume profile, since

$$L = \sum_{i=1}^{Q} \bar{\ell}_i \propto \sqrt{Q}, \tag{8.19}$$

which means that the integrated volume Q over a region of extension L away from the mid-price grows like L^2. The density of orders at distance L is therefore proportional to L.

8.7 The Continuum, Diffusive-Price Limit

Another route to understanding the shape of the LOB is to assume that the tick size is infinitesimal ($\vartheta \to 0$) so that the mid-price takes continuous values. However, working with the Santa Fe model under this assumption is difficult, because the motion of the mid-price is itself dictated by the interaction between the order flow and the instantaneous shape of the LOB. One possible way to make progress is to artificially decouple the motion of the mid-price from the state of the LOB, and to instead impose that it simply follows a random walk. As we noted in Section 8.5.3, this assumption is only valid on time scales larger than ν^{-1}, because substantial mean-reversion is present on shorter time scales.

Following this approach provides the following path to an approximate quantitative theory of the mean volume $V_{\mathrm{st.}}(d)$ at distance d from the mid-price. Buy (respectively, sell) orders at distance d from the current mid-price at time t

are those that were placed at a time $t' < t$ and have survived until time t. Between times t' and t, these orders:

(i) have not been cancelled; and
(ii) have not been crossed by the ask (respectively, bid) at any intermediate time t'', in the sense that the ask (respectively, bid) price has never been above (respectively, below) the price of the orders since the time that they were placed.

For the remainder of this section, we assume that the arrival rate of limit orders at a given price p depends on the distance d between p and the mid-price $m(t)$. Let $\lambda(d)$ denote the arrival rate of limit orders at a distance d from $m(t)$.

Consider a sell limit order in the LOB at some given time t. This order appeared in the LOB at some time $t' \leq t$, when its distance from the mid-price was $d' = d + m(t) - m(t')$. Therefore, its deposition occurred with some rate $\lambda(d')$. Using conditions (i) and (ii) above, the mean volume profile in the limit $t \to \infty$ can be written as:

$$V_{\mathrm{st.}}(d) = \lim_{t\to\infty} \int_{-\infty}^{t} \mathrm{d}t' \int_{-d}^{\infty} \mathrm{d}u \lambda(d+u) \mathbb{P}[u|t \to t'] \mathrm{e}^{-\nu(t-t')},$$

where $\mathbb{P}[u|t \to t']$ is the conditional probability that the time-evolution of the price produces a given value $u = m(t) - m(t')$ of the mid-price difference, given the condition that the path satisfies[8] $d + m(t) - m(t'') \geq 0$ at all intermediate times $t'' \in [t', t]$.

Evaluating $\mathbb{P}$ requires knowledge of the statistics of the price process, which we assume to be purely diffusive, with some diffusion constant D. In this case, $\mathbb{P}$ can be calculated using the standard **method of images**,[9] to yield:

$$\mathbb{P}[u|t \to t'] = \frac{1}{\sqrt{2\pi D\tau}} \left[\exp\left(-\frac{u^2}{2D\tau}\right) - \exp\left(-\frac{(2d+u)^2}{2D\tau}\right)\right],$$

where $\tau = t - t'$. After a simple computation, one finds (up to a multiplicative constant, which only affects the overall normalisation):

$$V_{\mathrm{st.}}(d) = \mathrm{e}^{-\kappa d} \int_{0}^{d} \mathrm{d}u \lambda(u) \sinh(\kappa u) + \sinh(\kappa d) \int_{d}^{\infty} \mathrm{d}u \lambda(u) \mathrm{e}^{-\kappa u}, \qquad (8.20)$$

where $\kappa^{-1} = \sqrt{D/2\nu}$ measures the typical range of price changes during the lifetime of an order ν^{-1}.

[8] We neglect here the size of the spread. The condition should in fact read $d + b(t) - b(t'') = d + m(t) - m(t'') - (s(t) - s(t''))/2 \geq 0$, for all $t'' \in [t', t]$.

[9] See, e.g., Redner, S. (2001). *A guide to first-passage processes*. Cambridge University Press.

Equation (8.20) depends on the statistics of the incoming limit order flow, which is modelled by the deposition rate $\lambda(d)$. Let us consider a simple case

$$\lambda(d) = e^{-\alpha d}$$

in which $\lambda(d)$ decays exponentially in d. In this case, it is possible to evaluate the integrals explicitly, to yield

$$V_{\text{st.}}(d) = V_0 \left[e^{-\alpha d} - e^{-\kappa d} \right],$$

where V_0 is some volume scale. In the original specification of the Santa Fe model, $\lambda(d)$ is constant in d. In this case, $\alpha = 0$, so $V_{\text{st.}}(d) = V_0 \left[1 - e^{-\kappa d} \right]$, which implies that the mean volume increases monotonically with increasing d. This is at odds with empirical observations of real LOBs, for which the mean volume profile first increases but then decreases with increasing d.

For general values of α, it is clear from the expression of $V_{\text{st.}}(d)$ that the mean volume:

- decays to 0 (linearly) when $d \to 0$. We will return to this important point in Chapter 19;
- reaches a maximum at some distance d^*; and
- decays back to 0 (exponentially) for large d.

Although only approximate, this simple framework provides insight into why the mean volume profile for small-tick stocks exhibits a universal *hump shape* (see Section 4.7 and Figure 4.8). In particular, the liquidity tends to be small close to the current mid-price. This will be a recurring theme in the following (see Chapter 18).

8.8 Conclusion: Weaknesses of the Santa Fe Model

The Santa Fe model makes many extreme assumptions that are clearly unrealistic. Some are probably innocuous, at least for a first account of the dynamics of the LOB, whereas others are more problematic, even at a qualitative level. To conclude our discussion of the model, we list some of these unrealistic assumptions and propose some possible cures.

(i) The model assumes that buy (respectively, sell) limit orders arrive at the same rate at all prices $p \leq m(t)$ (respectively, $p \geq m(t)$), whereas in real markets the rates of limit order arrivals vary strongly with relative price (see Figure 4.5). As we have seen in Section 8.7, allowing the arrival rate to vary as a function of price can be incorporated in the model relatively easily, and improves the model's ability to reproduce the behaviour observed in real LOBs, like the average volume profile.

(ii) The model assumes that all sources of temporal randomness are governed by Poisson processes with time-independent rates, whereas real order arrivals and cancellations cluster strongly in time and follow intra-day patterns (see Chapter 2). The problem of event clustering could be rectified by replacing the Poisson processes by self-exciting processes, such as Hawkes processes (see Chapter 9), and the problem of intra-day seasonalities can be eliminated by working in event-time.

(iii) The model assumes that all order flows are independent of the current state of the LOB and of previous order flow. However, as we have already seen for the dynamics of the best queues in Chapter 7, order-flow rates clearly depend both on the same-side and opposite-side best queues, and there is no reason to expect that order-flow rates will not also depend on other characteristics of the LOB. Two important dependencies can be accommodated relatively straightforwardly within an extended Santa Fe model: (a) the volume of incoming market orders tends to grow with the available volume and (b) the probability that a market order hits the ask rather than the bid increases with the volume imbalance I introduced in Section 7.2.

(iv) Perhaps even more importantly, the Santa Fe model assumes that all order flows are independent of each other. This modelling assumption is deeply flawed. As we will repeatedly emphasise in the coming chapters, there are several strong, persistent correlations between order flows in an LOB. For example, the signs of market orders are long-range autocorrelated.[10] Market order arrivals also tend to be followed by limit order arrivals that refill the depleted queue, and vice-versa, in a tit-for-tat dance between liquidity provision and consumption. This is important to produce a price that behaves approximately as a martingale.

Despite these simplifications, the model incorporates all of the concrete rules of trading via an LOB. Specifically, orders arrive and depart from the LOB in continuous time (due to particle arrivals, evaporations and annihilations), at discrete price levels (due to the discreteness of the pricing grid) and in discrete quantities (due to the discreteness of each particle's size). Whenever a particle arrives at a price already occupied by another particle of opposite type, the particles annihilate (which corresponds to the arrival of a market order). Hence, although the Santa Fe model does not seek to incorporate all empirical properties of order flow observed in real markets, it captures the essential mechanics of trading via an LOB. Therefore, the model can be regarded as a simple null model of price formation, without considering any strategic behaviour of the market participants. Interestingly, the model allows one to reproduce many empirical

[10] For models that include these long-range correlations, see: Mastromatteo, I., Tóth B., & Bouchaud, J-P. (2014). Agent-based models for latent liquidity and concave price impact. *Physical Review E*, 89(4), 042805.

quantities, such as the average or distribution of the bid–ask spread, with a remarkable accuracy, vindicating the idea that for some observables the influence of intelligence may be secondary to the influence of the simple rules governing trade. More sophisticated ingredients, such as modelling optimised strategies of market-makers/high-frequency traders or the fragmentation of large orders that lead to correlated order flows, can be included at a later stage.

Take-Home Messages

(i) Modelling the full LOB is difficult because the state space is large and investors' actions are extremely complex.
(ii) The so-called zero-intelligence approach simplifies the problem by assuming that investors' actions follow stochastic processes. Though oversimplified, this approach allows some empirical stylised facts to be recovered. This shows that some regularities of markets may simply be a consequence of their stochastic nature.
(iii) The Santa Fe model regards the LOB as a collection of queues that evolve according to a constant limit order arrival rate, a constant cancellation rate per existing limit order, and a constant market order arrival rate at the best queues.
(iv) By estimating the empirical rates for each such events, the Santa Fe model can be used to make predictions of several important statistical properties of real markets.
(v) The model makes reasonably good predictions about the mean bid–ask spread, and about the full distribution of the bid–ask spread.
(vi) The model makes reasonably good predictions about the increasing mean volume profile near to the best quotes, but does not capture the eventual decrease in mean volume deeper into the LOB.
(vii) The model slightly overestimates volatility for small-tick stocks, but grossly underestimates volatility for large-tick stocks.
(viii) In the model, the mean impact of an order is smaller than the half spread, creating profitable market-making opportunities. This is because the zero-intelligence approach misses a crucial ingredient of financial markets: feedback of prices on strategic behaviour.

8.9 Further Reading

Stylised Facts and Modelling

Bouchaud, J. P., Mézard, M., & Potters, M. (2002). Statistical properties of stock order books: Empirical results and models. *Quantitative Finance*, 2(4), 251–256.

Zovko, I., & Farmer, J. D. (2002). The power of patience: A behavioural regularity in limit-order placement. *Quantitative Finance*, 2(5), 387–392.
Weber, P., & Rosenow, B. (2005). Order book approach to price impact. *Quantitative Finance*, 5(4), 357–364.
Mike, S., & Farmer, J. D. (2008). An empirical behavioral model of liquidity and volatility. *Journal of Economic Dynamics and Control*, 32(1), 200–234.
Bouchaud, J. P., Farmer, J. D., & Lillo, F. (2009). How markets slowly digest changes in supply and demand. In Hens, T. & Schenk-Hoppe, K. R. (Eds.), *Handbook of financial markets: Dynamics and evolution*. North-Holland, Elsevier.
Eisler, Z., Bouchaud, J. P., & Kockelkoren, J. (2012). The price impact of order book events: Market orders, limit orders and cancellations. *Quantitative Finance*, 12(9), 1395–1419.
Gould, M. D., Porter, M. A., Williams, S., McDonald, M., Fenn, D. J., & Howison, S. D. (2013). Limit order books. *Quantitative Finance*, 13(11), 1709–1742.
Abergel, F., Chakraborti, A., Anane, M., Jedidi, A., & Toke, I. M. (2016). *Limit order books*. Cambridge University Press.

The Santa Fe Model

Daniels, M. G., Farmer, J. D., Gillemot, L., Iori, G., & Smith, E. (2003). Quantitative model of price diffusion and market friction based on trading as a mechanistic random process. *Physical Review Letters*, 90(10), 108102.
Smith, E., Farmer, J. D., Gillemot, L. S., & Krishnamurthy, S. (2003). Statistical theory of the continuous double auction. *Quantitative Finance*, 3(6), 481–514.
Farmer, J. D., Patelli, P., & Zovko, I. I. (2005). The predictive power of zero intelligence in financial markets. *Proceedings of the National Academy of Sciences of the United States of America*, 102(6), 2254–2259.

Agent-Based Models of LOBs

Bak, P., Paczuski, M., & Shubik, M. (1997). Price variations in a stock market with many agents. *Physica A: Statistical Mechanics and its Applications*, 246(3–4), 430–453.
Chiarella, C., & Iori, G. (2002). A simulation analysis of the microstructure of double auction markets. *Quantitative Finance*, 2(5), 346–353.
Challet, D., & Stinchcombe, R. (2003). Non-constant rates and over-diffusive prices in a simple model of limit order markets. *Quantitative Finance*, 3(3), 155–162.
Preis, T., Golke, S., Paul, W., & Schneider, J. J. (2006). Multi-agent-based order book model of financial markets. *EPL (Europhysics Letters)*, 75(3), 510.
Chiarella, C., Iori, G., & Perello, J. (2009). The impact of heterogeneous trading rules on the limit order book and order flows. *Journal of Economic Dynamics and Control*, 33(3), 525–537.
Tóth, B., Lemperiere, Y., Deremble, C., De Lataillade, J., Kockelkoren, J., & Bouchaud, J. P. (2011). Anomalous price impact and the critical nature of liquidity in financial markets. *Physical Review X*, 1(2), 021006.
Mastromatteo, I., Tóth, B., & Bouchaud, J. P. (2014). Agent-based models for latent liquidity and concave price impact. *Physical Review E*, 89(4), 042805.
Huang, W., Lehalle, C. A., & Rosenbaum, M. (2015). Simulating and analysing order book data: The queue-reactive model. *Journal of the American Statistical Association*, 110(509), 107–122.
Muni Toke, I. (2015). Stationary distribution of the volume at the best quote in a Poisson order book model. arXiv preprint arXiv:1502.03871.

Mathematical Models for the Dynamics of LOBs

Luckock, H. (2003). A steady-state model of the continuous double auction. *Quantitative Finance*, 3(5), 385–404.

Foucault, T., Kadan, O., & Kandel, E. (2005). Limit order book as a market for liquidity. *Review of Financial Studies*, 18(4), 1171–1217.

Rosu, I. (2009). A dynamical model of the limit order book. *The Review of Financial Studies*, 22(11), 4601–4641.

Cont, R., & De Larrard, A. (2012). Order book dynamics in liquid markets: Limit theorems and diffusion approximations. https://ssrn.com/abstract=1757861.

Abergel, F., & Jedidi, A. (2013). A mathematical approach to order book modeling. *International Journal of Theoretical and Applied Finance*, 16(05), 1350025.

Slanina, F. (2013). *Essentials of econophysics modelling*. Oxford University Press.

Horst, U., & Kreher, D. (2015). A weak law of large numbers for a limit order book model with fully state dependent order dynamics. arXiv preprint arXiv:1502.04359.

Gao, X., Dai, J. G., Dieker, T., & Deng, S. (2016). Hydrodynamic limit of order book dynamics. *Probability in the Engineering and Informational Sciences*, 1–30.

Huang, W., & Rosenbaum, M. (2015). Ergodicity and diffusivity of Markovian order book models: A general framework. arXiv preprint arXiv:1505.04936.

Muni Toke, I., & Yoshida, N. (2016). Modelling intensities of order flows in a limit order book. *Quantitative Finance*, 17, 1–19.

PART IV
Clustering and Correlations

Introduction

So far, we have restricted our attention to models with simple statistical properties, such as Brownian motions and time-homogeneous Poisson processes. These models have assumed that all events are independent (once they are conditioned on the current state of the world), and have produced simple, stationary outputs.

This picture is far from the truth, for at least two reasons. First, real order flow exhibits complex feedback loops. As empirical measurements demonstrate, a large fraction of market activity is generated endogenously, so assuming that order flows are conditionally independent is clearly problematic. Whether stabilising or destabilising, feedback loops in financial markets can (and do) fundamentally change the resulting global behaviour. Second, order flows show long-range autocorrelations, to the extent that future order flows are often highly predictable, given observations of the past.

In this part, we introduce a collection of tools to model and understand the complex dependencies between market events. As with the previous parts, we start this discussion by using empirical evidence to address several important questions on this topic. To what extent do past events influence future events? What part of market activity is triggered by exogenous events, like the arrival of news? And what part is triggered endogenously by past activity?

The first part of our discussion will focus on the important role played by the time-clustering of events. We will introduce a flexible class of processes called Hawkes processes, which are the continuous-time analogue of a discrete ARCH process. As we will discuss, the Hawkes framework is a simple yet powerful tool for estimating and modelling events that cluster in time. By fitting Hawkes processes to LOB data, we will show that a large fraction of market activity appears to be endogenous. This will help us to illustrate how clustering might be responsible for the fat-tailed returns distributions that we discussed in Section 2.2.1.

The second part of our discussion will focus on an important puzzle that we will attempt to solve throughout the remaining parts of the book: the highly predictable character of order-flow direction. Specifically, we will show that the sign of market orders is strongly autocorrelated in time, such that observing a buy market order now significantly increases the probability of observing another buy market order far in the future.

This last point deserves detailed investigation. Where do such autocorrelations come from? Do traders exhibit herding behaviour, such that they all buy or sell aggressively at the same time? Or do these regularities emerge from the behaviour of individual agents? As usual, we will use empirical data to settle the argument. We will show that although herding behaviours exist, they are actually quite weak. Instead, what really causes order-flow autocorrelations is the *order-splitting* behaviour of traders, who slice and dice so-called *metaorders* into many smaller *child orders*, which they execute incrementally over time periods of days or even weeks. Why do traders behave in this way? The answer lies in the ever-present shortage of liquidity in financial markets. This concept, which we will discuss extensively, will come to form one of the most important concepts of this book.

9
Time Clustering and Hawkes Processes

In economics, there can never be a "theory of everything". But I believe each attempt comes closer to a proper understanding of how markets behave.

(Benoît B. Mandelbrot)

When studying order flows in real markets, two striking phenomena are readily apparent:

- Market activity follows clear **intra-day patterns** that are related to predictable events, such as the start of a trading day, the end of a trading day, the opening of another market, the slowdown around lunch time, and scheduled macroeconomic news announcements. These intra-day patterns are clearly visible in Section 4.2. Such patterns induce strong 24-hour periodicities in several important market properties such as volatility (see, e.g., Figure 2.5). Other patterns also exist at the weekly, monthly, and yearly levels.
- Even after accounting for these intra-day patterns, market events still strongly cluster in time. Some periods have extremely high levels of market activity, with seemingly random durations, such that the level of market activity is "bursty" or "intermittent". In fact, financial markets are one among many other examples of **intermittent processes**, like neuronal activity or Barkhausen noise in disordered magnets (see references in Section 9.7). A well-known case is seismic activity, where one earthquake triggers aftershocks or even other major earthquakes elsewhere on the planet.

Throughout the previous four chapters, one of the key assumptions was that order flow can be described as a Poisson process with constant rate parameters. Although this assumption provides a convenient framework for building simple models, it clearly does not reflect these two important empirical facts.

In this chapter, we consider a class of models called *Hawkes processes*, which capture these properties via two key mechanisms. First, to account for exogenous

This chapter is not essential to the main story of the book.

intra-day patterns, they allow the Poissonian rate parameters to be explicitly time-dependent. Second, they incorporate a "self-exciting" feature, such that their local rate (or intensity) depends on the history of the process itself. As we illustrate throughout the chapter, these two core components make Hawkes processes powerful tools that strongly outperform homogeneous Poisson processes for replicating the true order-flow dynamics observed in real markets.

9.1 Point Processes

Throughout this chapter, our main object of study will be stochastic processes that create instantaneous temporal events. For $i \in \mathbb{N}$, we write t_i to denote the time at which the ith event occurs. These events could be, for example, market order arrivals, changes in the mid-price, or even any event that changes the state of the LOB. The set of arrival times defines a **point process** (PP).

For a given point process $(t_i)_{i\in\mathbb{N}}$, the **inter-arrival times** are defined as

$$\delta_i = (t_{i+1} - t_i),\ i = 0, 1, 2, \ldots. \tag{9.1}$$

Given a PP $(t_i)_{i\in\mathbb{N}}$, the associated **counting process**

$$N(t) := \sum_{i\in\mathbb{N}} \mathbf{1}_{t_i \le t} \tag{9.2}$$

counts the number of arrivals that have occurred up to (and including) time t, where $\mathbf{1}_{t_i \le t}$ denotes the indicator function

$$\mathbf{1}_{t_i \le t} := \begin{cases} 1, & \text{if } t_i \le t, \\ 0, & \text{otherwise.} \end{cases} \tag{9.3}$$

Intuitively, the counting process $N(t)$ is an increasing step function with a unit-sized jump discontinuity at each time t_i.

For an infinitesimal time increment $\mathrm{d}t$, we define the **counting increment**

$$\mathrm{d}N(t) := \begin{cases} 1, & \text{if there exists an } i \text{ such that } t \le t_i \le t + \mathrm{d}t, \\ 0, & \text{otherwise.} \end{cases} \tag{9.4}$$

9.1.1 Homogeneous Poisson Processes

A **homogeneous Poisson process** is a PP in which events occur independently of each other and with a fixed intensity φ per unit time, such that the probability that $\mathrm{d}N(t) = 1$ in an infinitesimal time interval $(t, t + \mathrm{d}t)$ is equal to $\varphi \mathrm{d}t$.

An important property of a homogeneous Poisson process is that the inter-arrival times δ_i are independent random variables with an exponential distribution:

$$\mathbb{P}[\delta_i > \tau] = \mathbb{P}[N(t_i + \tau) - N(t_i) = 0] = e^{-\varphi\tau}, \qquad i = 0, 1, 2, \ldots.$$

Because the events are independent, the number of events that occur in an interval $(t, t+\tau)$ obeys a Poisson distribution:

$$\mathbb{P}[N(t+\tau) - N(t) = n] = \frac{(\varphi\tau)^n}{n!} e^{-\varphi\tau}. \tag{9.5}$$

By standard results for the Poisson distribution, the mean and variance of the number of events that occur in an interval $(t, t+\tau)$ are both equal to $\varphi\tau$:

$$\mathbb{E}[N(t+\tau) - N(t)] = \varphi\tau, \tag{9.6}$$

$$\mathbb{V}[(N(t+\tau) - N(t)] = \varphi\tau. \tag{9.7}$$

9.1.2 The Clustering Ratio

When studying PPs, the **clustering ratio**

$$r(\tau) = \frac{\mathbb{V}[(N(t+\tau) - N(t)]}{\mathbb{E}[(N(t+\tau) - N(t)]} \tag{9.8}$$

is a useful tool for understanding the temporal clustering of events:

- The value $r = 1$ (which, by Equations (9.6) and (9.7), is the case for a homogeneous Poisson process) indicates that events do not cluster in time.
- Values $r > 1$ indicate that arrivals attract each other (so that many arrivals can occur in the same interval), and therefore indicate the presence of clustering.
- Values $r < 1$ indicate that arrivals repel each other, and therefore indicate the presence of inhibition (see Section 9.5).

As we will see below, events in financial time series are characterised by values of r larger (and often much larger) than 1.

9.1.3 Inhomogeneous Poisson Processes

An **inhomogeneous Poisson process** is a Poisson PP with a time-dependent intensity $\varphi(t)$, which describes the instantaneous arrival rate of events at time t. The intensity is defined as

$$\varphi(t) = \mathbb{E}\left[\frac{\mathrm{d}N(t)}{dt}\right] = \lim_{\epsilon \downarrow 0} \mathbb{E}\left[\frac{N(t+\epsilon) - N(t)}{\epsilon}\right]. \tag{9.9}$$

One simple application of an inhomogeneous Poisson process is to choose $\varphi(t)$ as a deterministic function that tracks the intra-day pattern of activity in financial markets. Another possible application is to consider the intensity $\varphi(t)$ to itself be a random variable, with a certain mean φ_0 and covariance $\mathbb{V}[\varphi(t)\varphi(t')] := C_\varphi(t, t')$. This "doubly stochastic"[1] point process is called a **Cox process**. Note

[1] The random nature of the intensity adds to the Poisson randomness.

that whenever $C_\varphi(t,t') > 0$ for $t' \neq t$, it follows that the clustering ratio satisfies $r > 1$, which indicates clustering.

9.2 Hawkes Processes

By allowing its rate parameter to vary as a function of time, an inhomogeneous Poisson process can address the first of the two empirical properties of order-flow series that we listed at the beginning of this chapter, namely predictable patterns. However, to also account for the second property (i.e. self-excitation) will require φ to depend not only on t, but on the past realisations of the process itself. This idea forms the basis of the Hawkes process.

9.2.1 Motivation and Definition

The inspiration for Hawkes processes originates not from financial markets, but from earthquake dynamics, where events seem to *trigger* other events, or replicas. This causes a point process of arrival times of earthquakes to exhibit a self-exciting structure, in which earthquakes are more likely to occur if other earthquakes have also occurred recently. The same phenomenon also occurs in financial markets. For example, trades occur frequently when many other trades have occurred recently, and occur infrequently when few other trades have occurred recently.

Hawkes processes are a specific family of inhomogeneous Poisson PPs where the temporal variation of $\varphi(t)$ has not only an **exogenous component** (such as the processes that we described in Section 9.1.3), but also an **endogenous component** that depends on the recent arrivals of the PP itself. More precisely, a **Hawkes process** is a point process whose arrival intensity is given by

$$\varphi(t) = \varphi_0(t) + \int_{-\infty}^{t} \mathrm{d}N(u)\Phi(t-u), \tag{9.10}$$

where $\varphi_0(t)$ is a deterministic base intensity and $\Phi(t) \geq 0$ is a non-negative **influence kernel** that describes how past events influence the current intensity. Often $\Phi(t)$ is chosen to be a strictly decreasing function of t, meaning that the influence of past events fades away with time.

Importantly, the integral in Equation (9.10) is calculated with respect to the counting process $N(u)$. The purpose of doing so is to sum the contributions to $\varphi(t)$ caused by all previous arrivals – which are precisely what $N(u)$ counts. Since $\mathrm{d}N(u)$ is equal to 1 if and only if one of the event-times t_i falls in the interval $[u, u + \mathrm{d}u]$, one can rewrite Equation (9.10) by replacing this integral with a sum that runs over the realised arrival times of the point process:

$$\varphi(t) = \varphi_0(t) + \sum_{t_i < t} \Phi(t - t_i). \tag{9.11}$$

As illustrated by Equation (9.11), the intensity function of a Hawkes process consists of two parts:

- The first part (i.e. the $\varphi_0(t)$ term) is deterministic, and describes the exogenous contribution to the rate dynamics. Its temporal variation might represent, for example, the intra-day periodicity of the activity.
- The second part (i.e. the $\sum_{t_i<t} \Phi(t - t_i)$ term) describes the endogenous contribution to the rate dynamics, which determines how previous arrivals affect the present arrival intensity. Because the second part of the intensity function depends on the history of the stochastic point process, the intensity function of a Hawkes process is itself a stochastic process.

Hawkes processes are widely used to model high-frequency market microstructure events in calendar time. As we will see throughout the remainder of this chapter, they offer several attractive benefits, including simplicity, flexibility, their ability to account for a wide range of different event types in a multivariate setting, their ability to account for non-stationarities through the exogenous $\varphi_0(t)$ term, and their appealing mathematical and statistical properties. Their parameters also have a simple and useful interpretation, and they thereby lead to a concise description of many complex aspects of market microstructure.

9.2.2 Basic Properties of Hawkes Processes

To simplify our discussion, we first consider some basic properties of Hawkes processes in the case where the exogenous intensity φ_0 is time-independent. We also assume that the process starts at $t = 0$ with no past (i.e. $N(0) = 0$). In this framework, it is possible to derive several interesting results about the statistical properties of the process.

First-Order Statistics

We first consider the mean intensity $\bar{\varphi}$ of the process, defined as the long-term growth rate of the counting process:

$$\bar{\varphi} := \lim_{t\to\infty} \frac{N(t)}{t}.$$

Provided that the process reaches a **stationary state** such that this quantity exists, $\bar{\varphi}$ must be equal to the long-term mean value of the intensity, after the influence of the initial state of the point process has died out:

$$\bar{\varphi} = \lim_{t\to\infty} \mathbb{E}[\varphi(t)].$$

Since $\mathbb{E}[\mathrm{d}N(u)] = \varphi(u)\mathrm{d}u$, one finds from Equation (9.10) that $\bar{\varphi}$ must obey the self-consistent equation

$$\bar{\varphi} = \varphi_0 + \bar{\varphi}\int_0^\infty \mathrm{d}u\, \Phi(u),$$

which only has a non-negative solution when the norm[2]

$$g := \int_0^\infty \mathrm{d}u\, \Phi(u) \tag{9.12}$$

of the kernel satisfies $g < 1$. Note that this condition requires $\Phi(t)$ to decrease sufficiently quickly for the integral to converge, and to be small enough for the process to reach a stationary state.

Let $g_c = 1$ denote the critical value for g. When $g < g_c$, the mean intensity of the Hawkes process is given by

$$\bar{\varphi} = \frac{\varphi_0}{1-g} \geq \varphi_0, \tag{9.13}$$

which indicates that the Hawkes self-exciting mechanism causes the mean intensity to increase. In the extreme case where $g \to g_c = 1$, this feedback is so strong that $\bar{\varphi}$ diverges. When $g > 1$, the process explodes, in the sense that $N(t)$ grows faster and faster with increasing t, so that no stationary state can ever be reached and the above mathematical formulas are meaningless. In the limit $g = 0$, there is no feedback, so $\bar{\varphi} = \varphi_0$ and we recover the homogeneous PP.

Second-Order Statistics

For a stationary Hawkes process with $g < 1$, the function

$$c(\tau) := \frac{1}{\bar{\varphi}^2}\mathrm{Cov}\left[\frac{\mathrm{d}N(t)}{dt}, \frac{\mathrm{d}N(t+\tau)}{dt}\right]; \qquad \tau > 0$$

describes the rescaled covariance of the arrival intensity. Note that since $\mathrm{d}N(t)^2 = \mathrm{d}N(t)$, there is a singular $\bar{\varphi}\delta(\tau)$ contribution to the equal-time covariance.

Observe that

$$\mathbb{E}[\mathrm{d}N(t)\mathrm{d}N(u)] = \mathbb{E}[\mathrm{d}N(t)|\mathrm{d}N(u) = 1]\mathbb{E}[\mathrm{d}N(u)].$$

Recalling also that $\mathbb{E}[\mathrm{d}N(u)] = \bar{\varphi}\mathrm{d}u$, it follows that the function $c(\cdot)$ is related to conditional expectation of the intensity function, as:

$$\mathbb{E}[\mathrm{d}N(t)|\mathrm{d}N(u) = 1] := \bar{\varphi}(1 + c(t-u))\,\mathrm{d}t + \delta(t-u). \tag{9.14}$$

Now, from the definition of the Hawkes process in Equation (9.10), one also has, for $t > u$:

$$\mathbb{E}[\mathrm{d}N(t)|\mathrm{d}N(u) = 1] = \left[\varphi_0 + \int_0^\infty \mathrm{d}v\, \Phi(v)\mathbb{E}[\mathrm{d}N(t-v)|\mathrm{d}N(u) = 1]\right]\mathrm{d}t.$$

[2] This quantity is also sometimes called the branching ratio (see Section 9.2.6).

Using Equation (9.14), the second term on the right-hand side can be transformed into

$$\int_0^\infty \mathrm{d}v\, \Phi(v)\mathbb{E}[\mathrm{d}N(t-v)|\mathrm{d}N(u)=1] = \bar{\varphi}g + \Phi(t-u) + \bar{\varphi}\int_0^\infty \mathrm{d}v\Phi(v)c(t-v-u).$$

By comparing the two expressions for $\mathbb{E}[\mathrm{d}N(t)|\mathrm{d}N(u)=1]$ and noting that $\varphi_0 + g\bar{\varphi} = \bar{\varphi}$, we finally arrive at the **Yule–Walker equation**, for $\tau > 0$:

$$c(\tau) = \frac{1}{\bar{\varphi}}\Phi(\tau) + \int_0^\infty \mathrm{d}u\,\Phi(u)c(\tau-u); \qquad c(-\tau) = c(\tau). \tag{9.15}$$

Introducing

$$\widehat{c}(z) = \int_{-\infty}^{\infty} \mathrm{d}u c(u) e^{-zu}, \quad \widehat{\Phi}(z) = \int_0^\infty \mathrm{d}u\,\Phi(u)e^{-zu},$$

this equation turns out to be equivalent to[3]

$$1 + \bar{\varphi}\widehat{c}(z) = \frac{1}{(1-\widehat{\Phi}(z))(1-\widehat{\Phi}(-z))}. \tag{9.16}$$

Equation (9.15) thus provides a way to compute the covariance of the arrival intensity $c(\cdot)$, given the Hawkes kernel Φ. Conversely, it also provides a way to reconstruct the kernel of the underlying Hawkes process from an empirical determination of the covariance c. In other words, the rescaled covariance of the intensity $c(\tau)$ fully characterises the Hawkes kernel $\Phi(\tau)$, while the average intensity $\bar{\varphi}$ allows one to infer the bare intensity φ_0. Much like the Gaussian process, the Hawkes process is entirely determined by its first- and second-order statistics.[4]

9.2.3 Some Useful Results

By noting that $\widehat{\Phi}(0) = g$, we obtain the identity

$$\bar{\varphi}\widehat{c}(0) := \bar{\varphi}\int_{-\infty}^{\infty} \mathrm{d}u\, c(u) = \frac{g(2-g)}{(1-g)^2}, \tag{9.17}$$

which is valid for any kernel shape. Importantly, this identity tells us that the integrated rescaled covariance is only determined by the branching ratio g of the Hawkes process.

[3] Deriving this result is non-trivial due to the infinite upper limit of the integral on the right-hand side of Equation (9.15). For a derivation, see Bacry, E., & Muzy, J.-F. (2016). First- and second-order statistics characterization of Hawkes processes and non-parametric estimation. *IEEE Transactions on Information Theory*, 62(4), 2184–2202.

[4] Note that by the same token, it is hard to distinguish correlation and causality, since two-point correlation functions are always invariant under time reversal. Therefore, fitting a point process using a Hawkes model in no way proves the existence of genuinely causal self-exciting effects. For an extended discussion, see Blanc, P., Donier, J., & Bouchaud, J. P. (2016). Quadratic Hawkes processes for financial prices. *Quantitative Finance*, 17, 1–18.

From the covariance of the arrival intensity, one can compute the variance of the number of events $N(t+\tau) - N(t)$ in an interval of size $\tau > 0$:

$$\mathbb{V}[N(t+\tau) - N(t)] = \int_t^{t+\tau} \int_t^{t+\tau} \mathbb{E}[\mathrm{d}N(u)\mathrm{d}N(u')]$$

$$= \bar{\varphi}\tau + 2\bar{\varphi}^2 \int_0^{\tau} \mathrm{d}u\,(\tau - u)c(u), \tag{9.18}$$

where the first term comes from the singular, equal-time contribution to c, and the second term is zero for a homogeneous Poisson process, for which $\Phi = c = 0$. It also follows from Equation (9.18) that the clustering ratio (9.8) is:

$$r(\tau) = 1 + 2\bar{\varphi} \int_0^{\tau} \mathrm{d}u \left(1 - \frac{u}{\tau}\right) c(u). \tag{9.19}$$

Note that for a pure Poisson process, $c(u) = 0$, and we recover the standard result $r = 1$. For a positive Hawkes kernel Φ, the value of $c(u)$ becomes positive as well, which leads to clustering (i.e. $r > 1$).

If we assume that $c(u)$ decreases faster than u^{-2} for large u, so that $\int_0^{\infty} \mathrm{d}u\, u c(u)$ is a finite integral, then Equation (9.19) can be simplified at large τ, as

$$r(\tau) \approx_{\tau\to\infty} 1 + 2\bar{\varphi} \int_0^{\infty} \mathrm{d}u\, c(u) = 1 + \bar{\varphi} \int_{-\infty}^{\infty} \mathrm{d}u\, c(u) = \frac{1}{(1-g)^2}, \tag{9.20}$$

where we have used the identity (9.17). This useful result can be used to estimate the feedback parameter g from empirical data, without having to specify the shape of the Hawkes kernel Φ.

9.2.4 Example I: The Exponential Kernel

To illustrate the application and interpretation of some of these results, we first consider the case of an **exponential kernel**

$$\Phi(t \geq 0) = g\omega e^{-\omega t}, \qquad t \geq 0, \tag{9.21}$$

where ω^{-1} sets the time scale beyond which past events can be regarded not to influence present intensity, and g is the norm of the kernel. Using Equation (9.15) in Laplace space, one readily derives:

$$c(t > 0) = \frac{\omega g(2-g)}{2\varphi_0} \omega e^{-(1-g)\omega t}. \tag{9.22}$$

This result is very interesting, because it tells us that the Hawkes feedback not only increases the mean intensity of the process (as intended), but also increases the memory time of the process. While the direct self-exciting mechanism dies out in a time ω^{-1}, the induced extra activity survives until a time $\omega^{-1}/(1-g)$, which diverges as $g \to 1$. We present an interpretation of this effect in terms of population dynamics, or branching process, in Section 9.2.6.

The clustering ratio r can also be calculated exactly, using Equations (9.19) and (9.22). One finds that

$$r = \frac{1}{(1-g)^2} - \frac{g(2-g)}{(1-g)^3\omega\tau}(1 - e^{-(1-g)\omega\tau}), \tag{9.23}$$

which rises from the Poisson value $r = 1$ when $\tau \ll \omega^{-1}$ to the asymptotic value $r = (1-g)^{-2}$ when $\tau \gg \omega^{-1}$.

Note that it is straightforward to simulate a Hawkes process with an exponential kernel. The trick is to simulate the process in event-time by calculating the (calendar) arrival times. Assume n events have already arrived at times $t_1, t_2, \cdots, t_n$. We need to determine the probability density of the arrival time of the $(n+1)^{\text{th}}$ event. The cumulative law of the inter-arrival times δ_n is given by

$$\mathbb{P}[\delta_n \le \tau] = 1 - \exp\left[-\varphi_0\tau - g(1 - e^{-\omega\tau})\sum_{k\le n} e^{-\omega(t_n - t_k)}\right]. \tag{9.24}$$

One can now generate the time series of arrival times iteratively from Equation (9.24) which can be achieved numerically by inverse-transform sampling.

9.2.5 Example II: The Near-Critical Power-Law Kernel

Consider now the case of a **power-law kernel**

$$\Phi(t \ge 0) = g\frac{\omega\beta}{(1+\omega t)^{1+\beta}}, \tag{9.25}$$

where the parameter $\beta > 0$ ensures that the integral defining the norm of Φ is convergent. In the limit $z \to 0$, the behaviour of $\widehat{\Phi}(z)$ depends on the value of β. For $\beta < 1$, asymptotic analysis leads to

$$\widehat{\Phi}(z) \approx g - g\Gamma[1-\beta](z/\omega)^{\beta}.$$

Inserting this behaviour into Equation (9.16) allows one to derive the behaviour of the correlation function $c(t)$. For $g \to 1$, the inversion of the Laplace transform gives a power-law behaviour for $c(t)$ at large times, as[5]

$$c(t) = c_\infty t^{-\gamma}; \qquad \gamma = 1 - 2\beta. \tag{9.26}$$

This result only makes sense if $\beta < 1/2$; otherwise, γ would be positive and $c(t)$ would grow without bound (which is clearly absurd). In fact, this is precisely

[5] See Appendix A.2. The prefactor identification gives $c_\infty^{-1} = 2g^2\omega^{-2\beta}\cos(\pi\beta)\Gamma^2[1-\beta]\Gamma[2\beta]$.

the result of Brémaud and Massoulié: the critical Hawkes process only exists for power-law kernels that decay with an exponent β in the range $0 < \beta < 1/2$, such that $\gamma = 1 - 2\beta$ satisfies $0 < \gamma < 1$. For $\beta > 1/2$ and $g = 1$, the Hawkes process never reaches a stationary state, or is trivially empty. Conversely, when the correlation function decays asymptotically as a power law with $\gamma < 1$, the Hawkes process is necessarily critical (i.e. $g = 1$).

When $\beta < 1/2$ and $g = 1 - \epsilon$ with $\epsilon \ll 1$, the correlation function $c(t)$ decays as $t^{-\gamma}$ up to a (long) crossover time $\tau^* \sim \epsilon^{-1/\beta}$, beyond which it decays much faster, such that $c(t)$ is integrable.

Finally, one can compute the large-τ behaviour of the clustering ratio r:

$$r(\tau) \sim \frac{2c_\infty}{(1-\gamma)(2-\gamma)}\tau^{1-\gamma}, \qquad (\tau \ll \tau^*),$$

which grows without bound when $\tau^* \to \infty$ (i.e. $g \to 1$).

9.2.6 Hawkes Processes and Population Dynamics

By regarding them as **branching processes** (also known as Watson–Galton models), Hawkes processes can also model population dynamics. In a branching process, exogenous "mother" (or "immigration") events occur at rate φ_0, and endogenously produce "child" (or "birth") events, which themselves also endogenously produce child events, and so on. A single mother event and all its child events are collectively called a *family*.

In such models, an important quantity is the expected number g of child events that any single (mother or child) event produces. This number g is called the **branching ratio**. If $g > 1$, then each event produces more than one child event on average, so the population explodes. If $g < 1$, then each event produces less than one child event on average, so the endogenous influence of any single event eventually dies out (although in a Hawkes process, this does not necessarily correspond to the whole process dying out, due to the exogenous immigration arrivals, which are driven by φ_0). The mean family size $\bar{S}$ is $1/(1-g)$.

The critical case $g = g_c = 1$ is extremely interesting, because for some special choices of power-law kernels Φ, the population can survive indefinitely and reach a stationary state even in the limit of no "ancestors", when $\varphi_0 \to 0$. In this case, the family sizes S are distributed as a power-law with a universal exponent $S^{-3/2}$ (i.e. independently of the precise shape of Φ). We will see below that this special case, first studied by Brémaud and Massoulié,[6] seems to corresponds to the behaviour observable in financial markets.

[6] See: Brémaud, P., & Massoulié, L. (2001). Hawkes branching point processes without ancestors. *Journal of Applied Probability*, 38(01), 122–135.

The analogy with population dynamics also allows us to understand why the memory time of the process increases as the expected number of child events increases: while mother events only have a certain fertility lifetime ω^{-1}, the lifetime of the whole lineage (which keeps the memory of the process) can be much longer – in fact, by a factor of the mean family size, $1/(1-g)$.

9.3 Empirical Calibration of Hawkes Processes

At the beginning of this chapter, we remarked that two features of real markets render homogeneous Poisson processes as unsuitable for modelling event arrival times. The first feature is the existence of seasonalities; the second feature is the existence of temporal clustering (even after the seasonalities have been accounted for). As we have discussed throughout the chapter, a Hawkes process (as defined in Equation (9.10)) can address both of these problems by incorporating seasonalities via its exogenous arrival rate $\varphi_0(t)$ and incorporating clustering via the Hawkes kernel. In this section, we explain how Hawkes processes can be calibrated on data. We first discuss how to account for the observed intra-day seasonalities, then we present a selection of methods for estimating the shape of the Hawkes kernel.

9.3.1 Addressing Intra-day Seasonalities

As we have discussed several times already, clear intra-day patterns influence activity levels in financial markets (see Figures 1.3 and 2.5 and Section 4.2). In the context of Hawkes processes, this feature can be addressed as follows. First, measure the intra-day pattern of activity $\bar{\varphi}(t)$ across many different trading days. Then, given $\bar{\varphi}(t)$, define **business time** as

$$\hat{t}(t) = \int_0^t \mathrm{d}u\, \bar{\varphi}(u).$$

In this re-parameterised version of calendar time, the speed of the clock is influenced by the average level of activity. This simple trick causes the activity profile to become flat in business time. We implement this technique (which may essentially be regarded as a pre-processing step) in the financial applications that we describe throughout the remainder of this chapter.

9.3.2 Estimating the Hawkes Kernel

We now turn to estimating the Hawkes kernel $\Phi(t)$. This kernel is characterised by two important features:

- its time dependence (e.g. an exponential kernel with some well-defined decay time ω^{-1}); and
- its norm $g = \int_0^\infty \mathrm{d}u\,\Phi(u)$, which quantifies the level of self-reflexivity in the activity.

As we stated in Section 9.3.1, we assume that we have re-parameterised time to address intra-day seasonalities, and therefore that φ_0, when measured in business time, is constant. We first present the simplest way to elicit and quantify non-Poissonian effects, based on the clustering ratio r. Within the context of Hawkes processes, estimating r immediately leads to an estimate of the feedback intensity (or kernel norm) g, independently of the form of the kernel itself (see Equation 9.19). In the context of microstructural activity, we will see that g is not only positive (which indicates that clustering effects are present beyond the mere existence of intra-day patterns) but, quite interestingly, is not far from the critical value $g_c = 1$. We then proceed to the calibration of Hawkes processes using two traditional calibration processes: maximum likelihood estimation and the method of moments.

9.3.3 Direct Estimation of the Feedback Parameter

Recall from Equation (9.20) that for large τ,

$$r(\tau) \approx_{\tau\to\infty} \frac{1}{(1-g)^2}.$$

Empirically, one can extract M (possibly overlapping) intervals of size τ from a long time series of events, and compute from these M sub-samples the mean and variance of the number of events in such intervals.[7] The theoretical prediction of Equation (9.23) is very well-obeyed by the surrogate series generated using an exponential kernel, whereas for the financial data, the clustering ratio $r(\tau)$ does not seem to saturate even on windows of $\tau = 10^5$ seconds, but rather grows as a power-law of time. This means that as the time scale of observations increases, the effective feedback parameter $g_{\text{eff}}(\tau) := 1 - 1/\sqrt{r(\tau)}$ increases towards $g_c = 1$, and does so as a power-law:

$$g_{\text{eff}}(\tau) \cong 1 - \frac{A}{\tau^\beta},$$

where A is a constant and $\beta \cong 0.35$ (see the inset of Figure 9.1).

Returning to Equation (9.19), this power-law behaviour suggests that the correlation function $c(u)$ itself decays as a slow power-law, as we mentioned in

[7] See: Hardiman, S. J., & Bouchaud, J. P. (2014). Branching-ratio approximation for the self-exciting Hawkes process. *Physical Review E*, 90(6), 062807.

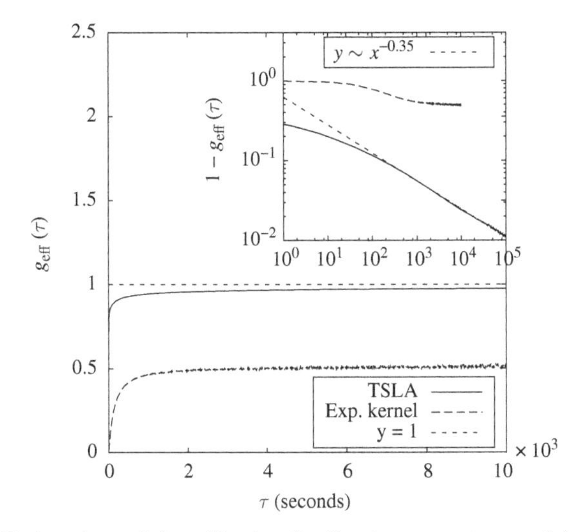

Figure 9.1. Estimation of the effective feedback parameter $g_{\rm eff}(\tau)$ as a function of τ, measured in seconds, for (solid curve) mid-price changes of TSLA, and (dashed) a simulation of a Hawkes process using an exponential kernel with norm $g = 1/2$. We confirm our estimation procedure by observing that the simulated $g_{\rm eff}$ converges exponentially fast to its theoretical asymptotic value, $g_{\rm eff}(\infty) \to g = 1/2$. The empirical time series exhibits criticality because $g_{\rm eff}(\tau)$ converges to 1, (inset) as a power-law with exponent $\beta \cong 0.35$ (the dotted line shows a power law with exponent 0.35).

Section 2.2.4. More precisely, assuming that $c(u) \sim_{u\to\infty} c_\infty/u^\gamma$ with $\gamma < 1$, one obtains

$$r(\tau) = \frac{1}{(1 - g_{\rm eff}(\tau))^2} \approx \frac{2c_\infty}{(1-\gamma)(2-\gamma)}\tau^{1-\gamma},$$

which indeed leads to $g_{\rm eff}(\tau)$ approaching 1 as a power law, with $\beta = (1-\gamma)/2$ (leading to $\gamma \cong 0.3$). This is as expected for a critical Hawkes process with a power-law kernel – see Equation (9.26). We will return to this observation in Section 9.3.5.

9.3.4 Maximum-Likelihood Estimation

If we assume that the kernel function $\Phi(t)$ can be written in a parametric form, with parameter vector $\boldsymbol{\theta}$, then we can use standard **maximum-likelihood estimation** (MLE) to estimate $\boldsymbol{\theta}$ from a realisation of the point process. If N arrivals occur during the interval $[0,T)$, then the likelihood function $\mathbb{L}$ is given by the joint density of all the observed intervals $[t_0 = 0, t_1]$, $[t_1, t_2]$, ... in $[0,T)$. First, note that the probability for an event to occur between t_i and $t_i + \mathrm{d}t$ but not at any intermediate times $t_{i-1} < t < t_i$ is given by:

$$\exp\left(-\int_{t_{i-1}}^{t_i} \mathrm{d}u\, \varphi(u|\boldsymbol{\theta})\right)\varphi(t_i|\boldsymbol{\theta})\mathrm{d}t.$$

The likelihood of the whole sequence of event times is then given by:

$$\mathbb{L}(t_1,t_2,\ldots,t_N|\boldsymbol{\theta}) = \left(\prod_{i=1}^{N} \varphi(t_i|\boldsymbol{\theta}) \exp\left(-\int_{t_{i-1}}^{t_i} \mathrm{d}u\, \varphi(u|\boldsymbol{\theta})\right)\right) \exp\left(-\int_{t_N}^{T} \mathrm{d}u\, \varphi(u|\boldsymbol{\theta})\right),$$

$$= \left(\prod_{i=1}^{N} \varphi(t_i|\boldsymbol{\theta})\right) \exp\left(-\int_{0}^{T} \mathrm{d}u\, \varphi(u|\boldsymbol{\theta})\right).$$

Taking logarithms of both sides yields the log likelihood

$$\ln \mathbb{L}(t_1,t_2,\ldots,t_N|\boldsymbol{\theta}) = \sum_{i=1}^{N} \ln \varphi(t_i|\boldsymbol{\theta}) - \int_{0}^{T} \mathrm{d}u\, \varphi(u|\boldsymbol{\theta}). \tag{9.27}$$

The MLE $\boldsymbol{\theta}^*$ is known to have nice properties: it is both *consistent* (i.e. it converges to the true $\boldsymbol{\theta}$ when $T \to \infty$) and asymptotically normally distributed around the true $\boldsymbol{\theta}$ with a known covariance matrix, with all elements of order T^{-1}. However, because the likelihood itself depends on the set of arrival times $t_1,t_2,\ldots,t_N$, there does not exist a simple "one-size-fits-all" closed-form solution for the MLE of $\boldsymbol{\theta}$. In many cases, the simplest approach is to use a numerical optimisation algorithm to maximise Equation (9.11) directly. However, some specific choices of Φ lead to sufficiently simple parametric forms for Equation (9.11) to allow easy computation of this integral, and therefore of the MLE. One such example is the exponential kernel, which we now investigate in more detail.

The Exponential Kernel

The exponential Hawkes process, which is defined by the kernel in Equation (9.21), depends on three parameters: $\boldsymbol{\theta} = (\varphi_0, \omega, g)$. With this kernel, the Hawkes dynamics obeys the stochastic differential equation

$$\mathrm{d}\varphi(t) = -\omega(\varphi(t) - \varphi_0)\mathrm{d}t + g\omega \mathrm{d}N(t). \tag{9.28}$$

We can interpret this as $\varphi(t)$ relaxing exponentially towards its base intensity φ_0, but being jolted upwards by an amount $g\omega$ each time an event occurs.

This exponential form is popular in many practical applications, mainly due to its Markovian nature: the evolution of $\varphi(t)$ only depends on the present state of the system, because all of the past is encoded into the value of $\varphi(t)$ itself. Numerically, this enables fast updates of the quantities needed to compute the likelihood function. Indeed, one can write:

$$\ln \mathbb{L}(t_1,t_2,\ldots,t_N|\boldsymbol{\theta}) = -T\varphi_0 + \sum_{i=1}^{N} \left[\ln(\varphi_0 + g\omega Z_i) - g\left(e^{-\omega(T-t_i)} - 1\right)\right] \tag{9.29}$$

where we have introduced the notation

$$Z_i := \sum_{k=1}^{i-1} e^{-\omega(t_i - t_k)}.$$

Expressed in this way, it seems that evaluating the likelihood function is an $O(N^2)$ operation. However, by using another trick, this can be reduced further, to $O(N)$. The trick is to notice that Z_i can be computed as a simple recursion, since:

$$Z_i = e^{-\omega(t_i - t_{i-1})} \sum_{k=1}^{i-1} e^{-\omega(t_{i-1} - t_k)} = (1 + Z_{i-1})\, e^{-\omega(t_i - t_{i-1})}.$$

The same trick can also be used when the kernel is a weighted sum of K exponentials:

$$\Phi(t) = g \sum_{k=1}^{K} a_k \omega_k e^{-\omega_k t}, \qquad \sum_{k=1}^{K} a_k = 1. \tag{9.30}$$

To use the trick in this case, one has to introduce K different objects Z_i^k that all evolve according to their own recursion:

$$Z_i^k = e^{-\omega_k(t_i - t_{i-1})}\left(1 + Z_{i-1}^k\right).$$

In this way, the computation of $\mathbb{L}$ and its gradients with respect to the three parameters $\boldsymbol{\theta} = (\varphi_0, \omega, g)$ (which are needed to locate the MLE $\boldsymbol{\theta}^*$) can all be performed efficiently, even on long time series. This is important because (as we discuss in the next section) power-law kernels can be approximated as sums of a finite number of exponentials.

The Power-Law Kernel

The power-law Hawkes process defined in Equation (9.25),

$$\Phi(t \geq 0) = g \frac{\omega\beta}{(1 + \omega t)^{1+\beta}},$$

depends on four parameters: $\boldsymbol{\theta} = (\varphi_0, \omega, g, \beta)$. Interestingly, exponential kernels can be seen as a special limit of power-law kernels, when $\beta \to \infty$ and $\omega \to 0$ with a fixed value of the product $\omega\beta$. Therefore, exponential Hawkes processes are a sub-family of power-law Hawkes processes.

Conversely, a power-law kernel can be accurately reproduced by a sum of exponential kernels with different time scales and different weights.[8] In particular,

[8] On this point, see, e.g., Bochud, T., & Challet, D. (2007). Optimal approximations of power laws with exponentials: Application to volatility models with long memory. *Quantitative Finance*, 7(6), 585–589, and Hardiman, S. J., Bercot, N., & Bouchaud, J. P. (2013). Critical reflexivity in financial markets: A Hawkes process analysis. *European Physical Journal B*, 86, 442–447.

choosing a discrete set of time scales as a geometric series, i.e. $\omega_k = \omega/b^k$, $b > 1$, and weights as $a_k = \omega_k^\beta$, Equation (9.30) allows one to reproduce quite reasonably a power-law decaying kernel $\Phi(t) \sim t^{-1-\beta}$, at least for t such that $\omega^{-1} \ll t \ll b^K \omega^{-1}$.

When using the MLE method for estimating the parameters of power-law kernels on financial data, special care must be given to errors in the time-stamping of the events. Such errors can substantially affect the MLE procedure and lead to spurious solutions, due to artefacts arising from short-time dynamics. In any case, an MLE of Hawkes parameters applied to a wide variety of financial contracts suggests very high values of g^*, close to the critical value $g_c = 1$, and yields a power-law exponent β^* in the range 0.1–0.2 (see references in Section 9.7). This clearly excludes the possibility of an exponential kernel, which corresponds to $\beta \gg 1$. Interestingly, the characteristic time scale ω^{-1}, below which the power-law behaviour saturates, has decreased considerably over time, from several seconds in 1998 to milliseconds in 2012. This is clearly contemporaneous to the advent of execution algorithms and high-frequency trading, which vindicates the idea that any kind of feedback mechanism has accelerated during that period. Note, however, that in contrast with the exponential case, a power-law kernel plausibly suggests the existence of a wide spectrum of reaction times in the market, from milliseconds to days or longer.

There is, however, a problem with using this approach to calibrate a Hawkes process to empirical data. The MLE method implicitly over-focuses on the short-time behaviour of the kernel and is quite insensitive to the shape of the kernel $\Phi(t)$ at large lags. Indeed, since the power-law kernel gives much greater weight to short lags than to long lags, the local intensity φ is mostly determined by the recent events and much less by the far-away past. Therefore, the MLE procedure will mostly optimise the short-time behaviour of the kernel, and will be quite sensitive to the detailed shape of this kernel for $t \to 0$, as well as on the quality of the data for short times. This is why the value of the exponent β^* obtained using MLE is significantly smaller than the value of $\beta \cong 0.35$ obtained from the direct estimation method used in the previous section: the two methods zoom in on different regions of the time axis, as we elaborate on in the next paragraph.

9.3.5 The Method of Moments

Another popular tool for calibrating Hawkes process is the **method of moments**, which seeks to determine the parameters $\boldsymbol{\theta}$ such that the theoretical moments of specified observables match their empirically determined counterparts. Restricting this method to first- and second-order statistics (which is enough to completely determine a Hawkes process, see Section 9.2.2), this method amounts to fitting φ_0 and $\Phi(t)$ such that:

- the mean intensity $\bar{\varphi} = \varphi_0/(1-g)$ matches the empirically determined mean; and

- the predicted intensity correlation function $c(t)$ (using the Yule–Walker equation (9.15)) reproduces the empirically determined intensity correlation.

The first condition simply amounts to counting the number of events in sufficiently large intervals, but the second condition requires extracting the shape of the kernel $\Phi(t)$ using the empirical correlation and Equation (9.15), which is much less trivial.

In fact, seen as an expression for $\Phi(t)$ *given* $c(t)$, Equation (9.15) is called a **Wiener–Hopf equation**. Solving it requires complicated numerical methods, but once this is done, the kernel $\Phi(t)$ can be determined in a wide region of time scales[9] (provided $c(t)$ can be measured with sufficient accuracy). Again, implementing this method on empirical data leads to a power-law kernel with a norm close to 1. An alternative method consists in fitting the measured correlation function $c(t)$ with simple mathematical expressions, and inverting the Wiener–Hopf equation analytically.

The method of moments requires measuring the function $c(t)$ accurately. This can prove difficult, as it requires choosing an adequate bin size Δt to define the count increments ΔN. Interestingly, the clustering ratio $r(\tau)$ allows one to bypass this step, and provides simple and intuitive results. Plotting r as a function of τ gives direct access to the time dependence of $\int_0^\tau \mathrm{d}u(1-u/\tau)c(u)$, and thus of $c(t)$ itself. As shown in Figure 9.1, $c(t)$ actually has not one but *two* power-law regimes: a short-time behaviour with $\beta \cong 0.2-0.25$ (compatible with MLE) and a long-time behaviour with $\beta \cong 0.35-0.4$. A similar crossover between the two regimes exists for, e.g., the S&P 500 futures contract, and occurs for τ of the order of several minutes.[10] Both regimes are characterised by exponents β in the Brémaud–Massoulié interval $0 < \beta < \frac{1}{2}$.

9.4 From Hawkes Processes to Price Statistics

In Section 2.2.4, we introduced a simple model in which changes of the mid-price $m(t)$ occur as a homogeneous Poisson process with rate φ, all have size ϑ (i.e. one tick), and occur upwards or downwards with equal probabilities. This model can be expressed concisely as:

$$\mathrm{d}m(t) = \varepsilon(t)\vartheta\,\mathrm{d}N(t), \tag{9.31}$$

where $N(t)$ is the counting process of mid-price changes and $\varepsilon(t) = \pm 1$ is an independent binary random variable. We now extend this model to the case where $\varphi(t)$ is time-dependent, such that $\mathrm{d}N(t)$ is given by a stationary Hawkes process with an influence kernel Φ.

[9] Note, however, that this method provides no guarantee that for a given $c(t)$, the kernel function Φ will be everywhere non-negative, as is required for the intensity φ.

[10] See: Hardiman et al. (2013).

The local volatility of the price process, measured on a time interval τ small enough to neglect the evolution of $\varphi(t)$, is given by

$$\sigma^2(t) := \frac{1}{\tau}\mathbb{V}\left[\int_t^{t+\tau} \mathrm{d}m(t')\right] = \frac{1}{\tau}\mathbb{E}[N(t+\tau) - N(t)] = \vartheta^2\varphi(t),$$

where we have used the fact that $\mathbb{E}[\varepsilon] = 0$ and

$$\mathbb{E}[\varepsilon(t)\varepsilon(t')] = 0 \text{ for all } t' \neq t.$$

In other words, as in Section 2.2.4, volatility σ and market activity φ are one and the same thing in this simple framework (see Equation (2.18)).

For a Hawkes process, the intensity $\varphi(t)$ is a random variable with some distribution and some temporal correlations described by the function $c(t)$. This is an enticing model of prices, since its volatility is random and clustered in time, which matches empirical observations. In fact, choosing a power-law kernel Φ allows one to generate the **long-range correlations** that we discussed in Chapter 2.

Is this power-law Hawkes process for prices a simple way to reconcile these empirical observations in a parsimonious theoretical framework? As it turns out, there are both pros and cons to this idea. The fact that $\varphi(t)$ has long-range memory that can be described by a near-critical power-law kernel $\Phi(t)$ (see Section 9.2.5) is a clear plus. However, the distribution of φ values generated by a Hawkes process has relatively thin tails, which decay exponentially for large values of φ.[11]

This **exponential tail** is transmitted to the distribution of price changes, $m(t+\tau) - m(t)$, which therefore does not exhibit the power-law tails that are universally observed in real markets. In other words, Hawkes processes do not produce sufficiently large fluctuations in the activity rate to reproduce the large price moves induced by large activity bursts.

There are two (possibly complementary) paths to solving these issues with the model. The first path is to argue that Equation (9.31) is too restrictive, and to allow the amplitude of individual jumps to have fat tails (rather than simply taking the two values $\pm\vartheta$). This approach is certainly plausible for small-tick assets, for which both the bid–ask spread and the gap between the best and second-best occupied price levels in the LOB typically undergo large fluctuations. However, it is less satisfactory for large-tick assets, for which prices seldom change by more than one tick at a time but for which the distribution of returns aggregated over a longer time interval τ (say five minutes) still exhibits power-law tails. Moreover, this approach is not very satisfying, because it requires us to simply input a power-law

[11] In the limit $g = 1 - \epsilon$, with $\epsilon \to 0$, the full distribution of φ can be obtained as

$$P_{\mathrm{st.}}(\varphi) \propto \varphi^{\frac{2\varphi_0}{\omega}} e^{-2\epsilon\varphi/\omega},$$

which shows explicitly the exponential behaviour of the tail. More generally, the tail decays as $e^{-z_c\varphi}$, where z_c is the solution of $g\omega z_c = \ln(1 + \omega z_c)$.

distribution for individual price changes without understanding where it comes from.

The second path is to extend the model and allow the level of market activity to be influenced by the amplitude of price changes themselves, and not only by their frequency. This framework produces a much richer structure, in which power-law tails are indeed generated. We will discuss this approach – of **generalised Hawkes processes** – in the next section.

9.5 Generalised Hawkes Processes

We now consider some generalisations to the standard Hawkes framework that we have developed so far in this chapter, and also present some examples to illustrate how these generalisations can help with practical applications.

Inhibition

The condition that a Hawkes process feedback kernel $\Phi(t)$ is positive for all lags is both necessary and sufficient to ensure that the rate $\varphi(t)$ always remains positive for any realisation of the underlying Poisson process. A positive kernel always leads to self-excitation and clustering.

In some circumstances, however, it is desirable to describe **inhibitory effects**, such that previous events produce negative feedback on $\varphi(t)$. In the presence of such inhibitory effects, one way to ensure that $\varphi(t)$ remains positive is to write

$$\varphi(t) = F\left[\varphi_0 + \int_0^t \mathrm{d}N(u)\Phi(t-u)\right], \tag{9.32}$$

where $F(x)$ is a non-negative function such as $F(x) = \Theta(x)$ or $F(x) = \ln(1 + e^x)$. Provided that the slope of the function $F(x)$ is always less than or equal to 1, the same stability condition ($g < 1$) can be established in these cases.[12] Such models can be simulated and calibrated numerically, but are much harder to characterise analytically than Hawkes models that do not include such inhibitory effects.

Exogenous Events

Sometimes, it is known in advance that certain events occurring at known times (such as the release of scheduled macroeconomic news announcements) will lead to an increase in the activity rate. This effect is not a self-excitation, but is instead truly exogenous. A simple way to include these **exogenous** events in the Hawkes description is to write

$$\varphi(t) = \varphi_0 + \int_0^t \mathrm{d}N(u)\Phi(t-u) + \sum_{t_i<t} a_i \Phi_{\text{exo.}}(t-t_i), \tag{9.33}$$

where t_i is the time of the ith such event, which has strength a_i, and $\Phi_{\text{exo.}}$ is another kernel describing how the impact of these events dies out with time.[13]

[12] See: Brémaud, P., & Massoulié, L. (1996). Stability of nonlinear Hawkes processes. *Annals of Probability*, 24, 1563–1588.

[13] See: Rambaldi, M., Pennesi, P., & Lillo, F. (2015). Modelling foreign exchange activity around macroeconomic news: Hawkes-process approach. *Physical Review E*, 91(1), 012819.

Multi-Event Hawkes Processes

The temporal evolution of an LOB is a multi-event problem, in which the different types of events can influence the future occurrences of both themselves and each other. For example, market order arrivals can trigger other market order arrivals, limit order arrivals, and cancellations (and vice-versa). The events that occur in other markets could also influence the activity in a given LOB.

The Hawkes framework can easily be extended to a *multivariate framework* with K event types, each with a time-dependent rate $\varphi_i(t)$, $i = 1, \ldots, K$. The multivariate form of Equation (9.10) is simply

$$\varphi_i(t) = \varphi_{0i} + \int_0^t \sum_{j=1}^{K} \mathrm{d}N_j(u)\Phi_{i,j}(t-u), \tag{9.34}$$

where $\Phi_{i,j}(t) \geq 0$ is a non-negative kernel function that describes how the past occurrence of an event of type j at time $t-u$ increases the intensity of the events of type i at time t.

The multivariate Hawkes framework is very versatile, because both the cross-event excitations and the temporal structures of the feedback can be chosen to reflect the activity observable in empirical data. The kernel $\Phi_{i,j}(t)$ need not be symmetric, so that the influence of type-i events on type-j activity need not be equal to the influence of type-j events on type-i activity.

In the case where the exogenous event rates are constant, the first- and second-order statistics can again be calculated exactly:

- The mean intensity of type-i events is given by a vector generalisation of the 1 component case, Equation (9.13):

$$\bar{\varphi}_i = \sum_{j=1}^{K} [(\mathbf{1}-\mathbf{g})^{-1}]_{i,j}\varphi_{0,j},$$

  where $\mathbf{g}$ is the matrix of kernel norms

$$g_{i,j} := \int_0^\infty \mathrm{d}u\, \Phi_{i,j}(u).$$

  The process is stable provided the spectral radius of the matrix $\mathbf{g}$ is strictly less than 1.
- For $t > u$, the conditional mean intensity is given by

$$\mathbb{E}[\mathrm{d}N_i(t)|\mathrm{d}N_j(u) = 1] = \bar{\varphi}_i(1 + c_{i,j}(t-u))\mathrm{d}t,$$

  where $\mathbf{c}(u)$ is the solution of a matrix Wiener–Hopf equation

$$\bar{\varphi}_i c_{i,j}(u) = \Phi_{i,j}(u) + \sum_{k=1}^{K} \int_0^\infty \mathrm{d}v\, \Phi_{ik}(u-v)\bar{\varphi}_k c_{kj}(v). \tag{9.35}$$

  Again, the lag-dependent matrix $\mathbf{c}$ is directly related to the covariance of the intensity as:

$$\mathrm{Cov}\left[\frac{dN_i(t)}{dt}, \frac{dN_j(u)}{du}\right] = \bar{\varphi}_i\bar{\varphi}_j c_{i,j}(t-u) + \bar{\varphi}_i\delta_{i,j}\delta(t-u).$$

Hawkes Processes with Price Feedback

In a Hawkes process, past events feed back into future activity. Taking a step back from formalism and thinking about financial markets, it seems reasonable to conclude that while all types of market activity might influence future trading, price moves are particularly salient events that affect traders' future behaviour and decisions. For example, local trends (such as sequences of consecutive price changes in the same direction) induce reactions from the market, either because traders believe that these trends reveal some hidden information, or because if prices reach a certain level, they can trigger other execution decisions (such as stop-losses).

From this point of view, it is natural to extend the Hawkes formalism by writing:[14]

$$\varphi(t) = \varphi_0 + \int_0^t \mathrm{d}N(u)\Phi(t-u) + \left[\int_0^t \mathrm{d}m(u)\Psi(t-u)\right]^2, \tag{9.36}$$

and

$$\mathrm{d}m(t) = \varepsilon(t)\vartheta \mathrm{d}N(t), \tag{9.37}$$

where Equation (9.36) describes the intensity, Equation (9.37) describes the mid-price change, and Ψ is a kernel that measures the influence of past returns. In the simplest case, where Ψ is an exponential kernel, the trend is an exponential moving average of the recent price changes over some time horizon. The last term is the square of the trend, in which case only the amplitude (but not the sign) of the recent price change is relevant. This means that a large price change in a single direction produces a much greater increase in activity than does a sequence of many prices changes in alternating directions.

In this extended model with **price feedback**, one can derive closed-form expressions for the first-, second- and third-order statistics. For example, the mean activity rate $\bar{\varphi}$ is given by

$$\bar{\varphi} = \frac{\varphi_0}{1 - g_\Phi - g_{\Psi^2}},$$

where g_Φ is the norm of Φ and g_{Ψ^2} is the norm of the square of Ψ. The model thus becomes unstable when $g_\Phi + g_{\Psi^2} > 1$.

The specific form of Equation (9.36) is in fact motivated by empirical data, and the resulting model can be seen as a natural continuous-time definition of the usual GARCH model. Remarkably, when both Φ and Ψ are exponential kernels, some analytical progress is possible, and one can show that the distribution of φ acquires power-law tails. The value of the tail exponent depends on the norms g_Φ and g_{Ψ^2}, and can be set in the range of empirical values, even with a small value of the norm of g_{Ψ^2}. In other words, while the pure Hawkes model has exponentially decaying tails (and is therefore a poor starting point for explaining the distribution of returns), incorporating a small trend feedback allows one to generate much more realistic price trajectories with fat tails.

9.6 Conclusion and Open Issues

In this chapter, we have introduced the popular family of Hawkes models, which help to address the difficult problem of understanding the highly clustered and intermittent evolution of financial markets in calendar time. In a Hawkes process, the event arrival intensity at a given time depends not only on an exogenous component, but also on an endogenous or "self-excitation" term.

[14] See: Blanc, P., Donier, J., & Bouchaud, J. P. (2016). Quadratic Hawkes processes for financial prices. *Quantitative Finance*, 17, 1–18.

Hawkes processes are extremely versatile tools for modelling univariate and multivariate point processes in calendar time. Like Gaussian processes, Hawkes processes are entirely determined by their first- and second-order statistics. It is therefore relatively easy to calibrate a Hawkes process on data. Another benefit of using Hawkes processes is the possibility to interpret their fitted parameter values. This allows one to quantify interesting aspects such as the temporal structure of the memory kernel and the total feedback strength g.

When Hawkes models are calibrated on financial data, one finds empirically that the memory kernel has a power-law structure, which shows that multiple different time scales are needed to account for how markets respond to past activity. This is in line with many other observations, such as the well-known long memory structure that we discuss in Chapter 10. Note however that calibrating a Hawkes process to data does not prove the existence of a genuine causal self-excitation mechanism: as often, two-point correlations (used to calibrate the model) do not imply causality.

Perhaps surprisingly, empirical calibration systematically suggests that the feedback strength g is close to the critical value $g_c = 1$, beyond which the feedback becomes so strong that the activity intensity diverges and the Hawkes process ceases to be well defined. There are two possible ways to interpret this observation: one is that markets operate in a regime that is very close to being unstable; the other is that the Hawkes framework is too restrictive and fails to capture the complexity of real financial markets. In fact, as we emphasised in Section 9.2.6, the only way that a Hawkes process can reproduce long memory in the activity is for it to be critical, which is somewhat suspicious. More general models, such as the Hawkes model with price feedback, are not so constrained. This hints of a warning that is more general in scope: even when they are enticing and provide extremely good fits to the data, models can lead to erroneous conclusions because they fail to capture the underlying reality.

Take-Home Messages

(i) Market activity is not time-homogeneous, but rather exhibits both intra-day patterns and endogenous intermittency.

(ii) Real-time market activity can be modelled using point processes, in which events occur in continuous time according to a given rate (intensity). The simplest PP is a homogeneous Poisson process, for which inter-arrival times are independent exponential random variables with a constant rate parameter.

(iii) Linear Hawkes processes are simple auto-regressive models in which the intensity has both an exogenous component and an endogenous component. The exogenous component can vary with time to account for known trends or seasonalities. The endogenous component is measured by a kernel that describes how past events influence the present intensity.

(iv) The larger the norm of the kernel, the more "self-exciting" the process. The branching ratio describes the expected number of future events that will originate from a given event in a Hawkes process.

(v) When calibrated on financial data, the branching ratio is found to be close to 1, which suggests that a large fraction of market activity is endogenous.

(vi) Linear Hawkes processes fail to reproduce the strong volatility clustering and fat-tailed returns distributions that occur in real data. The Hawkes model can be made to produce more realistic returns by including additional effects, such as including price returns in the feedback mechanism.

9.7 Further Reading

General

Daley, D. J., & Vere-Jones, D. (2003). An introduction to the theory of point processes (Vols. I–II). *Probability and its applications*. Springer, second edition.

Bauwens, L., & Hautsch, N. (2009). Modelling financial high frequency data using point processes. In Andersen, T. G., Davis, R. A., Kreiss, & J.-P., Mikosch, Th. V. (Eds.), *Handbook of financial time series* (pp. 953–979). Springer.

Bacry, E., Mastromatteo, I., & Muzy, J. F. (2015). Hawkes processes in finance. *Market Microstructure and Liquidity*, 1(01), 1550005.

Intermittent Dynamics (Physics)

Alessandro, B., Beatrice, C., Bertotti, G., & Montorsi, A. (1990). Domain-wall dynamics and Barkhausen effect in metallic ferromagnetic materials. I. Theory. *Journal of Applied Physics*, 68(6), 2901–2907.

Frisch, U. (1997). *Turbulence: The Kolmogorov legacy*. Cambridge University Press.

Fisher, D. S. (1998). Collective transport in random media: From superconductors to earthquakes. *Physics Reports*, 301(1), 113–150.

Sethna, J. P., Dahmen, K. A., & Myers, C. R. (2001). Crackling noise. *Nature*, 410(6825), 242–250.

Mathematical Properties of Hawkes Processes

Hawkes, A. G. (1971). Point spectra of some mutually exciting point processes. *Journal of the Royal Statistical Society*. Series B (Methodological), 33, 438–443.

Hawkes, A. G., & Oakes, D. (1974). A cluster process representation of a self-exciting process. *Journal of Applied Probability*, 11(03), 493–503.

Brémaud, P., & Massoulié, L. (2001). Hawkes branching point processes without ancestors. *Journal of Applied Probability*, 38(01), 122–135.

Bacry, E., Delattre, S., Hoffmann, M., & Muzy, J. F. (2013). Some limit theorems for Hawkes processes and application to financial statistics. *Stochastic Processes and their Applications*, 123(7), 2475–2499.

Jaisson, T., & Rosenbaum, M. (2015). Limit theorems for nearly unstable Hawkes processes. *The Annals of Applied Probability*, 25(2), 600–631.

Bacry, E., & Muzy, J. F. (2016). First-and second-order statistics characterization of Hawkes processes and non-parametric estimation. *IEEE Transactions on Information Theory*, 62(4), 2184–2202.

Estimation of Hawkes Processes

Ogata, Y. (1978). The asymptotic behaviour of maximum likelihood estimators for stationary point processes. *Annals of the Institute of Statistical Mathematics*, 30(1), 243–261.

Bacry, E., Dayri, K., & Muzy, J. F. (2012). Non-parametric kernel estimation for symmetric Hawkes processes: Application to high frequency financial data. *The European Physical Journal B-Condensed Matter and Complex Systems*, 85(5), 1–12.

Dassios, A., & Zhao, H. (2013). Exact simulation of Hawkes process with exponentially decaying intensity. *Electronic Communications in Probability*, 18(62), 1–13.

Hardiman, S. J., & Bouchaud, J. P. (2014). Branching-ratio approximation for the self-exciting Hawkes process. *Physical Review E*, 90(6), 062807.

Lallouache, M., & Challet, D. (2016). The limits of statistical significance of Hawkes processes fitted to financial data. *Quantitative Finance*, 16(1), 1–11.

Financial Applications of Hawkes Processes

Filimonov, V., & Sornette, D. (2012). Quantifying reflexivity in financial markets: Toward a prediction of flash crashes. *Physical Review E*, 85(5), 056108.

Hardiman, S., Bercot, N., & Bouchaud, J. P. (2013). Critical reflexivity in financial markets: A Hawkes process analysis. *The European Physical Journal B*, 86, 442–447.

Bacry, E., & Muzy, J. F. (2014). Hawkes model for price and trades high-frequency dynamics. *Quantitative Finance*, 14(7), 1147–1166.

Da Fonseca, J., & Zaatour, R. (2014). Hawkes process: Fast calibration, application to trade clustering, and diffusive limit. *Journal of Futures Markets*, 34(6), 548–579.

Achab, M., Bacry, E., Muzy, J. F., & Rambaldi, M. (2017). Analysis of order book flows using a nonparametric estimation of the branching ratio matrix. arXiv:1706.03411.

Extensions of Hawkes Processes

Brémaud, P., & Massoulié, L. (1996). Stability of nonlinear Hawkes processes. *The Annals of Probability*, 24, 1563–1588.

Embrechts, P., Liniger, T., & Lin, L. (2011). Multivariate Hawkes processes: An application to financial data. *Journal of Applied Probability*, 48(A), 367–378.

Bormetti, G., Calcagnile, L. M., Treccani, M., Corsi, F., Marmi, S., & Lillo, F. (2015). Modelling systemic price cojumps with Hawkes factor models. *Quantitative Finance*, 15(7), 1137–1156.

Rambaldi, M., Pennesi, P., & Lillo, F. (2015). Modelling foreign exchange market activity around macroeconomic news: Hawkes-process approach. *Physical Review E*, 91(1), 012819.

Blanc, P., Donier, J., & Bouchaud, J. P. (2016). Quadratic Hawkes processes for financial prices. *Quantitative Finance*, 17, 1–18.

Rambaldi, M., Bacry, E., & Lillo, F. (2016). The role of volume in order book dynamics: A multivariate Hawkes process analysis. *Quantitative Finance*, 17, 1–22.

10

Long-Range Persistence of Order Flow

An unfailing memory is not a very powerful incentive to the study of the phenomena of memory.

(Marcel Proust)

In the previous chapter, we noted that activity in financial markets tends to cluster in time. As we also noted, such clustering is not explained by local correlations of market activity, but instead suggests that market activity is a *long-memory process*. In this chapter, we discuss another type of long memory that is conceptually unrelated to this long memory of activity: the highly persistent nature of the sequence of binary variables ε_t that describe the direction of market orders. As we will see, buy orders tend to follow other buy orders and sell orders tend to follow other sell orders, both for very long periods of time.

More formally, let ε_t denote the sign of the tth market order, with $\varepsilon_t = +1$ for a buy market order and $\varepsilon_t = -1$ for a sell market order, where t is discrete and counts the number of market orders. In this event-time framework, one can characterise the statistical properties of the time series of signs via the market-order **sign autocorrelation function**

$$C(\ell) := \mathrm{Cov}[\varepsilon_t, \varepsilon_{t+\ell}]. \tag{10.1}$$

As we will see in this chapter, the surprising empirical result is that $C(\ell)$ decays extremely slowly with ℓ, and is well approximated by a power-law $\ell^{-\gamma}$ with $\gamma < 1$. Importantly, this effect is different from activity clustering in time, which we considered in Chapter 9. For example, a process with exponentially distributed inter-arrival times – as would occur if arrivals are described by a homogeneous Poisson process – can still have long-range persistence in order signs. Conversely, a process in which order signs are uncorrelated could still have long-range autocorrelations in inter-arrival times. Therefore, the underlying mechanisms explaining these two phenomena could be completely different.[1]

In this chapter, we first review the empirical evidence for long-range persistence in ε_t. We then consider some consequences of this fact, including the apparent **efficiency paradox**, which asks the question of how prices can remain unpredictable when order flow (which impacts the price) is so predictable. We then introduce two models that are capable of producing ε_t series with long-range autocorrelations, and consider how to calibrate such models on empirical data to make predictions of future order signs. Finally, we will discuss the possible origins of the long-range autocorrelations that we observe empirically.

10.1 Empirical Evidence

The order-sign autocorrelation function defined by Equation (10.1) has been studied by many authors and on many different asset classes, including equities, FX and futures (see references in Section 10.7). While it has been known for a long time that market order signs are positively autocorrelated, it came as a surprise that these autocorrelations decay extremely slowly. Figure 10.1 shows $C(\ell)$ for our four stocks on NASDAQ. Consistently with the existing empirical studies of other assets, the figure suggests that $C(\ell)$ decays as a power-law $\ell^{-\gamma}$ with $\gamma < 1$, at least up to very large lags (beyond which statistical precision is lost). Mathematically, the value $\gamma < 1$ means that $C(\ell)$ is non-integrable, which means that the order-sign process is a **long-memory** process.[2] Note that the initial decrease of $C(\ell)$ is often faster (exponential-like), followed by a slow, power-law regime.

The fact that $C(\ell)$ decays so slowly has many important consequences. For a long-memory process, the conditional expectation of $\varepsilon_{t+\ell}$ given that $\varepsilon_t = 1$ is

$$\mathbb{E}[\varepsilon_{t+\ell}|\varepsilon_t = 1] = C(\ell) \sim \frac{c_\infty}{\ell^\gamma}. \tag{10.2}$$

Numerically, with the realistic values $c_\infty = 0.5$ and $\gamma = \frac{1}{2}$, this gives $C(10000) \approx 0.005$. Therefore, if we observe a buy market order now, the probability that a market order 10,000 trades in the future is a buy order exceeds that for a sell order by more than 0.5%. In modern equities markets for liquid assets, a time-lag of 10,000 trades corresponds to a few hours or even a few days of trading. Where does this long-range predictability come from? We will return to this question later in this chapter.

[1] For example, using a Hawkes process to generate the arrival times of orders whose signs are uncorrelated would not create long-range autocorrelations in the order-sign series.

[2] For a detailed introduction to long-memory processes, see, e.g., Beran, J. (1994). *Statistics for long-memory processes*. Chapman & Hall.

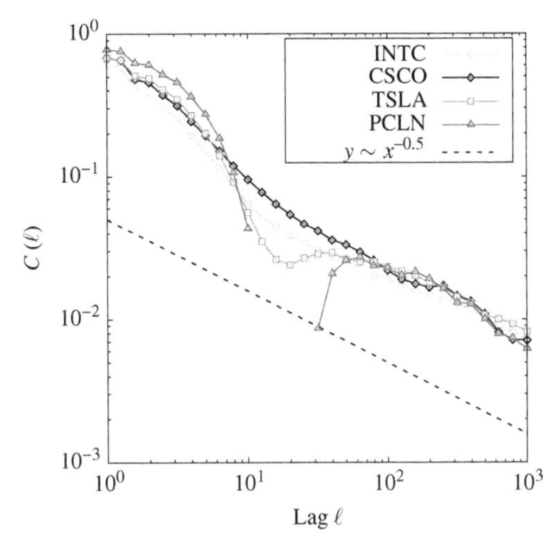

Figure 10.1. Autocorrelation function of the market order sign process for INTC, CSCO, TSLA and PCLN. The dashed line represents a power-law with exponent −0.5. The missing data points correspond to negative values.

10.2 Order Size and Aggressiveness

Until now, we have treated all market orders on an equal footing, independently of their volume. In theory, we could consider the time series of signed volumes $\varepsilon_t \upsilon_t$ (where υ_t denotes the volume of the market order arriving at time t). However, the distribution of order volumes has heavy tails, which complicates the estimation of the corresponding autocorrelation function. Furthermore, simply knowing that a market order is large only tells us half the story, because the impact caused by the arrival of a large buy (respectively, sell) market order depends on the volume available at the ask (respectively, bid) price.

A more useful idea is to label market orders with a simple binary **aggressiveness** variable $\pi_t \in \{0, 1\}$, which indicates whether or not the size of the market order is sufficiently large to trigger an immediate price change ($\pi_t = 1$) or not ($\pi_t = 0$). Put another way, a market order has $\pi_t = 1$ if and only if its size is greater than or equal to the volume available at the opposite-side best quote at its time of arrival.[3]

Throughout the book, we will write MO^1 to denote market orders for which $\pi_t = 1$ and MO^0 to denote market orders for which $\pi_t = 0$. We must then specify four correlation functions, depending on the type of events that we wish to consider. For clarity, we introduce an indicator variable $I(\pi_t = \pi)$, which is equal to 1 if the event at time t is of type π, and 0 otherwise. By standard properties of an indicator

[3] Note that in some cases, the emptied queue is then immediately refilled, reverting the initial price change. This is the case, for example, when hidden liquidity (e.g. iceberg orders) are present. Conversely, a market order smaller than the size of the queue can be immediately followed by a wave of cancellations, leading to a subsequent price change. Our definition is such that the first case corresponds to $\pi = 1$ and the second case to $\pi = 0$.

function, it follows that

$$\mathbb{P}(\pi) = \mathbb{E}[I(\pi_t = \pi)].$$

For a pair of events π and π' both in $\{\text{MO}^0, \text{MO}^1\}$, we define the **conditional correlation** of order signs as:

$$C_{\pi,\pi'}(\ell) := \mathbb{E}[\varepsilon_t \varepsilon_{t+\ell} | \pi_t = \pi, \pi_{t+\ell} = \pi'],$$
$$= \frac{\mathbb{E}[\varepsilon_t I(\pi_t = \pi) \cdot \varepsilon_{t+\ell} I(\pi_{t+\ell} = \pi')]}{\mathbb{P}(\pi)\mathbb{P}(\pi')}. \tag{10.3}$$

For $\ell > 0$, the first subscript of $C_{\pi,\pi'}(\ell)$ indicates the type of the event that happened first. The unconditional sign-correlation function is given by

$$C(\ell) = \sum_{\pi=0,1} \sum_{\pi'=0,1} \mathbb{P}(\pi)\mathbb{P}(\pi')C_{\pi,\pi'}(\ell).$$

By symmetry in the definition of C, it follows that

$$C_{\pi,\pi'}(\ell) = C_{\pi',\pi}(-\ell).$$

If $\pi \neq \pi'$, then in general $C_{\pi,\pi'}(\ell) \neq C_{\pi',\pi}(\ell)$. In other words, the correlation of the sign of an aggressive market order at time t with that of a non-aggressive market order at time $t + \ell$ has no reason to be equal to the corresponding correlation when the non-aggressive order arrives first.

Figure 10.2 shows the empirical values of these four conditional correlation functions. These results help to shed light on the origin of the slow decay of $C(\ell)$. For small-tick stocks, all four correlation functions behave similarly to each other. This makes sense, because an aggressive market order for a small-tick stock typically induces only a small relative price change, so there is little difference between market orders in MO^0 and those in MO^1 in this case. For large-tick stocks, by contrast, we observe significant differences between the behaviour for MO^0 and MO^1. Although the long-range power-law decay still occurs for pairs of market orders in MO^0, the sign of market orders in MO^1 is *negatively* correlated with that of the next market order. This also makes sense: after a substantial move up (respectively, down) induced by an aggressive buy (respectively, sell) market order, one expects that the flow inverts as sellers (respectively, buyers) are enticed by the new, more favourable price. Note finally that $C_{0,1}(\ell)$ decays faster than $C_{0,0}(\ell)$, but still approximately as a power-law.

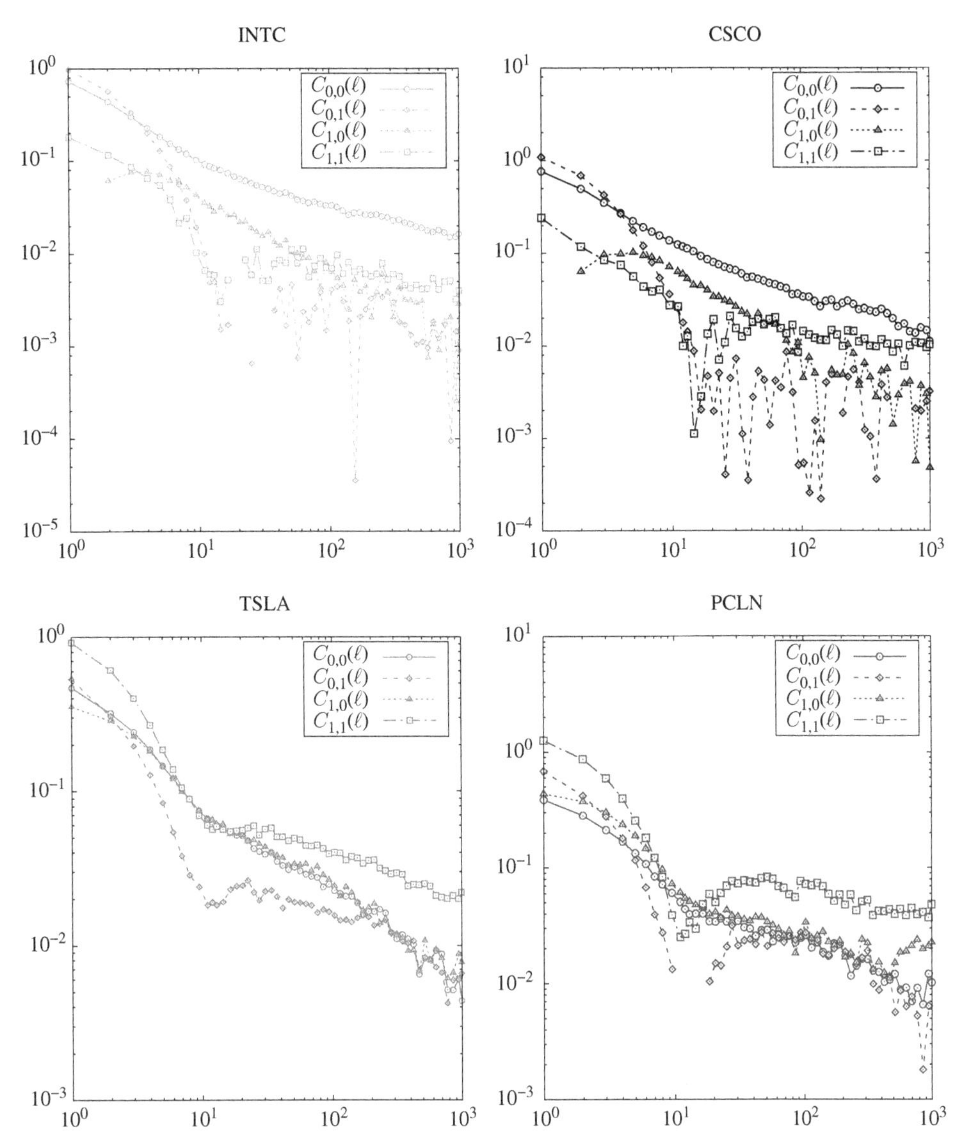

Figure 10.2. Conditional correlation functions (circles) $C_{0,0}$, (diamonds) $C_{0,1}$, (triangles) $C_{1,0}$ and (squares) $C_{1,1}$, of the market order sign process for (top left) INTC, (top right) CSCO, (bottom left) TSLA and (bottom right) PCLN. Missing points correspond to negative values.

10.3 Order-Sign Imbalance

For a finite event-time window of size N, consider the **order-sign imbalance**

$$\frac{1}{N}\sum_{t=1}^{N}\varepsilon_t.$$

The long-memory property of the ε_t series has an important consequence on the behaviour of this quantity. We will assume here and below that buy and sell market orders are such that, unconditionally, $\mathbb{E}[\varepsilon] = 0$. For short-memory processes, the amplitude of the fluctuations of the order imbalance in a window of size N scale as $1/\sqrt{N}$ for large N. More formally,

$$\mathbb{V}\left[\frac{1}{N}\sum_{t=1}^{N}\varepsilon_t\right] := \frac{1}{N} + \frac{2}{N}\sum_{\ell=1}^{N}\left(1 - \frac{\ell}{N}\right)C(\ell).$$

When $C(\ell)$ decays faster than $1/\ell$, the second term on the right-hand side also behaves as $1/N$, and we indeed recover that the variance of the order imbalance decays as $1/N$ for large N. If, however, $C(\ell) \sim c_\infty/\ell^\gamma$ with $\gamma < 1$, then the second term instead behaves as $N^{-\gamma}$, and therefore dominates at large N, leading to

$$\mathbb{V}\left[\frac{1}{N}\sum_{t=1}^{N}\varepsilon_t\right] \approx_{N\gg 1} \frac{2c_\infty}{(2-\gamma)(1-\gamma)}N^{-\gamma} \gg N^{-1} \qquad (\gamma < 1).$$

Hence, the fluctuations of order-sign imbalance decay very slowly with increasing N. This is a consequence of the persistence of order flow: although on average there are an equal number of buy and sell market orders, in any finite window there is likely to be an accumulation of one or the other. This is very important for market-makers, who strive to keep a balanced inventory, because these long-range autocorrelations make their job more difficult (see Sections 1.3 and 17.2).

Another way to highlight the problem caused by order-flow correlations is to assume that each market order on average pushes the price by a small amount $G \times \varepsilon_t$. As we will discuss in detail in Chapter 11, this is called **price impact**: buy orders tend to push the price up and sell orders tend to push the price down. If this impact was permanent, the mid-price m_ℓ would evolve according to

$$m_\ell = m_0 + G\sum_{t=0}^{\ell-1}\varepsilon_t + \sum_{t=0}^{\ell-1}\xi_t,$$

where ξ_t is an independent noise term (with variance Σ^2, say) that models all the other possible causes of price changes. For $\gamma < 1$, the variogram of price changes (see Section 2.1.1) would then be

$$\mathcal{V}(\ell) := \mathbb{V}[m_\ell - m_0] \approx_{\ell\gg 1} \frac{2c_\infty G^2}{(2-\gamma)(1-\gamma)}\ell^{2-\gamma} + \Sigma^2\ell.$$

Within our simple framework, the price would therefore be super-diffusive (i.e. the variogram would grow faster than linearly with ℓ for $\gamma < 1$). In other words, the long memory in the order signs would create trends in the price trajectory, because buyers (or sellers) consistently push the price in the same direction for long periods of time. This is at odds with empirical data, which suggests that the above picture is inadequate. As we discuss in Chapter 13, the resolution of this apparent **efficiency paradox** can be found within the framework of the **propagator model**. In other words, the removal of the order-flow correlations is the result of the counter-balancing role of liquidity providers (see also Section 16.2.1).

10.4 Mathematical Models for Persistent Order Flows

10.4.1 Herding versus Splitting

What causes the autocorrelations in market order signs? Intuitively, two possibilities come to mind. The first possibility is **herding**, in which different market participants submit orders with the same sign. For example, a group of market participants could observe the arrival of a buy market order and infer that there is an active buyer interested at the current price. The group of market participants may decide to join the bandwagon, and create many more buy market orders. This behaviour can cascade (similarly to the Hawkes mechanism that we discussed in Chapter 9) and thereby lead to long sequences of trades in the same direction.

The second possibility is **order-splitting**, in which market participants who wish to execute large trades split their intended volume into many smaller orders, which they then submit incrementally. We will discuss why this splitting should occur in real markets in Section 10.5. Clearly, such activity could also lead to long sequences of orders with the same signs, and could therefore cause autocorrelations in the order-sign series.

In the remainder of this section, we will consider two mathematical models: one of herding and one of order-splitting. We will discuss how to calibrate each of these models to empirical data, discuss the autocorrelation structures that each model predicts, then use these results to infer the relative adequacy of these models for explaining the autocorrelations that occur in real markets.

10.4.2 A Model for Herding

In this section, we introduce a family of models called **discrete auto-regressive** (DAR) processes,[4] which can be regarded as discrete analogues of Hawkes processes. A DAR process is constructed as follows. The sign ε_t at time t is thought of as the "child" of a previous sign at time $t - \ell$, where the distance ℓ is a random variable distributed according to a certain discrete distribution $\mathbb{K}(\ell)$, with

$$\sum_{\ell=1}^{\infty} \mathbb{K}(\ell) = 1. \tag{10.4}$$

[4] These processes are sometimes called integer-value auto-regressive (INAR) processes.

For a given value of $k > 0$, if $\mathbb{K}(\ell) \equiv 0$ for all $\ell > k$, the model is called DAR(k), and involves only k lags. Once the mother sign is chosen in the past, one posits that:

$$\begin{aligned} \varepsilon_t &= \varepsilon_{t-\ell} && \text{with probability} \quad p, \\ \varepsilon_t &= -\varepsilon_{t-\ell} && \text{with probability} \quad 1-p, \end{aligned} \tag{10.5}$$

for some constant $p < 1$.

This is a "copy-paste" model for trades: it assumes that a trader trading at time t selects a previous market order that occurred some ℓ steps in the past (where ℓ is chosen according to the probability distribution $\mathbb{K}(\ell)$), and either copies the sign of that market order (with probability p) or goes against the sign of that market order (with probability $1 - p$).

In the stationary state, the signs +1 and −1 are equiprobable, and the autocorrelation function $C(\ell)$ obeys the following **Yule–Walker equation**:[5]

$$C(\ell \geq 1) = (2p-1)\sum_{n=1}^{\infty} \mathbb{K}(n)C(\ell - n); \qquad C(0) = 1. \tag{10.6}$$

There is therefore a one-to-one relation between $\mathbb{K}(\ell)$ and $C(\ell)$, which allows one to calibrate the model, much as for Hawkes processes (see Section 9.3.5). Note that in the empirical case, where $C(\ell)$ decays as a power-law $\ell^{-\gamma}$ with exponent $\gamma < 1$, one can show (as in the corresponding Hawkes critical case) that $\mathbb{K}(\ell) \sim \ell^{(\gamma-3)/2}$ for large ℓ, together with $p \uparrow 1$ (compare with Equation (9.26)).

By construction, the conditional expectation of ε_t, given the past history of signs, is

$$\widehat{\varepsilon}_t = (2p-1)\sum_{\ell=1}^{\infty} \mathbb{K}(\ell)\varepsilon_{t-\ell}. \tag{10.7}$$

Therefore, the best predictor of the next sign in a DAR processes is a linear combination of the past signs. This result will turn out to be useful in Chapter 13.

10.4.3 A Model for Large Metaorders

An alternative model for the long-range correlation of order signs analyses the consequence of large, incrementally executed trading decisions. This model was originally proposed by Lillo, Mike and Farmer (LMF) in 2005.[6]

The LMF model assumes that the long memory in market order signs comes from very large orders that need to be split and executed slowly, over many individual trades. The underlying parent order is called a **metaorder**. It is well

[5] Note the similarity with the corresponding Equation (9.15) in the context of Hawkes processes.

[6] See: Lillo, F., Mike, S., & Farmer, J. D. (2005). Theory for long memory in supply and demand. *Physical Review E*, 71(6), 066122. Our presentation differs slightly in a few areas to simplify the mathematics, but is otherwise similar to theirs.

known that the amount of assets under management across different financial institutions is very broadly distributed, perhaps itself with a power-law tail. Therefore, if we assume that the size of a metaorder mostly depends on the size of the asset manager or mutual fund from which it originates, then we can reasonably expect the size of these metaorders to be power-law distributed as well.

The LMF model is a stylised model of a market, in which a fixed number M of independent metaorders are being executed at any instant of time. Each active metaorder $i = 1, \ldots, M$ is characterised by a certain sign ε_i and certain **termination rate** $\kappa_i \in (0, 1]$ drawn independently, for each metaorder, from a certain distribution $\rho(\kappa)$. At each discrete time step, one metaorder (say i) is chosen at random. Once chosen, the metaorder is either terminated (with probability κ_i), or has another unit executed (with probability $1 - \kappa_i$).[7] If terminated, the metaorder is replaced by a new metaorder with its own sign and termination rate. If it has a unit executed, the metaorder generates a market order with sign ε_i. Since the signs of metaorders are assumed to be independent, then for each $\ell \geq 1$, the value of the correlation function $C(\ell)$ is equal to the probability that a market order at time t and $t + \ell$ belong to the same metaorder. All other cases average to zero and do not contribute to $C(\ell)$.

Given that trade t belongs to a metaorder with termination rate κ, what is the probability that trade $t + \ell$ belongs to the same metaorder? Since the process is Markovian, this probability is given by

$$\left(1 - \frac{\kappa}{M}\right)^{\ell} \times \frac{1 - \kappa}{M},$$

because the probability for the metaorder to disappear is κ/M at each step, and the probability that the order at time $t + \ell$ belongs to the same metaorder as the order at time t is $1/M$. Similarly, the probability that a metaorder with termination rate κ remains active for a total number of time steps exactly equal to L is

$$\mathbb{P}(L|\kappa) = \left(1 - \frac{\kappa}{M}\right)^{L-1} \frac{\kappa}{M}.$$

Finally, the probability $\mathbb{P}(\kappa(\text{MO}_t) = \kappa)$ that a market order at a randomly chosen time belongs to a metaorder with termination rate κ and duration L is proportional to $\rho(\kappa)$ and to $L \times \mathbb{P}(L|\kappa)$, since the longer a metaorder is, the more likely it is to be

[7] In contrast to the original LMF model, we assume that when a trader submits a chunk of a metaorder, this action does not diminish the remaining size of their metaorder. Therefore, in our presentation, no metaorders ever diminish to size 0. Instead, we assume that traders simply stop executing their metaorders with a fixed rate. We make this simplification because it greatly simplifies the algebra involved in deriving the model's long-run behaviour.

encountered.[8] Hence

$$\mathbb{P}(\kappa(\mathrm{MO}_t) = \kappa) = \frac{1}{Z}\rho(\kappa) \sum_{L=1}^{\infty} L\mathbb{P}(L|\kappa),$$

where Z is a normalisation constant, such that

$$\int_0^1 \mathrm{d}\kappa \mathbb{P}(\kappa(\mathrm{MO}_t) = \kappa) = 1. \tag{10.8}$$

Putting everything together, one finally arrives at an exact formula for $C(\ell)$, as the probability for the order at time t and the order at time $t + \ell$ belong to the same metaorder:

$$C(\ell) = \frac{1}{MZ} \int_0^1 \mathrm{d}\kappa \rho(\kappa)(1 - \kappa) \sum_{L=1}^{\infty} L\mathbb{P}(L|\kappa)\left(1 - \frac{\kappa}{M}\right)^{\ell}. \tag{10.9}$$

To make sense of these formulas, we will first make some extra assumptions. First, we will assume that the number of active metaorders is large, implying $\kappa/M \ll 1$, which allows us to obtain the unconditional probability that a metaorder remains active for exactly L steps as:

$$\mathbb{P}(L) = \int_0^1 \mathrm{d}\kappa \rho(\kappa)\mathbb{P}(L|\kappa) \approx_{L\gg 1} \frac{1}{M} \int_0^1 \mathrm{d}\kappa\, \kappa\rho(\kappa)e^{-\kappa L/M}.$$

Second, we will assume that $\rho(\cdot)$ can be written in the form $\rho(\kappa) = \zeta\kappa^{\zeta-1}$ with $\zeta > 1$. This leads to:

$$\mathbb{P}(L) \approx_{L\gg 1} \zeta^2 \Gamma[\zeta] \frac{M^{\zeta}}{L^{1+\zeta}}.$$

Assuming that each market order has a size υ_0, this power law for L leads to a power-law decay of the distribution of the size $Q = \upsilon_0 L$ of metaorders, which is the initial motivation of the model and justifies our choice for $\rho(\kappa)$.

Within the same approximation,

$$\mathbb{E}[L] = \sum_{L=1}^{\infty} L\mathbb{P}(L|\kappa) \approx \frac{\kappa}{M} \int_0^{\infty} \mathrm{d}L\, Le^{-\kappa L/M} \approx \frac{M}{\kappa}, \tag{10.10}$$

and hence, using the normalisation condition from Equation (10.8),

$$Z = M \int_0^1 \mathrm{d}\kappa \frac{1}{\kappa}\rho(\kappa) = M\frac{\zeta}{\zeta - 1}. \tag{10.11}$$

Finally, from Equation (10.9), for $\ell \gg 1$,

$$C(\ell) \approx \frac{1}{Z} \int_0^1 \mathrm{d}\kappa \frac{1-\kappa}{\kappa} \rho(\kappa)e^{-\kappa\ell/M} \approx_{\ell\gg M} \Gamma[\zeta]\frac{M^{\zeta-2}}{\ell^{\zeta-1}},$$

[8] The sum over all L of $L \times \mathbb{P}(L|\kappa)$ must be convergent for the model to be well defined and stationary.

where the right-hand side is now in the form $c_\infty \ell^{-\gamma}$, as in Equation (10.2). Using this notation, this model produces a power-law decay for the autocorrelation of the signs with an exponent $\gamma = \zeta - 1$. Since the decay exponent γ must be positive, one concludes that $\zeta > 1$. In order to reproduce the empirical value of $\gamma \cong 1/2$, the model requires the underlying metaorder sizes to be distributed roughly as $L^{-1-\zeta}$ with $\zeta = 3/2$.

Clearly, the assumption that the number of active metaorders M is fixed and constant over time is not realistic. However, it is not crucial, because the relation $\gamma = \zeta - 1$ is robust and survives in situations where M fluctuates in time. A hand-waving argument for this is as follows: the probability that two orders separated by ℓ belong to the same metaorder (and contribute to $C(\ell)$) is proportional to the probability that they both belong to a metaorder of size larger than ℓ, which scales like $\int_\ell^\infty \mathrm{d}L\, L\mathbb{P}(L) \sim \ell^{1-\zeta}$.

10.4.4 Herding or Order-Splitting?

As we discussed in Section 10.4.1, herding and order-splitting could both provide plausible explanations for the autocorrelations in market order signs that occur empirically. Which is more likely to be the main reason for the observed phenomena? Although both explanations are likely to play a role, several empirical observations suggest that the influence of order-splitting is much stronger than that of herding.

One way to confirm this directly would be to analyse a data set that provides the identity of the initiator of each market order. This would allow detailed analysis of which market participant submitted which market order, and would thereby provide detailed insight into the relative roles of herding and order-splitting. Unfortunately, comprehensive data describing order ownership is very difficult to obtain. However, fragmented data from brokers, consultants and proprietary sources all confirm that order-splitting is pervasive in equity markets, futures markets, FX markets, and even on Bitcoin markets, with a substantial fraction of institutional trades taking several days or even several weeks to complete (see references in Section 9.7).

In some cases (such as the London Stock Exchange (LSE), the Spanish Stock Exchange (SSE), the Australian Stock Exchange (ASE) and the New York Stock Exchange (NYSE)), partial information about the identity of participants can be obtained through the code number associated with the broker who executes the trade. This data suggests that most of the long-range autocorrelations in $C(\ell)$ originate from the same broker submitting several orders with the same sign, rather than from other brokers joining the bandwagon. In fact, although some herding can be detected on short time lags (e.g. for ℓ less than about 10), the behaviour of other brokers at large lags is actually contrarian, and contributes *negatively* to $C(\ell)$.

After a short spree of copy-cat behaviour, other market participants react to a flow of buy market orders by sending sell market orders (and vice-versa), presumably because buy orders move the price up, and therefore create more opportunities for sellers.[9] This indicates that on the time scale over which $C(\ell)$ decays as a power-law, herding is not a relevant factor.

Is there any empirical evidence to suggest that the size of metaorders is power-law distributed, as assumed by the LMF model? Without identification codes, this is again difficult. However, a power-law distribution with an exponent $\zeta \cong 3/2$ has indeed been reported in the analysis of block trades (traded off-book in the upstairs market) on the LSE, in the reconstruction of large metaorders via brokerage codes on the SSE, and in a set of large institutional metaorders executed at Alliance Bernstein's buy-side trading desk in the US equities market. Bitcoin data allows a precise reconstruction of large metaorders, and suggests a smaller exponent $\zeta \cong 1$.[10]

10.5 Liquidity Rationing and Order-Splitting

In Section 10.4.4, we noted that a variety of empirical evidence suggests that order-splitting is the primary cause of the observed long-term autocorrelations in market order signs. This raises an interesting question: why would investors split their orders in the first place, rather than submitting their desired trades as quickly as possible, before the information edge they have (or believe they have) becomes stale?

As we noted at the end of Chapter 1, market participants face a quandary. Many buyers want to buy (and many sellers want to sell) quantities that are, in aggregate, very substantial. For example, the daily traded volume for a typical stock in an equities markets usually amounts to roughly 0.1%–1% of its total market capitalisation. Volumes in some commodity futures or FX markets are breathtaking. However, as we will discuss in this section, the total volume offered for an immediate transaction at any given instant of time is typically very small. Why is this the case? As we will argue, the answer is that most market participants work very hard to hide their real intentions to trade. An important consequence is then that the available liquidity at a given moment in time is rather like an iceberg:

[9] See; Tóth, B., Palit, I., Lillo, F., & Farmer, J. D. (2015). Why is equity order flow so persistent? *Journal of Economic Dynamics and Control*, 51, 218–239 and Tóth, B., Eisler, Z., & Bouchaud, J. P. (2017). https://ssrn.com/abstract=2924029, for the same analysis on CFM proprietary data.

[10] See: Vaglica, G., Lillo, F., Moro, E., & Mantegna, R. (2008). Scaling laws of strategic behavior and size heterogeneity in agent dynamics. *Physical Review E*, 77, 0036110; Bershova, N., & Rakhlin, D. (2013). The non-linear market impact of large trades: Evidence from buy-side order flow. *Quantitative Finance*, 13, 1759–1778; Donier, J., & Bonart, J. (2015). A million metaorder analysis of market impact on the Bitcoin. *Market Microstructure and Liquidity*, 1(02), 1550008.

it reveals only a small fraction of the huge underlying supply and demand. Most of the liquidity remains *latent*.

10.5.1 The Buyer's Conundrum

Imagine that after careful examination of some information, you estimate that the price of some given stock XYZ should increase by 20% in the coming year. After considering your portfolio and performing a suitable risk analysis, you decide to buy 1% of the market capitalisation of XYZ. You also know that the mean daily traded volume of XYZ is 0.5% of its market capitalisation. Even if you decide to participate with $\frac{1}{4}$ of that daily volume with your own trades (which is already quite large), it will take you at least eight days to complete your desired total volume. If we assume that daily volatility is 2%, then over eight days, the value of the stock typically fluctuates by $2\% \times \sqrt{8} \approx 6\%$. Therefore, about one-third (6/20) of your expected profit might evaporate during this execution period. This provides a strong incentive to hurry through your trades as much as possible. How should you proceed?

Frustratingly, there probably also exists a handful of other market participants who are actively trying to offload a position in stock XYZ, and who would be ready to sell you the full 1% at the current market price. However, you do not know who these people are – or where to find them. Moreover, you cannot tell them that you want to buy such a large quantity (for fear they will try to negotiate a much higher price), and they cannot tell you that they want to sell such a large quantity (for fear you will try to negotiate a much lower price).[11] This is precisely the **buyer's conundrum** illustrated by the quote at the beginning of Chapter 1.

If 0.5% of the market cap is traded over the course of a whole day, then roughly 0.005% is traded every five minutes (neglecting intra-day activity patterns and the opening auction and closing auction). This is roughly the volume typically available at the best bid or ask quote for small-tick stocks; the corresponding volume for large-tick stocks is somewhat larger, but only by a factor of about 10. Sending a market order whose volume is much larger than these numbers is clearly a bad idea, because it would mean paying a much worse price than the ask-price, and could possibly wreak havoc in the market (remember that your desired metaorder size is between 20 and 200 times larger than the volume available at the ask!). Similarly, sending a very rapid succession of smaller market orders would presumably send a strong signal that there is a hurried buyer in the market, and would likely cause many sellers to increase their prices.

[11] In the old days, traders seeking to perform a large buy (respectively, sell) trade often commissioned a broker to find sellers (respectively, buyers). Such traders hoped that the broker was smart enough to get a good price while hiding their true intentions to trade. This was often a long, drawn-out and possibly costly process, with little transparency. However, even when using a broker, information leakage remained a costly consideration.

Rather than removing liquidity, how about instead contributing to liquidity by posting a very large limit order at or close to the bid-price? Unfortunately, this does not work either, because observing an unusually large limit order also signals a large buying interest. This influences both buyers (who are now tempted to buy at the ask-price rather than hoping to achieve a better price by placing their own limit orders) and sellers (who think that it might be a bad idea to sell now if the price is likely to go up, as suggested by the new limit order arrival). In fact, limit orders – which are often described as "passive" because they provide liquidity – can impact prices considerably.

The so-called **flash crash** of 6 May 2010 provides an example of the possible dangers of submitting unusually large limit orders. On this day, an asset manager decided to sell 75,000 S&P E-mini contracts (representing about 8% of the typical daily volume) using several huge sell limit orders.[12] The appearance of such an enormous volume in an already agitated market created a sudden drop of liquidity on the buy-side of the LOB and a rapid decline of prices that reached −9% after a few minutes, while also spilling over to equity markets. This is an extreme but vivid example of how signalling a desire to buy or sell a large quantity can impact and possibly destabilise prices, even with a purely passive behaviour. One clearly sees how self-reinforcing liquidity droughts can appear in financial markets. In other words, even limit orders that do not directly remove any liquidity from the market can indirectly generate a **liquidity crisis**.

In summary, as a buyer who seeks to purchase a very large quantity of an asset, you have no other realistic choice than to split your desired trade into many small pieces and execute them incrementally, over a period which might span several days or even months. Intuitively, the probability of revealing information increases with the size of a (limit or market) order, because smaller orders are more likely to go unnoticed while larger orders are more likely to attract attention. Therefore, the **information leakage cost** of an order is expected to increase with its volume (relative to the available liquidity). Of course, the same arguments hold for sellers as well. In both cases, these actions are consistent with the idea that market participants' execution of large metaorders can cause long-range autocorrelations in the observed order flow.

10.5.2 Metaorder Execution

In practice, it is difficult to manually execute a large metaorder over a long time period in an efficient way. Therefore, execution algorithms are now routinely used to perform this task. These algorithms are either built in-house by asset managers,

[12] See: Kirilenko, A. A., Kyle, A. S., Samadi, M., & Tuzun, T. (2017). The flash crash: High frequency trading in an electronic market. *The Journal of Finance*. doi:10.1111/jofi.12498.

or proposed by brokers who sell execution as a service (with a fee!). We now provide a brief description of what these algorithms or brokers attempt to achieve. We also return to this question, in the context of optimal execution, in Chapter 21.

Some common execution schedules are:

- **The time-weighted average price (TWAP) benchmark:** TWAP execution aims to achieve an average execution price that is as close as possible to the time-weighted average price available in the market during a specified period (typically one trading day).
- **The volume-weighted average price (VWAP) benchmark:** VWAP execution aims to achieve an average execution price that is as close as possible to the volume-weighted average price available in the market during a specified period (typically one trading day).
- **The Almgren–Chriss optimal schedule:** This algorithm aims to find an execution strategy that minimises a combination of trading costs and the variance of the difference between the execution price and a given reference price (such as the price at the open). We discuss this algorithm in more detail in Chapter 21.

Intuitively, VWAP reflects that large volumes should be more representative of the fair price paid during the day, whereas TWAP is insensitive to the size of the trades. Note that the TWAP and VWAP benchmarks are somewhat misleading since they hide the impact of the executed metaorder: in the limit of a very large metaorder dominating the market, its average execution price is very close to the VWAP, since the VWAP is computed using the trades of the metaorder itself. This misleadingly suggests a high-quality execution, when these benchmark prices themselves are adversely impacted by the metaorder and actually quite far from the decision price (such as the price at the beginning of the day). The Almgren–Chriss algorithm attempts to correct this drawback.

One can also devise more sophisticated algorithms that take into account the local liquidity fluctuations or short-term predictability in the price, or even the simultaneous execution of different metaorders on different instruments. In any case, the important conclusion is that all such execution algorithms slice metaorders into small pieces that are executed incrementally, either as market orders or as limit orders, and thereby result in a long-range autocorrelated order flow.

10.6 Conclusion

The long memory of market order signs is a striking stylised fact in market microstructure. At first sight, the effect is extremely puzzling, because it appears to contradict the near-absence of predictability in price series. How can it be that one can make good predictions of the sign of a market order far in the future without being able to predict that the price will increase or decrease over the same time horizon? As we have discussed in this chapter, it must be the case that the market somehow reacts to the correlated order flow in such a way that the price becomes (approximately) *statistically efficient*. We will return to our discussion of this *efficiency paradox* several times in the coming chapters.

In this chapter, we have also summarised evidence to support that the long memory of order flow is a consequence of metaorder-splitting. Even in so-called "liquid" markets, such as US large-cap stocks, investors are not faced with plentiful liquidity. As we have argued, the volume available in the LOB is only a very small fraction of the total volume desired for trade in the market. Therefore, the only sensible possibility for market participants who wish to execute large trades is to slice and dice their desired metaorders into small quantities, which they execute incrementally over long periods of time.

An important conclusion is that at any instant of time, there are huge chunks of metaorders that still await execution. At odds with Walras' picture of price formation, where the price instantaneously clears supply and demand (see Chapter 18), markets in fact only slowly resolve the imbalance between buyers and sellers. Therefore, the revealed liquidity in an LOB far from illuminates the true buying and selling intentions in the market. Instead, most of the liquidity remains latent, as we will explore in detail in Chapter 18.

Take-Home Messages

(i) The signs of arriving market orders have long-range autocorrelations. This makes the signs of future market orders predictable, which seems to be at odds with the (nearly) uncorrelated nature of price returns.

(ii) Empirical studies suggest that the main cause of these autocorrelations is single investors splitting large metaorders, rather than different investors herding.

(iii) Traders seek to minimise the information leakage that occurs when they make their trading intentions public. Therefore, the liquidity in an LOB is typically much smaller than the sum of all latent trading intentions.

(iv) Due to this shortage of liquidity, investors cannot execute large orders quickly without destabilising the market, and therefore need to split them using execution strategies such as VWAP, TWAP and Almgren–Chriss execution.

10.7 Further Reading

Long-Memory Processes, DAR(p) and Other Models

Jacobs, P. A., & Lewis, P. A. (1978). Discrete time series generated by mixtures. I: Correlational and runs properties. *Journal of the Royal Statistical Society*. Series B (Methodological), 40, 94–105.

Jacobs, P. A., & Lewis, P. A. (1978). Discrete time series generated by mixtures. III. Autoregressive processes (DAR (p)). *Naval Postgraduate School Technical Report* (Monterey, CA).

Jacobs, P. A., & Lewis, P. A. (1983). Stationary discrete autoregressive-moving average time series generated by mixtures. *Journal of Time Series Analysis*, 4(1), 19–36.

Raftery, A. E. (1985). A model for high-order Markov chains. *Journal of the Royal Statistical Society*. Series B (Methodological), 47, 528–539.

Beran, J. (1994). *Statistics for long-memory processes*. Chapman & Hall.

Berchtold, A., & Raftery, A. E. (2002). The mixture transition distribution model for high-order Markov chains and non-Gaussian time series. *Statistical Science*, 17, 328–356.

Lillo, F., Mike, S., & Farmer, J. D. (2005). Theory for long memory in supply and demand. *Physical Review E*, 71(6), 066122.

Taranto, D. E., Bormetti, G., Bouchaud, J. P., Lillo, F., & Tóth, B. (2016). Linear models for the impact of order flow on prices II. The Mixture Transition Distribution model. https://ssrn.com/abstract=2770363.

Long-Range Correlation of Order Flow

Bouchaud, J. P., Gefen, Y., Potters, M., & Wyart, M. (2004). Fluctuations and response in financial markets: The subtle nature of random price changes. *Quantitative Finance*, 4(2), 176–190.

Lillo, F., & Farmer, J. D. (2004). The long memory of the efficient market. *Studies in Nonlinear Dynamics & Econometrics*, 8(3), 1.

Bouchaud, J. P., Kockelkoren, J., & Potters, M. (2006). Random walks, liquidity molasses and critical response in financial markets. *Quantitative Finance*, 6(02), 115–123.

Bouchaud, J. P., Farmer, J. D., & Lillo, F. (2009). How markets slowly digest changes in supply and demand. In Hens, T. & Schenk-Hoppe, K. R. (Eds.), *Handbook of financial markets: Dynamics and evolution*. North-Holland, Elsevier.

Yamamoto, R., & Lebaron, B. (2010). Order-splitting and long-memory in an order-driven market. *The European Physical Journal B-Condensed Matter and Complex Systems*, 73(1), 51–57.

Tóth, B., Palit, I., Lillo, F., & Farmer, J. D. (2015). Why is equity order flow so persistent? *Journal of Economic Dynamics and Control*, 51, 218–239.

Taranto, D. E., Bormetti, G., Bouchaud, J. P., Lillo, F., & Tóth, B. (2016). Linear models for the impact of order flow on prices I. Propagators: Transient vs. history dependent impact. https://ssrn.com/abstract=2770352.

Large Institutional Trades

Chan, L. K., & Lakonishok, J. (1993). Institutional trades and intra-day stock price behavior. *Journal of Financial Economics*, 33(2), 173–199.

Chan, L. K., & Lakonishok, J. (1995). The behavior of stock prices around institutional trades. *The Journal of Finance*, 50(4), 1147–1174.

Gabaix, X., Ramalho, R., & Reuter, J. (2003). *Power laws and mutual fund dynamics*. MIT mimeo.

Gabaix, X., Gopikrishnan, P., Plerou, V., & Stanley, H. E. (2006). Institutional investors and stock market volatility. *The Quarterly Journal of Economics*, 121(2), 461–504.

Vaglica, G., Lillo, F., Moro, E., & Mantegna, R. (2008). Scaling laws of strategic behavior and size heterogeneity in agent dynamics. *Physical Review E*, 77, 0036110.

Schwarzkopf, Y., & Farmer, J. D. (2010). Empirical study of the tails of mutual fund size. *Physical Review E*, 81(6), 066113.

Bershova, N., & Rakhlin, D. (2013). The non-linear market impact of large trades: Evidence from buy-side order flow. *Quantitative Finance*, 13(11), 1759–1778.

Donier, J., & Bonart, J. (2015). A million metaorder analysis of market impact on the Bitcoin. *Market Microstructure and Liquidity*, 1(02), 1550008.

Kyle, A. S., & Obizhaeva, A. A. (2016). Large bets and stock market crashes. https://ssrn.com/abstract=2023776.

Kirilenko, A., Kyle, A. S., Samadi, M., & Tuzun, T. (2017). The flash crash: High frequency trading in an electronic market. *The Journal of Finance*, 72, 967–998.

PART V
Price Impact

Introduction

In this part, we jump into the core topic of this book: understanding price dynamics. More specifically, we will consider the important question of how order flows interact to form a price. In particular, how can the *predictable* order flow elicited in the previous chapter produce *unpredictable* prices? This question will be our core motivation throughout the remainder of the book.

As ever, we start with two fundamentals: concepts and data. We begin by zooming out and asking questions about how market participants' actions impact prices, and, more broadly, the state of the market. We are immediately faced with a difficulty: since history cannot be replayed, how can we know how an action *really* affected the market? Put another way, how can we know that what we observed was not going to happen anyway? This conundrum leads us to the necessity of decomposing impact into two components: the *reaction impact,* which describes how an action directly affects future prices, and the *prediction impact,* which describes what would have happened anyway, even in the absence of the action. The sum of reaction impact and prediction impact will correspond to the *observed impact,* as measurable by an outside observer.

Because it does not require any information about the circumstances in which trades occurred, observed impact often seems to be a natural quantity of study for an outside observer who only has access to public data. However, upon deeper reflection, observed impact reveals itself to be rather non-trivial. On the one hand, some aspects behave as we would expect. For example, buy market orders push the price up on average, sell market orders push the price down on average, and observed impact first increases rapidly before reaching a stable plateau. On the other hand, other aspects are much more surprising. One prominent example is that impact scales as a concave function of order size, such that the seemingly natural idea that this relationship should be linear actually breaks down.

In the quest for reliable measurements of impact, private data is a prerequisite. By allowing a deconvolution of the prediction impact at the scale of the metaorder – and thus getting rid of most spurious conditioning effects – private data offers a reliable way to estimate how order flow dynamically affects the market. As we will discuss, this understanding provides a solid starting point for price-formation models, which indicate that prices move with the *square root* of metaorder volumes.

This widely reported empirical phenomenon is extremely surprising, and holds the key to unlocking several mysteries surrounding order flow and price formation. Does such impact remain forever imprinted on the market and the price? What does it mean for investors' trading strategies? And how does it relate to the idea of market efficiency? This part opens these and many other questions, which the remainder of this book will seek to address.

11
The Impact of Market Orders

To measure is to know. (Lord Kelvin)

In this chapter, we address the seemingly obvious notion of **price impact** (which we first discussed in Section 1.3.2): buy trades tend to push the price up and sell trades tend to push the price down. Expressed in the notation and language that we have subsequently developed, we might also express this notion by saying that price impact refers to the positive correlation between the sign ($+1$ for a buy order and -1 for a sell order) of an incoming market order and the subsequent price change that occurs upon or after its arrival. As we discuss in Section 11.3, these seemingly obvious statements are indeed verified by empirical data.

Price impact is an all-too-familiar reality for traders who need to buy or sell large quantities of an asset. To these traders, price impact is tantamount to a cost, because the impact of their earlier trades makes the price of their subsequent trades worse on average. Therefore, monitoring and controlling impact costs is one of the most active and rapidly expanding domains of research in both academic circles and trading firms.

Understanding and assessing price impact entails considering two different but related topics. The first is how *volume* creates impact: how much more impact does a larger trade cause? The second is the *temporal behaviour* of impact: how much of a trade's impact is permanent, how much decays over time, and how does this transient behaviour unfold? We will consider both of these topics in detail throughout this chapter. As we will discuss, both of these aspects of price impact are far from trivial. Is a transaction not a fair deal between a buyer and a seller? If so, which of the two is really impacting the price?

11.1 What Is Price Impact?

In much of the existing literature, there are two strands of interpretation for price impact, which reflect the great divide between efficient-market enthusiasts (who believe that the price is always close to its fundamental value) and sceptics (who believe that the dynamics of financial markets is primarily governed by order flow). At the two extremes of this spectrum are the following stories:

(i) *Agents successfully forecast short-term price movements, and trade accordingly.* This is the **efficient-market point of view**, which asserts that a trader who believes that the price is likely to rise will buy in anticipation of this price move. This clearly results in a positive correlation between the sign of the trade and the subsequent price change(s), even if the trade by itself has no effect on prices. In this framework, a noise-induced trade that is based on no information at all should have no long-term impact on prices – otherwise, prices could end up straying very far from their fundamental values, which cannot be the case if markets are efficient. By this interpretation, if the price was meant to move due to information, it would do so even *without* any trades.

(ii) *Price impact is a reaction to order-flow imbalance.* This is the **efficient-market sceptic view**, which asserts that the fundamental value is irrelevant, at least on short time scales, and that even if a trade reflected no information in any reasonable sense, then price impact would still occur.
As an illustration of this viewpoint, recall the Santa Fe model (see Chapter 8), in which all order-flow events are described by independent, homogeneous Poisson processes. All else being held constant, then the mid-price will on average be higher (respectively, lower) conditional on the arrival of an extra buy (respectively, sell) market order than it would be conditional on that market order not arriving. This effect is readily apparent in Figure 11.2, where we plot the mean impact of a buy trade in the Santa Fe model. Clearly, there is a well-defined and measurable price impact, even though there is no notion of fundamental price or information in this zero-intelligence model.

Although both of the above explanations result in a positive correlation between trade signs and price movements, they are conceptually very different.[1] In the first story, as emphasised by J. Hasbrouck, *"orders do not impact prices. It is more accurate to say that orders forecast prices."*[2] Put another way, trades reveal private information about the fundamental value, creating a so-called **price discovery** process. In the second story, the act of trading itself impacts the price. In this case, one should remain agnostic about the information content of the trades, and should therefore speak of **price formation** rather than price discovery. If market

[1] On this point, see Lyons, R. (2001). *The microstructure approach to exchange rates.* MIT Press. Lyons writes: *Consider an example that clarifies how economist and practitioner worldviews differ. The example is the timeworn reasoning used by practitioners to account for price movements. In the case of a price increase, practitioners will assert "there were more buyers than sellers". Like other economists, I smile when I hear this. I smile because in my mind the expression is tantamount to the "price had to rise to balance demand and supply".*

[2] Hasbrouck, J. (2007). *Empirical market microstructure.* Oxford University Press.

participants believe that the newly established price is the "right" price and act accordingly, "information revelation" might simply be a self-fulfilling prophecy.

As mentioned above, the Santa Fe model (see Chapter 8) provides an illustration of the second story. In this model, the mechanism that generates impact can be traced back to the modelling assumption that at any given time, agents submitting orders always use the current mid-price as a reference. Any upwards (respectively downwards) change in mid-price therefore biases the subsequent order flow in an upwards (respectively downwards) direction. This causes the model to produce a diffusive mid-price in the long run, but only resulting from the permanent impact of a purely random order flow, in a purely random market.

Whether prices are formed or discovered remains a topic of much debate. At this stage, there is no definitive answer, but because trades in modern markets are anonymous, and because the line between real information and noise is so blurry, reality probably lies somewhere between these two extremes. Since some trades may contain real private information, and since other market participants do not know which trades do and do not contain such information, it follows that all trades must (on average) impact the price, at least temporarily. The question of how much real information is revealed by trades is obviously crucial in determining whether markets are closer to the first picture or the second picture. Several empirical results suggest that the impact of random trades is similar to that of putative informed trades (at least on the short run), and that the amount of information per trade is extremely small (see the discussions in Chapters 13, 16 and 20).

11.2 Observed Impact, Reaction Impact and Prediction Impact

From a scientific (but slightly ethereal) point of view, one would ideally like to assess the impact of a market order by somehow measuring the difference between the mid-price in a world where the order is executed and the mid-price in a world where all else is equal but where the given order is not executed. For a buy market order, for example,

$$\mathcal{I}^{\text{react.}}_{t+\ell}(\text{exec}_t \,|\, \mathcal{F}_t) := \mathbb{E}[m_{t+\ell} \,|\, \text{exec}_t, \mathcal{F}_t] - \mathbb{E}[m_{t+\ell} \,|\, \text{no exec}_t, \mathcal{F}_t], \tag{11.1}$$

where "exec_t" and "no exec_t" denote, respectively, the execution or non-execution of the market order at time t. We call this quantity the **reaction impact**, because it seeks to quantify how the market price reacts to the arrival of a given order. In this formulation, $\mathcal{F}_t$ represents the state of the world at time t. In particular, $\mathcal{F}_t$ contains all information that may have triggered the given buy order, but not whether the trade is executed or not. To aid readability, we will sometimes omit the conditioning on $\mathcal{F}_t$, but it is always implicitly present in our arguments.

The definition of reaction impact in Equation (11.1) is close in spirit to what natural scientists would like to consider: an experiment where the system is perturbed in a controlled manner, such that the result of that perturbation can be cleanly observed and quantified. Unfortunately, this definition cannot be implemented in a real financial system, because the two situations (i.e. the market order arriving or not arriving) are mutually exclusive, and history cannot be replayed to repeat the experiment in the very same conditions. Instead, what can be measured in a real financial market is the **observed impact**:

$$\mathcal{I}^{\text{obs.}}_{t+\ell}(\text{exec}_t) := \mathbb{E}[m_{t+\ell} \mid \text{exec}_t] - m_t, \tag{11.2}$$

where m_t is the observed mid-price just before the execution occurred.

Most studies of impact focus on measuring and studying observed impact, which is readily available ex-post in a given data set. If prices were martingales, then it would follow that $\mathbb{E}[m_{t+\ell} \mid \text{no exec}_t, \mathcal{F}_t]$ is equal to m_t, so observed impact would be precisely equal to reaction impact. In real markets, however, this equality does not hold (because $\mathcal{F}_t$ contains the information available to the trader, so one should expect the price to increase on average even in the absence of his or her trade, as the prediction motivating the trade is revealed). The amount of information contained in $\mathcal{F}_t$ can be written as:

$$\mathcal{I}^{\text{pred.}}_{t+\ell} := \mathbb{E}[m_{t+\ell} \mid \text{no exec}_t, \mathcal{F}_t] - m_t. \tag{11.3}$$

We call this quantity the **prediction impact**, in the spirit of Hasbrouck's view (recalled above).[3]

By Equations (11.1) and (11.2), the difference between observed impact and reaction impact is given by prediction impact:

$$\begin{aligned}\mathcal{I}^{\text{obs.}}_{t+\ell}(\text{exec}_t) - \mathcal{I}^{\text{react.}}_{t+\ell}(\text{exec}_t) &= \mathbb{E}[m_{t+\ell} \mid \text{exec}_t] - m_t - \mathbb{E}[m_{t+\ell} \mid \text{exec}_t] \\ &\quad + \mathbb{E}[m_{t+\ell} \mid \text{no exec}_t], \\ &= \mathbb{E}[m_{t+\ell} \mid \text{no exec}_t] - m_t.\end{aligned}$$

To recap: we have introduced three types of impact – namely, observed impact, reaction impact, and prediction impact – which are related by the equality:

$$\mathcal{I}^{\text{obs.}} = \mathcal{I}^{\text{react.}} + \mathcal{I}^{\text{pred.}}. \tag{11.4}$$

Prediction impact is very difficult to estimate empirically, because the full information set $\mathcal{F}_t$ used by market participants to predict future prices is extremely

[3] Note that we prefer here the term "prediction" to the term "information", to avoid any confusion with "fundamental information". The latter term suggests some knowledge of the fundamental price of the asset, whereas we prefer the agnostic view that some market participants successfully predict the future evolution of prices, whether or not this is justified by fundamentals. For an extended discussion on this point, see Chapter 20.

large and difficult to quantify. Reaction impact is somewhat easier to estimate, via one of the following methods:

(i) by performing experiments where trading decisions are drawn at random, such that $\mathcal{I}^{\text{pred.}}_{t+\ell} = 0$ by construction (up to statistical noise);
(ii) by choosing at random whether or not to execute an order with a given prediction signal (so as to measure both $\mathbb{E}[m_{t+\ell} \mid \text{exec}_t, \mathcal{F}_t]$ and $\mathbb{E}[m_{t+\ell} \mid \text{no exec}_t, \mathcal{F}_t]$ for the same strategy but at different times t), then subtracting the latter from the former;
(iii) ex-post, by conditioning on the order-sign imbalance of the rest of the market between t and $t + \ell$, and using this imbalance as a proxy for the presence of informed trading, i.e. for whether $\mathcal{I}^{\text{pred.}}$ is zero or not.[4]

None of these approaches are perfect. To measure anything meaningful, the first idea requires generating a large number of random trades, which is a costly and time-consuming experiment! A handful of studies have attempted to determine $\mathcal{I}^{\text{react.}}$ empirically, either by actually performing random trades or by carefully mining existing data sets to identify specific orders for which $\mathcal{I}^{\text{pred.}}$ can be regarded to be zero (such as trades initiated for cash-inventory purposes only). These studies all conclude that *on short time scales*, the mechanical impact estimated from random trades is to a good approximation identical to the mechanical impact estimated from proprietary (allegedly informed) trades, or from all trades in a given data set.[5] This shows that the prediction component $\mathcal{I}^{\text{pred.}}$ (if any) is only expected to show up at longer times, when the prediction signal that initiated the trade is realised (see Chapter 20).

11.3 The Lag-1 Impact of Market Orders

Although all types of order-flow events (market order arrivals, limit order arrivals, and cancellations) can impact prices, it is conceptually and operationally simpler to first study only the impact of market orders (see Chapter 14 for an extended discussion of the other events). One reason for doing so is that this analysis requires only trades-and-quotes data (i.e. the time series of bid-prices b_t, ask-prices a_t and trade prices p_t). Because we consider only market orders, in the following we count time t in market-order time, in which we increment t by 1 for each market

[4] This was suggested in Donier, J., & Bonart, J. (2015). A million metaorder analysis of market impact on the Bitcoin. *Market Microstructure and Liquidity*, 1(02), 1550008.

[5] See Section 14.5.2 and, e.g., Gomes, C., & Waelbroeck, H. (2015). Is market impact a measure of the information value of trades? Market response to liquidity vs. informed metaorders. *Quantitative Finance*, 15(5), 773–793, and Tóth, B., Eisler, Z., & Bouchaud, J.-P. (2017). The short-term price impact of trades is universal. https://ssrn.com/abstract=2924029.

order arrival, and in which b_t and a_t denote the values of the bid- and ask-prices immediately *before* the arrival of the tth market order.

In high-quality trades-and-quotes data, there is an exact match between the transaction price p_t and either the bid- or the ask-price at the same time. If $p_t = a_t$, then the transaction is due to an incoming buy market order (which we label with the order sign $\varepsilon_t = +1$); if $p_t = b_t$, then the transaction is due to an incoming sell market order (which we label with the order sign $\varepsilon_t = -1$). Each entry in a trades-and-quotes data set also specifies the trade volume υ_t. If an incoming buy (respectively, sell) market order's size does not exceed the volume at the best ask (respectively, bid) quote, then υ_t is the full size of the incoming market order. If the volume of the incoming market order exceeds the volume at the ask (respectively, bid) quote, further transactions will occur at higher (respectively, lower) prices, within the limits of the market order volume and price. Any unmatched part of the market order will remain as a limit order in the LOB, at the price at which it was sent. Trades-and-quotes data sets typically report such activity as different, successive transactions with identical or very similar time stamps.

11.3.1 Unconditional Impact

The simplest measure of price impact is the mean difference between the mid-price just before the arrival of a given market order and the mid-price just before the arrival of the next market order.[6] To align activity for buy and sell market orders, our general definition of impact must also incorporate the order sign ε_t. Recalling that m_t denotes the mid-price immediately before the arrival of the tth market order, we define the **lag-1 unconditional impact** as

$$\mathcal{R}(1) := \langle \varepsilon_t \cdot (m_{t+1} - m_t) \rangle_t, \tag{11.5}$$

where the empirical average $\langle \cdot \rangle_t$ is taken over all market orders regardless of their volume and regardless of the state of the world (including the LOB) just before the transaction. We could of course perform more precise measurements of price impact in specific situations by also conditioning on extra variables, but for now we consider the general definition in Equation (11.5).

Table 11.1 lists several statistics related to price impact. For all stocks, it is clear that $\mathcal{R}(1) > 0$ with strong statistical significance.[7] This demonstrates that

[6] The choice of the mid-price m_t as the relevant reference price is the simplest, but is not necessarily the most adequate. For large-tick stocks, in particular, we have seen in Section 7.2 that the volume imbalance I is a strong predictor of the sign of the future price change, so a better reference price could be defined as $\widetilde{m}_t = p_+(I)a_t + p_-(I)b_t$. Throughout the book, however, we stick with the mid-price m_t.

[7] Indeed, from the last column of Table 11.1, the total number of events used to compute $\mathcal{R}(1)$ is of the order of 10^6, which leads to a relative standard error of less than 1%.

Table 11.1. *The average spread just before a market order, $\langle s \rangle$; the lag-1 response functions, $\mathcal{R}(1)$ (all market orders), $\mathcal{R}^1(1)$ (price-changing market orders) and $\mathcal{R}^0(1)$ (non-price-changing market orders); the standard deviation of price fluctuations around the average price impact of a market order, $\Sigma_{\mathcal{R}} = \sqrt{\mathcal{V}(1) - \mathcal{R}(1)^2}$, all measured in dollar cents; the fraction of market orders that immediately change the price, $P[\mathrm{MO}^1]$; and the number of market orders observed between 10:30 and 15:00 during each trading day, N_{MO}, for 10 small- and large-tick stocks during 2015.*

	$\langle s \rangle$	$\mathcal{R}(1)$	$\mathcal{R}^1(1)$	$\mathcal{R}^0(1)$	$\Sigma_{\mathcal{R}}$	$P[\mathrm{MO}^1]$	N_{MO}
SIRI	1.06	0.058	0.516	0.006	0.213	0.112	623
INTC	1.08	0.246	0.769	0.029	0.422	0.293	4395
CSCO	1.09	0.206	0.735	0.022	0.386	0.256	3123
MSFT	1.09	0.276	0.769	0.039	0.441	0.322	7081
EBAY	1.10	0.348	0.745	0.059	0.502	0.419	3575
FB	1.21	0.481	0.818	0.124	0.674	0.514	10703
TSLA	12.99	2.59	3.79	0.403	4.49	0.649	3932
AMZN	21.05	3.63	5.57	0.597	6.38	0.618	4411
GOOG	26.37	3.97	6.65	0.644	7.62	0.557	3710
PCLN	94.68	15.30	24.97	2.17	28.77	0.579	1342

a buy (respectively, sell) market order is on average followed by an immediate increase (respectively, decrease) in m_t. We also point out three other interesting observations:

(i) For small-tick stocks, the value of $\mathcal{R}(1)$ is proportional to the mean spread $\langle s \rangle_t$ (see also Figure 11.1). In other words, the scale of mid-price changes induced by market order arrivals is of the same order as the bid–ask spread s_t. This turns out to capture a profound truth that we already alluded to in Section 1.3.2, and on which we will expand in Chapter 17 below. In a nutshell, this linear relation follows from the argument that market-making strategies must be roughly break-even: market-makers attempt to earn the bid–ask spread s, but face impact costs due to the adverse price move after a market order. The relation $\mathcal{R}(1) \propto \langle s \rangle_t$ means that, to a first approximation, adverse selection is compensated by the spread (see Chapters 16 and 17 for an extended discussion of this point).

(ii) For large-tick stocks, the average spread is bounded below by one tick, so the linear relationship saturates. However, $\mathcal{R}(1)$ itself is not bounded, because the proportion of trades that result in a one-tick price change may become arbitrarily small, resulting in a small average impact. In this situation, the market-making problem becomes more subtle and requires study of the full queuing systems (see Chapter 17).

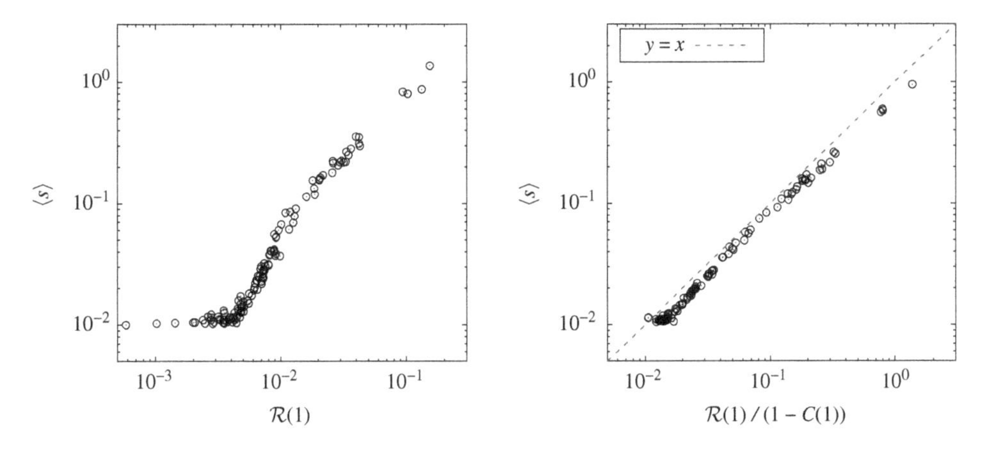

Figure 11.1. (Left panel) Average spread $\langle s \rangle$ versus the lag-1 impact $\mathcal{R}(1)$, for 120 stocks traded on NASDAQ, and (right panel) similar plot but with $\mathcal{R}(1)$ adjusted for the lag-1 correlation $C(1)$ in the market order flow (see Chapter 17 for a theoretical justification).

(iii) There is a substantial amount of noise around the mean impact $\mathcal{R}(1)$. One way to measure this dispersion is to calculate the lag-1 variogram of the mid-price:

$$\Sigma_{\mathcal{R}} := \mathcal{V}(1) - [\mathcal{R}(1)]^2; \qquad \mathcal{V}(1) = \langle (m_{t+1} - m_t)^2 \rangle.$$

As is clear from Table 11.1, the magnitude of the fluctuations is larger than the mean impact itself. As noted in (ii) above, some market orders do not change the price at all, others trigger large cancellations and hence have a very large impact, and some are even followed by a price change in the opposite direction! This highlights that $\mathcal{R}(1)$ does not simply measure the simple mechanical effect of the market order arrival, but instead incorporates the full sequence of other limit order arrivals and cancellations that occur between two successive market order arrivals.

There is no reason to limit our definition of price impact to the lag-1 case. For any $\ell > 0$, we can easily extend the definition from Equation (11.5) to the general case:

$$\mathcal{R}(\ell) := \langle \varepsilon_t \cdot (m_{t+\ell} - m_t) \rangle_t. \tag{11.6}$$

The function $\mathcal{R}(\cdot)$ is called the **response function**.

Figure 11.2 shows the shape of the response function for four stocks in our sample. In each case, $\mathcal{R}(\ell)$ rises from an initial value $\mathcal{R}(1)$ to a larger value $\mathcal{R}_\infty = \mathcal{R}(\ell \to \infty)$, which is 2–5 times larger than the initial response $\mathcal{R}(1)$. This occurs as a result of the autocorrelations in market order signs, which tend to push the price in the same direction for a while (see Section 13.2.1). This illustrates an important point, which we will return to in the next chapter: one should not confuse the response function with the mechanical impact of an isolated random trade. In the

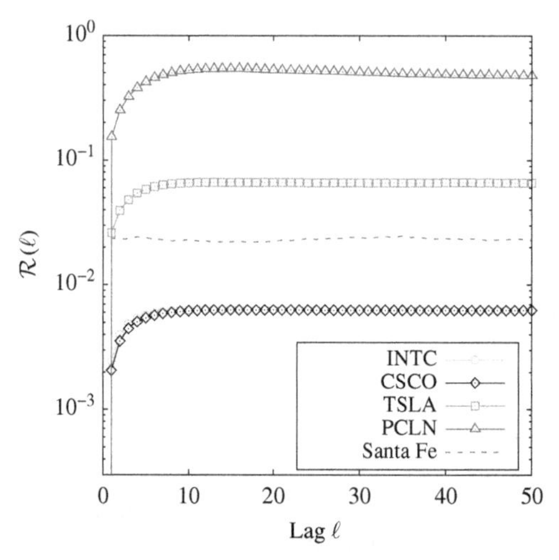

Figure 11.2. Response function $\mathcal{R}(\ell)$ in market order event-time, for (circles) INTC, (diamonds) CSCO, (squares) TSLA, (triangles) PCLN and (dashed curve) in the simulated Santa Fe model. The parameters of the Santa Fe model correspond to TSLA.

language of the previous section, there is an information contribution $\mathcal{I}^{\text{pred.}}$ that comes from the autocorrelation of the market order signs. Put another way, the response function $\mathcal{R}(\ell)$ also contains the reaction impact of future trades, which, as we saw in Chapter 10, are correlated with the present trade.

11.3.2 Conditioning on Trade Volume

So far, we have considered the impact of a market order irrespective of its volume. However, it seems natural that large market orders should somehow impact prices more than small market orders. As we discussed in Section 10.5.1, one possible reason that this should be the case is information leakage: if market orders reveal information, larger trades may indeed lead to larger subsequent price moves. Another possible reason is the purely statistical observation that a larger market order is more likely to consume all the available volume at the opposite-side best quote, and is therefore more likely to lead to a price change both instantaneously (by directly changing the state of the LOB) and subsequently (by causing other traders to modify their subsequent order flow).

Both of these intuitive arguments are indeed confirmed by empirical data. However, the effect of volume on impact is much weaker than might be naively anticipated. In fact, after normalising the volume of a market order υ by the mean volume at the opposite-side best quote $\bar{V}_{\text{best}}$, one finds that the volume-dependence

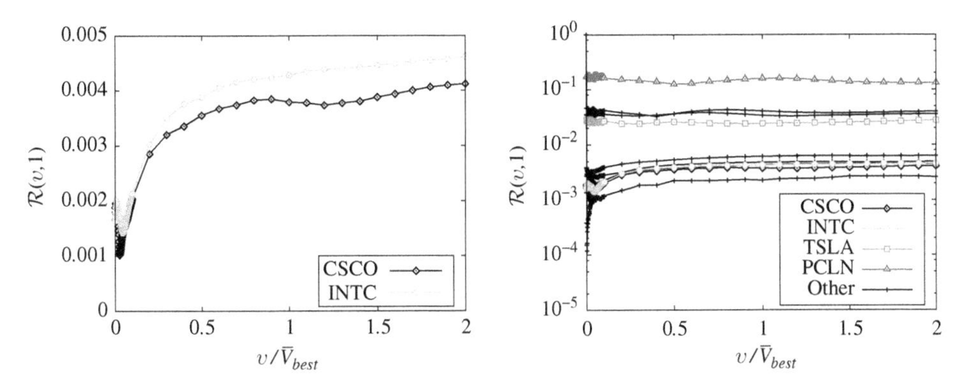

Figure 11.3. (Left panel) Lag-1 impact response function $\mathcal{R}(v,1)$ for INTC and CSCO, as a function of normalised market order volume (where we normalise by the average volume at the same-side queue). Observe the non-monotonic effect for small v. (Right panel) The same plot on a semi-logarithmic scale, for the ten stocks in Table 11.1. Small-tick stocks (with a large $\mathcal{R}(v,1)$) do not exhibit any significant dependence on v.

of the immediate impact is actually strongly sub-linear:

$$\mathcal{R}(v,1) := \langle \varepsilon_t \cdot (m_{t+1} - m_t) | v_t = v \rangle_t \cong A \left(\frac{v}{\bar{V}_{\text{best}}} \right)^{\zeta} \langle s \rangle_t, \qquad (11.7)$$

where A is a constant of order 1 and ζ is an exponent that takes very small values. For the stocks in our sample, we find values $\zeta \cong 0 - 0.3$ (see Figure 11.3). In other words, the lag-1 impact of a single market order is a strongly concave function of its volume, and perhaps even a constant for small-tick stocks. As we discuss in the next section, this concavity mostly comes from a conditioning bias called selective liquidity taking.

Note a curious feature for large-tick stocks: the impact curve is non-monotonic for small volumes, such that small market orders seem to have an anomalously high impact. This is related to volume imbalance effects, as noted in Section 7.2: when the volume in one of the two queues is very small compared to the volume in the opposite queue, it is highly probable that a small market order will grab the small remaining quantity, resulting in a one-tick price jump.

11.3.3 Selective Liquidity Taking

Whether or not a given market order has instantaneous impact depends not only on the size of the market order, but also on the state of the LOB at its time of arrival. Therefore, attempting to analyse impact only as a function of market order size could be misleading. At the very least, one should distinguish between **aggressive market orders** and **non-aggressive market orders**.

Recall from Section 10.2 that we use the notation MO^1 to denote market orders that consume all volume available at the best opposite quote, which leads to an *immediate* price move, and MO^0 to denote market orders that do not. We introduce a similar notation for impact:

$$\mathcal{R}^1(1) := \langle \varepsilon_t \cdot (m_{t+1} - m_t) | \pi_t = \mathrm{MO}^1 \rangle_t,$$
$$\mathcal{R}^0(1) := \langle \varepsilon_t \cdot (m_{t+1} - m_t) | \pi_t = \mathrm{MO}^0 \rangle_t.$$

Table 11.1 lists the empirical values of $\mathcal{R}^1(1)$ and $\mathcal{R}^0(1)$ for the stocks in our sample. As might be expected, $\mathcal{R}^1(1) > \mathcal{R}^0(1)$, but note that even for MO^0 events, the response of the market is strictly positive: $\mathcal{R}^0(1) > 0$. This is due to the fact that there is a non-zero probability for the non-executed limit orders at the opposite-side best quote to be cancelled before the next market order arrival, and thereby to produce a price change in the direction of the initial trade. Clearly, one has

$$\mathcal{R}(1) = \mathbb{P}[\mathrm{MO}^0]\mathcal{R}^0(1) + \mathbb{P}[\mathrm{MO}^1]\mathcal{R}^1(1),$$

where $\mathbb{P}[\mathrm{MO}^1]$ is the probability that the market order is aggressive and $\mathbb{P}[\mathrm{MO}^0]$ is the probability that it is not (such that $\mathbb{P}[\mathrm{MO}^1] = 1 - \mathbb{P}[\mathrm{MO}^0]$).

It is interesting to study the distribution of market order volumes, conditioned to the volume at the opposite-side best quote V_{best} at their time of arrival.[8] Figure 11.4 shows the mean size of market order arrivals for given values of V_{best}. The plot suggests that these mean order sizes grow sub-linearly, and appear to be well described by the power-law

$$\langle \upsilon | V_{\text{best}} \rangle \propto V_{\text{best}}^{\chi}; \qquad \chi \cong 0.6.$$

Figure 11.5 shows the distribution $f(x)$ of the ratio $x = \upsilon / V_{\text{best}}$. We observe that $f(x)$ mostly decreases with x and behaves qualitatively similarly for large- and small-tick stocks. Note also the important round-number effects: $f(x)$ has spikes when the market order size is equal to simple fractions of V_{best}. These round-number effects persist even when the market order is larger than the available volume (corresponding, for example, to $V_{\text{best}} = 100$, $\upsilon = 200$). The largest spike occurs for $x = 1$, which means that traders submit a significant number of market orders with a size that exactly matches the available volume at the best. Finally, only few market orders are larger than the available volume at the best. In other words, when traders submit market orders, they often adapt their order volumes to the volume available at the opposite-side best quote. This phenomenon is known as **selective liquidity taking**: the larger the volume available at the opposite-side best quote, the larger the market orders that tend to arrive.

[8] As is clear from Section 4.2, V_{best} has a strong intra-day pattern. It can also fluctuate considerably from one trading day to the next.

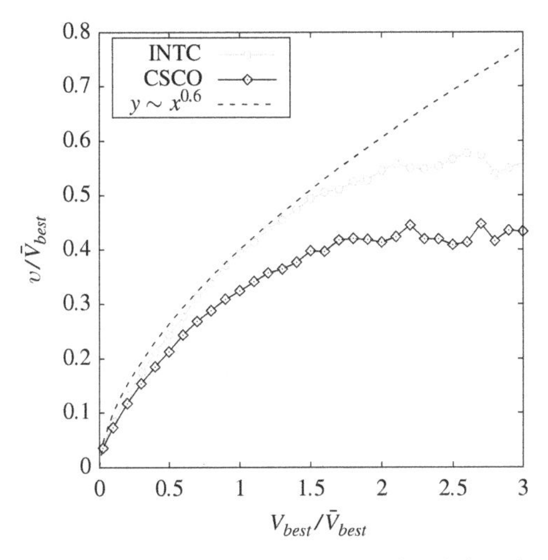

Figure 11.4. Average market order size (normalised by the average queue volume) versus the available volume at the same-side best quote (normalised by the average queue volume) for INTC and CSCO. For small queue volumes, the relationship roughly follows a power-law with exponent 0.6 (dashed curve), and saturates for large queue volumes.

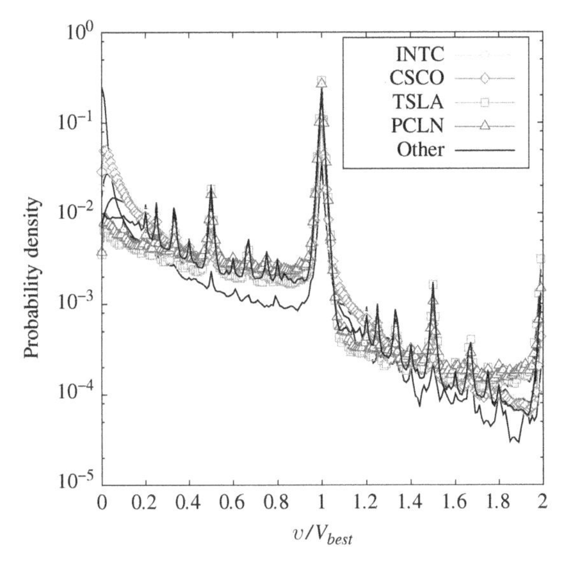

Figure 11.5. Empirical distribution of the fraction of executed queue volume q/V_{best} for the ten stocks in our sample.

One can make use of a simple caricature to understand the strong concavity of $\mathcal{R}(\upsilon, 1)$ as a function of υ, as we reported in the previous section (see Equation 11.7). Assume for simplicity that $\mathcal{R}^0(\upsilon, 1) \approx 0$ and $\mathcal{R}^1(\upsilon, 1) \approx \mathcal{I}$, such that market orders with a volume less than that of the opposite-side best quote have no impact at all, while market orders with a volume that matches that of the opposite-side best quote impact the price by a fixed quantity $\mathcal{I}$ (which we also assume to be independent of volume and equal to some fixed fraction of the spread). In other words, we assume that market orders never eat more than one level, and that the impact of new limit orders and cancellations can be neglected. If this were the case in real markets, then it would follow that

$$\mathcal{R}(\upsilon,1) \approx \mathbb{P}[\mathrm{MO}^1|\upsilon]\mathcal{I}.$$

In real markets, $\mathbb{P}[\mathrm{MO}^1|\upsilon]$ is an increasing function of υ, is zero when $\upsilon = 0$ and converges to 1 when $\upsilon \to \infty$. The resulting $\mathcal{R}(\upsilon,1)$ is therefore a concave function. In fact, using Bayes rule,

$$\mathbb{P}[\mathrm{MO}^1|\upsilon] = \int_0^{\upsilon} \mathrm{d}V_{\mathrm{best}}\, P(V_{\mathrm{best}}|\upsilon) = \frac{\int_0^{\upsilon} \mathrm{d}V_{\mathrm{best}}\, P(\upsilon|V_{\mathrm{best}})P(V_{\mathrm{best}})}{\int_0^{\infty} \mathrm{d}V_{\mathrm{best}}\, P(\upsilon|V_{\mathrm{best}})P(V_{\mathrm{best}})}.$$

Now suppose that the conditional distribution $P(\upsilon|V_{\mathrm{best}})$ is an arbitrary function of $x = \upsilon/V_{\mathrm{best}}$ and $P(V_{\mathrm{best}})$ is itself a power-law. In this case, one finds that $\mathbb{P}[\mathrm{MO}^1|\upsilon]$ is *independent* of υ, leading to a volume-independent impact $\mathcal{R}(\upsilon,1)$, as indeed observed for small-tick stocks (see Figure 11.3).

The conclusion of this toy calculation is that the strong concavity of lag-1 impact as a function of market order volume is (at least partially) a conditioning effect caused by the fact that most large market order arrivals happen when there is a large volume available at the opposite-side best quote.

11.3.4 Conditioning on the Sign of the Previous Trade

As we have shown throughout this chapter, market orders clearly impact the price in the direction of the trade. However, in Chapter 10 we showed that market order signs are autocorrelated. As illustrated by Table 11.2, the lag-1 autocorrelation coefficient $C(1) = \langle \varepsilon_t \varepsilon_{t-1} \rangle$ is quite large. Given that market orders impact prices, it might seem reasonable to expect that the autocorrelation in market order signs should also lead to some predictability in price moves.

If the lag-1 impact $\mathcal{R}(1)$ was independent of the past, then conditional on the last trade being a buy, the next price change would also be on average positive. Conditional on the last trade being a buy, the probability that the next trade is also a buy is $p_+ = (1 + C(1))/2$, and the probability that the next trade is a sell is $p_- = (1 - C(1))/2$. Therefore, it follows that:

$$\langle m_{t+1} - m_t | \varepsilon_{t-1} \rangle_{\mathrm{naive}} = p_+ \mathcal{R}(1) - p_- \mathcal{R}(1) = C(1)\mathcal{R}(1). \tag{11.8}$$

In this naive view, the presence of sign autocorrelations should thus lead to price predictability. However, price changes in financial markets are difficult to predict, even at high frequencies, so this naive picture is likely to be incorrect. To

Table 11.2. *The values of* $C(1)$, f, $\mathcal{R}_+(1)$ *and* $\mathcal{R}_-(1)$ *for the ten stocks in our sample. Impact is measured in dollar cents.*

	$C(1)$	$\mathcal{R}_+(1)$	$\mathcal{R}_-(1)$	f
SIRI	0.93	0.064	0.027	0.90
INTC	0.59	0.26	0.22	0.61
CSCO	0.66	0.22	0.17	0.71
MSFT	0.57	0.29	0.25	0.56
EBAY	0.48	0.36	0.33	0.47
FB	0.37	0.47	0.51	0.31
TSLA	0.69	2.53	2.84	0.52
AMZN	0.72	3.61	4.17	0.57
GOOG	0.75	3.94	4.87	0.60
PCLN	0.78	15.67	16.99	0.63

quantify this small level of predictability, let f denote the actual lag-1 mid-price predictability as a fraction of the above naive predictability. We define the fraction f as:

$$\langle \varepsilon_{t-1} \cdot (m_{t+1} - m_t) \rangle := f C(1) \mathcal{R}(1).$$

Table 11.2 shows the values of f for the stocks in our sample. In all cases, the value of f is smaller than 1, which is the value that the above naive picture would suggest.

How should we understand this empirical result? A first step is to note that a buy trade in fact impacts the mid-price less if it follows another buy trade than if it follows a sell trade. More formally, we define *two* impacts, as follows:

$$\begin{aligned} \mathcal{R}_+(1) &:= \mathbb{E}[\varepsilon_t \cdot (m_{t+1} - m_t) | \varepsilon_t \varepsilon_{t-1} = +1]; \\ \mathcal{R}_-(1) &:= \mathbb{E}[\varepsilon_t \cdot (m_{t+1} - m_t) | \varepsilon_t \varepsilon_{t-1} = -1]. \end{aligned} \tag{11.9}$$

By taking the product of successive trade signs, the conditioning selects successive trades in the same direction for $\mathcal{R}_+(1)$ and in the opposite direction for $\mathcal{R}_-(1)$.

As shown by Table 11.2, it holds that $\mathcal{R}_+(1) < \mathcal{R}_-(1)$ for small-tick stocks.[9] This illustrates an important empirical fact: the most likely outcome has the smallest impact. For example, if the previous trade is a buy, then due to the autocorrelation of market order signs, the next trade is more likely to also be a buy. When the next trade occurs, its impact will, on average, be smaller if it is indeed a buy than if it is

[9] For large-tick stocks, the situation is inverted as a consequence of the influence of volume imbalance on future price changes. The unpredictability argument is no longer about the mid-price m_t but instead about the modified mid-price $\widetilde{m}_t$ defined in Footnote 6.

a sell. This mechanism, which is a crucial condition for market stability, is called **asymmetric dynamical liquidity**.

Of course, considering the lag-1 autocorrelation is only the tip of the iceberg. As we saw in Chapter 10, market order signs actually have long-range autocorrelations that decay very slowly. Therefore, the sign of the next trade can be better predicted by looking at the whole history of trades, and not only the single most recent trade, as would be the case if trade signs were Markovian.[10] In Chapter 13, we extend the analysis from this section to incorporate autocorrelations at larger lags and to build a full theory of the delayed impact of market orders.

11.4 Order-Flow Imbalance and Aggregate Impact

To reduce the role of microstructural idiosyncrasies and conditioning biases (such as selective liquidity taking), price impact is often measured not at the trade-by-trade level (as we have done so far in this chapter), but instead over some coarse-grained time scale T. In many practical applications, choices of T range from about five minutes to a full trading day. In this framework, the goal is to characterise the positive correlations between aggregate signed order flow and contemporaneous price returns.

Consider all transactions that occur in a given time interval $[t, t+T)$, for some $T > 0$ and where t and T are expressed in either event-time or calendar-time. Throughout this section, we again choose to work in market order time, whereby we advance t by 1 for each market order arrival. For each $n \in [t, t+T)$, let ε_n denote the sign of the nth market order and let v_n denote its volume. The **order-flow imbalance** is:

$$\Delta V = \sum_{n \in [t,t+T)} \varepsilon_n v_n.$$

If $\Delta V > 0$, then more buy volume arrives in the given interval than does sell volume, so one expects the price to rise.[11]

The **aggregate impact** is the price change over the interval $[t, t+T)$, conditioned to a certain volume imbalance:

$$\mathbb{R}(\Delta V, T) := \mathbb{E}\left[m_{t+T} - m_t | \sum_{n \in [t,t+T)} \varepsilon_n v_n = \Delta V \right]. \tag{11.10}$$

This quantity can be studied empirically using only public trades-and-quotes data. If $T = 1$, then ΔV is simply the volume of the single market order at time t, and

[10] This Markovian assumption is precisely the starting point of the MRR model that we discuss in Section 16.2.1.

[11] This market scenario is often described as there being "more buyers than sellers"; in reality, of course, there is always an equal number of buy orders and sell orders for each transaction. What is usually meant by the phrase is that there are more buy market orders due to aggressive buyers, but even this is only a rough classification because buyers might also choose to use limit orders to execute their trades (see the discussion in Section 10.5.1 and Chapter 21).

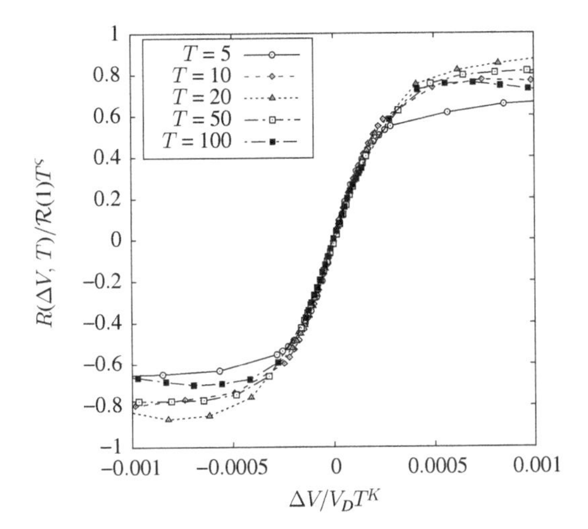

Figure 11.6. Aggregate impact scaling function $\mathscr{F}(x)$ for TSLA. By rescaling $\mathbb{R}(\Delta V, T)$ for $T = 5, 10, 20, 50$ and 100, with $\chi = 0.65$ and $\varkappa = 0.95$, the curves approximately collapse onto each other. Note that $\mathcal{R}$ and ΔV are rescaled each day by the corresponding values of $\mathcal{R}(1)$ and the daily volume $V_{\rm D}$. A similar rescaling works similarly well for all stocks and futures.

one recovers the definition of lag-1 impact

$$\mathbb{R}(\Delta V, 1) \equiv \mathcal{R}(\upsilon = \Delta V, 1).$$

Therefore, by the same arguments as in Section 11.3.2, $\mathbb{R}(\Delta V, 1)$ is a strongly concave function of ΔV.

How does the behaviour of $\mathbb{R}(\Delta V, T)$ change for larger values of T? Empirically, when T increases, the dependence of $\mathbb{R}(\Delta V, T)$ on the volume imbalance becomes closer to a linear relationship for small ΔV, while retaining its concavity for large $|\Delta V|$. In fact, the following empirical **scaling law** appears to hold for a large variety of stocks and futures contracts:[12]

$$\mathbb{R}(\Delta V, T) \cong \mathcal{R}(1) T^{\chi} \times \mathscr{F}\left(\frac{\Delta V}{V_{\rm D} T^{\varkappa}}\right), \tag{11.11}$$

where $V_{\rm D}$ is the daily volume and $\mathscr{F}(u)$ is a scaling function (see Figure 11.6) that is linear for small arguments ($\mathscr{F}(u) \sim_{|u| \ll 1} u$) and concave for large arguments[13] and where empirically the exponents are given by $\chi \cong 0.5 - 0.7$ and $\varkappa \cong 0.75 - 1$. We will attempt to rationalise the values of these exponents in Section 13.4.3.

The slope of the linear region of $\mathbb{R}(\Delta V, T)$ is usually called **Kyle's lambda**, in reference to the Kyle model, which we discuss in Chapter 15. The value of this

[12] See: Patzelt, F., & Bouchaud, J. P. (2018). Universal scaling and nonlinearity of aggregate price impact in financial markets. *Physical Review E*, 97, 012304.

[13] A possible example is $\mathscr{F}(u) \sim_{|u| \gg 1} \text{sign}(u)|u|^{\zeta}$ with a small exponent ζ, or other functions with a similar behaviour but that saturate for large u.

slope is often regarded as a measure of a market's (il-)liquidity. Using the scaling result in Equation (11.11), it follows that Kyle's lambda *decreases* with the time interval over which it is measured:[14]

$$\mathbb{R}(\Delta V, T) \sim_{|\Delta V| \to 0} \Lambda(T)\Delta V; \qquad \Lambda(T) \cong \Lambda(1)T^{-(\varkappa-\chi)}.$$

Interestingly, the empirical determination of $\varkappa - \chi \cong 0.25\text{–}0.3$ is more robust and stable across assets than χ and $\varkappa$ independently.

The saturation for larger values of ΔV is also interesting, but is likely to be related to selective liquidity taking (see Section 11.3.3) for single market orders, which is expected to persist on longer time scales $T \gg 1$. For example, imagine that there is an excess of sell limit orders in the LOB during the whole time interval $[t, t+T)$. This will likely attract buy market orders that can be executed without impacting the price too much. Therefore, one expects that these situations will bias $\mathbb{R}(\Delta V, T)$ downwards for large ΔV, following the very same conditioning argument that led to a concave shape for $\mathcal{R}(\Delta V, 1)$.

11.5 Conclusion

The primary conclusion of this chapter is that there exists a clear empirical correlation between the signs of trades and the directions of price moves. Trivial as it may sound, the mechanism underlying this behaviour is not immediately clear. Does this phenomenon occur because information is revealed? Or does the very occurrence of a trade itself impact prices? Generally speaking, one expects that both effects should contribute. However, empirical data suggests that the reaction part of impact is predominant on short to medium time scales, and is identical for all trades irrespective of whether they are informed or not. Since trades in modern financial markets are anonymous, it would be surprising if any distinction (between whether or not a given trade was informed) could be made before any genuine prediction contained in some of these trades reveals itself. By a similar argument, one expects that the impact of trades could also explain a large fraction of the volatility observable in real markets (see the discussion in Chapter 20).

From the point of view of this book, this is very good news! It suggests that there is hope for modelling the reaction part of the impact, which should follow reasonably simple principles, as we shall discuss in Chapters 13 and 19. The prediction part of impact is of course much harder to model, since it depends on a huge information set and on the specific strategies used to exploit mispricings. In any case, the information content of individual trades must be very small (see

[14] In the Kyle model (see Chapter 15), aggregate impact is linear and additive, so $\Lambda(T)$ must be constant and therefore independent of T. Since order flow is assumed to be completely random in this model, the value of the exponents corresponds to $\chi = \varkappa = \frac{1}{2}$ in that case, and the scaling function is linear: $\mathscr{F}_{\text{Kyle}}(x) = x$.

Chapters 13, 16 and 20). Furthermore, as we will see, the decay of the reaction impact is so slow that the distinction between transient impact (often associated with noise trades) and permanent impact (often associated with informed trades) is fuzzy.

Several empirical conclusions from this chapter will now trickle throughout much of the rest of the book. For example, we saw in Section 11.3.1 that the lag-1 impact of a market order is of the same order as the bid–ask spread. We will investigate this from a theoretical perspective in Chapter 17. We also saw in Section 11.3.1 that the response function $\mathcal{R}(\ell)$ is an increasing function of ℓ that saturates at large lags (see Figure 13.1). This can be traced back to the autocorrelation of the trade-sign series (see Sections 13.2 and 16.2.4, and Equation (16.22)), from which we will construct a theory for the reaction impact of trades (see Chapter 13). In Section 11.3.4, we saw that the impact of two trades in the same direction is less than twice the unconditional impact of a single trade. This is due to liquidity providers buffering the impact of successive trades in the same direction, and is a crucial condition for market stability, as we will see in Sections 13.3 and 14.4.

Finally, we have discussed the impact of trades at an aggregate level. We saw that small price returns are linearly related to small volume imbalances, with a slope (called "Kyle's lambda") that is found to decrease with the aggregation time scale T. We will return to this discussion in Section 13.4.3. As we discuss in the next chapter, however, this linear impact law, obtained by a blind aggregation of all trades, is of little use for understanding the price impact of a *metaorder* executed over a time horizon T. This necessary distinction is among the most striking results in market microstructure, and will require a detailed, in-depth discussion (see Section 13.4.4 and Part VIII).

Take-Home Messages

(i) Trade directions are positively correlated with subsequent price changes. There are two competing views that attempt to explain this phenomenon: that trades *forecast* prices or that trades *impact* prices.
(ii) The observed impact of a trade can be decomposed into two components: a reaction component, which describes how the market reacts to the trade itself, and a prediction component, which describes all other dynamics not directly related to the trade (e.g. exogenous information).
(iii) Measuring the reaction impact of a given order is difficult, because doing so would require replaying history to consider a world in

which all else was equal except that the given order was not submitted.

(iv) The response function describes the mean price trajectory at a given time lag after a trade. Empirically, the response function is found to be an increasing function of lag, starting from 0 and reaching a plateau with a scale that is of the same order as the bid–ask spread. Importantly, the response function must not be confused with the reaction impact of a market order, because it also contains the impact of future orders.

(vi) Because market order signs are strongly autocorrelated, if trades impacted prices linearly, then prices would be strongly predictable. This is not observed empirically. Therefore, the impact of trades must be history-dependent. For example, a buy trade that follows a buy trade impacts the price less than a buy trade that follows a sell trade.

(vii) It is non-trivial to assess the dependence of impact on volumes. At the scale of individual trades, conditioning effects lead to a strongly concave dependence.

(viii) The aggregate impact of volume imbalance is linear for small imbalances and saturates for large imbalances. The slope of the linear part decreases as a power-law of the aggregation time. However, this naive averaging only shows part of the story – as we will see in the next chapter.

11.6 Further Reading

Price Impact

Weber, P., & Rosenow, B. (2005). Order book approach to price impact. *Quantitative Finance*, 5(4), 357–364.

Gerig, A. (2007). *A theory for market impact: How order flow affects stock price*. PhD thesis, University of Illinois, available at: arXiv:0804.3818.

Farmer, J. D., & Zamani, N. (2007). Mechanical vs. informational components of price impact. *The European Physical Journal B-Condensed Matter and Complex Systems*, 55(2), 189–200.

Hasbrouck, J. (2007). *Empirical market microstructure: The institutions, economics, and econometrics of securities trading*. Oxford University Press.

Bouchaud, J. P., Farmer, J. D., & Lillo, F. (2009). How markets slowly digest changes in supply and demand. In Hens, T. & Schenk-Hoppe, K. R. (Eds.), *Handbook of financial markets: Dynamics and evolution*. North-Holland, Elsevier.

Bouchaud, J. P. (2010). Price impact. In Cont, R. (Ed.), *Encyclopedia of quantitative finance*. Wiley.

Eisler, Z., Bouchaud, J. P., & Kockelkoren, J. (2012). The price impact of order book events: Market orders, limit orders and cancellations. *Quantitative Finance*, 12(9), 1395–1419.

Hautsch, N., & Huang, R. (2012). The market impact of a limit order. *Journal of Economic Dynamics and Control*, 36(4), 501–522.
Cont, R., Kukanov, A., & Stoikov, S. (2014). The price impact of order book events. *Journal of Financial Econometrics*, 12(1), 47–88.
Donier, J., & Bonart, J. (2015). A million metaorder analysis of market impact on the Bitcoin. *Market Microstructure and Liquidity*, 1(02), 1550008.
Gomes, C., & Waelbroeck, H. (2015). Is market impact a measure of the information value of trades? Market response to liquidity vs. informed metaorders. *Quantitative Finance*, 15(5), 773–793.
Tóth, B., Eisler, Z. & Bouchaud, J.-P. (2017). The short-term price impact of trades is universal. https://ssrn.com/abstract=2924029.

(Weak) Volume Dependence of Impact

Hasbrouck, J. (1991). Measuring the information content of stock trades. *The Journal of Finance*, 46(1), 179–207.
Jones, C. M., Kaul, G., & Lipson, M. L. (1994). Transactions, volume, and volatility. *Review of Financial Studies*, 7(4), 631–651.
Chen, Z., Stanzl, W., & Watanabe, M. (2002). *Price impact costs and the limit of arbitrage*. Yale ICF Working Paper No. 00–66.
Lillo, F., Farmer, J. D., & Mantegna, R. N. (2003). Econophysics: Master curve for price-impact function. *Nature*, 421(6919), 129–130.
Potters, M., & Bouchaud, J. P. (2003). More statistical properties of order books and price impact. *Physica A: Statistical Mechanics and its Applications*, 324(1), 133–140.
Zhou, W. X. (2012). Universal price impact functions of individual trades in an order-driven market. *Quantitative Finance*, 12(8), 1253–1263.
Taranto, D. E., Bormetti, G., & Lillo, F. (2014). The adaptive nature of liquidity taking in limit order books. *Journal of Statistical Mechanics: Theory and Experiment*, 2014(6), P06002.
Gomber, P., Schweickert, U., & Theissen, E. (2015). Liquidity dynamics in an electronic open limit order book: An event study approach. *European Financial Management*, 21(1), 52–78.

History Dependence of Impact

Bouchaud, J. P., Gefen, Y., Potters, M., & Wyart, M. (2004). Fluctuations and response in financial markets: The subtle nature of random price changes. *Quantitative Finance*, 4(2), 176–190.
Bouchaud, J. P., Kockelkoren, J., & Potters, M. (2006). Random walks, liquidity molasses and critical response in financial markets. *Quantitative Finance*, 6(02), 115–123.
Farmer, J. D., Gerig, A., Lillo, F., & Mike, S. (2006). Market efficiency and the long-memory of supply and demand: Is price impact variable and permanent or fixed and temporary? *Quantitative Finance*, 6(02), 107–112.
Taranto, D. E., Bormetti, G., Bouchaud, J. P., Lillo, F., & Tóth, B. (2016). Linear models for the impact of order flow on prices I. Propagators: Transient vs. history dependent impact. Available at SSRN: https://ssrn.com/abstract=2770352.

Aggregate Impact

Kempf, A., & Korn, O. (1999). Market depth and order size. *Journal of Financial Markets*, 2(1), 29–48.
Plerou, V., Gopikrishnan, P., Gabaix, X., & Stanley, H. E. (2002). Quantifying stock-price response to demand fluctuations. *Physical Review E*, 66(2), 027104.

Chordia, T., & Subrahmanyam, A. (2004). Order imbalance and individual stock returns: Theory and evidence. *Journal of Financial Economics*, 72(3), 485–518.

Evans, M. D., & Lyons, R. K. (2002). Order flow and exchange rate dynamics. *Journal of Political Economy*, 110(1), 170–180.

Gabaix, X., Gopikrishnan, P., Plerou, V., & Stanley, H. E. (2006). Institutional investors and stock market volatility. *The Quarterly Journal of Economics*, 121(2), 461–504.

Lillo, F., Farmer J. D., & Gerig A. (2008). *A theory for aggregate market impact*. Technical report, Santa Fe Institute, unpublished research.

Hopman, C. (2007). Do supply and demand drive stock prices? *Quantitative Finance*, 7, 37–53.

12

The Impact of Metaorders

It doesn't matter how beautiful your theory is, it doesn't matter how smart you are. If it doesn't agree with experiment, it's wrong.

(Richard P. Feynman)

In the previous chapter, we considered how the arrival of a single market order impacts the mid-price. However, as we noted in Section 10.5.1, most traders do not execute large trades via single market orders, but instead split up their trades into many small pieces. These pieces are executed incrementally, using market orders, limit orders, or both, over a period of several minutes to several days. As we saw in the last chapter, the chaining of market orders greatly affects their impact. Therefore, understanding the impact of a single market order is only the first step towards understanding the impact of trading more generally. To develop a more thorough understanding, we must also consider the impact of **metaorders** (defined in Section 10.4.3).

The empirical determination of metaorder impact is an important experiment whose results, when measured properly, are of great interest to academics, investors and market regulators alike. From a fundamental point of view, how does a metaorder of size Q contribute to price formation? From the point of view of investors, what is the true cost of performing such a trade? How does it depend on market conditions, execution strategies, time horizons, and so on? From the point of view of regulators, can large metaorders destabilise markets? Is marked-to-market accounting wise when, as emphasised above, the market price is (at best) only meaningful for infinitesimal volumes?

Naively, it might seem intuitive that the impact of a metaorder should scale linearly in its total size Q. Indeed, as we will discuss in this chapter, many simple models of price impact predict precisely a linear behaviour. Perhaps surprisingly, empirical analysis reveals that in real markets, this scaling is not linear, but rather is approximately square-root. Throughout this chapter, we present this square-root law of impact and discuss several of its important consequences.

12.1 Metaorders and Child Orders

Assume that a trader decides to buy or sell some quantity Q of a given asset. Ideally, the trader would like to buy or sell this whole quantity immediately, at the market price. However, as we discussed in Section 10.5.1, unless Q is smaller than the volume available for immediate purchase or sale at the opposite-side best quote, then conducting the whole trade at the bid- or ask-price is not possible. Therefore, there is no such thing as a "market price" for a trade, because such a price can only be guaranteed to make sense for a very small Q. For the volumes typically executed by large financial institutions, there is rarely enough liquidity in the whole LOB to match the required quantity Q all at once. Therefore, traders must split their desired metaorder (i.e. the full quantity Q) into many smaller pieces, called **child orders**, which they submit gradually over a period of minutes, hours, days and even months.

It is common practice for traders to decompose their trading activity into two distinct stages:

(i) The *investment-decision stage,* during which the trader determines the sign ε (i.e. buy/sell) and volume Q of the metaorder, and the desired time horizon T over which to execute it, usually based on some belief about the future price of the asset.

(ii) The *execution stage,* during which the trader conducts the relevant trades to obtain the required quantity at the best possible price within the time window T.

The execution stage is sometimes delegated to a broker, who seeks to achieve specified execution targets such as VWAP or TWAP (see Section 10.5.2) by performing incremental execution of the metaorder in small chunks. As we will discuss further in Chapter 19, this incremental execution seeks to make use of the gradual refilling of the LOB from previously undisclosed orders. In principle, this should considerably improve the price obtained for the metaorder execution when compared to submitting large market orders that penetrate deep into the LOB (and that could even destabilise the market).

Importantly, the direction, volume and time horizon of a metaorder are determined and fixed during the investment-decision stage. Any decision to stop, extend or revert a metaorder comes from a new investment decision. As we discuss later in this chapter, it is important to keep this in mind to avoid spurious conditioning effects when measuring the impact of a metaorder.

12.2 Measuring the Impact of a Metaorder

Consider a metaorder with volume Q and sign ε. Let N denote the number of constituent child orders of this metaorder. For $i = 1,2,\ldots,N$, let t_i, υ_i and p_i denote, respectively, the execution time, volume and execution price of the ith child order. By definition, it follows that

$$\sum_{i=1}^{N} v_i = Q.$$

Because each of the N child orders are part of the same metaorder, it follows that the sign of each child order is also equal to ε.[1]

As we discussed (in the context of market order impact) in Section 11.2, the ideal experiment for measuring the impact of a metaorder would involve comparing two different versions of history: one in which a given metaorder arrived, and one in which it did not, but in which all else was equal. In reality, however, this is not possible, so we must instead make do with estimating impact by measuring quantities that are visible in empirical data.

12.2.1 The Ideal Data Set

To gain a detailed understanding of the impact of a metaorder, one would ideally have access to some form of proprietary data or detailed broker data that lists:

- which child orders belong to which metaorders;
- the values of t_i, p_i, and v_i for each child order; and
- whether each child order was executed via a limit order or a market order.

With access to this ideal data, it would be straightforward to construct a detailed execution profile of a metaorder by recording the values of t_i, p_i and v_i for each child order. In reality, however, it is rare to have access to such rich and detailed data, so it is often necessary to cope with less detailed data, and to impose additional assumptions about metaorder execution.

12.2.2 Less-Detailed Data

Even in the absence of the full information described in Section 12.2.1, it is still possible to gain insight into the impact of a metaorder, given only the following information:

- the sign ε and total quantity Q of the metaorder;
- the time of the first trade t_1 and the corresponding mid-price price m_1;[2]
- the time of the last trade t_N and the corresponding mid-price price m_N. The execution horizon T is then given by $T = t_N - t_1$.

[1] We assume here that the investor does not send child orders to sell (respectively buy) when s/he wants to buy (respectively sell). This is reasonable as any round-trip is usually costly – see Section 19.5.

[2] In principle, t_1 is different from the time t_0 at which the investment decision is made. However, we neglect here the short-term predictability that would lead to a systematic price change between these two times.

In the absence of more detailed information about the execution profile of a metaorder, we will rely on an important assumption: that, given t_1 and t_N, the execution profile of the metaorder's profile is approximately linear, such that the cumulative volume $q(t)$ executed up to time t is

$$q(t) \approx \frac{t - t_1}{T} Q, \qquad q(t_N) \equiv Q. \tag{12.1}$$

In Section 12.2.3, we list several interesting properties of impact that can be measured within this simple framework. However, it is important to stress that measuring metaorder impact still requires relatively detailed data that indicates which child orders belong to which metaorders. This information is not typically available in most publicly available data, which is anonymised and provides no explicit trader identifiers. Using such data only allows one to infer the aggregate impact as in Section 11.4. Identifying aggregate impact with metaorder impact is misleading, and in most cases leads to a substantial underestimation of metaorder impact.

12.2.3 Impact Path and Peak Impact

Given metaorder data of the type described in Section 12.2.2, and making the assumptions described in that section, one can define several quantities of interest for characterising the impact of a metaorder of total volume Q and horizon T:

(i) The average **impact path** is the mean price path between the beginning and the end of a metaorder (see Figure 12.1):

$$\mathfrak{I}^{\text{path}}(q, t - t_1 | Q, T) = \langle \varepsilon \cdot (m_t - m_1) \mid q; Q, T \rangle, \tag{12.2}$$

where q is the quantity executed between t_1 and t (which, by Equation (12.1), we assume to grow approximately linearly with t). We will see later that the conditioning on Q and T can in fact be removed if the execution profile is linear. We introduce the notation $\mathfrak{I}$ for the impact of a metaorder to distinguish it from the impact $\mathcal{I}$ of a single market order, which we discussed in Chapter 11.

(ii) The average **peak impact** of a metaorder of size Q executed over a time horizon T is:

$$\mathfrak{I}^{\text{peak}}(Q, T) := \langle \varepsilon \cdot (m_N - m_1) \mid Q \rangle. \tag{12.3}$$

As alluded to above, for a metaorder executed at a constant rate, it is reasonable that there cannot be any difference between the mechanical impact of the volume q of a partially executed metaorder and a fully executed metaorder of volume q, since the rest of the market does not know whether the metaorder is continuing or not. Hence, neglecting any prediction contribution coming from short-term signals, we expect that

$$\mathfrak{I}^{\text{path}}(q, t - t_1 | Q, T) \approx \mathfrak{I}^{\text{peak}}(q, t - t_1), \qquad \text{for all } q \leq Q, t \leq T.$$

Empirical data confirms that this equality does indeed hold (see Section 12.5).

(iii) The **execution shortfall** $\mathscr{C}$ (also called execution cost or "slippage") is the average difference between the price paid for each subsequent child order and the decision price (which we assume to be the price paid for the first child order; see previous footnote). Neglecting the spread contribution, one writes

$$\mathscr{C}(Q,T) = \left\langle \sum_i \varepsilon v_i \cdot (m_i - m_1) \middle| Q \right\rangle. \tag{12.4}$$

This is the volume-weighted average premium paid by the trader executing the metaorder.

(iv) **Impact after-effects** describe the mean price path for $t > t_N$ (i.e. after the metaorder has been fully executed; see Figure 12.1). At any given $t > t_N$, the impact after-effect of a metaorder can be decomposed into a **transient component** $\mathfrak{I}^{\text{trans.}}(Q,t)$ and a **permanent component** $\mathfrak{I}^{\infty}(Q)$, such that, when $t > T$,

$$\mathfrak{I}^{\text{path}}(Q,t) = \mathfrak{I}^{\text{trans.}}(Q,t) + \mathfrak{I}^{\infty}(Q), \tag{12.5}$$

with

$$\mathfrak{I}^{\text{trans.}}(Q, t \to \infty) = 0; \qquad \mathfrak{I}^{\infty}(Q) = \lim_{t \to \infty} \mathfrak{I}^{\text{path}}(Q,t).$$

Note that the permanent component $\mathfrak{I}^{\infty}(Q)$ receives two types of contribution:

(a) One coming from the prediction signal at the origin of the metaorder, called the prediction impact in Section 11.2;

(b) The other coming from the possibly permanent reaction of the market to all trades, even uninformed. This contribution is nicely illustrated by the zero-intelligence Santa Fe model, for which indeed $\mathfrak{I}^{\infty}(Q) > 0$.

For non-linear execution profiles, one should expect the impact path to depend not only on q, but possibly on the whole execution schedule $\{q(t)\}$.

12.3 The Square-Root Law

At the heart of most empirical and theoretical studies of **metaorder impact** lies a very simple question: how does the impact of a metaorder depend on its size Q? Many models, including the famous Kyle model (see Chapter 15), predict this relationship to be linear. Although this answer may appear intuitive, there now exists an overwhelming body of empirical evidence that rules it out, in favour of a

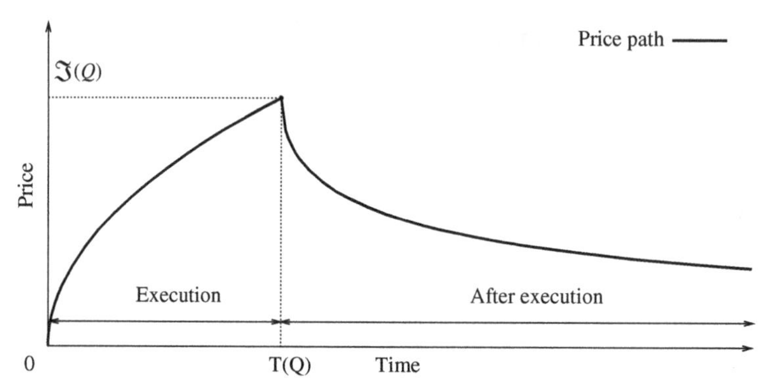

Figure 12.1. Average shape of the impact path. Over the course of its execution, a buy metaorder pushes the price up, until it reaches a peak impact. Upon completion, the buying pressure stops and the price reverts abruptly. Some impact is however still observable long after the metaorder execution is completed, and sometimes persists permanently. For real data, one should expect a large dispersion around the average impact path.

concave, and apparently square-root dependence on Q. In this section, we discuss the empirical basis of this so-called **square-root impact law**.

12.3.1 Empirical Evidence

Since the early 1980s, a vast array of empirical studies[3] spanning both academia and industry have concluded that the impact of a metaorder scales approximately as the square-root of its size Q. This result is reported by studies of different markets (including equities, futures, FX, options, and even Bitcoin), during different epochs (including pre-2005, when liquidity was mostly provided by market-makers, and post-2005, when electronic markets were dominated by HFT), in different types of microstructure (including both small-tick stocks and large-tick stocks), and for market participants that use different underlying trading strategies (including fundamental, technical, and so on) and different execution styles (including using a mix of limit orders and market orders or using mainly market orders).

In all of these cases, the peak impact of a metaorder with volume Q is well described by the relationship

$$\mathfrak{I}^{\text{peak}}(Q,T) \cong Y\sigma_T \left(\frac{Q}{V_T}\right)^{\delta}, \qquad (Q \ll V_T), \tag{12.6}$$

where Y is a numerical coefficient of order 1 ($Y \cong 0.5$ for US stocks), δ is an exponent in the range 0.4–0.7, σ_T is the contemporaneous volatility on the time horizon T, and V_T is the contemporaneous volume traded over time T. Note that

[3] We list a wide range of such studies in Section 12.7.

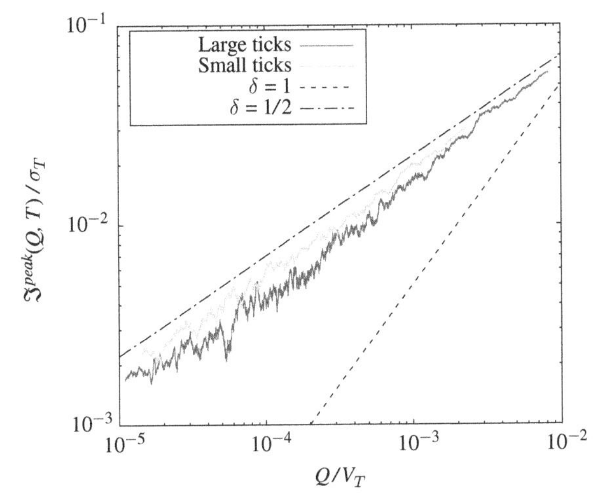

Figure 12.2. The impact of metaorders for Capital Fund Management proprietary trades on futures markets, during the period from June 2007 to December 2010 (see Tóth et al. (2011)). We show $\Im^{\text{peak}}(Q,T)/\sigma_T$ versus Q/V_T on doubly logarithmic axes, where σ_T and V_T are the daily volatility and daily volume measured on the day the metaorder is executed. The black curve is for large-tick futures and the grey curve is for small-tick futures. For comparison, we also show a (dash-dotted) line of slope $\frac{1}{2}$ (corresponding to a square-root impact) and a (dotted) line of slope 1 (corresponding to linear impact).

Equation (12.6) is dimensionally consistent, in the sense that both the left-hand side and the right-hand side have the dimension [% of price].

To illustrate the relationship in Equation (12.6), Figure 12.2 shows $\Im^{\text{peak}}(Q,T)/\sigma_T$ vs. Q/V_T for the data published in the paper by Tóth et al. in 2011,[4] which corresponds to nearly 500,000 metaorders executed between June 2007 and December 2010 on a variety of liquid futures contracts. The data suggests a value of $\delta \cong 0.5$ for small-tick contracts and $\delta \cong 0.6$ for large-tick contracts, for values of Q/V_T ranging from about 10^{-5} to a few per cent. Several other empirical studies of different markets have drawn similar conclusions, with $\delta \cong 0.6$ for US and international stock markets, $\delta \cong 0.5$ for Bitcoin and $\delta \cong 0.4$ for volatility markets (see Section 12.7 for references). In all published studies, the exponent δ is around $\frac{1}{2}$, hence the name "square-root impact" that we will use henceforth.

12.3.2 A Very Surprising Law

The square-root law of metaorder impact is well established empirically, but there are several features that make it extremely surprising theoretically. In this section,

[4] Tóth, B., Lemperiere, Y., Deremble, C., De Lataillade, J. Kockelkoren, J., & Bouchaud, J. P. (2011). Anomalous price impact and the critical nature of liquidity in financial markets. *Physical Review X*, 1(2), 021006.

we summarise some of these surprising features, and highlight some important lessons that the square-root law teaches us about financial markets.

The first surprising feature of Equation (12.6) is that metaorder impact does not scale linearly with Q – or, said differently, that metaorder impact is not additive. Instead, one finds empirically that the second half of a metaorder impacts the price much less than the first half. This can only be the case if there is some kind of **liquidity memory time** T_m, such that the influence of past trades cannot be neglected for $T \ll T_m$ but vanishes for $T \gg T_m$, when all memory of past trades is lost. We will hypothesise in Chapter 18 that T_m is in fact imprinted in the "latent" LOB that we alluded to above.

The second surprising feature of Equation (12.6) is that Q appears not (as might be naively anticipated) as a fraction of the total market capitalisation $\mathcal{M}$ of the asset,[5] but instead as a fraction of the total volume V_T traded during the execution time T. In modern equities markets, $\mathcal{M}$ is typically about 200 times larger than V_T for $T = 1$ day. Therefore, the impact of a metaorder is much larger than if the Q/V_T in Equation (12.6) was instead $Q/\mathcal{M}$. The square-root behaviour for $Q \ll V_T$ also substantially amplifies the impact of small metaorders: executing 1% of the daily volume moves the price (on average) by $\sqrt{1\%} = 10\%$ of its daily volatility. The main conclusion here is that even relatively small metaorders cause surprisingly large impact.

The third surprising feature of Equation (12.6) is that the time horizon T does not appear explicitly. To consider this in more detail, let

$$\mathcal{L}_T := \sqrt{V_T}/\sigma_T \tag{12.7}$$

denote the **liquidity ratio**, which measures the capacity of the market to absorb incoming order flow. If prices are exactly diffusive, one has $\sigma_T = \sqrt{T}\sigma_1$, and if traded volume grows linearly with time, i.e. $V_T = TV_1$ (where "1" denotes, say, one trading day), then one finds

$$\mathcal{L}_T = \frac{\sqrt{TV_1}}{\sqrt{T}\sigma_1} = \frac{\sqrt{V_1}}{\sigma_1}.$$

Therefore, $\mathcal{L}_T$ is independent of T. If we rewrite Equation (12.6) to include $\mathcal{L}_T$, we arrive at (for $\delta = 0.5$)

$$\mathfrak{I}^{\text{peak}}(Q,T) = \frac{Y}{\mathcal{L}_T}\sqrt{Q}.$$

[5] The idea that trading 1% of a stock's total market capitalisation should move its price by about 1% was common lore in the 1980s, when impact was deemed totally irrelevant for quantities representing a few basis points of $\mathcal{M}$. Neglecting the potential impact of trades representing 100% of V_T, but (at the time) about 0.25% of the market capitalisation, is often cited as one of the reasons for the 1987 crash, when massive portfolio insurance trades created havoc in financial markets (see, e.g., Treynor, J. L. (1988). Portfolio insurance and market volatility. *Financial Analysts Journal*, 44(6), 71–73).

Hence, the square-root impact can be written as:

$$\mathfrak{I}^{\text{peak}}(Q,T) \cong Y\sigma_1 \sqrt{\frac{Q}{V_1}}, \qquad (Q \ll V_T), \tag{12.8}$$

which illustrates that the impact of a metaorder is only determined by Q, not the time T that it took to be executed. A possible explanation, from an economic point of view, is that the market price has to adapt to a change of global supply/demand εQ, independently of how this volume is actually executed. We provide a more detailed, Walrasian view of this non-trivial statement in Chapter 18.

12.3.3 Domain of Validity

As with all empirical laws, the square-root impact law only holds within a certain domain of parameters. In this section, we discuss this domain and highlight some of the limitations that one should expect on more general grounds.

First, as we noted in Section 12.3.2, the square-root law should only hold when the execution horizon T is shorter than a certain memory time T_m, which is related to the time after which the underlying latent liquidity of the market has evolved significantly (see Chapter 18). The memory time T_m is difficult to estimate empirically, but one may expect that it is of the order of days. Therefore, the impact of a metaorder of size Q that is executed over a horizon of several weeks or months should become approximately proportional to Q, if the permanent component of impact is non-zero.

Second, the ratio Q/V_T should be small, such that the metaorder constitutes only a small fraction of the total volume V_T and such that the impact is small compared to the volatility. In the case where Q/V_T is substantial, one enters a different regime, because the metaorder causes a large perturbation to the ordinary course of the market. In this large Q/V_T regime, a metaorder with a sufficiently long execution horizon T for the underlying latent liquidity to reveal itself is expected to have a completely different impact than a metaorder that is executed extremely quickly, with the risk of destabilising the market and possibly inducing a crash.

More formally, let τ_{liq} be the time needed for the LOB liquidity to refill, which is typically on the scale of seconds to minutes. The two above-mentioned regimes are:

- **Slow execution**, such that $\tau_{\text{liq}} \ll T \ll T_m$. For small Q/V_T, this is the square-root regime discussed in the last section. As Q/V_T increases beyond (say) 10%, one may actually expect that the square-root behaviour becomes even more concave, because more and more sellers (respectively, buyers) eventually step in to buffer the increase (respectively, decrease) in price (see Equation (18.10) for a concrete example).[6]
- **Fast execution**, such that $T \lesssim \tau_{\text{liq}}$. In this case, impact becomes a convex (rather than concave) function of Q for large Q. This is clear in the limit $T \to 0$, where the metaorder simply consumes the liquidity immediately available in the LOB. Since the mean volume profile in the LOB first increases then decreases with increasing distance from the mid-price (see Section 4.7), the immediate impact of a volume

[6] This seems consistent with empirical findings, see, e.g., Zarinelli et al. (2015). Beyond the square root: Evidence for logarithmic dependence of market impact on size and participation rate. *Market Microstructure and Liquidity*, 1(02), 1550004. One should however keep in mind the possible "Implementation Bias 1" discussed in the next section.

Q will first be a concave, and then a convex, function of Q.[7] This is the reason that some financial markets implement circuit breakers, which aim to slow down trading and allow time for liquidity to replenish (see Section 22.2).

Finally, the regime where Q is smaller than the typical volume at best V_{best} is again expected to depart from a square-root behaviour and recover a linear shape. This is suggested both by some empirical results[8] and by the theoretical analysis of Chapters 18 and 19.

In summary, the square-root impact law holds (approximately) in the regime of intermediate execution horizons T and intermediate volume fractions Q/V_T:

$$\tau_{\text{liq}} \ll T \ll T_m; \qquad \frac{V_{\text{best}}}{V_T} \ll \frac{Q}{V_T} \lesssim 0.1,$$

which is the regime usually adopted by investors in normal trading conditions. Different behaviours should be expected outside of these regimes, although data to probe these regions is scarce, and the corresponding conclusions currently remain unclear.

12.3.4 Possible Measurement Biases

The peak impact of a metaorder $\mathfrak{I}^{\text{peak}}(Q,T)$ can be affected by several artefacts and biases. To obtain reproducible and understandable results, one should thus stick to a well-defined experimental protocol, and be aware of the following possible difficulties:

(i) **Prediction bias 1**: The larger the volume Q of a metaorder, the more likely it is to originate from a stronger prediction signal. Therefore, a larger part of its impact may be due to short-term predictability, and not reveal any structural regularity of the reaction part of impact.

(ii) **Prediction bias 2**: Traders with strong short-term price-prediction signals may choose to execute their metaorders particularly quickly, to make the most of their signal before it becomes stale or more widely-known. Therefore, the strength of a prediction signal may itself influence the subsequent impact path (this is the prediction impact), in particular when the prediction horizon is comparable to the execution horizon. This bias is also likely to affect the long-term, permanent component $\mathfrak{I}^{\infty}(Q)$ of the impact (see Section 12.2).

(iii) **Synchronisation bias**: The impact of a metaorder can change according to whether or not other traders are seeking to execute similar metaorders at the same time. This can occur if different traders' prediction signals are correlated, or if they trade based on the same piece of news. (Note that this bias overlaps with the prediction biases discussed in (i) and (ii)).

(iv) **Implementation bias 1**: Throughout this chapter, we have assumed that both the volume Q and execution horizon T are fixed before a metaorder's execution begins. In reality, however, some traders may adjust these values during execution, by conditioning on the price path. In these cases, understanding metaorder impact is much more difficult. For example, when examining a buy metaorder that is only executed if the price goes down, and abandoned if the price goes up, impact will be negative. This implementation bias is expected to be stronger for large volumes, since the price is more likely to have

[7] Indeed, the immediate impact $\mathcal{I}$ is set by the condition:

$$\int_{m_1}^{m_1+\mathcal{I}} \mathrm{d}p'\, V_+(p') = Q,$$

where $V_+(p)$ is the density of available orders at price p (see Section 3.1.7). Solving for $\mathcal{I}$ as a function of Q leads to a convex shape at large Q when $V_+(p)$ is a decreasing function of p.

[8] Zarinelli et al. (2015).

moved adversely in these cases. This conditioning may result in a systematic underestimation of $\Im^{\text{peak}}(Q,T)$ at large Q's.

(v) **Implementation bias 2**: The impact path can become distorted for metaorders that are not executed at an approximately constant rate. For example, if execution is front-loaded, in the sense that most of the quantity Q is executed close to the start time t_1, the impact at time t_N will have had more time to decay, and may therefore be lower than for a metaorder executed at constant rate.

(vi) **Issuer bias**: Another bias may occur if a trader submits several dependent metaorders successively. If such metaorders are positively correlated and occur close in time to one another, the impact of the first metaorder will be different to the impact of the subsequent metaorders. Empirically, the impact of the first metaorder is somewhat greater than that of the second, and so on (much as for single market orders; see Section 11.3.4). Reasons for this will become clear in Chapter 19.

All these biases are pervasive. Fortunately, most of them can be avoided when one has access to proprietary trading data that contains full information about each child order, as described in Section 12.2.1. However, even less-ideal data sets still lead to similar conclusions on the dependence of the peak impact $\Im^{\text{peak}}(Q,T)$ on Q. Remarkably, many different studies (based on different assets, epochs, trading strategies and microstructure) all converge to a universal, square-root-like dependence of $\Im^{\text{peak}}(Q,T)$ on Q, and a weakly decreasing dependence on T.

12.3.5 Common Misconceptions

We now discuss some misconceptions and confusions about the impact of metaorders that exist in the literature.

- The impact of a metaorder of volume Q is *not* equal to the aggregate impact of order imbalance ΔV, which is linear, and not square-root-like, for small Q (see Section 11.4).[9] Therefore, one cannot measure the impact of a metaorder without being able to ascribe its constituent trades to a given investor.
- The square-root impact law applies to slow metaorders composed of several individual trades, but *not* to the individual trades themselves. Universality, if it holds, can only result from some mesoscopic properties of supply and demand that are insensitive to the way markets are organised at the micro-scale. This would explain why the square-root law holds equally well in the pre-HFT era (say, before 2005) as today, why it is insensitive to the tick size, etc.
- Conversely, at the single-trade level, microstructure effects (such as tick size and market organisation) play a strong role. In particular, the impact of a single market order, $\mathcal{R}(\upsilon, 1)$, does not behave like a square-root, although it behaves as a concave function of υ. This concavity has no immediate relation with the concavity of the square-root impact for metaorders.
- A square-root law for the *mean impact* of a metaorder is not related to the fact that the average *squared* price difference on some time interval grows linearly with the total exchanged volume:

$$\mathbb{E}\left[(m_{t+T} - m_t)^2 \middle| \sum_{n \in [t,t+T)} \upsilon_n = V\right] \propto V. \tag{12.9}$$

[9] In particular, the assumption that the aggregate order-flow imbalance ΔV in a given time window is mostly due to the presence of a single trader executing a metaorder of size Q, such that $\Delta V = Q+$ noise, is not warranted. For example, ΔV likely results from the superposition of *several* (parts of) overlapping metaorders with different sizes, signs and start times.

Equation (12.9) is a trivial consequence of the diffusive nature of prices when the total traded volume V scales like T. Cursorily, this relationship reads as "price differences grow as the square-root of exchanged volumes", but it tells us nothing about the average *directional* price change in the presence of a directional metaorder.

The square-root law is a genuinely challenging empirical fact. Most models in the literature (such as the famous Kyle model; see Chapter 15) predict linear impact. We review some possible scenarios that could explain the concavity of metaorder impact in Section 12.6.

12.4 Impact Decay

At time t_N, the pressure exerted by the metaorder stops and the price reverts back (on average) towards its initial value. Empirical data again suggests a universal behaviour, at least shortly after t_N, when impact relaxes quite abruptly. The decay then slows down considerably, and impact appears to reach a plateau.

For times beyond t_N, empirical data becomes increasingly noisy, because the variance of price changes itself increases linearly with t. The issue of long-term impact is thus extremely difficult to settle, in particular because one expects impact to decay as a slow power-law of time to compensate the long memory of order signs (see Section 13.2 and Chapter 19). Therefore, although the transient, square-root impact $\mathfrak{I}^{\text{trans.}}(Q,t)$ is most likely universal, the permanent impact component $\mathfrak{I}^{\infty}(Q)$ is a combination of a reaction component (as would happen in the Santa Fe model) and a genuine prediction component (resulting from the fact that some agents do successfully predict the future price).[10] Perhaps tautologically, one expects $\mathfrak{I}^{\infty}(Q)$ to be higher for agents with a high predictive power than for noise trades (who trade without information, e.g. for risk or cash-flow purposes).

Since metaorders themselves tend to be autocorrelated in time (the same decision to buy or sell might still be valid on the next day, week or month), it is difficult to separate the mechanical contribution from the informational contribution, especially in view of the slow decay of impact. An apparent permanent impact could well be due to the continuing pressure of correlated metaorders. Again, measuring the amount of true information revealed by trades is very difficult (see Chapter 20).

12.5 Impact Path and Slippage Costs

How can we be sure that the square-root law is really universal, and is not a consequence of measurement biases (such as those listed in Section 12.3.4) or conditioning by traders, perhaps by some sophisticated optimisation program (see

[10] See references in Section 12.7.

Section 21.2.3)? Thankfully, high-quality proprietary data allows us to dismiss most of these concerns and to check that the square-root law is actually extremely robust. Indeed, one finds that the impact of the initial $\phi\%$ of a metaorder of total size Q *also obeys the square-root law:*[11]

$$\mathfrak{I}^{\text{path}}(\phi Q, \phi T | Q, T) \cong \mathfrak{I}^{\text{peak}}(\phi Q) = \sqrt{\phi}\,\mathfrak{I}^{\text{peak}}(Q). \tag{12.10}$$

Hence, a partial metaorder consisting of the first $\phi\%$ of a metaorder of size Q behaves as a *bona fide* metaorder of size ϕQ, and also obeys the square-root impact law. This implies that the square-root impact is not a mere consequence of how traders choose the size and/or the execution horizon of their metaorders as a function of their prediction signal (see Section 21.2.3 for a detailed discussion of this point).

Interestingly, Equation (12.10) allows one to relate the execution shortfall $\mathscr{C}$ (or impact-induced slippage) to the peak impact

$$\mathscr{C}(Q) = \left\langle \sum_i \varepsilon \upsilon_i \cdot (m_i - m_1) \middle| Q \right\rangle = \sum_i \upsilon_i \mathfrak{I}^{\text{path}}\left(\sum_{j<i} \upsilon_j \right). \tag{12.11}$$

Using Equation (12.10) and taking the continuous limit $\upsilon_i \to Q\mathrm{d}\phi$,

$$\mathscr{C}(Q) \approx Q \int_0^1 \mathrm{d}\phi\, \mathfrak{I}^{\text{path}}(\phi Q, \phi T | Q, T) \approx Q \mathfrak{I}^{\text{peak}}(Q) \int_0^1 \mathrm{d}\phi\, \sqrt{\phi} = \frac{2}{3} Q \mathfrak{I}^{\text{peak}}(Q).$$

A square-root impact therefore leads to a square-root execution shortfall. Importantly, the execution shortfall per unit volume corresponds to $\frac{2}{3}$ of the peak impact. If impact was linear, the execution shortfall per unit volume would only be $\frac{1}{2}$ of the peak impact.

12.6 Conclusion

The most important conclusion of this chapter is that, contrarily to intuition, the average impact of a metaorder does not scale linearly with its volume Q. Instead, impact obeys the square-root law from Equation (12.6). This behaviour is contrary to what is predicted by many models, such as the Kyle model (see Chapter 15), which predict linear behaviour. This is an interesting case where empirical data compelled the finance community to accept that reality was fundamentally different from mainstream theory.

Since the mid-nineties, several stories have been proposed to account for the square-root impact law. The first attempt, due to the Barra Group and Grinold

[11] See, e.g., Moro, E., Vicente, J., Moyano, L. G., Gerig, A., Farmer, J. D., Vaglica, G., & Mantegna, R. N. (2009). Market impact and trading profile of hidden orders in stock markets. *Physical Review E*, 80(6), 066102; Donier, J., & Bonart, J. (2015). A million metaorder analysis of market impact on the Bitcoin. *Market Microstructure and Liquidity*, 1(02), 1550008.

and Kahn (1999),[12] argues that the square-root behaviour is a consequence of market-markers being compensated for their inventory risk. The reasoning is as follows. Assume that a metaorder of volume Q is absorbed by market-makers who will need to slowly offload their position later on. The amplitude of a potentially adverse move of the price during this unwinding phase is of the order of $\sigma_1 \sqrt{T_{\text{off}}}$, where T_{off} is the time needed to offload an inventory of size Q. It is reasonable to assume that T_{off} is proportional to Q and inversely proportional to the trading rate of the market V_1 (see Section 12.3.2). If market-makers respond to the metaorder by moving the price in such a way that their profit is of the same order as the risk they take, then it would follow that $\mathfrak{I} \propto \sigma_1 \sqrt{Q/V_1}$, as found empirically. However, this story assumes no competition between market-makers. Indeed, inventory risk is diversifiable over time, and in the long run averages to zero. Charging an impact cost compensating for the inventory risk of each metaorder would lead to formidable profits and would necessarily attract competing liquidity providers, eventually leading to a Y-ratio much smaller than 1.

Another theory, proposed by Gabaix et al. (2003), ascribes the square-root impact law to the fact that the optimal execution horizon T^* for informed metaorders of size Q grows like $T^* \sim \sqrt{Q}$ (see Section 21.2.3 for a detailed derivation). This theory argues that since the price is expected to move linearly in the direction of the trade during T^* (as information is revealed), the peak impact itself behaves as $\sqrt{Q}$. However, this scenario would imply that the impact path is linear in the executed quantity q, which is at odds with empirical data. As we noted in Section 12.5, the full impact path also behaves as a square-root of q (at least when the execution schedule is flat).

Recently, Farmer et al. (2013) proposed yet another theory that is very reminiscent of the Glosten–Milgrom model, which argues that the size of the bid–ask spread is actually set competitively (see Section 16.1). The theory assumes that metaorders arrive sequentially, with a volume Q distributed according to a power-law. Market-makers attempt to guess whether the metaorder will continue or stop at the next time step, and set the price such that it is a martingale and such that the average execution price compensates for the information contained in the metaorder (this condition is sometimes called "fair-pricing"). If the distribution of metaorder sizes behaves as $Q^{-5/2}$, these two conditions lead to a square-root impact law (see Sections 13.4.5 and 16.1.6 for more details). Although enticing, this theory has difficulty explaining why the square-root impact law appears to be much more universal than the distribution of the size of metaorders or of the autocorrelation of trade signs.[13] For example, the square-root law holds very precisely in Bitcoin markets, where the distribution of metaorder sizes behaves

[12] See also Section 12.7, in particular Zhang, Y.-C. (1999).
[13] On this last point, see Mastromatteo, I., Tóth, B., & Bouchaud, J. P. (2014). Agent-based models for latent liquidity and concave price impact. *Physical Review E*, 89(4), 042805.

as Q^{-2}, rather than $Q^{-5/2}$, and at a time where market-making was much less competitive.

Closer to the spirit of the present book, Tóth et al. (2011) proposed an alternative theory based on a dynamical description of supply and demand. This approach provides a natural statistical interpretation for the square-root law and its apparent universality. We provide a detailed overview of this idea in Chapters 18 and 19.

Take-Home Messages

(i) By studying the impact of metaorders, it is possible to quantify how market participants' actions impact prices at a mesoscopic scale.
(ii) According to the square-root law, the immediate price impact of a metaorder behaves as a square root of the order's volume.
(iii) The square-root law holds in a wide range of market scenarios, and can be observed empirically for both informed and uninformed metaorders.
(iv) When studying metaorder impact, it is important to avoid the many biases that could influence measurements.
(v) After a metaorder stops, the price reverts (on average) towards its pre-trade value. How much of the impact remains long after the execution (i.e. the permanent impact) depends on the predictive power or informational content of the metaorder.

12.7 Further Reading

Treynor, J. L. (1988). Portfolio insurance and market volatility. *Financial Analysts Journal*, 44(6), 71–73.

Empirical Evidence for a Square-Root Law

Loeb, T. F. (1983). Trading cost: The critical link between investment information and results. *Financial Analysts Journal*, 39, 39–44.
Torre, N., & Ferrari, M. (1997). *Market impact model handbook*. BARRA Inc. Available at www.barra.com/newsletter/nl166/miminl166.asp.
Almgren, R., Thum, C., Hauptmann, E., & Li, H. (2005). Direct estimation of equity market impact. *Risk*, 18(5752), 10.
Kissel, R., & Malamut, R. (2006). Algorithmic decision-making framework. *Journal of Trading*, 1(1), 12–21.
Moro, E., Vicente, J., Moyano, L. G., Gerig, A., Farmer, J. D., Vaglica, G., & Mantegna, R. N. (2009). Market impact and trading profile of hidden orders in stock markets. *Physical Review E*, 80(6), 066102.
Tóth, B., Lemperiere, Y., Deremble, C., De Lataillade, J., Kockelkoren, J., & Bouchaud, J. P. (2011). Anomalous price impact and the critical nature of liquidity in financial markets. *Physical Review X*, 1(2), 021006.

Engle, R., Ferstenberg, R., & Russell, J. (2012). Measuring and modelling execution cost and risk. *The Journal of Portfolio Management*, 38(2), 14–28.

Mastromatteo, I., Tóth, B., & Bouchaud, J. P. (2014). Agent-based models for latent liquidity and concave price impact. *Physical Review E*, 89(4), 042805.

Donier, J., & Bonart, J. (2015). A million metaorder analysis of market impact on the Bitcoin. *Market Microstructure and Liquidity*, 1(02), 1550008.

Zarinelli, E., Treccani, M., Farmer, J. D., & Lillo, F. (2015). Beyond the square root: Evidence for logarithmic dependence of market impact on size and participation rate. *Market Microstructure and Liquidity*, 1(02), 1550004.

Kyle, A. S., & Obizhaeva, A. A. (2016). Large bets and stock market crashes. https://ssrn.com/abstract=2023776.

Bacry, E., Iuga, A., Lasnier, M., & Lehalle, C. A. (2015). Market impacts and the life cycle of investors orders. *Market Microstructure and Liquidity*, 1(02), 1550009.

Tóth, B., Eisler, Z., & Bouchaud, J.-P. (2017). The short-term price impact of trades is universal. https://ssrn.com/abstract=2924029.

Several internal bank documents have also reported such a concave impact law, e.g.: Ferraris, A. (2008). *Market impact models*. Deutsche Bank internal document, http://dbquant.com/Presentations/Berlin200812.pdf.

Decay of Metaorder Impact

Bershova, N., & Rakhlin, D. (2013). The non-linear market impact of large trades: Evidence from buy-side order flow. *Quantitative Finance*, 13(11), 1759–1778.

Brokmann, X., Serie, E., Kockelkoren, J., & Bouchaud, J. P. (2015). Slow decay of impact in equity markets. *Market Microstructure and Liquidity*, 1(02), 1550007.

Gomes, C., & Waelbroeck, H. (2015). Is market impact a measure of the information value of trades? Market response to liquidity vs. informed metaorders. *Quantitative Finance*, 15(5), 773–793.

see also: Moro, E. et al.; Zarinelli, E. et al. in the previous subsection.

Theories About the Square-Root Law

Zhang, Y. C. (1999). Toward a theory of marginally efficient markets. *Physica A: Statistical Mechanics and Its Applications*, 269(1), 30–44.

Grinold, R. C., & Kahn, R. N. (2000). *Active portfolio management*. McGraw-Hill.

Gabaix, X., Gopikrishnan, P., Plerou, V., & Stanley, H. E. (2003). A theory of power-law distributions in financial market fluctuations. *Nature*, 423(6937), 267–270.

Barato, A. C., Mastromatteo, I., Bardoscia, M., & Marsili, M. (2013). Impact of meta-order in the Minority Game. *Quantitative Finance*, 13(9), 1343–1352.

Farmer, J. D., Gerig, A., Lillo, F., & Waelbroeck, H. (2013). How efficiency shapes market impact. *Quantitative Finance*, 13(11), 1743–1758.

Mastromatteo, I., Tóth, B., & Bouchaud, J. P. (2014). Anomalous impact in reaction-diffusion financial models. *Physical Review Letters*, 113(26), 268701.

Donier, J., Bonart, J., Mastromatteo, I., & Bouchaud, J. P. (2015). A fully consistent, minimal model for non-linear market impact. *Quantitative Finance*, 15(7), 1109–1121.

Donier, J., & Bouchaud, J. P. (2016). From Walras auctioneer to continuous time double auctions: A general dynamic theory of supply and demand. *Journal of Statistical Mechanics: Theory and Experiment*, 2016(12), 123406.

Pohl, M., Ristig, A., Schachermayer, W., & Tangpi, L. (2017). The amazing power of dimensional analysis: Quantifying market impact. arXiv preprint arXiv:1702.05434.

see also: K. Rodgers, https://mechanicalmarkets.wordpress.com/2016/08/15/price-impact-in-efficient-markets/.

PART VI

Market Dynamics at the Micro-Scale

Introduction

In the previous parts of this book, we have addressed the empirical and conceptual aspects of the relationship between order flow and price formation. Let us recap what we have so far: a highly predictable order flow, which is a consequence of traders splitting their metaorders into small chunks due to liquidity shortage, and some observations of how trades impact prices. In this part, we will begin our modelling efforts of market dynamics at the micro-scale by introducing a class of simple models of price formation, often dubbed *propagator models.* When fitted on data, these models provide a route to computing many different quantities of interest, including the aggregate impact of individual orders and the impact of metaorders.

We will use propagator models to explore two interesting topics. The first topic is how to infer the reaction impact from the observed impact of real trades. By making explicit the relationship between these two types of impact, propagator models make it possible to estimate the reaction impact from the empirical *response function,* which we have already encountered in Chapter 11. In this view, the prediction component of impact is simply the reaction component of *other* (past and future) trades. Correlations in the order flow therefore play a central role. As we will discuss in this part, the outcome of such empirical calculations is quite intriguing: the reaction impact of orders is neither permanent (as it decays to zero in the long run) nor transient (as this decay is so slow that it takes an extremely long time to vanish).

The second topic is understanding how prices can remain unpredictable when order flows exhibit strong, long-range autocorrelations. As we discussed in the first part of this book, the signature plot of market prices is nearly flat on all scales, which indicates that price returns are linearly unpredictable. How is this possible when the order flow itself is so predictable? As we will see, the answer stems from the dynamic nature of price impact. By performing detailed calculations, we will show that the slowly decaying reaction impact *exactly* compensates for the autocorrelation in order flow, so as to produce diffusive prices. We will also present another angle for understanding this fine balance: if the impact of an order is proportional to its *surprise*, and not just its direction, then by definition prices must be unpredictable. As we will show, such history-dependent impact models are equivalent to propagator models in some particular cases.

Thanks to their simplicity, propagator models are powerful and versatile tools for modelling markets. In addition to the basic modelling framework, we will consider a selection of variants and extensions, including considering several types of orders with different reaction impacts. We will also discuss how one can compare the impact between different investors and analyse the cross-asset impact of trades. As we will discuss in the last part of this book, propagator models can also form the basis of simple trading-optimisation schemes. Therefore, in addition to their usefulness as theoretical tools for understanding complex market behaviour, propagator models are also helpful for market practitioners who seek to optimise the execution of their trades.

13
The Propagator Model

It is better to be roughly right than precisely wrong.

(John Maynard Keynes)

In the previous two chapters, we have considered how individual market orders and whole metaorders impact the mid-price. In both cases, our study of impact has primarily revolved around measuring the change in price that occurs contemporaneously with the specified market order or metaorder. However, as we briefly discussed in Sections 11.3.1 and 12.4, impact is not only felt immediately, but rather evolves dynamically. In fact, the "reactional" component of impact is expected to be mostly transient. In this chapter, we introduce a model that seeks to describe how impact evolves over time, in such a way to avoid the **efficiency paradox** created by the long memory of order signs.

Similarly to the previous chapters, we restrict our attention to market order time, in which we advance the clock by a single unit for each market order arrival. In this framework, we introduce and study a simple, linear model called the **propagator model**, which expresses price moves in terms of the influence of past trades. We demonstrate that to avoid creating statistical arbitrage opportunities and ensure **market efficiency**, the propagator model must possess certain properties, which we subsequently use to refine the model towards a more specific form. Despite its simplicity, the propagator model provides deep insight into the nature of price impact in financial markets. To conclude the chapter, we frame the propagator model in a wider context and discuss various conceptual points and extensions.

13.1 A Simple Propagator Model

In Chapter 11, we discussed how strongly, on average, the arrival of a market order of a given volume υ impacts the mid-price. After the arrival occurs, there is no reason why the impact of a market order should remain visible indefinitely, because that would mean that the order's arrival shifted the entire supply and demand curves forever. As we will show in this section, if that was the case, then the long-range autocorrelations of market order signs would create strong autocorrelations in returns. This is at odds with widespread empirical observations of signature plots (see the discussion in Section 10.3), which are nearly flat. Therefore, market efficiency imposes that a large fraction of a market order's impact must relax over time.

To illustrate why this is the case, we first consider a naive model of impact, in which every trade causes a fixed, constant lag-1 impact that persists permanently. To consider this in detail, let us consider the mathematical properties of the return series that such a process would generate. Let

$$r_t := m_{t+1} - m_t \tag{13.1}$$

denote the change in mid-price from immediately before the tth event until immediately before the $(t+1)^{\text{th}}$ event. We assume that each trade has a mean permanent impact equal to G_1.[1] Using this notation, our simple model translates into

$$r_t = G_1 \varepsilon_t + \xi_t, \tag{13.2}$$

where ξ_t is a random noise term that models fluctuations around the average impact contribution and that also captures price changes not directly attributed to trading (e.g. quote changes that occur due to the release of public news, even without any trades occurring). We will assume here that ξ_t is independent of the order flow ε_t, with $\mathbb{E}[\xi] = 0$ and $\mathbb{E}[\xi^2] = \Sigma^2$.

Given the mid-price m_{t_0} at some previous time $t_0 \leq t$, the mid-price m_t is given by

$$m_t = m_{t_0} + G_1 \sum_{t_0 \leq n < t} \varepsilon_n + \sum_{t_0 \leq n < t} \xi_n. \tag{13.3}$$

Equation (13.3) highlights the non-decaying nature of impact in this model: whatever happened at all previous times is permanently imprinted on the price at time t.

When the signs of the trades are independent random variables with zero mean, this simple model predicts that the response function (see Equation (11.6)) is simply constant:

$$\begin{aligned} \mathcal{R}(\ell) &= \mathbb{E}[\varepsilon_t \cdot (m_{t+\ell} - m_t)], \\ &= \mathbb{E}\left[\varepsilon_t \cdot [(m_{t+\ell} - m_{t+1}) + (m_{t+1} - m_t)]\right], \\ &= G_1. \end{aligned} \tag{13.4}$$

[1] In the present context, G_1 coincides with the quantity $\mathcal{R}(1)$ considered in the previous chapters, but this correspondence no longer holds in the presence of sign autocorrelations, which we will address in the subsequent sections.

The model also makes the following prediction for the mid-price variogram (see Equation (2.1)):

$$\begin{aligned}\mathcal{V}(\ell) &:= \mathbb{V}[m_{t+\ell} - m_t], \\ &= \left(G_1^2 + \Sigma^2\right)\ell.\end{aligned} \tag{13.5}$$

Therefore, the model predicts that the mid-price obeys a pure diffusion with lag-independent diffusion coefficient (or squared volatility)

$$\mathcal{D}(\ell) = \frac{\mathcal{V}(\ell)}{\ell} = G_1^2 + \Sigma^2, \tag{13.6}$$

as is observed empirically in liquid markets (except perhaps at small ℓ).

Despite the appealing simplicity of these results, this naive model ignores an important ingredient of real markets: the autocorrelation of the order-sign series ε_t. As we discussed in Chapter 10, real market order sign series have strong **long memory**, with an autocorrelation function $C(\ell)$ that decays approximately according to a power law. Incorporating this autocorrelation structure into the model, we arrive at the following expression for the autocovariance of price returns for $\ell > 0$:

$$\begin{aligned}\mathbb{E}[r_t r_{t+\ell}] &= G_1^2 \mathbb{E}[\varepsilon_t \varepsilon_{t+\ell}], \\ &= G_1^2 C(\ell).\end{aligned}$$

Therefore, if we incorporate the autocorrelation structure of market order signs, then our simple model predicts that price returns become strongly autocorrelated. This would violate market efficiency, because price returns would be strongly predictable. If prices are to be efficient, then we must conclude that the empirically observed long memory of order flow rules out the above assumption of impact being a simple constant (and thus permanent).

How might we resolve this conundrum? In the next section, we will consider two possibilities: relaxing the assumption that price impact is permanent, or relaxing the assumption that lag-1 impact is independent of the past order flow. As we will see from our analysis, these two seemingly different approaches are actually closely related.

13.2 A Model of Transient Impact and Long-Range Resilience

As we have noted several times, the long-range autocorrelation of market order signs is an interesting and puzzling property of financial markets. The fact that returns remain uncorrelated despite this long-range autocorrelation in market order signs suggests that markets are somehow resilient to these clusters of highly predictable market orders in the same direction. To understand this interesting property more clearly, let's consider a more general version of Equation (13.3), in which we regard the impact not as a constant G_1, but instead as a function $G(\ell)$ that describes how the impact of the trade propagates through time. We call the function $G(\cdot)$ the **propagator**.

Using this notation, we can generalise Equation (13.3) as:

$$m_t = m_{t_0} + \sum_{t_0 \le n < t} G(t-n)\varepsilon_n + \sum_{t_0 \le n < t} \xi_n, \tag{13.7}$$

where ξ_n again captures all price moves not directly induced by trades (e.g. due to quote revisions during news announcements).

Clearly, the functional form of the propagator $G(\cdot)$ influences the values of m_t in the model. For $G(\ell)$ to capture the transient component of impact, it seems reasonable that $G(\ell)$ should decrease with increasing ℓ. The long-time limit

$$G_\infty := \lim_{\ell \to \infty} G(\ell)$$

captures the permanent reaction impact of a market order. The simple, constant propagator that we considered in Equation (13.3) corresponds to the function $G(\ell) = G_1$ for all lags.

For the sake of simplicity, we will assume that impact is independent of trade volume, time and market conditions. All three of these assumptions are gross oversimplifications that do not accurately reflect the behaviour observed in real markets (see, e.g., Sections 11.3.1 and 11.3.2), but by adopting them we greatly simplify the mathematics associated with formulating a first model. We will revisit and relax these assumptions within a more general model in Chapter 14.

Equation (13.7) defines a family of linear, **transient impact models** (TIM). Within this family of models, reaction impact is parameterised by the propagator $G(\ell)$, which encodes how a market order executed at time t impacts the price at a later time $t + \ell$.

We can use the model in Equation (13.7) to write an expression for the return series:

$$r_t = \underbrace{G_1 \varepsilon_t}_{\text{immediate impact}} + \underbrace{\sum_{n<t} K(t-n)\varepsilon_n}_{\text{impact decay}} + \underbrace{\xi_t}_{\text{noise}}, \tag{13.8}$$

where $G_1 := G(1)$, and for $\ell \ge 1$,

$$K(\ell) := G(\ell+1) - G(\ell),$$

which is the discrete derivative of $G(\ell)$. Therefore, the model predicts the return at time t is due not only to the most recent trade (whose contribution is $G_1 \varepsilon_t$), but also to *all* previous trades. Introducing the convention $G(0) = 0$, Equation (13.8) can be written in a more compact form:

$$r_t = \sum_{n \leq t} K(t-n)\varepsilon_n + \xi_t. \tag{13.9}$$

As we will now demonstrate, the model allows us to compute the response function $\mathcal{R}(\ell)$ (or observed impact) and the price variogram $\mathcal{V}(\ell)$ for an arbitrary function $G(\ell)$ and an arbitrary correlation between trades $C(\ell)$.

13.2.1 The Response Function

By plugging Equation (13.7) into Equation (13.4), one readily obtains the following expression for the **response function**:

$$\mathcal{R}(\ell) = G(\ell) + \sum_{0<n<\ell} G(\ell-n)C(n) + \sum_{n>0} [G(\ell+n) - G(n)]\, C(n). \tag{13.10}$$

If the terms in the market order sign series were uncorrelated, it would follow that $C(n) = 0$ for all $n > 0$, so only the first term of the expression would be non-zero, leading to $\mathcal{R}(\ell) = G(\ell)$. In this case, the observed impact $\mathcal{R}(\ell)$ and the bare reaction impact $G(\ell)$ would be one and the same thing. In reality, however, market order signs are strongly autocorrelated, so we need to consider the full expression.

In this model, the lag-1 impact $\mathcal{R}(1)$ (which we measured in Chapter 11) includes the decay of the impact of all previous correlated trades, and is given by

$$\mathcal{R}(1) = \sum_{n \geq 0} K(n)C(n) = G_1 + \sum_{n>0} K(n)C(n). \tag{13.11}$$

Because the impact of an individual trade should decay with time, one expects $K(n > 0) \leq 0$. If trade signs are positively autocorrelated, one concludes that $\mathcal{R}(1) \leq G_1$.

The asymptotic response function is given by

$$\mathcal{R}_\infty = G_\infty + \lim_{\ell \to \infty} \left\{ \sum_{0<n<\ell} G(\ell-n)C(n) + \sum_{n>0} [G(\ell+n) - G(n)]\, C(n) \right\}, \tag{13.12}$$

which in fact requires some special conditions for the result to be finite (i.e. non-divergent). We return to this discussion in Section 13.2.3.

For any $\ell \geq 0$, Equation (13.10) relates the bare reaction impact $G(\ell)$, which is *not* directly observable, to two quantities that are: the observed impact $\mathcal{R}(\ell)$ and the autocorrelation of trade signs $C(\ell)$. One could therefore use this equation to estimate $G(\ell)$ from empirical data. However, this direct method is very sensitive to finite-size effects, and therefore provides poor estimates of $G(\ell)$: in practice, ℓ

must be smaller than the maximum lag L that allows a reasonable estimation of $\mathcal{R}$ and C, beyond which it becomes difficult to separate the signal from noise.

An alternative approach to this problem is to use the lagged **sign-return correlation**, which is defined as

$$S(\ell) := \mathbb{E}[r_{t+\ell} \cdot \varepsilon_t] = \mathcal{R}(\ell + 1) - \mathcal{R}(\ell). \tag{13.13}$$

The function $S(\cdot)$ can be regarded as the discrete derivative of $\mathcal{R}(\cdot)$. It is straightforward to find the analogue of Equation (13.10) for $S(\cdot)$:

$$S(\ell) = \sum_{n \geq 0} C(|n - \ell|)K(n), \tag{13.14}$$

where we have again used the convention $G(0) = 0$. This equation can also be written in matrix form, as

$$\vec{S} = \mathbf{C}\vec{K}, \tag{13.15}$$

where the matrix $\mathbf{M}$ is given by

$$\mathbf{C}_{n,m} = C(|n - m|); \qquad 0 \leq n, m \leq L.$$

By inverting the matrix $\mathbf{C}$, Equation (13.15) can be rewritten as

$$K(\ell) = \sum_{n=0}^{L} (\mathbf{C}^{-1})_{\ell,n} S(n).$$

Observe that calculating $K(\ell)$ only requires knowledge of the values of S for positive lags.

In Figure 13.1, we plot $\mathcal{R}(\ell)$ and $G(\ell)$ for one large-tick stock and one small-tick stock. As expected, $G(\ell)$ decays with increasing ℓ. Upon closer scrutiny, one finds that the long-time behaviour of $G(\ell)$ decays approximately according to a power-law,

$$G(\ell) \approx_{\ell \gg 1} \frac{\Gamma_\infty}{\ell^\beta}, \qquad \beta < 1,$$

(see the inset of Figure 13.1). This is the precise definition of the "resilience" property that we discussed at the beginning of Section 13.2: the propagator $G(\ell)$ decays all the way to zero, but so slowly that its sum over all ℓ is divergent. We call this property **long-range resilience**, because it parallels the definition of long memory (see Section 10.1) when the decay exponent γ of $C(\ell)$ is less than 1.

In summary, the impact of a trade dissipates over time, but does so very slowly. As we will now show by considering the price variogram $\mathcal{V}(\ell)$, this slow decay turns out to play a crucial role in ensuring that the mid-price remains diffusive in the face of the long-range autocorrelations of market order signs.

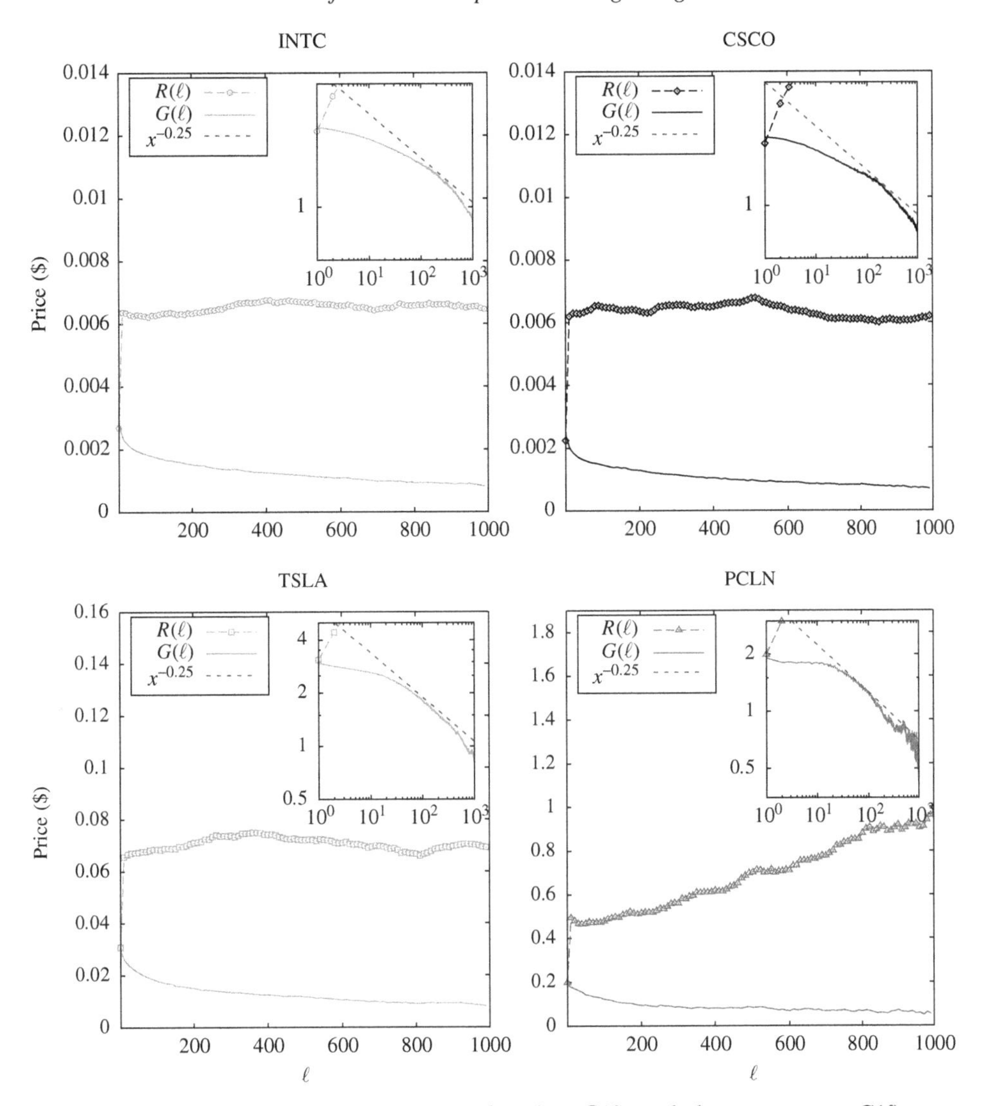

Figure 13.1. The empirical response function $\mathcal{R}(\ell)$ and the propagator $G(\ell)$, calculated from the empirical $\mathcal{R}(\ell)$ and $C(\ell)$, for (top left) INTC, (top right) CSCO, (bottom left) TSLA and (bottom right) PCLN. Inset: The same plots in doubly logarithmic axes. The dashed black lines depict a power-law with exponent −0.25.

13.2.2 The Variogram

Once $G(\ell)$ is determined, the propagator model provides a definite prediction for the **variogram**, shown in Figure 13.2. In full generality, the equation relating $\mathcal{V}(\ell)$ to $G(\ell)$ and $C(\ell)$ reads

$$\mathcal{V}(\ell) = \sum_{0 \le n < \ell} G^2(\ell - n) + \sum_{n > 0} [G(\ell + n) - G(n)]^2 + 2\Psi(\ell) + \Sigma^2 \ell, \tag{13.16}$$

where $\Psi(\ell)$ is the correlation-induced contribution to $\mathcal{V}$:

$$\Psi(\ell) = \sum_{0\le j<k<\ell} G(\ell-j)G(\ell-k)C(k-j) + \sum_{0<j<k} [G(\ell+j)-G(j)]\,[G(\ell+k)-G(k)]\,C(k-j) + \sum_{0\le j<\ell}\sum_{k>0} G(\ell-j)\,[G(\ell+k)-G(k)]\,C(k+j).$$

Let us analyse the large-ℓ behaviour of $\mathcal{V}(\ell)$, in the case where $C(\ell)$ decays as $c_\infty \ell^{-\gamma}$ with $\gamma < 1$ (which, as we saw in Chapter 10, is a good approximation of the large-ℓ behaviour of $C(\ell)$ for real markets). Motivated by the numerical result that we obtained for $G(\ell)$, we assume that for large ℓ, the propagator $G(\ell)$ decays as a power law $\Gamma_\infty \ell^{-\beta}$. When $0<\beta<1/2$ and $0<\gamma<1$, asymptotic analysis (with $\ell \gg 1$) of the first two terms of Equation (13.16) reveals that they both grow sub-linearly with ℓ:

$$\sum_{0\le n<\ell} G^2(\ell-n) \approx \sum_{0<n\le\ell} \frac{\Gamma_\infty^2}{\ell^{2\beta}} \approx \frac{\Gamma_\infty^2}{1-2\beta}\ell^{1-2\beta} \ll \Sigma^2\ell,$$

and

$$\sum_{n>0} [G(\ell+n)-G(n)]^2 \approx \Gamma_\infty^2 \ell^{1-2\beta} \int_0^\infty \mathrm{d}u \left(\frac{1}{(1+u)^\beta} - \frac{1}{u^\beta}\right)^2 = A(\beta)\Gamma_\infty^2 \ell^{1-2\beta} \ll \Sigma^2\ell,$$

where A is a finite number,

$$A = \frac{1}{1-2\beta}\left(\frac{2\Gamma[2\beta]\Gamma[1-\beta]}{\Gamma[\beta]} - 1\right); \qquad 0<\beta<\frac{1}{2}.$$

Therefore, only the contribution $\Psi(\ell)$ in Equation (13.16) may compete with the long-term diffusion behaviour of the price induced by the public news term ξ_t. Asymptotic analysis of $\Psi(\ell)$ yields the following contribution to the diffusion coefficient:

$$D_\Psi(\ell) := \frac{\Psi(\ell)}{\ell} \approx \Gamma_\infty^2 c_\infty I(\gamma,\beta)\, \ell^{1-2\beta-\gamma},$$

where $I(\gamma,\beta) > 0$ is a (complicated but finite) numerical integral.

We first examine what would happen if the propagator $G(\ell)$ did not decay at all ($\beta = 0$). This case corresponds to the assumption that impact is permanent. In this case, in the presence of long memory of the order flow $0<\gamma<1$, one finds that $D_\Psi(\ell) \propto \ell^{1-\gamma}$, which grows with increasing ℓ. This corresponds to a super-diffusive price – or, in financial language, the presence of persistent trends in the price series. This is the **efficiency paradox** created by the long-range correlation of market order signs.

As the propagator decays more quickly (i.e. as β increases), super-diffusion is less pronounced (i.e. $D_\Psi(\ell)$ grows more slowly), until

$$\beta = \beta_c := (1-\gamma)/2, \tag{13.17}$$

for which $D_\Psi(\ell)$ is a constant, independent of ℓ. Adding the variance of the "news" contribution Σ^2, the long-term volatility reads

$$\mathcal{D}_\infty = \Gamma_\infty^2 c_\infty I(\gamma,\beta_c) + \Sigma^2.$$

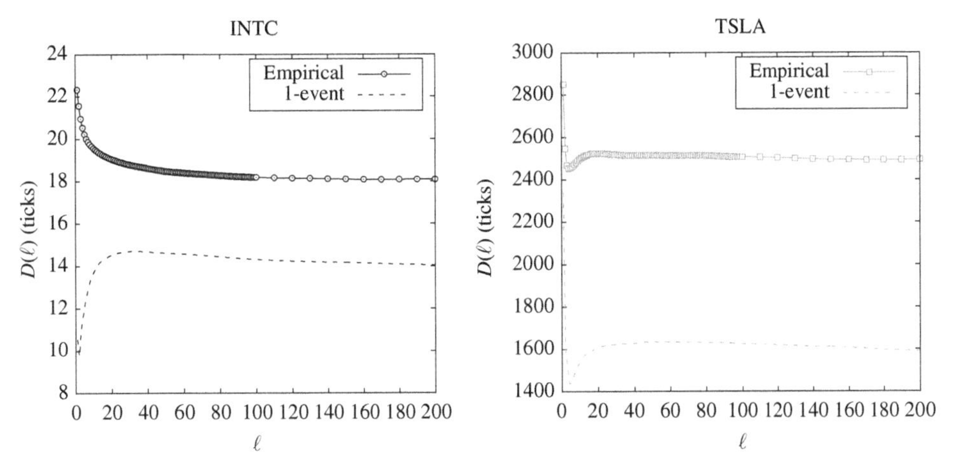

Figure 13.2. (Markers and solid curves) The empirical values of the function $\mathcal{D}(\ell)$ and (dashed curve) its approximation with the 1-event (market order) propagator model, Equation (13.16), with $\Sigma^2 = 0$. The left panel shows the results for INTC, which is a large-tick stock, and the right panel shows the results for TSLA, which is a small-tick stock. Remarkably, the trade-only contribution accounts for 0.65–0.8 of the long-term squared volatility. However, the model completely fails to capture the short-term structure of $\mathcal{D}(\ell)$, which suggests that one should also take into account some time-dependence of $G(\ell)$ (which might depend on event type and past history; see Chapter 14).

When $\beta > \beta_c$, $D_\Psi(\ell)$ decreases with ℓ, as do the contributions of the first two terms of Equation (13.16).[2] The lag-dependent volatility $\mathcal{D}(\ell)$ is in this case enhanced at short lags by the transient impact contribution, resulting in sub-diffusive prices at short lags.

Based on the calculations we have performed throughout this section, we can now reach an important conclusion: provided that the impact of single trades is transient (i.e. $\beta > 0$) and with a decay that precisely offsets the long-range autocorrelations in market order signs (i.e. $\beta = \beta_c := (1-\gamma)/2$), then the long-range autocorrelation of market order signs *is* compatible with a random-walk behaviour of the mid-price, and thus with statistical efficiency.

13.2.3 A Fine Balance

When the "fine balance" condition $\beta = \beta_c$ holds, we are in a rather odd situation where impact is not permanent (since in the long-time limit $\ell \to \infty$, the propagator $G(\ell)$ is zero) but is not really transient either, because the decay is extremely slow. The convolution of this **semi-permanent impact** balances exactly the slow decay of trade correlations to produce a finite contribution to the long-term volatility.

[2] Since $\gamma < 1$, it follows that $2 - 2\beta - \gamma > 1 - 2\beta$, and therefore at large ℓ the term $D_\Psi(\ell)$ always dominates the first two contributions of Equation (13.16) to $\mathcal{D}(\ell)$.

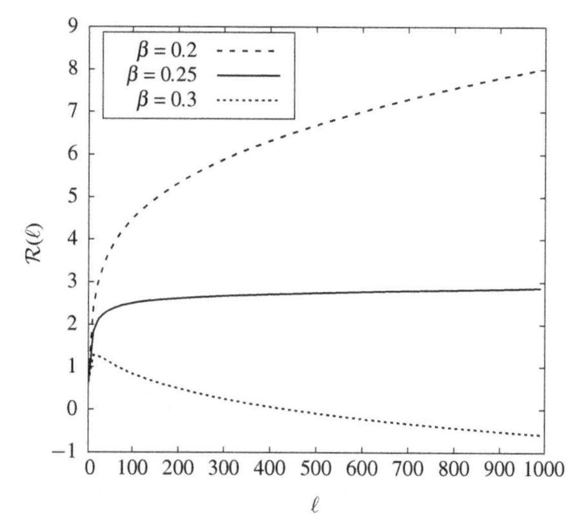

Figure 13.3. Calculation of $\mathcal{R}(\ell)$ for a synthetic propagator $G(\ell) \sim \ell^{-\beta}$ and a synthetic correlation function $C(\ell) \sim \ell^{-1/2}$, with $\beta \in \{0.2, 0.25, 0.35\}$. Observe that after an initial rapid increase, $\mathcal{R}$ is approximately flat for the parameter choice $\beta = \beta_c = 0.25$.

In fact, plugging a power law for $G(\ell)$ into Equation (13.10) leads to a response function $\mathcal{R}(\ell)$ that for large ℓ behaves as

$$\mathcal{R}(\ell) \approx_{\ell \gg 1} \Gamma_\infty c_\infty \frac{\Gamma[1-\gamma]}{\Gamma[\beta]\Gamma[2-\beta-\gamma]} \left[\frac{\pi}{\sin \pi\beta} - \frac{\pi}{\sin \pi(1-\beta-\gamma)} \right] \ell^{1-\beta-\gamma} \tag{13.18}$$

provided $\beta + \gamma < 1$. We display the explicit numerical prefactor to highlight that when $\ell \to \infty$, the value of $\mathcal{R}(\ell)$ diverges to $+\infty$ whenever $\beta < \beta_c$, and diverges to $-\infty$ whenever $\beta > \beta_c$. If $\beta = \beta_c$, then

$$\frac{\pi}{\sin \pi\beta_c} = \frac{\pi}{\sin \pi(1-\beta_c-\gamma)},$$

so the leading term in $\mathcal{R}(\ell)$ vanishes. In this case, one must also estimate the sub-leading term, which saturates to a finite value $\mathcal{R}(\infty)$. Hence, for $\beta = \beta_c$, the total asymptotic response to a trade (including all future correlated trades) is non-zero, even though the impact of each individual trade $G(\ell)$ decays to zero!

Figure 13.3 summarises this rather peculiar situation. Interestingly, we see that financial markets operate in a fragile regime where liquidity providers and liquidity takers offset each other, such that most of the predictable patterns that would otherwise be created by their predictable actions are removed from the price trajectory. Only an unpredictable contribution remains, even at the highest frequencies. We will return to the question of how this occurs in Chapter 14, where we will attempt to model the impact of all LOB events, not only that of market orders.

13.3 History-Dependent Impact Models

An alternative way to ensure the statistical efficiency of prices is to assume that the price impact of each order is permanent but history-dependent. Indeed, one way to rephrase the fact that price returns are difficult to predict is that most of the available information at time t is already included in the price, such that price moves only occur if something truly unexpected happens. In other words, since the signs of future trades are predictable, only the **surprise component** of the next sign should have an impact on the price. Why should this be the case? One possible explanation is that high-frequency participants act in such a way as to remove any predictability – e.g. by providing liquidity on the side where the next trade is most likely to happen (see Section 11.3.4).

13.3.1 The Impact of Surprises

We extend the expression in Equation (13.2) to formulate a version of the propagator model in which returns are unpredictable, but still linear in ε_t. There are two main ingredients to this approach. The first is to replace the (highly predictable) order-sign term ε_t term with the corresponding sign surprise $(\varepsilon_t - \widehat{\varepsilon}_t)$, where $\widehat{\varepsilon}_t$ is the conditional expectation of ε_t:

$$\widehat{\varepsilon}_t := \mathbb{E}_{t-1}[\varepsilon_t].$$

In this way, we replace the ε_t term, whose behaviour depends only on the market order arrival at time t, with a term that also reflects the previous market order arrivals. The second ingredient to replace the constant term G_1 in Equation (13.2) with a term that may depend on t (to reflect, for example, the spread dynamics). Together, these ingredients lead to the following **history-dependent impact model** (HDIM):

$$r_t = G_{1,t} \times (\varepsilon_t - \widehat{\varepsilon}_t) + \xi_t. \tag{13.19}$$

Since neither the sign surprise $(\varepsilon_t - \widehat{\varepsilon}_t)$ nor ξ_t can be predicted, it follows that for any immediate impact $G_{1,t}$,

$$\mathbb{E}_{t-1}[r_t] = G_{1,t}\mathbb{E}_{t-1}\left[\varepsilon_t - \mathbb{E}_{t_1}[\varepsilon_t]\right] + \mathbb{E}_{t-1}[\xi_t] = 0.$$

Therefore, prices in the model are a martingale, even when the sign of the next trade is highly predictable. This is a plausible first approximation to how prices behave in real markets, although one observes that real prices deviate from the martingale property at very high frequencies (see Figure 13.2 and Section 16.3).

Now, either the sign of the tth transaction matches the sign of $\widehat{\varepsilon}_t$, or it does not. Let $\mathbb{E}_t^+[r_t]$ denote the expected ex-post value of the return of the tth transaction, given that ε_t matches that of $\widehat{\varepsilon}_t$, and let $\mathbb{E}_t^-[r_t]$ denote the expected ex-post value of the return of the tth transaction, given that ε_t does not match that of $\widehat{\varepsilon}_t$. By definition, the most likely outcome is that ε_t does match the sign of its predictor

$\widehat{\varepsilon_t}$, which happens with a probability given by $(1 + |\widehat{\varepsilon_t}|)/2$ (see Equation (13.21), below).

The absence-of-predictability criterion $\mathbb{E}_{t-1}[r_t] = 0$ can be rewritten as:

$$\frac{1+|\widehat{\varepsilon_t}|}{2}\mathbb{E}_t^+[r_t] + \frac{1-|\widehat{\varepsilon_t}|}{2}\mathbb{E}_t^-[r_t] = 0 \Rightarrow \left|\frac{\mathbb{E}_t^+[r_t]}{\mathbb{E}_t^-[r_t]}\right| = \frac{1-|\widehat{\varepsilon_t}|}{1+|\widehat{\varepsilon_t}|} \leq 1.$$

Therefore, the most likely outcome has the smallest impact. This generalises the result on $\mathcal{R}_\pm(1)$ given in Section 11.3.4, which was based on a qualitative argument. In the present picture, each trade has a permanent impact, but the impact depends on the past order flow and on its predictability.

13.3.2 The DAR Model

We now assume that the time series of market order signs is well modelled by a **discrete auto-regressive** process (see Section 10.4.2). In this framework, the best predictor $\widehat{\varepsilon_t} = \mathbb{E}_{t-1}[\varepsilon_t]$ of the next trade sign, just before it happens, can be written as

$$\widehat{\varepsilon_t} = \sum_{k=1}^{p} \mathbb{K}(k)\varepsilon_{t-k}. \qquad (13.20)$$

The backward-looking kernel $\mathbb{K}(k)$ can be inferred from the sign autocorrelation function $C(\ell)$ using the Yule–Walker equation (10.6). The probability that the next sign is ε_t is then given by:

$$\mathbb{P}_{t-1}(\varepsilon_t) = \frac{1+\varepsilon_t\widehat{\varepsilon_t}}{2}. \qquad (13.21)$$

Because ε_t is a binary random variable, Equation (13.21) is equivalent to $\widehat{\varepsilon_t} = \mathbb{E}_{t-1}[\varepsilon_t]$.

Inserting Equation (13.20) into Equation (13.19), we arrive at

$$r_t = G_{1,t}\varepsilon_t - G_{1,t}\sum_{k=1}^{p} \mathbb{K}(k)\varepsilon_{t-k} + \xi_t. \qquad (13.22)$$

In the case where $G_{1,t}$ does not depend on time, and can therefore be written as a constant G_1, we see (by comparing this expression with that in Equation (13.8)) that this framework is equivalent to the linear propagator model, with

$$K(\ell) := G(\ell+1) - G(\ell) \equiv -G_1\mathbb{K}(\ell); \qquad (\ell \leq p); \qquad K(\ell > p) = 0. \qquad (13.23)$$

Therefore, a slowly decaying $G(\ell) \sim \ell^{-\beta}$ can only be reproduced by a high-order DAR(k) model for which $\mathbb{K}(\ell)$ itself decays as a power-law (in fact, precisely as $\ell^{-\beta-1}$). The case where

$$\mathbb{K}(k) = \begin{cases} \rho, & \text{for } k = 1, \\ 0, & \text{for } k > 1 \end{cases}$$

leads to an exponential decay of the sign correlation, $C(\ell) = \rho^\ell$. This corresponds exactly to the MRR model that we will discuss in Section 16.2.1.

Throughout this section, we have greatly simplified all calculations by neglecting the influence of trade volume. We present an extended version of this model, including trade volumes, in Appendix A.3. Interestingly, even in cases where volumes fluctuate considerably, the final results and conclusions are qualitatively the same as those that we have presented above.

13.4 More on the Propagator Model

We now provide some advanced remarks about the propagator model, concerning in particular the impact of metaorders, the aggregate impact of volume imbalance, and its close kinship with Hasbrouck's VAR model.

13.4.1 Hasbrouck's VAR Model

In the wider literature on market microstructure, the so-called **vector autoregressive model** (VAR), as proposed by J. Hasbrouck in 1991, is a standard and widely used tool for predicting the present price return r_t and signed volume $v_t = \varepsilon_t \upsilon_t$ from their previous realisations. In this section, we illustrate how the propagator framework, as encapsulated in Equation (13.8), can be regarded as a special case of Hasbrouck's VAR model.

The VAR model is the joint linear regression:

$$\begin{aligned} r_t &= G_1 v_t + \sum_{-\infty<n<t} K(t-n)(t-n)v_n + \sum_{-\infty<n<t} \mathcal{H}_{rr}(t-n)r_n + \xi_t, \\ v_t &= \sum_{-\infty<n<t} \mathcal{H}_{vv}(t-n)v_n + \sum_{-\infty<n<t} \mathcal{H}_{rv}(t-n)r_n + \xi'_t. \end{aligned} \tag{13.24}$$

The propagator model can be seen as a special case of Equation (13.24), where:

- In the propagator model, the signed volume v_t in Equation (13.24) is replaced by the sign ε_t. This is justified by the fact that trade volumes carry much less information than a linear model would imply.
- In the propagator model, there is no direct impact of past returns on the present return (i.e. $\mathcal{H}_{rr} = 0$). This choice corresponds to a mechanistic interpretation of trade impact, where one assumes that past price changes by themselves do not influence present returns.
- In the propagator model, the dynamics of signed volumes (or signs) is summarised by the autocorrelation function $C_v(\ell)$ but is not described explicitly, although the DAR model is one possible specification, with the Yule–Walker relation:

$$C_v(\ell) = \sum_{k=1}^{\infty} \mathcal{H}_{vv}(k) C_v(|\ell - k|).$$

- In the propagator model, there is no direct feedback of the price change on the order flow (i.e. $\mathcal{H}_{rv} = 0$). The order flow is therefore assumed to be "rigid", independent of the price trajectory. This is clearly an approximation, which is partially addressed by the generalised propagator model (see Chapter 14).

In the VAR model, the **information content** I_H of a trade is defined as the long-term average price change, conditioned on a trade occurring at time $t = 0$:

$$I_H = \mathbb{E}[m_\infty - m_0 \mid v_0 = v].$$

This is very close to the asymptotic response function $\mathcal{R}_\infty$ in the propagator model, with the only difference being that volume fluctuations are neglected in the latter.

In other words, I_H is given by the same expression as $\mathcal{R}_\infty$ (see Equation (13.10)), with $C(\ell)$ replaced by $C_v(\ell)$ and with a prefactor proportional to the volume v. In Hasbrouck's initial specification, the information content of a trade is therefore proportional to its volume, whereas we know now that there is much less information in volume than initially anticipated (see Section 11.3.2).

13.4.2 The Propagator Model as a Reduced Description of All Order Flow

The propagator model that we have explored in this chapter considers price returns only at the times of market order arrivals. In a real LOB, by contrast, several other limit order arrivals and cancellations may occur between the arrivals of market orders. In Chapter 14, we extend the propagator model to include these other events explicitly. However, there are also good reasons for wishing to exclude these other events from the model specification, such as not having access to them in market data or simply wishing to keep the model formulation parsimonious. Therefore, we now explore the extent to which the basic propagator model that we have introduced in this chapter effectively includes the influence of these other LOB events.

Imagine that two types of events are important for price dynamics, but that we only observe one of them. Observable events (e.g. market order arrivals) are characterised by a random variable ε_t, which is simply the trade sign in the propagator model. Unobservable events (e.g. limit order arrivals and cancellations) are characterised by another random variable $\widehat{\xi}_t$. If all events were observed, the full propagator model for the mid-price, generalising Equation (13.7), would read

$$m_t = m_{t_0} + \sum_{t_0 \le n < t} G(t-n)\varepsilon_n + \sum_{t_0 \le n < t} H(t-n)\widehat{\xi}_n + \sum_{t_0 \le n < t} \xi_t. \tag{13.25}$$

Assume for simplicity that the ε_t and $\widehat{\xi}_t$ are correlated Gaussian random variables. One can then always express the unobserved $\widehat{\xi}$ terms as a linear superposition of past ε terms, plus noise. Inserting this decomposition into the equation above, one finds an effective propagator model in terms of the ε only, plus an additional noise component ξ' coming from the unobserved events and from ξ:

$$m_t = m_{-\infty} + \sum_{-\infty < n < t} G(t-n)\varepsilon_n + \sum_{-\infty < n < t} H(t-n) \sum_{-\infty < m < n} \Xi(n-m)\varepsilon_m + \sum_{0 \le n < t} \xi'_t, \tag{13.26}$$

where Ξ is the linear filter for expressing the $\widehat{\xi}$s in terms of the past εs. This kernel is the result of the classical *Wiener filter*, and can be expressed in terms of the correlation function between the two sets of variables.[3]

Equation (13.26) can in fact be recast in the form of the single-event propagator model, Equation (13.7), with an **effective propagator** $\widetilde{G}$ that contains both the direct effect modelled by G and the indirect influence of the unobserved events:

$$\widetilde{G}(\ell) = G(\ell) + \sum_{n=1}^{\ell} H(\ell')\Xi(\ell - \ell').$$

From this equation, it is clear that a non-trivial temporal dependence of $\widetilde{G}$ can arise, even when the true propagators G and H are lag-independent. In other words, the decay of a single market order's impact should in fact be interpreted a consequence of the interplay between market orders and limit orders. As a trivial example, suppose both propagators are equal and constant in time, such that $G(\ell) = H(\ell) = G_0$, but that $\varepsilon_n = -\widehat{\xi}_n$ for all n. This means that both types of events impact the price, but that

[3] See, e.g., Levinson, N. (1947). The Wiener RMS error criterion in filter design and prediction. *Journal of Mathematical Physics*, 25, 261–278, and Appendix A.3 for an explicit example.

they cancel each other out exactly. Then $\Xi(\ell) = -\delta_{\ell,0}$ leading to $\widetilde{G}(\ell) = 0$ for all ℓ. Therefore, the effective impact of events of the first type is zero in the effective model (and not G_0). In reality, limit orders only partially "screen" the impact of market orders, resulting in a non-trivial, time-decaying effective propagator.

13.4.3 Aggregate Impact

Within the propagator model, it is interesting to compute the **aggregate impact** of a given order-flow imbalance in a time interval containing T trades (see Section 11.4). For simplicity, we assume that the signed volume $v_n = \varepsilon_n \upsilon_n$ of each transaction n is a Gaussian variable with zero mean and unit variance, and long-range autocorrelations given by $C(\ell) \sim \ell^{-\gamma}$. In this framework, we aim to compute the average value of the mid-price change $m_{t+T} - m_t$, conditioned on a specified order-flow imbalance $\Delta V = \sum_{j=0}^{T-1} v_j$. In the following, we choose for convenience the origin of time such that $t = 0$.

Let us first compute the average value of v_n when such a condition is imposed:

$$\mathbb{E}[v_n|\Delta V] = \frac{1}{Z}\int \prod \mathrm{d}v_j\, v_n e^{-\frac{1}{2}\left(\sum_{j,k} v_j C^{-1}_{j,k} v_k\right)} \delta\left(\sum_{j=0}^{T-1} v_j - \Delta V\right),$$

where $C_{j,k} = C(|j-k|)$ is the correlation matrix, $\delta(\cdot)$ is the Dirac delta function, and Z is the same expression as in the numerator but without the v_n term.

Using the standard representation of the δ function in terms of its Fourier transform:

$$\mathbb{E}[v_n|\Delta V] = \frac{\partial}{\partial w_n} \ln\left(\int \mathrm{d}z e^{-iz\Delta V} \int \prod \mathrm{d}v_i e^{-\frac{1}{2}\sum_{ij} v_i C^{-1}_{ij} v_j + iz\sum_{j=0}^{T-1} v_j + w_n v_n}\right)\Bigg|_{w_n=0}.$$

Performing the Gaussian integral, and keeping only terms that survive when $w_n \to 0$:

$$\mathbb{E}[v_n|\Delta V] = \frac{\partial}{\partial w_n} \ln\left(\int \mathrm{d}z e^{-iz\Delta V} e^{-\frac{z^2}{2}\sum_{j,k=0}^{T-1} C_{jk} + izw_n \sum_{j=0}^{T-1} C_{nj}}\right)\Bigg|_{w_n=0}.$$

Performing the Gaussian integration over z:

$$\mathbb{E}[v_n|\Delta V] = -\frac{\partial}{\partial w_n} \frac{\left(\Delta V - w_n \sum_{j=0}^{T-1} C_{nj}\right)^2}{2\sum_{j,k=0}^{T-1} C_{jk}}\Bigg|_{w_n=0},$$

$$= \frac{\sum_{j=0}^{T-1} C_{nj}}{\sum_{j,k=0}^{T-1} C_{jk}} \Delta V. \tag{13.27}$$

In the case where there are no correlations (i.e. when $C_{i,j} = \delta_{i,j}$), one can check this expression directly:

$$\mathbb{E}[v_n|\Delta V] = \begin{cases} \frac{1}{T}\Delta V, & \text{for } 0 \le n \le T-1; \\ 0, & \text{otherwise.} \end{cases}$$

In the absence of correlations, the constraint that $\Delta V = \sum_{j=0}^{T-1} v_j$ is only effective when $n \in [1, T]$. Within this interval, all of the n's play a symmetric role.

The result is even more interesting in the limit of long-range autocorrelations $C(\ell) \sim_{\ell \gg 1} c_\infty \ell^{-\gamma}$, with $0 < \gamma < 1$. When $T \gg 1$, one finds

$$\mathbb{E}[v_n|\Delta V] \approx \Delta V \frac{2-\gamma}{2T}\left[\left(1 - \frac{n}{T}\right)^{1-\gamma} + \left(\frac{n}{T}\right)^{1-\gamma}\right], \qquad 1 \le n \le T-1. \tag{13.28}$$

The propagator model predicts that the average change in mid-price between $t = 0$ and $t = T-1$ is given by

$$\mathbb{R}(\Delta V, T) = \mathbb{E}[m_T - m_0|\Delta V],$$

$$= \sum_{0 \le n \le T-1} G(T-n)\mathbb{E}[\varepsilon_n|\Delta V] + \sum_{n>0} [G(T+n) - G(n)]\,\mathbb{E}[\varepsilon_n|\Delta V],$$

where ε_n is the sign of v_n (i.e. the sign of a Gaussian variable with a given conditional mean and variance, as computed in Equation (13.27)).

If $\mathbb{E}[v_n|\Delta V] \ll 1$ and $T \gg 1$, then $\mathbb{E}[\varepsilon_n|\Delta V]$ simplifies to

$$\mathbb{E}[\varepsilon_n|\Delta V] \approx \sqrt{\frac{2}{\pi}}\mathbb{E}[v_n|\Delta V].$$

Using also that for $n \gg 1$, the propagator can be approximated as $G(n) \approx \Gamma_\infty/n^\beta$, one finally finds that in the large-T limit, the aggregated impact is given by

$$\mathbb{E}[m_T - m_0|\Delta V] \approx A(\beta,\gamma)\Gamma_\infty \frac{\Delta V}{T^\beta},$$

where $A(\beta,\gamma)$ is a numerical constant that depends on β and γ.

When ΔV reaches values of order T, one finds that $\mathbb{E}[\varepsilon_n|\Delta V] \to 1$ and $\mathbb{E}[m_T - m_0|\Delta V] \propto T^{1-\beta}$. The propagator model therefore yields the following result for the aggregate impact:

$$\mathbb{R}(\Delta V, T) = \Gamma_\infty T^{1-\beta} \times \mathscr{F}_{\text{prop}}\left(\frac{\Delta V}{T}\right), \tag{13.29}$$

where $\mathscr{F}_{\text{prop}}(u)$ is a scaling function that is linear for small arguments and that saturates for large arguments. Observe that Equation (13.29) has the same functional form as Equation (11.11), which describes empirical observations and that we recall here for convenience:

$$\mathbb{R}(\Delta V, T) \cong \langle s\rangle T^\chi \times \mathscr{F}\left(\frac{\Delta V}{\bar{V}_{\text{best}} T^\varkappa}\right).$$

However, the propagator prediction makes incorrect predictions for the exponents χ and $\varkappa$. Empirical data suggests $\chi \approx 0.5 - 0.65$ and $\varkappa \approx 0.75 - 0.9$ (see Section 11.4), while the propagator model predicts $\chi = 1 - \beta \approx 0.75$ and $\varkappa = 1$. Note, however, that the model correctly predicts the difference $\chi - \varkappa = 0.25$ which governs the scaling of Kyle's lambda with T (see Chapter 15).

13.4.4 Metaorder Impact

We now illustrate how the propagator model can also help to illuminate the peak impact and impact path of a metaorder. Throughout this section, we also make some additional assumptions about metaorder execution. Specifically, we assume that a metaorder of total volume Q is executed in N separate child orders, each of size υ, and that the (trade) time between each child order is equal to n. We describe such a metaorder as having a *participation rate* $f = 1/n$. In the present context, we assume that all trades have unit size, so the participation rate of a given metaorder is simply equal to the fraction of the total number of trades that it comprises. We also assume that the sign of the metaorder is uncorrelated with the sign of all other transactions.

We arbitrarily set the time of the first trade to be $t_0 = 0$. Due to our assumptions, trades occur at time $0, n, 2n, \ldots, T = (N-1)n$, such that the total execution time of the metaorder is $T = (N-1)n$. Conditional on the presence of a buy metaorder, the propagator model predicts that the average price change between t_0 and some $t \leq T$ is given by

$$\mathfrak{I}^{\text{path}}(t) = \mathbb{E}[m_t - m_0|\text{metaorder}] = \sum_{k=0}^{\lfloor t/n \rfloor} I_1(\upsilon)G(t-kn),$$

where $I_1(\upsilon)$ is the lag-1 impact of a child order of size υ. In what follows, we assume that lag-1 impact is constant, so we can absorb the $I_1(\upsilon)$ term into the definition of G, to lighten notation.

Approximating the discrete sum as an integral, and using the Euler–McLaurin formula, one finds:

$$\mathfrak{I}^{\text{path}}(t) \approx \int_0^{t/n} \mathrm{d}k\, G(t-kn) + \frac{1}{2}\left[G(1)+G(t)\right] + \ldots.$$

If we also assume that the propagator $G(t)$ decays as $\Gamma_\infty t^{-\beta}$ for large t, with $\beta < 1$, then to leading order

$$\mathfrak{I}^{\text{path}}(t) \approx \frac{\Gamma_\infty}{n(1-\beta)} t^{1-\beta}.$$

The impact of such a metaorder peaks at time $t = T$, with a value given by:

$$\mathfrak{I}^{\text{peak}}(Q, f) \approx \frac{\Gamma_\infty}{(1-\beta)} f^{\beta} \left(\frac{Q}{\upsilon}\right)^{1-\beta}, \qquad (N = Q/\upsilon \gg 1). \tag{13.30}$$

This **peak impact** is concave in the volume of the metaorder Q, and increases with the participation rate f (or, alternatively, decreases with the execution horizon $T \approx Q/(\upsilon f)$). In other words, expressing Equation (13.30) in terms of T instead of f:

$$\mathfrak{I}^{\text{peak}}(Q, T) \approx \frac{\Gamma_\infty}{(1-\beta)} T^{-\beta} \frac{Q}{\upsilon}, \qquad (N = Q/\upsilon \gg 1). \tag{13.31}$$

This corresponds to a linear impact, with a slope that decays as $T^{-\beta}$, in agreement with Equation (13.29). This result is not surprising: the propagator model is linear, so impacts add up.

Therefore, although for a fixed f it is qualitatively similar to the square-root impact law for $\beta < 1/2$, the metaorder impact law in the propagator framework is fundamentally *linear*. Furthermore, the model predicts that the impact of a metaorder scales like f^{β} or $T^{-\beta}$, which is not consistent with experimental data: as we emphasised in Section 12.3, the impact of a metaorder is in fact only weakly dependent on the time to completion T. Hence, while the propagator model captures some aspects of the square-root law, it needs to be amended in some way to reflect the behaviour of real markets, as observable in empirical data. We will examine two such amendments in Section 13.4.5 and in Chapter 19.

13.4.5 The LMF Surprise Model

In Chapter 10, we saw that a plausible explanation for the long-range autocorrelation of market order signs is the power-law distribution of metaorder volumes. In the context of the Lillo, Mike and Farmer (LMF) model, we showed that in order to account for an autocorrelation function $C(\ell)$ that behaves as $\ell^{-\gamma}$, the probability $\mathbb{P}(L)$ that a sequence of market order signs (chosen uniformly at random among all such sequences) has length exactly L should decay as $L^{-2-\gamma}$ for large L.

We will now show that in the extreme case where there is only a single active metaorder at each time step, it is straightforward to construct the best predictor for the next sign. This expression turns out to be very different from the linear combination of past signs that we previously obtained within a DAR description in Equation (13.20).

Consider a situation in which the previous ℓ market orders have all had sign $\varepsilon_t = +1$. If it is known that only one metaorder is active at this time, then the probability that the next market order sign is also $+1$ is given by

$$P_+(\ell) = \frac{\sum_{L=\ell+1}^{\infty} \mathbb{P}(L)}{\sum_{L=\ell}^{\infty} \mathbb{P}(L)} \approx_{\ell \gg 1} 1 - \frac{\gamma+1}{\ell}.$$

Within the LMF framework, the best predictor of the next sign is

$$\widehat{\varepsilon}_t = +\left(1 - \frac{\gamma+1}{\ell}\right) - \frac{\gamma+1}{\ell} = 1 - 2\frac{\gamma+1}{\ell}.$$

The history-dependent (or "surprise") model, which generalises the linear propagator model from Section 13.3, then reads

$$r_t = m_{t+1} - m_t = G_1(\varepsilon_t - \widehat{\varepsilon}_t) + \xi_t, \tag{13.32}$$

with

$$r_t = \begin{cases} 2G_1\left(\frac{\gamma+1}{\ell}\right) + \xi_t, & \text{if } \varepsilon = +1, \\ -2G_1 + 2G_1\left(\frac{\gamma+1}{\ell}\right) + \xi_t, & \text{if } \varepsilon = -1. \end{cases}$$

The impact of a sequence of $\ell = Q$ consecutive trades in the same direction is then

$$\mathfrak{I}_{\text{LMF}}(Q) = \mathbb{E}[m_\ell - m_0] \approx_{Q \gg 1} 2G_1(1+\gamma)\ln Q.$$

Note that this behaviour is quite different from the prediction of the propagator model (see Equation (13.30)), which instead leads to $\mathfrak{I}_{\text{prop}}(Q) \sim Q^{1-\beta}$.

In the LMF model, the average price paid to execute the Q child orders is given by

$$p_{\text{ex.}} = m_0 + \frac{2G_1(1+\gamma)}{Q}\sum_{L=1}^{Q} \ln L.$$

For $Q \gg 1$, this expression can be approximated as

$$\begin{aligned} p_{\text{ex.}} &\approx m_0 + 2G_1(1+\gamma)(\ln Q - 1), \\ &\approx m_\ell - 2G_1(1+\gamma). \end{aligned}$$

At the end of a long metaorder, one does not need to wait long before a trade in the unexpected direction takes place. This trade has a large price impact of approximately $-2G_1$. Therefore, soon after the end of the metaorder, the price reverts to

$$m_\infty \approx m_\ell - 2G_1 > p_{\text{ex.}},$$

which exceeds the average paid price by an amount $2\gamma G_1$. This highlights a problem with this model: the *price is a martingale* thereafter, so the metaorder would be profitable on average. This represents an arbitrage opportunity that is absent from real markets.

How might we alter the surprise model in Equation (13.32) to address this weakness? One possible approach is to allow the impact parameter G_1 to be dependent on time and/or market conditions. Then, one way to remove the above arbitrage is to assume that the longer the metaorder has lasted, the larger the coefficient G_1:

$$G_1(\ell) := G_0\ell^\delta, \qquad \delta > 0. \tag{13.33}$$

One now finds that the impact of a sequence of Q trades in the same direction is

$$\mathfrak{I}_\delta(Q) \approx 2G_0(1+\gamma)\frac{Q^\delta}{\delta},$$

and that the average price paid to execute the Q trades is

$$p_{\text{ex.}} \approx p_0 + 2G_0\frac{(1+\gamma)}{\delta(1+\delta)}Q^\delta.$$

Long after the metaorder execution ends, the price reverts to

$$m_\infty \approx m_\ell - 2G_0 Q^\delta \qquad (Q \gg 1).$$

To make further progress with our study of this model, we now make the additional assumption that the **execution price is fair**, in the sense that the average execution price is equal to the long-term price. This leads to the condition

$$\begin{aligned} p_{\text{ex.}} = m_\infty &\Rightarrow 2G_0\frac{(1+\gamma)}{\delta(1+\delta)}\ell^\delta = 2G_0(1+\gamma)\frac{\ell^\delta}{\delta} - 2G_0\ell^\delta, \\ &\Rightarrow \frac{(1+\gamma)}{(1+\delta)}\left(1 - \frac{1}{\delta}\right) = 1, \\ &\Rightarrow \delta = \gamma. \end{aligned}$$

This idea, promoted by Farmer, Gerig, Lillo and Waelbrouck (FGLW), leads to a concave impact for metaorders, $\mathfrak{I}_\delta(Q) \sim Q^\delta$, with an exponent δ that is equal to the decay exponent of the autocorrelation of the trade-sign series, $\delta = \gamma < 1$. In this framework, the impact decay is nearly immediate, because the market is assumed to detect when the metaorder has ended. Interestingly, the FGLW calculation can be framed using slightly different language, in the context of the Glosten–Milgrom model (see Section 16.1.6).

The FGLW model also suggests a non-linear generalisation of the linear propagator model, where

$$G_1(\widehat{\varepsilon}_t) = G_0\left(1 + A\widehat{\varepsilon}_t^2 + \dots\right),$$

where A is a constant and by symmetry the term linear in $\widehat{\varepsilon}_t$ is absent. In this set-up, the value of G_1 increases as the level of predictability increases, similarly to Equation (13.33). Setting $A = 0$ recovers the linear propagator model with $G_1 = G_0$. To date, the properties of this extended propagator model still remain to be explored. We present another (arguably better-motivated) non-linear generalisation of the propagator model in Chapter 19.

13.5 Conclusion

In this chapter, we have presented a simple linear model that expresses price moves in terms of the influence of all past trades. The model assumes that each trade should be treated on the same footing (independently of its size or how aggressive it is) and leads to a reaction impact in the direction of the trade, with dynamics encoded in a lag-dependent function $G(\cdot)$, called the propagator. In the model, the immediate reaction impact of a trade is given by $G(1)$ and the long-term reaction impact of a trade is given by $G(\infty)$.

To avoid creating statistical arbitrage opportunities (such as trends), the fact that real trade signs are long-ranged autocorrelated imposes that $G(\ell)$ has some special shape. In Section 13.2, we showed that if the autocorrelation of trade signs decays as $\ell^{-\gamma}$ with $\gamma < 1$, then $G(\ell)$ must decay to zero as $\ell^{-\beta}$, with $\beta = (1-\gamma)/2$. This very slow decay, which we call long-range resilience, means that the distinction between "transient" and "permanent" impact is necessarily fuzzy, at least empirically. Furthermore, even when the permanent impact of a single trade G_∞ is zero, the cumulative observed impact $\mathcal{R}_\infty$ of that trade and all of its correlated kinship can be non-zero, as a result of the subtle compensation between trade correlations and impact decay (see Equation (13.18)). Intuitively, this compensation reflects the way that liquidity providers and liquidity takers offset each other, to remove most of the linearly predictable patterns that their (predictable) actions would otherwise create.

Despite the fact that it only deals with market orders, we have argued that the propagator model is actually an effective reduced-form description of the full interplay between market orders, limit orders and cancellations. We have also noted that it is a simplified, and arguably more intuitive, specification of Hasbrouck's classic VAR model.

Although very illuminating, the propagator model suffers from several drawbacks. For example, the linearity of the model is at odds with empirical findings such as the square-root impact of metaorders. Resolving these problems requires a more sophisticated, non-linear version of the propagator model, which we introduce in Chapter 19.

Take-Home Messages

(i) The simple, linear propagator model assumes that all market orders have an identical, time-dependent impact on the price, consisting of both an immediate reaction and a lagged reaction.
(ii) Because market order signs are positively autocorrelated, if impact was constant and permanent, then the mid-price would be strongly super-diffusive, and therefore predictable. To ensure that the mid-price series is diffusive, we must assume that the impact of a market order decays over time.
(iii) To balance the power-law (long-range) autocorrelations of trade signs, the propagator must decay slowly as well. We call this behaviour, in which the effect of an action takes a long time to vanish, the *long-range resilience of markets*.
(iv) The propagator kernel G can be fitted empirically from the two-point correlation function of market order signs C and the response function $\mathcal{R}$, through a linear equation.
(v) In history-dependent impact models, impact is permanent but depends on the history in such a way that prices are unpredictable. In a DAR hypothesis for the order flow, one recovers exactly the propagator model.
(vi) Despite its many appealing properties, the linear propagator model is too simple to reproduce some empirical regularities observable in real markets. Addressing these deficiencies of the model involves considering a more complex, non-linear framework, which we introduce in Chapter 19.

13.6 Further Reading

See also Section 10.7.

The Propagator Model and History-Dependent Impact Models

Bouchaud, J. P., Gefen, Y., Potters, M., & Wyart, M. (2004). Fluctuations and response in financial markets: The subtle nature of random price changes. *Quantitative Finance*, 4(2), 176–190.

Bouchaud, J. P., Kockelkoren, J., & Potters, M. (2006). Random walks, liquidity molasses and critical response in financial markets. *Quantitative Finance*, 6(02), 115–123.

Farmer, J. D., Gerig, A., Lillo, F., & Mike, S. (2006). Market efficiency and the long-memory of supply and demand: Is price impact variable and permanent or fixed and temporary? *Quantitative Finance*, 6(02), 107–112.

Bouchaud, J. P., Farmer, J. D., & Lillo, F. (2009). How markets slowly digest changes in supply and demand. In Hens, T. & Schenke-Hoppe, K. R. (Eds.), *Handbook of financial markets: Dynamics and evolution*. North-Holland, Elsevier.

Eisler, Z., J. Kockelkoren, J., & Bouchaud, J. P. (2012). Models for the impact of all order book events. In Abergel, F., Bouchaud, J.-P., Foucault, T., Lehalle, C.-A., & Rosenbaum, M. (Eds.), *Market microstructure: Confronting many viewpoints*. Wiley

Taranto, D. E., Bormetti, G., Bouchaud, J. P., Lillo, F., & Tóth, B. (2016). Linear models for the impact of order flow on prices I. Propagators: Transient vs. history dependent impact. Available at SSRN: https://ssrn.com/abstract=2770352.

Tóth, B., Eisler, Z., & Bouchaud, J.-P. (2017). The short-term price impact of trades is universal. https://ssrn.com/abstract=2924029.

Hasbrouck's Model

Hasbrouck, J. (1991). Measuring the information content of stock trades. *The Journal of Finance*, 46(1), 179–207.

Jones, C. M., Kaul, G., & Lipson, M. L. (1994). Transactions, volume, and volatility. *Review of Financial Studies*, 7(4), 631–651.

Hasbrouck, J. (2007). *Empirical market microstructure: The institutions, economics, and econometrics of securities trading*. Oxford University Press.

Metaorders and Propagators

Farmer, J. D., Gerig, A., Lillo, F., & Waelbroeck, H. (2013). How efficiency shapes market impact. *Quantitative Finance*, 13(11), 1743–1758.

Donier, J., Bonart, J., Mastromatteo, I., & Bouchaud, J. P. (2015). A fully consistent, minimal model for non-linear market impact. *Quantitative Finance*, 15(7), 1109–1121.

14
Generalised Propagator Models

When I want to understand what is happening today or try to decide what will happen tomorrow, I look back.

(Omar Khayyám)

14.1 Price Micro-Mechanics

In the previous chapter, we studied a class of propagator models that consider the change in mid-price that occurs between subsequent market order arrivals. The models that we studied in that chapter consider all market order arrivals on an equal footing, irrespective of their size and how aggressive they are.

In the present chapter, we extend those models in two different ways. First, we consider a propagator model that partitions market order arrivals according to whether or not they consume the entire opposite-side best queue upon arrival. We introduce a mathematical framework that enables us to distinguish the impact of orders partitioned in this way, and we show that this extension helps to solve some of the problems with the one-event-type propagator models from the previous chapter.

Second, we introduce a generalised propagator model that considers not only market order arrivals, but also some limit order arrivals and cancellations. In this framework, we are able to track all events that cause price changes, and we are therefore able to monitor the evolution of impact on a more microscopic scale. However, we also argue that performing such a granular analysis of order flow requires working with rather complex models, which can be difficult to calibrate. We then turn to a more intuitive formulation of these multi-event propagator models that naturally encompasses the idea of history-dependent liquidity.

14.2 Limitations of the Propagator Model

The propagator model that we introduced in Chapter 13 is a reduced-form description of LOB dynamics. The model assumes that all market orders lead

This chapter is not essential to the main story of the book.

to the same impact dynamics, characterised by the propagator $G(\ell)$, and does not explicitly track other LOB events that occur between market order arrivals. Clearly, this approach neglects many effects that could be useful for understanding or modelling price changes in real LOBs. For example, partitioning market orders according to whether or not they consume all the available liquidity at the opposite-side best quote is very useful, not only for assessing whether such market orders will cause a change in mid-price immediately upon arrival, but also for predicting the string of other LOB events that are likely to follow.

As we detailed in Section 13.2, calibrating the one-event propagator $G(\ell)$ only requires access to data describing the response function $\mathcal{R}(\ell)$ for $\ell \geq 0$ (or alternatively $S(\ell \geq 0)$) and the autocorrelation function for market order signs $C(\ell)$. This information is readily available in many empirical data sets.

Once $G(\ell)$ has been calibrated, the propagator model leads to a precise prediction for the lag-dependent volatility $\mathcal{D}(\ell) = \mathcal{V}(\ell)/\ell$ in terms of $G(\ell)$ and $C(\ell)$, with the only further adjustable parameters arising from the variance Σ^2 of the news (or other types of events) contribution ξ_t (see Equation (13.7)). Figure 13.2 shows the result of performing this analysis. As the figure reveals, the outcome is quite disappointing: the model fails to capture all the interesting structure of the signature plot $\mathcal{D}(\ell)$, and only reproduces the long-term value of $\mathcal{D}(\ell)$ when choosing a suitable value for Σ^2.

In this way, the propagator model makes a falsifiable prediction about the detailed structure of the **signature plot** (or *lag-dependent volatility* $\mathcal{D}(\ell)$). Recall that for a purely diffusive process, $\mathcal{D}(\ell)$ is lag-independent, such that volatility does not depend on the time scale chosen to measure it. If $\mathcal{D}(\ell)$ increases with lag, then returns are positively autocorrelated, so prices trend; if $\mathcal{D}(\ell)$ decreases with lag, then returns are negatively autocorrelated, so prices mean-revert (see Section 2.1.1).

Another weakness of the propagator model concerns the shape of the **negative-lag response function**, i.e. $\mathcal{R}(\ell)$ for $\ell < 0$. As emphasised above, only the positive side of $\mathcal{R}(\ell)$ is needed to calibrate $G(\ell)$. Once this is known, $\mathcal{R}(\ell < 0)$ can be predicted without any additional parameters. Figure 14.1 shows the results for the stocks in our sample. In all cases, the empirical $\mathcal{R}(\ell < 0)$ lies above the theoretical prediction of the propagator model. The discrepancy is quite important and shows that the simple propagator model fails to grasp an important aspect of the dynamics of markets.[1]

[1] The discrepancy tends to be smaller for small-tick assets than for large-tick assets; see Taranto, D. E., Bormetti, G., Bouchaud, J. P., Lillo, F., & Tóth, B. (2016). Linear models for the impact of order flow on prices I. Propagators: Transient vs. history dependent impact. https://ssrn.com/abstract=2770352.

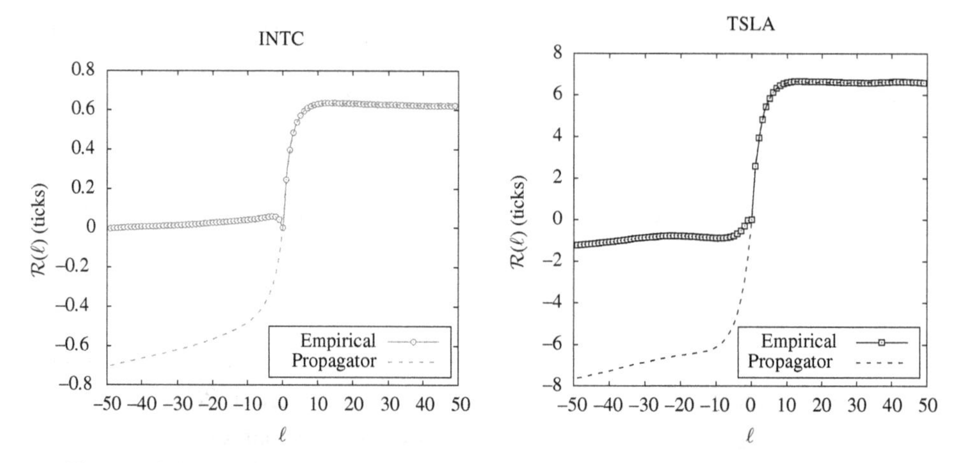

Figure 14.1. (Dashed curves) The response function calculated from the propagator model, fitted using only positive lags, for (left panel) INTC and (right panel) TSLA. The markers denote the corresponding empirical values.

Why does this discrepancy occur? To address this question, we first rewrite the response function at negative lags:

$$\begin{aligned}\mathcal{R}(-\ell) &= \langle \varepsilon_t \cdot (m_{t-\ell} - m_t) \rangle, \\ &= -\langle (m_{t+\ell} - m_t) \cdot \varepsilon_{t+\ell} \rangle,\end{aligned}$$

where the second line is obtained by shifting t to $t+\ell$. The negative-lag response function therefore tells us how the price change between t and $t+\ell$ is correlated with the sign of the market order arriving at time $t+\ell$. Intuitively, one would expect that price changes themselves directly influence the sign of future trades (e.g. that a price increase motivates more sell orders, everything else being equal, contributing positively to $\mathcal{R}(-\ell)$, and vice-versa). However, this mechanism is not present in the simple propagator model, which considers order flow to have no reaction to price changes.[2] As we show below, most of the effect can simply be captured by treating the non-price-changing market orders MO^0 and price-changing market orders MO^1 separately.

14.3 Two Types of Market Orders

As an initial step towards addressing the full complexity of LOB dynamics, we first extend the propagator model to distinguish between the set of market orders MO^1 (i.e. those that cause a non-zero lag-1 price change) and MO^0 (i.e. those that do not).

[2] This mechanism can however be included within Hasbrouck's VAR model (see Section 13.4.1), and corresponds to a negative kernel $\mathcal{H}_{rv}$.

In this section, we still work in transaction time (such that each time a market order arrives, we increment t by a single unit), and we use the notation $\pi_t \in \{\mathrm{MO}^0, \mathrm{MO}^1\}$ to indicate the market order type at time t. The generalisation that we introduce now is to allow the propagator function to be different for events of type MO^0 than for events of type MO^1, such that the mid-price dynamics obey

$$m_t = m_{t_0} + \sum_{t_0 \leq t' < t} \left(\sum_{\pi = \mathrm{MO}^0, \mathrm{MO}^1} \mathbb{I}(\pi_{t'} = \pi) G_\pi(t - t') \varepsilon_{t'} + \xi_{t'} \right), \tag{14.1}$$

where $\mathbb{I}$ is an indicator function, such that for each t', one selects the propagator $G_{\pi_{t'}}$ corresponding to the particular event type that occurred at that time.

To calibrate this model, we simply generalise the response function $\mathcal{R}(\ell)$ to account for whether a given market order belongs to MO^0 or MO^1:

$$\mathcal{R}_\pi(\ell) = \langle \varepsilon_t \cdot (m_{t+\ell} - m_t) | \pi_t = \pi \rangle := \frac{\langle \varepsilon_t \mathbb{I}(\pi_t = \pi) \cdot (m_{t+\ell} - m_t) \rangle}{\mathbb{P}(\pi)}. \tag{14.2}$$

The function $\mathcal{R}_\pi(\ell)$ measures the lagged covariance (measured in market order time) between the signed event $\varepsilon_t \mathbb{I}(\pi_t = \pi)$ at time t and the mid-price change between times t and $t + \ell$, normalised by the stationary probability of the event π,

$$\mathbb{P}(\pi) = \langle \mathbb{I}(\pi_t = \pi) \rangle.$$

The normalised response function in Equation (14.2) gives the expected (signed) price change after an event π.

Figure 14.2 shows the behaviour of $\mathcal{R}_\pi(\ell)$ for $\pi = \mathrm{MO}^0$ and $\pi = \mathrm{MO}^1$. The figure paints a more detailed picture of the same phenomenon as we reported for lag $\ell = 1$ in Table 11.1, where we saw that $\mathcal{R}^1(1)$ (which is equal to $\mathcal{R}_{\mathrm{MO}^1}(1)$) is larger than $\mathcal{R}^0(1)$ (which is equal to $\mathcal{R}_{\mathrm{MO}^0}(1)$). Figure 14.2 shows that $\mathcal{R}^0(\ell)$ grows with ℓ for both small-tick and large-tick assets, whereas $\mathcal{R}^1(\ell)$ converges almost immediately to a plateau value for large-tick assets.

We also define the **signed-event correlation function** as

$$C_{\pi,\pi'}(\ell) = \frac{\langle \mathbb{I}(\pi_t = \pi) \varepsilon_t \mathbb{I}(\pi_{t+\ell} = \pi') \varepsilon_{t+\ell} \rangle}{\mathbb{P}(\pi)\mathbb{P}(\pi')}, \tag{14.3}$$

where the event π occurs before the event π'. In general, there is no reason to expect **time-reversal symmetry** (i.e. that $C_{\pi,\pi'}(\ell)$ is equal to $C_{\pi',\pi}(\ell) \equiv C_{\pi,\pi'}(-\ell)$). We have already discussed this extended correlation matrix in Section 10.2; see Figure 10.2.

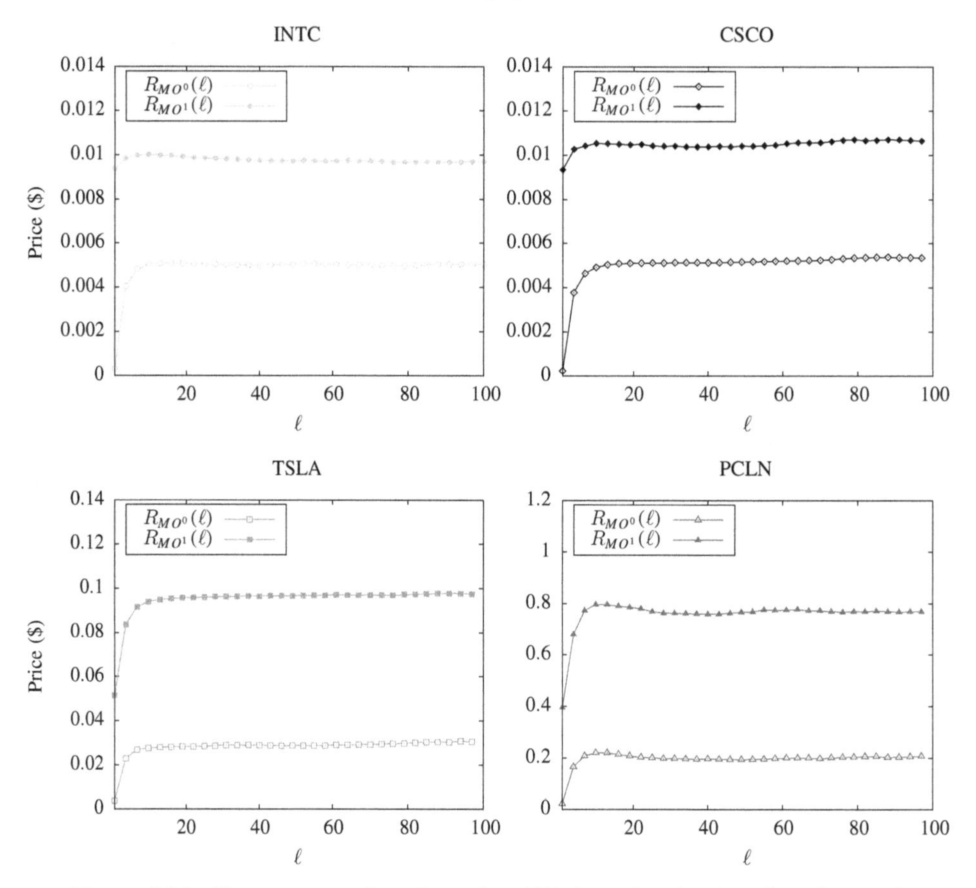

Figure 14.2. The response function of a (filled markers) price-changing and (hollow markers) non-price-changing market order for (top left) INTC, (top right) CSCO, (bottom left) TSLA and (bottom right) PCLN.

Using some straightforward algebra, the return response function in Equation (14.2) can be expressed as a generalisation of Equation (13.14):

$$S_\pi(\ell) = \sum_{\pi'} \mathbb{P}(\pi') \left[\sum_{0 \le n \le \ell} C_{\pi,\pi'}(\ell - n) K_{\pi'}(n) + \sum_{n > \ell} C_{\pi',\pi}(n - \ell) K_{\pi'}(n) \right], \qquad (14.4)$$

where $K_\pi(\ell) := G_\pi(\ell + 1) - G_\pi(\ell)$.

Similarly to Section 13.2.1, one can invert the system of equations in (14.4) to evaluate the (un-observable) G_π in terms of the (observable) S_π and $C_{\pi,\pi'}$. Figure 14.3 shows the G_π-propagators for INTC and TSLA. The price-changing propagator is, as expected, larger in amplitude than the non-price-changing propagator, but reveals a substantially stronger impact decay.

It is also possible to generalise the calculation of the price variogram $\mathcal{V}(\ell)$ to this framework, but the calculation is messy, so we present it in Appendix A.4. In any case, once the $G_\pi(\ell)$ are known, one can again (as in Equation (13.16))

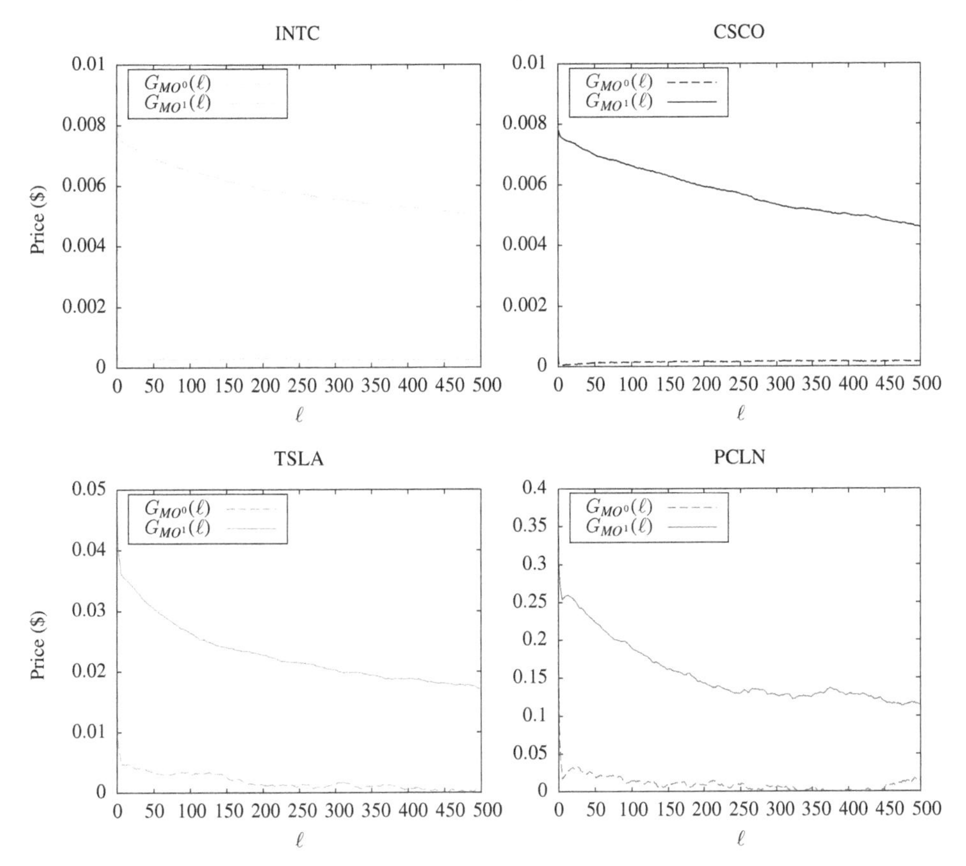

Figure 14.3. Propagators of (solid curve) price-changing market orders and (dashed curve) non-price-changing market orders for (top left) INTC, (top right) CSCO, (bottom left) TSLA and (bottom right) PCLN.

predict $\mathcal{D}(\ell) = \mathcal{V}(\ell)/\ell$ up to an additive constant Σ^2, which corresponds to the contribution of all events other than market orders. Figures 14.4 and 14.5 show the results of the two-state propagator model for the negative-lag response function $\mathcal{R}(\ell) = \sum_\pi \mathbb{P}(\pi)\mathcal{R}_\pi(\ell)$ and the signature plot $\mathcal{D}(\ell)$. The discrepancies noted for the one-state propagator are clearly reduced. Of particular interest is the fact that most of the long-term volatility is now explained solely in terms of market order impact.

14.4 A Six-Event Propagator Model

Compared to the simple propagator model from Chapter 13, the extended model in Section 14.3 fares much better at reproducing both $\mathcal{D}(\ell)$ and the negative part of the response function $\mathcal{R}(\ell)$. However, this extended model is still imperfect and suffers from the problem of only considering market orders, while treating limit order arrivals and cancellations in an indirect, implicit way.

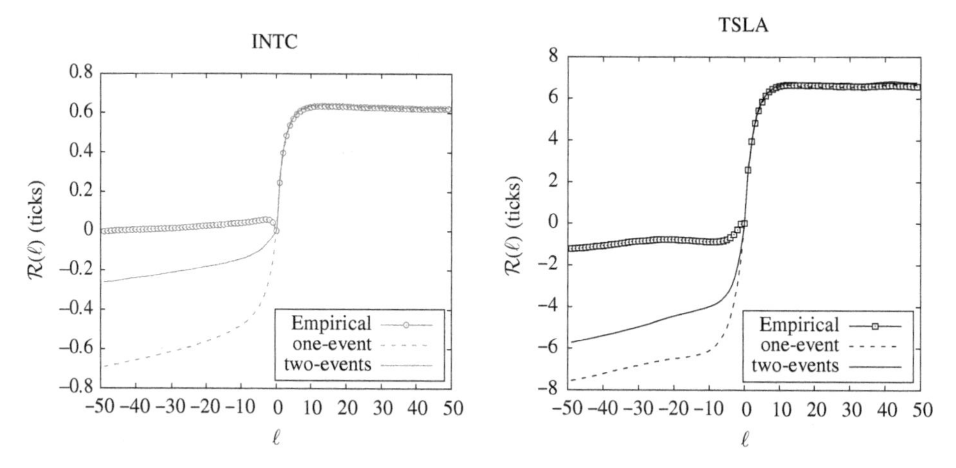

Figure 14.4. Predictions of the response function calculated from the propagator model, fitted by using only positive lags, for (left panel) INTC and (right panel) TSLA. The markers depict the empirical values, the dashed curves show the results from the one-event propagator model, and the solid curves show the results from the two-event propagator model.

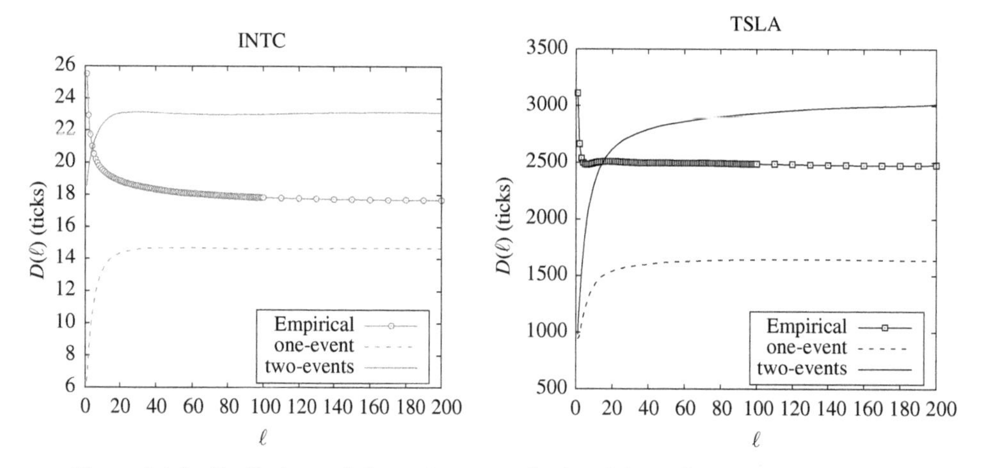

Figure 14.5. Predictions of the variogram calculated from the propagator model, fitted using only positive lags, for (left panel) INTC and (right panel) TSLA. The markers depict the empirical values, the dashed curves show the results from the one-event propagator model, and the solid curves show the results from the two-event propagator model.

In this section, we turn to a more granular formulation of the propagator model. If desired, we could formulate a version of the propagator model that considers all LOB events at all prices. However, such a model would involve processing a huge number of events, some of which occur very far away from the best quotes. Instead, we restrict our attention to LOB events that occur at or inside the best quotes, because only those event types can cause a change in the mid-price. Such events could be market order arrivals, cancellations at the best quotes, or limit order arrivals at the existing best quotes or inside the spread.

In this extended propagator model, we advance the clock by one unit whenever any of these events occurs. We again use a superscript 0 to indicate events that do not create an immediate (in this updated version of event-time) change in the mid-price, and we use a superscript 1 to indicate events that lead to an immediate change in the mid-price. We therefore partition the events that we consider into six different types of events: $\pi \in \{\mathrm{MO}^0, \mathrm{CA}^0, \mathrm{LO}^0, \mathrm{MO}^1, \mathrm{CA}^1, \mathrm{LO}^1\}$. For market order arrivals and limit order arrivals, we write $\varepsilon = +1$ to denote buy orders and $\varepsilon = -1$ to denote sell orders. For cancellations, we use the *opposite* signs, such that we write $\varepsilon = -1$ for the cancellation of a buy order and $\varepsilon = +1$ for the cancellation of a sell order. For a price-changing event, we call the associated change in mid-price the **step size** Δ_t. We provide a detailed description of these event types and definitions in Table 14.1.

It can also be useful to introduce another sign, $\varepsilon^\dagger = \pm 1$, which identifies the side of the LOB where each event occurs. We write $\varepsilon^\dagger = +1$ for any event happening on the ask side (i.e. buy market order arrivals, sell limit order arrivals and sell limit order cancellations) and $\varepsilon^\dagger = -1$ for any event happening on the bid side (i.e. sell market order arrivals, buy limit order arrivals and buy limit order cancellations).

14.4.1 Price Changes in the Six-Event Propagator Model

As in previous chapters, let m_t denote the mid-price immediately *before* the event that occurs at time t. In this framework, the generalised propagator description of mid-price changes can be written as

$$m_t = m_{t_0} + \sum_{t_0 \le t' < t} \left(\sum_{\pi} \mathbb{I}(\pi_{t'} = \pi) G_\pi(t - t') \varepsilon_{t'} + \xi_{t'} \right), \tag{14.5}$$

where the sum over π now covers all six event types, defining six different propagators $G_\pi(\ell)$.

To build a statistical theory of price changes, and in particular to understand how a certain market event impacts the price, one needs to study the properties of the time series of events and signs. To do so, we can follow a similar approach as in Equation (14.1), except that π_t can now take six different types. Correspondingly, the correlation matrix becomes a 6×6 matrix, with each entry being a function of time.

How does this correlation matrix behave for real stocks? Plotting the full matrix graphically is difficult, so we instead summarise its salient features:[3]

[3] For a generalisation of the DAR model of Sections 10.4.2 and 13.3.2 to multi-event-time series, see Taranto D. E., Bormetti, G., Bouchaud, J. P., Lillo, F., & Tóth, B. (2016). Linear models for the impact of order flow on prices I. Propagators: Transient vs. history dependent impact. https://ssrn.com/abstract=2770352.

Table 14.1. *Summary of the* 6 *possible event types* $\pi \in \{\mathrm{MO}^0, \mathrm{CA}^0, \mathrm{LO}^0, \mathrm{MO}^1, \mathrm{CA}^1, \mathrm{LO}^1\}$, *with the corresponding event signs and step sizes.*

Type	Event	Event Signs	Step Size (Δ)
MO^0	arrival of a market order with volume less than the outstanding volume at the opposite-side best quote	$\varepsilon = \varepsilon^\dagger = 1$ for buy market orders; $\varepsilon = \varepsilon^\dagger = -1$ for sell market orders	0
CA^0	partial cancellation of the bid/ask-queue	$\varepsilon = \varepsilon^\dagger = -1$ for buy limit order cancellation; $\varepsilon = \varepsilon^\dagger = +1$ for sell limit order cancellation;	0
LO^0	arrival of a limit order at the current best bid/ask	$\varepsilon = -\varepsilon^\dagger = +1$ for buy limit order arrivals; $\varepsilon = -\varepsilon^\dagger = -1$ for sell limit order arrivals	0
MO^1	arrival of a market order with volume greater than or equal to the outstanding volume at the opposite-side best quote	$\varepsilon = \varepsilon^\dagger = +1$ for buy market orders; $\varepsilon = \varepsilon^\dagger = -1$ for sell market orders;	half of the first gap behind the ask ($\varepsilon = 1$) or bid ($\varepsilon = -1$)
CA^1	cancellation of the whole best bid/ask-queue	$\varepsilon = \varepsilon^\dagger = -1$ for buy limit order cancellation; $\varepsilon = \varepsilon^\dagger = +1$ for sell limit order cancellation	half of the first gap behind the ask ($\varepsilon = 1$) or bid ($\varepsilon = -1$)
LO^1	arrival of a limit order inside the spread	$\varepsilon = -\varepsilon^\dagger = +1$ for buy limit order arrivals; $\varepsilon = -\varepsilon^\dagger = -1$ for sell limit order arrivals	half the distance of the limit order from the previous same-side best quote

- Autocorrelations of the "side" variable $\varepsilon_t^\dagger$ (see Figure 14.6) decay as a power law. In other words, LOB activity persists being either bid-side or ask-side for long periods of time. This is a more general statement of the same phenomenon that we observed for market order signs in Chapter 10.
- All non-price-changing correlation functions of the form $C_{\pi^0,\pi^0}(\ell)$ (with $\pi^0 = \text{MO}^0, \text{LO}^0$ or CA^0) are positive and decay as a slow power law. For price-changing events ($\pi^1 = \text{MO}^1, \text{LO}^1$ or CA^1), this correlation becomes short-ranged for large-tick stocks.
- When considering all market order arrivals, limit order arrivals and cancellations together (i.e. mixing between the different event types), autocorrelations of ε_t decay exponentially, which indicates that this series is short-range autocorrelated. Therefore, although each of the separate order-flow sign series are (separately) long-range autocorrelated, their intertwined series is not. This is the mechanism of the "tit-for-tat" dance that makes prices diffusive, as we described in Section 13.2.
- After a market order of either type MO^0 or MO^1, the flow of limit orders and cancellations first pushes the price in the same direction for about ten events, then reverses and opposes the market order flow, particularly through LO-type events, which correspond to **liquidity refill**.
- Newly posted price-improving limit orders LO^1 attract market orders. More precisely, LO^1 events rapidly trigger a strong opposite flow of MO^1 orders and, for large-tick stocks, of MO^0 orders as well. By contrast, LO^0 events are initially followed by market orders in the same direction, before the flow of these market orders inverts.

In a nutshell, the correlation matrix is consistent with the story that market participants who seek to execute large volumes split their trades into many different orders, which they execute via a mix of both market orders and limit orders. This explains why the diagonal correlation functions are positive and long-ranged. Liquidity providers step in rather quickly to counteract the correlated flow of market orders (via liquidity refill).[4]

Endowed with the knowledge of all correlation functions $C_{\pi,\pi'}(\ell)$ and all response functions $\mathcal{R}_\pi(\ell)$, one can invert Equations (14.4) to obtain the six propagators $G_\pi(\ell)$. Using the expression in Appendix A.4, one can then reconstruct the lag-dependent volatility $\mathcal{D}(\ell)$ from $C_{\pi,\pi'}(\ell)$ and $G_\pi(\ell)$. This recipe works reasonably well for small-tick stocks, but works much less well for large-tick stocks.[5]

[4] At the very highest frequencies, the reaction of the market is actually to first reduce liquidity, before this refill occurs. This could be explained by some liquidity providers cancelling their previous limit orders and possibly replacing them slightly further from the spread, in an attempt to buy or sell at a slightly better price immediately after the trade, before the refill occurs. See Bonart, J., & Gould, M. D. (2017). Latency and liquidity provision in a limit order book. *Quantitative Finance*, 1–16.

[5] On this point, see Eisler, Z., J. Kockelkoren, J., & Bouchaud, J. P. (2012). Models for the impact of all order book events. In Abergel, F. et al. (Eds.), *Market microstructure: Confronting many viewpoints*. Wiley; and Patzelt, F., & Bouchaud, J.-P. (2017). Nonlinear price impact from linear models. *Journal of Statistical Mechanics*, 12, 123404.

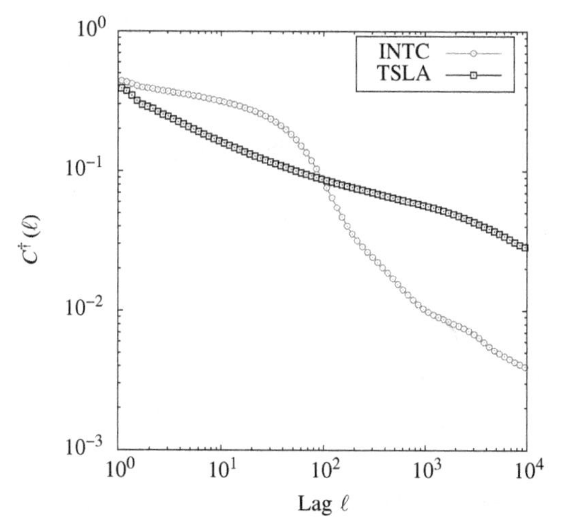

Figure 14.6. Autocorrelation function of the "side" variable $\varepsilon^\dagger$ for TSLA and INTC, plotted in doubly logarithmic coordinates. The plot suggests that the "side" process is a long-memory process.

Why is this so? In the next subsection, we argue that although generalising the transient impact model seems natural at first sight, attempting to do so naively can lead to erroneous results. Instead, we argue that an approach of history-dependent impact makes much more sense when generalised to all LOB events, and helps control the unavoidable noise that affects the determination of the propagators, especially for large-tick stocks. While the HDIM and TIM approaches are equivalent in the single-event case (see Section 13.3), they are actually distinct modelling strategies in the general case.

14.4.2 A Generalised, History-Dependent Impact Model

If we consider a model that includes all LOB events, then the dynamics of the mid-price can be *exactly* represented by the expression

$$r_t = \varepsilon_t \Delta(\pi_t, \varepsilon_t, t), \tag{14.6}$$

where $\Delta(\pi_t, \varepsilon_t, t)$ is the amplitude of the price change at time t for an event of type π_t and sign ε_t. By definition, $\Delta = 0$ for any event $\pi \in \{\text{MO}^0, \text{CA}^0, \text{LO}^0\}$. Therefore, any price change must be due to one event $\pi \in \{\text{MO}^1, \text{CA}^1, \text{LO}^1\}$. The definition of the non-zero Δ's depend on the type of event that occurs at time t (see Table 14.1). For example, if $\pi \in \text{MO}^1$, then Δ is one-half of the gap behind the best ask (for $\varepsilon = +1$) or behind the best bid (for $\varepsilon = -1$). In contrast to Equations (13.8) and (13.19), the present model has no need for a term ξ_t to account for "news" contributions, because it already includes all LOB events.

Large-Tick Stocks: The Case of Constant Δ

We first consider Equation (14.6) as a model for large-tick stocks. For such stocks, the LOB is densely populated, and the bid–ask spread and the gap behind the best quote are usually equal to one tick ϑ. To model this case, we assume that Δ does not fluctuate at all, such that

$$\begin{aligned} \Delta(\pi^0, \varepsilon, t) &= 0, \\ \Delta(\pi^1, \varepsilon, t) &= \Delta_{\pi^1} = \vartheta/2. \end{aligned} \tag{14.7}$$

In the propagator framework, this amounts to setting

$$G_{\pi^0}(\ell) = 0; \qquad G_{\pi^1}(\ell) = \Delta_{\pi^1},$$

such that all price-changing events have constant impact. In this case, the expression for $\mathcal{R}_\pi(\ell)$ simplifies to

$$\mathcal{R}_\pi(\ell) \approx \Delta_\pi + \sum_{0<t'<\ell} \sum_{\pi^1} \Delta_{\pi^1} \mathbb{P}(\pi^1) C_{\pi,\pi^1}(t'). \tag{14.8}$$

Therefore, the price response to a given event can be understood as its own "mechanical" impact (possibly zero) plus the sum of the average impact of all future events correlated with this initial event. Within the same model, the expression for $\mathcal{V}(\ell)$ simplifies to

$$\mathcal{V}(\ell) \approx \sum_{0 \le t',t'' < \ell} \sum_{\pi^1} \sum_{\pi^{1'}} \mathbb{P}(\pi^1)\mathbb{P}(\pi^{1'}) C_{\pi^1,\pi^{1'}}(t'-t'') \Delta_{\pi^1} \Delta_{\pi^{1'}}. \tag{14.9}$$

For large-tick stocks, these predictions are extremely accurate. In principle, this model is equivalent to the six-event propagator model with a constrained form for G_{π^0} and G_{π^1}. The reason it fares much better than the unconstrained model is that, after matrix inversion, noisy estimates of $\mathcal{R}_\pi(\ell)$ and $C_{\pi,\pi'}(\ell)$ lead to propagators that have some non-trivial (but spurious) structure, which in turn considerably pollutes the determination of $\mathcal{V}(\ell)$.

Small-Tick Stocks: The Case of History-Dependent Δ

We next consider Equation (14.6) as a model for small-tick stocks. For such stocks, the LOB is sparsely populated, and the bid–ask spread and the gap behind the best quote fluctuate over time. The propagator model does not address these fluctuations, since $G_\pi(\ell)$ only depends on the event type, not on the particular state of the LOB when the event π took place. This provides motivation for considering alternative approaches to the one that we have developed so far in this chapter. One possible alternative is to consider a model in which the history of the order flow feeds back into the present values of $\Delta(\pi, \varepsilon, t)$, to reflect how past order flow affects present liquidity conditions.[6]

Assuming that all the consequences of buys are the same as those of sells (up to a sign change), we can extend the model in Equation (14.7) to

$$\Delta(\pi^1, \varepsilon, t) = \Delta_{\pi^1} + \sum_{t'<t} \sum_{\pi'} \mathbb{I}(\pi_{t'} = \pi') \kappa_{\pi',\pi^1}(t-t') \varepsilon_{t'} \varepsilon + \widetilde{\xi}_t, \tag{14.10}$$

where $\kappa_{\pi,\pi'}$ are kernels that model the dependence of gaps on past order flow, and $\widetilde{\xi}$ describes the part of the evolution of the spreads and gaps that is not explained by the past order flow. Observe that the entries whose second index corresponds to a non-price-changing event, for which $\Delta_{\pi^0} = 0$ by definition, so κ_{π,π^0} must be zero for all π. For large-tick stocks, the influence kernels $\kappa_{\pi,\pi'}$ are all extremely small and can be neglected, leading back to Equations (14.8) and (14.9).

To illustrate what this model encodes, imagine that the event taking place at time t' is a buy market order with type MO^1. This market order exerts a buy pressure on the LOB. If this pressure scares away sellers, then the gap behind the ask could increase;

[6] For more details on the content of this section, and empirical calibration, see Eisler, Z., J. Kockelkoren, J., & Bouchaud, J. P. (2012). The price impact of order book events: Market orders, limit orders and cancellations. *Quantitative Finance*, 12(9), 1395–1419; Eisler et al. (2012) as listed in Footnote 5; and Patzelt, F., & Bouchaud, J.-P. (2017). Nonlinear price impact from linear models. *Journal of Statistical Mechanics*, 12, 123404.

if this pressure attracts other sellers to submit limit orders, then the gap behind the ask could decrease. The first case would correspond to $\kappa_{\mathrm{MO}^1,\mathrm{MO}^1} > 0$ and the second case would correspond to $\kappa_{\mathrm{MO}^1,\mathrm{MO}^1} < 0$.

As we now show, the model in Equation (14.10) corresponds precisely to a generalisation of the model in Equation (13.8). By substituting Equation (14.10) into the exact evolution equation (14.6), we arrive at

$$r_t = \Delta_{\pi_t}\varepsilon_t + \sum_{t'<t} \kappa_{\pi_{t'},\pi_t}(t-t')\varepsilon_{t'} + \xi_t; \qquad (\xi_t := \widetilde{\xi}_t\varepsilon_t). \tag{14.11}$$

It is interesting to compare Equation (14.11) with the analogue for the transient impact model in Equation (14.5):

$$r_t = G_{\pi_t}(1)\varepsilon_t + \sum_{t'<t}\left[G_{\pi_{t'}}(t-t'+1) - G_{\pi_{t'}}(t-t')\right]\varepsilon_{t'} + \xi_t. \tag{14.12}$$

By comparing Equation (14.1) and (14.12), one sees that the two models are equivalent if and only if

$$G_\pi(1) = \Delta_\pi$$

and

$$\kappa_{\pi,\pi'}(\ell) = G_\pi(\ell+1) - G_\pi(\ell), \qquad \text{for all } \pi', \tag{14.13}$$

which can occur if and only if $\kappa_{\pi,\pi'}$ does not depend on the value of π'. This constraint cannot hold in general since when the second event is non-price-changing, $\kappa_{\pi,\pi^0} = 0$, which would imply that κ is zero in all cases. The only exception is when there is one possible type of event, in which case we recover the equivalence between transient impact models (TIM) and history-dependent impact models (HDIM), as we saw in Section 13.3.

When calibrating $\kappa_{\pi,\pi'}$ empirically, one in fact finds a significant dependence on the second event type, absent in transient impact models. The model fares quite well at reproducing the negative-lag response function and the signature plot of small-tick stocks. One furthermore finds that for ℓ large enough, and when π' is a price-changing event

$$\kappa_{\pi^1,\pi'}(\ell) \leq 0, \tag{14.14}$$

Equation (14.14) recovers the **asymmetric dynamical liquidity** phenomenon in a wider context (see Section 11.3.4): past price-changing events tend to reduce the impact of future events of the same sign, and increase the impact of future events of opposite sign (recall the content of Equation (14.10)). Note however that $\kappa_{MO^1,\pi'}(\ell)$ is positive for short lags: as we mentioned in Section (14.4.1), the knee-jerk reaction of markets is first to reduce liquidity upon the arrival of aggressive market orders, before the liquidity refill phenomenon takes place.

14.5 Other Generalisations

In this section, we present other ways to tag different types of LOB events with a π_t term, different from the MO, LO, CA types discussed above. One extremely interesting case is when some trade identification is possible, allowing one to distinguish the price impact of different institutions or individual traders. Another situation where tagging is important is cross-impact, where trading one asset may impact another correlated asset in a different market.

14.5.1 Trade-Ownership Data

One very interesting case arises when the owner of each trade can be identified. For example, some data sets include an identification number for each market order, which indicates the financial institution (or even individual trader) that initiated the trade.[7] In this case, one could tag each market order with this information, and define a propagator $G_\pi(\ell)$ for each institution (or trader) π. By fitting such a model empirically, it could be possible to gain insight into how the impact of specified actions differs according to who conducted them.

14.5.2 Proprietary Trades versus Market Trades

Another situation where partial identification is possible is when a trading firm knows about its own trades, but cannot identify those of other firms. In this situation, the trading firm can tag market orders with a binary variable $\pi \in$ {own, other}. From the empirical determination of the cross-correlation function $C_{\text{own,other}}(\ell)$, the trading firm can then determine how its own trades tend to be anticipated or followed by the rest of the market.

Together with the two empirical response functions $\mathcal{R}_{\text{own}}(\ell)$ and $\mathcal{R}_{\text{other}}(\ell)$, it is again possible to obtain two propagators, $G_{\text{own}}(\ell)$ and $G_{\text{other}}(\ell)$, by inverting a relation similar to Equation (14.4). When this analysis is possible, one finds that the two propagators are identical, up to statistical fluctuations.[8] Since $G_{\text{other}}(\ell)$ must reflect a large fraction of noise trades, the similarity between $G_{\text{own}}(\ell)$ and $G_{\text{other}}(\ell)$ is compatible with the idea that these propagators describe the reaction impact, which is independent of the information content of the trades (which only shows up via the correlations with future orders, and on longer time scales). This is in line with the order-flow view of price formation (see the discussion in Sections 11.1 and 11.2).

14.5.3 Cross-Impact

Another natural extension of the propagator model is to consider how trades for a given asset j can (directly or indirectly) impact the price of another asset i. Restricting to market orders, and neglecting order volumes, Equation (13.8) can be generalised to describe both self- and cross-impact, as follows:[9]

$$r_{i,t} = m_{i,t+1} - m_{i,t} = \sum_{t' \leq t} K_{ij}(t-n)\varepsilon_{j,t'} + \xi_{i,t}, \tag{14.15}$$

where $r_{i,t}$ is the return of asset i at time t, $\varepsilon_{j,t'}$ is the sign of the market order for asset j at time t', and $\xi_{i,t}$ are residuals that capture the component of returns not directly related to trading (and that are possibly correlated between different assets). If one considers the joint dynamics of N stocks, then the propagator is

[7] Sometimes, only the identification of the executing broker is possible, not the final buyer/seller of the asset.

[8] On this point, see Tóth, B., Eisler, Z., & Bouchaud, J. P. (2017). The short-term price impact of trades is universal. https://ssrn.com/abstract=2924029.

[9] Note that because the market order time has no reason to correspond across different assets, t is here calendar-time and not event-time.

a lag-dependent $N \times N$ matrix. The diagonal terms of this matrix correspond to **self-impact**, whereas the off-diagonal terms correspond to **cross-impact**.

This model can be calibrated by inverting a relation similar to Equation (14.4). The detailed discussion of this topic is beyond the scope of this book, so we instead refer to recent papers cited in Section 14.7. The two important empirical conclusions from these papers are:

- Both the on-diagonal and off-diagonal elements of the propagator matrix decay as a power-law of the lag, $G(\ell) \sim \ell^{-\beta}$, with $\beta < 1$, as in the single-asset case.
- Perhaps surprisingly, most of the cross-correlations between price moves are mediated by trades themselves (i.e. through a cross-impact mechanism) rather than through the cross-correlation of the residual terms $\xi_{i,t}$, which are not directly related to trading.

Yet another related situation arises in fragmented markets, when trading on different venues can impact prices differently, corresponding to different propagators.

14.6 Conclusion

We conclude our discussion of propagator models by recapping the thread of ideas that we have explored. We started our discussion in Chapter 13 with a simple picture in which we interpreted price moves in terms of the impact of market orders and an extra contribution not related to trades. In this description, we saw that the long memory of trade signs imposes constraints on how price impact must decay with time, to ensure that the mid-price remains approximately diffusive. In the present chapter, we developed our discussion towards a more complete picture of LOB dynamics, where we interpreted price moves in terms of complex, intertwined flows (of market orders, limit orders and cancellations) that statistically coordinate and respond to each other.

We have also seen that the transient nature of price impact, which is described by the decaying propagator $G(\ell)$, could equivalently be interpreted as a permanent but history-dependent impact. In this interpretation, past trades themselves shape present liquidity in a way that decreases the impact of expected market orders and increases the impact of surprising market orders (see Section 13.3).

In generalising the propagator model to describe all LOB events, it becomes apparent that transient impact models (TIM) and history-dependent impact models (HDIM) are not equivalent, but rather that transient impact models are a special subclass of HDIMs (see Equation (14.11)). The compelling idea behind HDIMs is that the current spread and liquidity in the LOB are affected by all previous LOB events. This allows one to capture in detail how past order flow shapes (on average) the LOB in the vicinity of the best quotes, and therefore how the next event is likely to impact the mid-price.

Similarly to in Section 11.3.4, we again find that liquidity is dynamically asymmetric in this extended framework. More precisely, past events tend to reduce the impact of future events of the same sign and increase the impact of future events of opposite sign, as is required if markets are to be stable and prices are to be statistically efficient. This is one of the most important messages of this chapter, and indeed of the whole book: financial markets operate in a kind of "tit-for-tat" mode, where liquidity providers react to the actions of liquidity takers, and vice-versa. During the normal-functioning of financial markets, these retroactions have a stabilising effect that allows markets to function in an orderly fashion. Any breakdown of the asymmetric liquidity mechanism may lead to crises.

Finally, we emphasise that the generalised propagator models we have presented in this chapter are still linear models. When restricted to market orders, they boil down to the effective propagator model from Chapter 13 (see Section 13.4.2). Therefore, this family of models cannot account for non-linear effects such as the square-root impact of metaorders. Reproducing these more complex effects requires genuinely new ingredients, which we turn to in Chapter 19.

Take-Home Messages

(i) Single-event propagator models are too simple to reproduce some empirical regularities of real markets. Accounting for the more complex dynamics observable in empirical data requires extending this basic framework.

(ii) One possible generalisation is to partition market orders according to whether or not they change the price. For large-tick stocks, this partitioning improves the propagator model's predictions.

(iii) Another possible generalisation is to include all LOB events. Fitting such a model requires large amounts of data and can lead to expensive computations, but allows one to measure the impact of limit orders and cancellations, on top of the impact of market orders.

(iv) In the propagator framework, there are two main approaches to including all LOB events: transient impact models (TIM), which assume that all events are characterised by a different (decaying) propagator, and history-dependent impact models (HDIM), which describe how each event changes the future impact of all other events – i.e. of future liquidity.

(v) The propagator framework can also be generalised to many other situations, including comparing the impact of different traders' actions or measuring cross-impact across different assets.

14.7 Further Reading

Generalised Propagator Models

Eisler, Z., Bouchaud, J. P., & Kockelkoren, J. (2012). The price impact of order book events: Market orders, limit orders and cancellations. *Quantitative Finance*, 12(9), 1395–1419.

Eisler, Z., J. Kockelkoren, J., & Bouchaud, J. P. (2012). Models for the impact of all order book events. In Abergel, F., Bouchaud, J.-P., Foucault, T., Lehalle, C.-A., & Rosenbaum, M. (Eds.), *Market microstructure: Confronting many biewpoints*. Wiley

Taranto, D. E., Bormetti, G., Bouchaud, J. P., Lillo, F., & Tóth, B. (2016). Linear models for the impact of order flow on prices I. Propagators: Transient vs. history dependent impact. Available at SSRN: https://ssrn.com/abstract=2770352.

Impact of Limit Orders and Other Events

Hautsch, N., & Huang, R. (2012). The market impact of a limit order. *Journal of Economic Dynamics and Control*, 36(4), 501–522.

Bershova, N., Stephens, C. R., & Waelbroeck, H. (2014). The impact of visible and dark orders. https://ssrn.com/abstract=2238087.

Cont, R., Kukanov, A., & Stoikov, S. (2014). The price impact of order book events. *Journal of Financial Econometrics*, 12(1), 47–88.

Gençay, R., Mahmoodzadeh, S., Rojcek, J., & Tseng, M. C. (2016). *Price impact of aggressive liquidity provision* (No. 16-21). Swiss Finance Institute.

Patzelt, F., & Bouchaud, J.-P. (2017). Nonlinear price impact from linear models. *Journal of Statistical Mechanics*, 12, 123404.

Propagator Models with Trader Identification

Tóth, B., Eisler, Z., Lillo, F., Kockelkoren, J., Bouchaud, J. P., & Farmer, J. D. (2012). How does the market react to your order flow? *Quantitative Finance*, 12(7), 1015–1024.

Tóth, B., Eisler, Z. & Bouchaud, J.-P. (2017). The short-term price impact of trades is universal. https://ssrn.com/abstract=2924029.

Propagators and Cross-Impact

Wang, S., Schäfer, R., & Guhr, T. (2015). Price response in correlated financial markets: Empirical results. arXiv preprint arXiv:1510.03205.

Schneider, M., & Lillo, F. (2016). Cross-impact and no-dynamic-arbitrage. arXiv:1612.07742.

Wang, S., & Guhr, T. (2016). Microscopic understanding of cross-responses between stocks: A two-component price impact model. https://ssrn.com/abstract=2892266.

Benzaquen, M., Mastromatteo, I., Eisler, Z., & Bouchaud, J. P. (2017). Dissecting cross-impact on stock markets: An empirical analysis. *Journal of Statistical Mechanics: Theory and Experiment*, 023406.

Mastromatteo, I., Benzaquen, M., Eisler, Z., & Bouchaud, J. P. (2017). Trading lightly: Cross-impact and optimal portfolio execution. arXiv:1702.03838. *Risk Magazine*, July 2017.

PART VII
Adverse Selection and Liquidity Provision

Introduction

Throughout the book, we have pieced together several clues that suggest that the unpredictable nature of prices emerges from a fine balance between a strongly autocorrelated order flow and a dynamic price impact. At the same time, however, we have seen that so-called surprise models can also reproduce this balance with very few assumptions. It therefore seems that there might be another side to the coin, which statistical models fail to capture. Are we missing something?

So far, we have regarded order flows as simple stochastic processes with specified statistical properties. In doing so, we have excluded the *strategic behaviours* implemented by real market participants. In reality, however, investors' actions are clearly influenced by their desire to make profits. Therefore, it seems logical that strategic behaviour should play an important role in real markets.

In this part, we will take the next logical step in our journey by exploring how including strategic considerations can shed light on some otherwise surprising features of financial markets. In doing so, we will examine how markets can remain in a delicate balance, maintained by ongoing competition between rational, profit-maximising agents. We will discuss the seminal Kyle model, which provides a beautiful explanation of how impact arises from liquidity providers' fears of adverse selection from informed traders. The model illustrates how impact allows information to be reflected in the price, and makes clear the important role played by noise traders – as we alluded to in the very first part of this book. More generally, the Kyle model, albeit not very realistic, is a concrete example of how competition and arbitrage can produce diffusive prices, even in the presence of private information.

Economic models also provide important insights on the challenges faced by market-makers. We will discuss the work of Glosten and Milgrom, which shows how the bid–ask spread must compensate market-makers for adverse selection in a competitive market. The relationship between a metaorder's slippage and its permanent impact will appear as another manifestation of the same idea. This framework paves the way for richer models of liquidity provision, where inventory risk, P&L skewness and finite tick-size effects all contribute to the challenge. We will also discuss how the Glosten–Milgrom model teaches us an important lesson: that the apparent profit opportunities from "buying low, selling high" around the spread are completely misleading.

15
The Kyle Model

Economists think about what people ought to do. Psychologists watch what they actually do.

(Daniel Kahneman)

As we discussed in Section 11.1, price impact can be interpreted in two different ways: as a statistical effect caused by a local order-flow imbalance in a market otherwise in equilibrium, or as the process by which revealed information is incorporated into the price. In the previous chapters, we have mainly focused on the first of these interpretations. In the present chapter, we turn our attention to the second interpretation. Specifically, we introduce the classic **Kyle model**, which seeks to shed light on the mechanism by which private information is gradually incorporated into prices, via trading. Kyle's original paper is often cited as the foundation of the field of market microstructure. The Kyle model shows how price impact can arise when market-makers anticipate adverse selection in an auction setting. As we discuss throughout the chapter, the model also makes clear why even informed traders should act cautiously and use small orders to ensure that they only reveal their information very slowly.

15.1 Model Set-Up

Consider a single-asset market populated by an **informed trader** (Alice), a **market-maker** (Bob), and a set of **noise traders**, who behave as follows:

- **Alice**: At time $t = 0$, Alice discovers some private information about the price p_F that the asset will have at $t = 1$. For example, Alice may be an insider trader who knows that at time $t = 1$, there will be a public announcement about a takeover bid at price p_F. Alice is the only market participant with access to this (private) information, so that Alice does not have to worry about other market participants trying to exploit the same information. Based on her private information, Alice chooses a volume Q of the asset to buy ($\varepsilon = +1$) or sell ($\varepsilon = -1$), in such a way as to maximise her expected profit, when discounting the expected impact cost of her own trade. Alice has no risk constraints that would limit her position.

- **Noise traders**: These uninformed traders do not have access to private information, but rather simply trade for idiosyncratic reasons (for a full discussion of the role of noise traders in the market ecology, see Chapter 1). In doing so, the noise traders generate a random order flow with a net volume of V_{noise}, whose sign and amplitude is independent from p_{F}.
- **Bob**: Bob is a market-maker. He clears the market by matching the net total buy or sell volume $\Delta V = V_{\text{noise}} + \varepsilon Q$ with his own inventory, at a clearing price $\widehat{p}$. Bob's choice of $\widehat{p}$ is rule-based, and such that he breaks even on average, in a sense that we make precise below. Bob has no inventory constraints that would limit his position.

At time $t = 1$, the price p_{F} is revealed. At this time, Alice's asset is exactly worth p_{F}. This assumes that she can buy or sell any quantity of the asset at price p_{F} without causing impact.

In the tradition of theoretical economics, one then looks for an equilibrium between Alice and Bob, such that:

(i) **Profit maximisation**: given Bob's price-clearing policy, Alice's signed volume εQ must maximise her expected gain. She buys/sells at the clearing price $\widehat{p}$, but the asset will in fact be worth p_{F}, so her gain is given by

$$\mathcal{G} = \varepsilon Q(p_{\text{F}} - \widehat{p}). \tag{15.1}$$

Note again that there is no risk constraint that would limit Alice's position in the asset.

(ii) **Market efficiency**: Bob's clearing price must be such that $\mathbb{E}_{\text{Bob}}[p_{\text{F}}|\Delta V] = \widehat{p}$, where the expectation is taken using the information available to Bob only (i.e. without the information known to Alice). This corresponds to a situation where Bob breaks even on average, given an incoming volume ΔV.

Given that Alice knows all of the above (i.e. she knows how both the noise traders and Bob will act in a given situation), how should she choose Q to maximise her expected profit $\mathbb{E}[\mathcal{G}]$ at time $t = 1$? For Alice to choose an optimal value of Q to maximise $\mathcal{G}$, she must consider Bob's clearing price. Specifically, Alice knows that Bob has a mechanical rule for choosing $\widehat{p}$ as a function of ΔV, so she must use this knowledge when deciding how to act.

Bob observes the net volume ΔV, but does not know the value of p_{F}. Since orders are anonymous, he also does not know the values of ε or Q. However, Bob knows that Alice knows his price-clearing rule, and also knows that she acts so as to optimise her gain with the information at her disposal (which he will try to infer from the knowledge of ΔV).

We further assume that Bob knows that the net volume executed by the noise traders is a Gaussian random variable with zero mean and standard deviation equal to Σ_V, and that the mispricing $p_F - p_0$ is similarly a Gaussian random variable with zero mean (over time) and standard deviation σ_F. The latter quantity is related to the typical amount of information available to insiders at time 0 but not yet included in the price.

15.2 Linear Impact

What are Alice's and Bob's optimal actions in this market? Consider the case where all random variables are Gaussian and where Bob's price-fixing rule is linear in the order imbalance, such that

$$\widehat{p} = p_0 + \Lambda \Delta V, \tag{15.2}$$

for some impact parameter Λ called **Kyle's lambda**. In this framework, one can show that the solution to this problem (or "market equilibrium") is self-consistent and can be fully determined. Indeed, profit maximisation on Alice's part leads to

$$\widehat{Q} = \text{argmax}_Q \mathbb{E}[\mathcal{G}]; \qquad \mathcal{G} = \varepsilon Q \times (p_F - p_0 - \Lambda \Delta V). \tag{15.3}$$

Since the expected value of the random imbalance V_{noise} is zero, one has $\mathbb{E}[\Delta V] = \varepsilon Q$. This leads to a quadratic maximisation problem, with the solution

$$\widehat{Q} = \frac{1}{2} \frac{(p_F - p_0)}{\Lambda}. \tag{15.4}$$

This result means that Alice should choose $\widehat{Q}$ to be proportional to the mispricing $p_F - \widehat{p}$. Since Bob knows that Alice will do this, he attempts to estimate the value of $\varepsilon\widehat{Q}$, when he only observes ΔV. Estimating $\varepsilon\widehat{Q}$ allows him, using Equation (15.4), to guess the value of p_F used by Alice, and choose Λ in such a way that his clearing price $\widehat{p}$ is an unbiased estimate of the future price.

Using **Bayes' theorem**, the conditional probability that Alice's volume is $\varepsilon\widehat{Q}$, given ΔV, is

$$\begin{aligned}
\mathbb{P}_{\text{Bob}}(\varepsilon\widehat{Q} \mid \Delta V) &\propto \mathbb{P}(\Delta V \mid \varepsilon\widehat{Q}) \times \mathbb{P}(\varepsilon\widehat{Q}), \\
&\propto \exp\left[-\frac{(\Delta V - \varepsilon\widehat{Q})^2}{2\Sigma_V^2}\right] \times \exp\left[-\frac{2(\varepsilon\widehat{Q})^2 \Lambda^2}{\sigma_F^2}\right],
\end{aligned}$$

where in the second exponential we have used the relation between $\widehat{Q}$ and $p_F - p_0$ from Equation (15.4). By merging the two exponentials together, we see that Bob's inferred distribution for $\varepsilon\widehat{Q}$ is Gaussian, with the following conditional mean:

$$\mathbb{E}_{\text{Bob}}[\varepsilon\widehat{Q} \mid \Delta V] = \sigma_F^2 \frac{\Delta V}{\sigma_F^2 + 4\Lambda^2 \Sigma_V^2}. \tag{15.5}$$

This in turn converts into Bob's best estimate of Alice's view on the future price, again using Equation (15.4)

$$\widehat{p} := \mathbb{E}_{\text{Bob}}\left[p_{\text{F}}|\Delta V\right] = p_0 + 2\Lambda\sigma_{\text{F}}^2 \frac{\Delta V}{\sigma_{\text{F}}^2 + 4\Lambda^2\Sigma_V^2}. \tag{15.6}$$

Identifying this expression with Equation (15.2), it also follows that

$$\Lambda = \frac{2\Lambda\sigma_{\text{F}}^2}{\sigma_{\text{F}}^2 + 4\Lambda^2\Sigma_V^2}.$$

Simplifying this expression provides the following solution for Kyle's lambda:

$$\Lambda = \frac{\sigma_{\text{F}}}{2\Sigma_V}. \tag{15.7}$$

By choosing this value of Λ, Bob thus makes sure that the strategic behaviour of Alice and the stochastic nature of the noise traders combine in such a way that the realised price $\widehat{p}$ is an unbiased estimate of the fundamental price p_{F}, given the publicly available information at time $t = 0$.

15.3 Discussion

The Kyle model raises several interesting points for discussion. First, the mean impact of an order grows linearly with Q: the Kyle model leads to a **linear impact law**. The linear scale factor Λ grows with the typical amount of private information present in the market (measured by σ_{F}) but decreases with the typical volume of uninformed trades (measured by Σ_V). This captures the basic intuition that market-makers protect themselves against adverse selection from informed traders by increasing the cost of trading, and benefit from the presence of noise traders to reduce price impact.

Second, using the result for $\widehat{Q}$ in Equation (15.4), the result for $\widehat{p}$ in Equation (15.6), and Equation (15.7), it follows that Alice's gain is given by

$$\begin{aligned}
\mathbb{E}[\varepsilon\widehat{Q}\cdot(p_{\text{F}} - \widehat{p})] &= \frac{\sigma_{\text{F}}^4}{2\Lambda(\sigma_{\text{F}}^2 + 4\Lambda^2\Sigma_V^2)}, \\
&= \frac{1}{2}\sigma_{\text{F}}\Sigma_V.
\end{aligned}$$

Therefore, the conditional expectation of Alice's gain increases with the amount of private information and with the overall liquidity of the market, which is (unwittingly) provided by the noise traders. For typical values of the predictor (i.e. for values of $(p_{\text{F}} - p_0)$ of the order of σ_{F}), it follows that $\widehat{Q} \sim \Sigma_V$, so Alice contributes a substantial fraction of the total traded volume. This is not very realistic in practice: as a common-sense precautionary measure, informed traders tend to limit their trading to a small fraction of the total volume. Out-sized trades risk destabilising the market, resulting in a much larger impact cost (see discussion in Section 10.5).

Finally, the pricing error at $t = 1$ can be measured as:

$$\mathbb{V}[\widehat{p} - p_{\mathrm{F}}] = \frac{1}{2}\mathbb{V}[p_{\mathrm{F}} - p_0] = \frac{1}{2}\sigma_{\mathrm{F}}^2. \tag{15.8}$$

Therefore, Bob is only able to reduce by one-half the variance of the uncertainty of the fundamental price known to Alice.

The Kyle model provides a clear picture of the origin of price impact. In the model, market-makers fear that someone in the market is informed, and therefore react by increasing the price when they observe a surplus of buyers and decreasing the price when they observe a surplus of sellers. The model also illustrates that even though the model permits Alice to enter an infinitely large position, her private information only leads to bounded profits, because of impact costs.

Note that when moving slightly away from its core assumptions, the Kyle model becomes a self-fulfilling mechanism. Suppose that market-makers overestimate σ_{F} (i.e. they overestimate the quality of information available to insiders). In reality, because there is no "terminal time" when a "true price" p_{F} is revealed, such market-makers will over-react to the order flows when setting the value of $\widehat{p}$. In the efficient-market picture, these pricing errors should self-correct through arbitrage from the informed trader (i.e. with a signal to trade in the opposite direction at the next time steps). This would lead to excess high-frequency volatility. However, signature plots of empirical data are rather flat (see Section 2.1.4), which instead suggests that the whole market shifts its expectations around the new traded price, much as assumed in the Santa Fe model in Chapter 8 (for which $\sigma_{\mathrm{F}} = 0$, which corresponds to the situation of no information but non-zero impact).

15.4 Some Extensions

There are several simple ways to extend the Kyle model:

(i) One possible extension, due to Kyle himself, is to consider a multi-step set-up in which the terminal price p_{F} known to Alice only reveals itself at time T. In the continuous-time limit, in which there are infinitely many steps between the initial time and the terminal time T, Alice's optimal trade at any time $t < T$ is still linear in the mispricing:

$$\varepsilon(t)\,\mathrm{d}\widehat{Q}(t) = \frac{\Sigma_V T}{\sigma_{\mathrm{F}}(T-t)}(p_{\mathrm{F}} - p(t))\,\mathrm{d}t, \tag{15.9}$$

while the impact parameter is independent of time and given by $\Lambda = \sigma_{\mathrm{F}}/\Sigma_V$. The resulting price volatility is also constant in time.[1] Quite remarkably,

[1] The fact that $\Lambda(t)$ is a constant in a continuous auction equilibrium implies that trading prices have constant volatility over time and therefore that information is gradually incorporated into prices at a constant rate.

due to the market-maker's clearing rule, even in the presence of Alice's systematic trading in the direction of the true price p_F, the price $p(t)$ remains a martingale for Bob. A similar property will hold in the Glosten–Milgrom model (see Section 16.1.5). Contrarily to empirical observations, however, the multi-time-step Kyle model's order-sign series is not autocorrelated at all!

In this model, Alice's aggressiveness increases as $t \to T$. In order to maximise her profit, Alice's impact on the price ensures that $p(t) \to p_F$ as $t \to T$. This is intuitively obvious if Alice's information is certainly true, because any mispricing would lead to unexploited profit opportunities.

Strangely, however, this convergence is actually driven by Alice's trading strategy, Equation (15.9). In other words, if Alice firmly believed that the price at time T should be equal to some arbitrary value p_{Alice} and Bob again acted as if Alice was truly informed, then the price would indeed converge to p_{Alice} at time T.

(ii) Another possible extension is to assume that Alice is risk-averse, so adds a risk-penalty term of the form $-\zeta Q^2$ to her objective function, such that Equation (15.3) instead becomes

$$\mathcal{G} = \varepsilon Q \times (p_F - p_0 - \varepsilon \Lambda Q) - \zeta Q^2.$$

To leading order in ζ, the inclusion of this term decreases the value of Kyle's lambda to $\Lambda - \zeta^2/2\Lambda$. Since Alice's new risk-aversion term constrains her from taking big positions, the adverse selection risk is reduced and Bob can offer more liquidity to the market.

(iii) A third simple extension is to remove the hypothesis that V_{noise} and $p_F - p_0$ are Gaussian random variables. In this case, one can still solve Kyle's model in the small-volume limit $Q \ll \Sigma_V$. Provided that:

- the distribution of the random component of the order flow has a quadratic maximum around zero, i.e. it behaves as

$$\mathbb{P}(V_{\text{noise}} \to 0) = P_0 - P_0'' V_{\text{noise}}^2/2 + \ldots;$$

- the distribution of $p_F - p_t$ has a finite variance σ_F^2,

then the main conclusions of the Kyle model still hold, in particular that price impact is linear when $Q \ll \Sigma_V$, with a coefficient proportional to σ_F.

Besides these three, many other extensions and generalisations of the Kyle model have been considered, for example the role of inventory risk constraints for the market-maker.[2]

15.5 Conclusion

The Kyle model elegantly elicits some deep truths about how markets function, but also fails to capture some important empirical properties of real markets. The most interesting outcomes of the model are:

- Trades impact prices. In the model, the mechanism that creates price impact is the market-maker's attempts to guess the amount of information contained in the order-flow, and to adjust the price up or down accordingly.

[2] See: Cetin, U., & Danilova, A. (2016). Markovian Nash equilibrium in financial markets with asymmetric information and related forward-backward systems. *The Annals of Applied Probability*, 26(4), 1996–2029.

- Impact is linear (i.e. price changes are proportional to order-flow imbalance) and permanent (i.e. there is no decay of impact). In the context of an LOB, linear impact is a generic consequence of having a finite density of buy/sell orders in the vicinity of the price, and permanent impact is a consequence of liquidity immediately refilling the gap left behind the incoming market order.[3] We will return to this discussion in the context of Walrasian auctions in Chapter 18.
- The impact of trading is inversely proportional to the total volume traded by noise traders. In other words, the existence of uninformed traders is essential for the market to function. In the next chapter, we will see that the Glosten–Milgrom model reaches a similar conclusion: a dearth of uninformed trading can lead to market breakdown.
- Because of impact, the informed trader must limit her trading volume to optimise her gains. Therefore, the amount of profit that can be made using private (insider) information is limited.

When confronted with empirical data, the Kyle model suffers from important drawbacks. The most obvious one is that the sign of order flow is found to be completely uncorrelated. This is a consequence of impact being permanent in the Kyle model. Any correlation in signs would lead to predictable returns, as emphasised in Sections 10.3 and 13.1. Therefore, in order to capture realistic market dynamics, the Kyle framework must be extended to accommodate sign correlations – a topic that we discussed in Chapter 13 and to which we will return in Section 16.2.1 below.

The fact that impact in the Kyle model is linear and permanent is at odds with the square-root impact law of Chapter 12. It also leads to the conclusion that returns and aggregate volume imbalance are related through the very same constant Λ, independently of the time scale T over which they are computed. This is in strong contrast with the empirical data shown in Section 11.4, which suggests that $\Lambda(T)$ in fact decays as $\sim T^{-0.25}$. Clearly, additional features must be included in Kyle's model to make it a realistic model of price impact.

Take-Home Messages

(i) The Kyle model is a simple model of price impact with three classes of agents. The informed trader has private information about the future price, and chooses a trade volume to optimise her profit. The noise traders submit a random trade volume. The market-maker acts as a counterpart for the sum of the trading volumes submitted by the informed trader and the noise traders, and chooses his clearing price

[3] See also Obizhaeva, A., & Wang, J. (2013). Optimal trading strategy and supply/demand dynamics. *Journal of Financial Markets*, 16, 1–32.

to equal his expectation of the fundamental price, given the volumes he observes – such that his expected profit is zero.

(ii) The informed trader anticipates the market-maker's clearing price and therefore optimises her volume to maximise her expected profit.

(iii) In the model, volumes impact the price because of their expected informational content. This exposes the market-maker to adverse selection. By adjusting the price (negatively) to order-flow imbalance, the market-maker ensures that on average, impact exactly compensates for this adverse selection.

(iv) When all distributions are Gaussian, the impact scales linearly with the informed trader's volume. The proportionality coefficient is often called Kyle's lambda.

(v) The value of Kyle's lambda measures market (il-)liquidity. The larger the coefficient, the more a given volume impacts the price and the more expensive trading is.

(vi) The larger the number of noise traders, the more liquid the market is. In this sense, a market needs uninformed participants to function smoothly.

15.6 Further Reading

The Kyle Model and Some Generalisations

Kyle, A. S. (1985). Continuous auctions and insider trading. *Econometrica: Journal of the Econometric Society*, 53, 1315–1335.

Back, K. (1992). Insider trading in continuous time. *Review of Financial Studies*, 5(3), 387–409.

Caballe, J., & Krishnan, M. (1994). Imperfect competition in a multi-security market with risk neutrality. *Econometrica*, 695–704.

Nöldeke, G., & Tröger, T. (2001). Existence of linear equilibria in the Kyle model with multiple informed traders. *Economics Letters*, 72(2), 159–164.

Corcuera, J. M., Farkas, G., Di Nunno, G., & Oksendal, B. (2010). Kyle-Back's model with Lévy noise. Preprint series. Pure mathematics. http://urn. nb. no/URN: NBN: no-8076.

Cetin, U., & Danilova, A. (2016). Markovian Nash equilibrium in financial markets with asymmetric information and related forward-backward systems. *The Annals of Applied Probability*, 26(4), 1996–2029.

16

The Determinants of the Bid–Ask Spread

Another issue brought to the fore by the crisis is the need to better understand the determinants of liquidity in financial markets. The notion that financial assets can always be sold at prices close to their fundamental values is built into most economic analysis...

(Ben Bernanke)

As we discussed in Chapter 1, organising a market to ensure fair and orderly trading is by no means a trivial task. As we also discussed in Chapter 1, transactions can only take place if some market participants post binding quotes to the rest of the market, in the sense that they specify prices at which they agree to buy or sell a specified quantity of an asset. By posting these quotes, liquidity providers put themselves at risk, because liquidity takers can decide whether or not they want to accept these offers to trade – and will only do so if they believe that the price is favourable. Liquidity providers are therefore exposed to a systematic, adverse bias: while some trades are uninformed and innocuous, other trades may be informed and be followed by large price moves in the direction of the trade, to the detriment of the liquidity provider.[1]

Given this seemingly unfavourable position, why do any market participants provide liquidity at all? The answer is that many liquidity-provision strategies, including the popular strategy of market-making, can be profitable in the long run because a large fraction of trades are in fact non-informed (or very weakly informed). As we noted in Section 1.3.2, the fundamental consideration for implementing these strategies in the long run is the balance between the mean size of the bid–ask spread and the mean strength of adverse impact.

[1] As noted by Perold, A. F. (1988). The implementation shortfall. Paper versus reality. *The Journal of Portfolio Management*, 14(3), 4–9: "*You do not know whether having your limit order filled is a blessing or a curse – a blessing if you have just extracted a premium for supplying liquidity; a curse if you have just been bagged by someone who knows more than you do.*"

In older financial markets, only a select few market participants were able to act as market-makers. Due to the large spreads that were typical of these markets, market-making was a highly profitable business for these individuals. In modern electronic markets, by contrast, any market participant can act as a market-maker. As we discuss in the chapter, this important change has made market-making a highly competitive business that is typically only marginally profitable.

In this chapter, we introduce and study some simple models that help to make our previous discussions of market-making and the bid–ask spread more precise. As we will see, modern trade-and-quote data confirms the existence of a close correspondence between impact, volatility and the bid–ask spread, enforced by competition between liquidity providers.

16.1 The Market-Maker's Problem

Market-makers attempt to earn profit from exploiting the difference between the bid- and the ask-price using (primarily) limit orders. Understanding how to choose the value of the bid–ask spread is a question of paramount importance for a market-maker: choosing a value that is too small will leave a market-maker under-compensated for the **adverse selection** risk inherent in implementing the strategy; choosing a value that is too large will prevent the market-maker from conducting any trades, as other liquidity providers will offer trades at better prices.

We begin by examining one of the first models to formalise the issue of adverse selection in an LOB framework. This model was originally due to a paper from Glosten and Milgrom in 1985, and is close in spirit to the Kyle model that we discussed in Chapter 15. As in the Kyle model, one assumes that some market participants are informed, in the sense that they have access to private information about the price of the asset at some future **terminal time**. Other market participants are either uninformed (even if they may believe otherwise!) or trade for other reasons, without any view on the future price.

16.1.1 Break-Even Quotes

In the model, the informed and uninformed agents trade with a single market-maker, who we again call Bob. Bob chooses a bid-price b, at which he places a unit volume for purchase, and an ask-price a, at which he places a unit volume for sale. We assume that Bob chooses the values of b and a so as to ensure he has **no ex-post regrets**, in the sense that the true price of the asset (revealed after the trade) is on average equal to b if an agent sold (and Bob bought) and is on average equal to a if an agent bought (and Bob sold). This quote-setting rule allows Bob to break even on average. In other words, we assume that Bob is risk-neutral and that he does not increase the bid–ask spread to compensate for the variance and skewness of his payoff (see Section 1.3.2). This is justified in a competitive situation, which pushes any risk-compensating premium to small values.

In addition to the market-maker, the **Glosten–Milgrom model** assumes that the market is populated by a set of liquidity takers. Each liquidity taker i maintains a private (idiosyncratic) valuation $\widehat{p}_i$, which is a random variable. If i is an informed trader, then $\widehat{p}_i$ is positively correlated with the terminal value p_F; if i is an uninformed trader, then $\widehat{p}_i$ is independent of p_F. Each liquidity taker i buys or sells the asset depending on $\widehat{p}_i$ and on Bob's quotes. Specifically, if $\widehat{p}_i > a$, then agent i buys the asset from Bob at price a; if $\widehat{p}_i < b$, then agent i sells the asset to Bob at price b. If $b < \widehat{p}_i < a$, the liquidity taker does nothing.

Within this framework, the conditions for Bob to have no ex-post regrets are given by

$$\begin{aligned} a &= \mathbb{E}_{\mathrm{Bob}}[p_\mathrm{F} | \widehat{p} > a], \\ b &= \mathbb{E}_{\mathrm{Bob}}[p_\mathrm{F} | \widehat{p} < b]. \end{aligned} \tag{16.1}$$

Equation (16.1) says that conditional on a buy trade occurring at the ask, the expected future ask-price (from Bob's standpoint) is equal to the price paid by the buyer. Symmetrically, conditional on a sell trade occurring at the bid, the expected future bid-price (from Bob's standpoint) is equal to the price paid by the seller. In both cases, the expectation is taken with respect to the information available to Bob. As in the Kyle model (see Chapter 15), this is not the full set of information available in the market. If Bob knew the terminal value p_F, then his break-even condition would simply be $a = b = p_\mathrm{F}$, since he would not be subject to any adverse selection, and all trades would be noise. In general, however, $b < a$ since a sell order brings a negative piece of information while a buy order brings a positive piece of information.

16.1.2 A Model with Well-Informed Traders

The simplest model for Bob's view of the world is that a fraction of agents are perfectly informed and know the value of p_F, while others are uninformed and their expectation of the future price is symmetrically distributed around the current price p_0:

$$\mathbb{P}(\widehat{p} | p_\mathrm{F}) = \underbrace{(1-\phi) f(\widehat{p} - p_0)}_{\text{uninformed}} + \underbrace{\phi \delta(\widehat{p} - p_\mathrm{F})}_{\text{informed}}, \tag{16.2}$$

where $f(\cdot)$ is a certain symmetric distribution function, ϕ is the fraction of informed traders, and the δ-function reflects that informed traders perfectly forecast the terminal price.[2]

[2] In fact, adding some uncertainty around p_F (i.e. fattening the δ-function) does not change the qualitative conclusions of the model.

In order to set his quotes using Equation (16.1), Bob needs to estimate the distribution of p_{F}, given that someone trades. Bayes' rule allows him to compute

$$\mathbb{P}(p_{\mathrm{F}}|\widehat{p}) = \frac{\mathbb{P}(\widehat{p}|p_{\mathrm{F}})\mathbb{Q}_0(p_{\mathrm{F}})}{\int_{-\infty}^{+\infty} \mathrm{d}p_{\mathrm{F}}\, \mathbb{P}(\widehat{p}|p_{\mathrm{F}})\mathbb{Q}_0(p_{\mathrm{F}})}, \tag{16.3}$$

where $\mathbb{Q}_0(p_{\mathrm{F}})$ denotes the prior distribution of p_{F} at time $t = 0$. As within the Kyle framework, we assume that Bob knows $\mathbb{Q}_0(p_{\mathrm{F}})$, but of course not the value of p_{F} itself. Bob can now use Equation (16.3) to compute his required conditional expectations (see Equations (16.1)):

$$a = \mathbb{E}_{\mathrm{Bob}}[p_{\mathrm{F}}|\widehat{p} > a] = \frac{\int_{-\infty}^{+\infty} \mathrm{d}p_{\mathrm{F}}\, p_{\mathrm{F}}\, \mathbb{P}(\widehat{p} > a|p_{\mathrm{F}})\mathbb{Q}_0(p_{\mathrm{F}})}{\int_{-\infty}^{+\infty} \mathrm{d}p_{\mathrm{F}}\, \mathbb{P}(\widehat{p} > a|p_{\mathrm{F}})\mathbb{Q}_0(p_{\mathrm{F}})},$$

$$b = \mathbb{E}_{\mathrm{Bob}}[p_{\mathrm{F}}|\widehat{p} < b] = \frac{\int_{-\infty}^{+\infty} \mathrm{d}p_{\mathrm{F}}\, p_{\mathrm{F}}\, \mathbb{P}(\widehat{p} < b|p_{\mathrm{F}})\mathbb{Q}_0(p_{\mathrm{F}})}{\int_{-\infty}^{+\infty} \mathrm{d}p_{\mathrm{F}}\, \mathbb{P}(\widehat{p} < b|p_{\mathrm{F}})\mathbb{Q}_0(p_{\mathrm{F}})}.$$

To give some flesh to these equations, we choose some specific form for the uninformed distribution f (in Equation (16.2)) and for $\mathbb{Q}_0$. Our choice, motivated by the simplicity of the resulting calculations, reads:

$$f(u) = \frac{e^{-|u|/w}}{2w}; \qquad \mathbb{Q}_0(p_{\mathrm{F}}) = \frac{e^{-|p_f - p_0|/\sigma}}{2\sigma},$$

where w is a parameter that controls the dispersion of the uninformed price expectations and σ is proportional to the volatility of the fundamental price. This choice of distribution allows us to obtain explicit results, but the resulting conclusions still hold qualitatively for a wide class of distributions.

By performing elementary integrals of exponential functions in the expressions for a and b, the final result reads:

$$\begin{aligned} a &= p_0 + s/2, \\ b &= p_0 - s/2, \end{aligned} \tag{16.4}$$

with

$$s = \frac{\phi(2\sigma + s)e^{s/w}}{\phi e^{s/w} + (1 - \phi)e^{s/\sigma}}. \tag{16.5}$$

Equation (16.5) is quite interesting and exhibits different regimes depending on the ratio w/σ, as we now detail.

16.1.3 An Orderly Market

In the case where $w \geq \sigma$, Equation (16.5) has a single solution, which, in the limit of a small fraction of informed traders (i.e. $\phi \to 0$) reads:[3]

$$s \approx 2\phi\sigma + O(\phi^2), \qquad (\phi \ll 1). \tag{16.6}$$

In other words, when uninformed traders have sufficiently dispersed signals with an amplitude w that is greater than the width σ of the informed signal, Bob should set s to be proportional to the product of the price volatility σ and the fraction of informed trades ϕ. Note that in the limit $\phi \ll 1$ one has $w \gg s$: uninformed agents trade (almost) unconditionally.

Since in reality the ratio of the bid–ask spread s to the daily volatility σ is very small (say, 0.01 to 0.1), Equation (16.6) suggests that the fraction of trades that are informed about daily price moves must indeed be small. In fact, the value of σ for informed traders increases with prediction horizon T as the price volatility itself (i.e. as $\sqrt{T}$). The corresponding fraction of informed traders (or the quality of their information; see Equation (16.9)) must therefore be bounded from above by $1/\sqrt{T}$ for the bid–ask spread to remain bounded, independent of the horizon of informed traders. In a nutshell, the fraction of the volatility captured by informed traders must decrease with the time horizon.

It is also interesting to consider the **skewness** ς of the distribution of future price changes. Conditional on a buy trade occurring, the updated distribution of price changes is $\mathbb{P}(p_F|\widehat{p} > a)$. In this case, what is the third moment ς of the distribution of $p_F - a$? To first-order in ϕ (assumed to be small):

$$\varsigma \approx \varsigma_0 + \phi \frac{\int_0^{\infty} \mathrm{d}p_F \, (p_F - p_0)^3 \mathbb{Q}_0(p_F)}{\left[\int_{-\infty}^{\infty} \mathrm{d}p_F \, (p_F - p_0)^2 \mathbb{Q}_0(p_F)\right]^{3/2}},$$

where ς_0 is the unconditional skewness of price changes, and the second term is strictly positive. Hence, even for a symmetric unconditional distribution of price changes $\mathbb{Q}_0(p_F)$, observing a buy (respectively, sell) transaction should lead to a positively (respectively, negatively) skewed distribution of realised price changes $p_F - a$.

One can in fact easily remove the distracting contribution of ς_0 by studying the skewness of the random variable

$$U := \mathbb{I}_{\widehat{p}>a}(p_F - a) + \mathbb{I}_{\widehat{p}<b}(b - p_F),$$

which simply represents the P&L of market orders. Again, to first-order in ϕ, it follows that $\varsigma(U) \propto \phi$. Figure 16.1 shows the (low-moment) skewness of U, estimated as the difference between the median and the mean (scaled by the r.m.s.), as a function of the time horizon T, and measured in market order event-time. We estimate p_F as the mid-price T market orders later,

$$\widetilde{U}(T) := \mathbb{I}_{\widehat{p}_t>a_t}(m_{t+T} - a_t) + \mathbb{I}_{\widehat{p}_t<b_t}(b_t - m_{t+T}). \tag{16.7}$$

For the small-tick stock (TSLA), the skewness is positive and decays roughly as $1/\sqrt{T}$, as predicted by the Glosten–Milgrom model. The skewness is however very

[3] This can easily be seen by assuming that s can be expanded as $c\phi + c'\phi^2 + o(\phi^2)$, then plugging this expression into Equation (16.5). Identifying the two sides of the equation order by order in ϕ gives Equation (16.6).

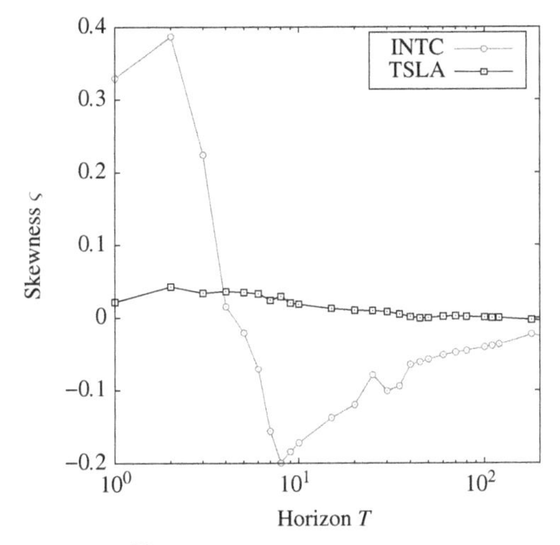

Figure 16.1. Skewness of $\widetilde{U}(T)$ (defined in Equation (16.7)), estimated with the low-moment difference between its median and its mean, as a function of the time horizon T, measured in market order event-time for INTC and TSLA.

small, even on short time scales, suggesting again that the fraction of informed trades is itself very small. For the large-tick stock (INTC), the skewness quickly becomes negative, in contradiction with the Glosten–Milgrom framework. In the case of large-ticks, however, we know that the bid–ask spread is bounded from below and the Glosten–Milgrom argument must be amended.

16.1.4 Market Breakdown

We now consider the more interesting case of Equation (16.5), where $w < \sigma$. In this case, uninformed agents do not predict very large price moves, so their signal rarely exceeds the bid–ask spread and they do not trade much when the spread is large. Figure 16.2 shows that a second, large-s solution appears in the interval $w^* < w < \sigma$, where w^* is a threshold that does not have a simple analytical form. Competition between market-makers enforces the solution with smallest s.[4] But when $w \downarrow w^*$, the large- and small-s solutions converge before they both disappear for smaller values of w.

The case with no solutions is particularly interesting: it suggests that the market breaks down because there is no longer any way for Bob to fix a bid–ask spread to break even. Put simply, the uninformed traders do not provide enough potential gains to compensate Bob for the adverse selection that he experiences from the informed traders, whatever the value of the spread he chooses. In the Kyle model, where traders send orders *before* knowing the transaction price, reducing the number of noise traders results in an increase of the market-impact parameter Λ,

[4] Still, one may expect in that case that the spread "hesitates" between the two solutions, leading to interesting regime shifts.

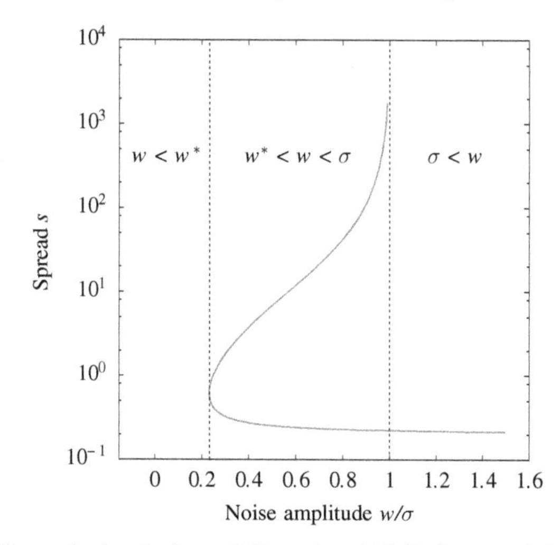

Figure 16.2. Numerical solution of Equation (16.5) for $\phi = 0.1$ and $\sigma = 1$ and different values of w. Note that there is a unique solution for $w > \sigma$, two solutions for $w^* < w < \sigma$ and no solutions for $w < w^* \approx 0.22\sigma$.

but without the market breaking down. In both cases, the presence of noise traders is crucial to ensure orderly trading.

This emergence of **market breakdown** is one of the most interesting features of the Glosten–Milgrom model. The effect would still exist even if informed traders had only partial information about the future price (i.e. when the δ-function in Equation (16.2) is replaced by a wider distribution). In fact, one can consider the case where all traders are partially informed, with some uncertainty about the true future price. For example, we might assume that Equation (16.2) is replaced by a logistic distribution

$$\mathbb{P}(\widehat{p}|p_{\mathrm{F}}) = \frac{\epsilon e^{-\epsilon(\widehat{p}-p_{\mathrm{F}})}}{\left(1 + e^{-\epsilon(\widehat{p}-p_{\mathrm{F}})}\right)^2}, \tag{16.8}$$

where ϵ describes the amount of information on the future price that all traders (except the market-maker) have at their disposal. In the limit of weak information (i.e. $\epsilon \to 0$), solving Equation (16.8) leads to

$$s \approx \epsilon\sigma. \tag{16.9}$$

This solution is very similar to the one in Equation (16.6). Again, the solution disappears when ϵ exceeds a critical value ϵ_c, which corresponds to the case when market-making becomes impossible because the information available to traders is too precise.

In summary, markets can only operate smoothly if the number of uninformed traders is sufficiently high. When market-makers fear that other market participants have too much information, they are unable to fix any spread that enables them to break even, which can lead to a **liquidity crisis**.

16.1.5 A Dynamical Version of the Model: Quote Updates

The setting of the above discussion was entirely static. What happens after the first trade has occurred? Throughout this section, we assume that a solution for s always exists (i.e. $w > \sigma$) and that the fundamental price (which Bob attempts to guess from the order flow) is fixed in time.

Assume that a trade has taken place at the ask a_t at time t. Bob should now use the information that the trade has occurred to update his unconditional distribution of the future price p_F, so that at time $t+1$:

$$\mathbb{Q}_{t+1}(p_\text{F}) = \mathbb{P}_t(p_\text{F}|\widehat{p}_t > a_t).$$

The right-hand side of this expression is simply given by Equation (16.3), with $\mathbb{Q}_0$ replaced by $\mathbb{Q}_t$ to reflect that we now consider a dynamical model. This, in turn, allows Bob to fix the next bid- and ask-prices, again imposing a break-even constraint. The detailed analysis of the induced price dynamics is beyond the scope of this book, but one can show that the resulting sequence of trade prices is a martingale that converges to p_F at the terminal time when the number of steps goes to infinity, while the bid–ask spread exponentially decreases in time, because transactions gradually reveal the signal of the informed agents.

The **martingale** property (using public information available to the market-maker) is in fact quite simple to prove. Observe that

$$p_t = \begin{cases} a_t, & \text{if the trade at time } t \text{ is a buy,} \\ b_t, & \text{if the trade at time } t \text{ is a sell.} \end{cases}$$

From Equation (16.1), one has that immediately after each transaction t, it must hold that

$$\mathbb{E}_t[p_\text{F}] = p_t.$$

The expected transaction price at the next trade can therefore be written as:

$$\begin{aligned} \mathbb{E}_t[p_{t+1}] &= \mathbb{E}_t\Big[\mathbb{E}_{t+1}[p_\text{F}]\Big], \\ &= \mathbb{E}_t[p_\text{F}], \\ &= p_t, \end{aligned}$$

which is precisely the martingale property.

Is the **exponential spread decay** (in time) predicted by Glosten–Milgrom also observable in empirical data? Figure 4.2 indeed shows that the spread tends to decay as the day proceeds. In the Glosten–Milgrom framework, this means that overnight shocks introduce some uncertainty on the fundamental price, and that this uncertainty gets progressively resolved throughout the day as smart traders work out the consequence of these shocks. However, the plot suggests that the precise dynamics in real markets is more complex than the prediction of the model. In particular, an exponential fit of the decay is not adequate. Clearly, the idea of a fixed fundamental price p_F that the market strives to discover throughout the day is grossly over-simplified. In fact, the target price itself changes throughout the day, making the dynamics somewhat more intricate.

We conclude this section with a remark that is similar to the one that we made in the context of the Kyle model in Section 15.3. The dynamical update process that we have considered in this section requires Bob to have an unbiased estimate of the quantity of information available to informed traders at each instant of time. If Bob tends to overestimate the information content of trades (perhaps in an attempt to protect himself against selection bias), then his quote updates are not justified by the fundamental valuation of the asset. However, if other traders adjust to these new quotes and change their views on the price, self-fulfilling prophecies and excess volatility may emerge.[5]

16.1.6 Metaorders in the Glosten–Milgrom Framework

In the Glosten–Milgrom framework, the permanent impact of a trade is the expected value of the price, given that the specified trade has taken place. As we discussed in Section 10.5, however, many traders in real markets do not simply submit individual market orders, but instead seek to execute large metaorders. To minimise the impact of their actions, traders typically submit their child orders for a single metaorder over a period of several days. In the context of the Glosten–Milgrom model, it is interesting to measure the permanent impact of not just single market orders, but also of metaorders.

To address this case, we follow Farmer et al.[6] and make the additional assumptions that metaorder executions occur one-after-the-other (and therefore do not overlap), and that Bob is notified (or is able to infer) every time that a new metaorder execution begins.

Recall from Section 12.2 that a metaorder has a direction ε and total volume Q, and is split into child orders. We now assume that each child order has a volume equal to the lot size υ_0 (see Section 3.1.5). Let q denote the volume of the metaorder executed so far, and let $p(q)$ denote the price set by Bob at that point. Let $p_\infty(Q)$ denote the expected price after a metaorder of total volume Q has terminated and impact has relaxed.

As in the previous section, we assume that Bob sets his price such that each time he receives a market order, he has no ex-post regrets and breaks even on average. As we discussed in Section 16.1.5, this can be seen as a martingale condition for the price. In the present context, these two conditions read:

(i) **Martingale condition**: Bob's action makes the price process p a martingale. Therefore, when a volume q has been executed, the price is such that

$$p(q) = \mathbb{E}\left[p_\infty(Q) \mid Q \geq q\right]. \tag{16.10}$$

[5] For more on this scenario, see also Section 20.3.

[6] Farmer, J. D., Gerig, A., Lillo, F., & Waelbroeck, H. (2013). How efficiency shapes market impact. *Quantitative Finance*, 13, 1743–1758.

(ii) **Break-even condition**: Irrespective of the metaorder size Q, the price must be such that

$$\frac{1}{Q}\int_0^Q p(q)\mathrm{d}q = \mathbb{E}[p_\infty(Q)], \qquad (16.11)$$

(where for simplicity we consider volumes and prices to be continuous, not discrete).

In words, the second condition simply stipulates that

$$\text{impact cost} = \mathbb{E}[\text{permanent impact}].$$

This is a metaorder-equivalent condition to the Glosten–Milgrom condition for single market orders, which reads

$$\text{half spread} = \mathbb{E}[\text{adverse selection}].$$

Together with the distribution of metaorder sizes $F(Q) := \mathbb{P}[q > Q]$, these two conditions allow one to determine the temporary impact $\Im(q) = p(q) - p_0$ and the permanent impact $\Im_\infty(Q) = p_\infty(Q) - p_0$ of a metaorder. Noting that the market-maker receives a volume flow that stops only when the metaorder is completed, one finds

$$\begin{aligned} p(q) &= \frac{1}{F(q)}\int_q^\infty \mathrm{d}Q f(Q)p_\infty(Q), \\ p_\infty(Q) &= \frac{1}{Q}\int_0^Q \mathrm{d}q p(q), \end{aligned} \qquad (16.12)$$

where $f(q) := -F'(q)$. By multiplying each of these equations by the denominator term on the right-hand side, then taking derivatives (using the product rule), one finds a pair of coupled differential equations:

$$F(q)p'(q) - f(q)p(q) = -f(q)p_\infty(q), \qquad (16.13)$$

$$p_\infty(q) + qp'_\infty(q) = p(q). \qquad (16.14)$$

By substituting the expression for p_∞ obtained from Equation (16.13) into Equation (16.14), we arrive at the following first-order differential equation for $p(q)$:

$$q\left(\frac{F}{f}p'\right)' + \left(\frac{F}{f} - q\right)p' = 0.$$

This equation can be solved by introducing an auxiliary function

$$g(q) := \frac{F}{f}p'(q),$$

leading to

$$g(q) = \frac{A}{qF(q)},$$

where A is a constant.

Finally, using the fact that $\Im(0) = 0$, it follows that

$$\Im(q) = A\int_0^q \mathrm{d}q' \frac{f(q')}{q'F(q')^2}. \qquad (16.15)$$

Hence, if metaorder volumes are distributed according to a power-law $F(q) \sim q^{-1-\gamma}$, where $0 < \gamma < 1$, then for large q, mean impact is asymptotically concave:

$$\Im(q) \propto q^\delta, \qquad \delta = \gamma. \qquad (16.16)$$

The mean permanent impact then follows from the fair-pricing condition:

$$\Im_\infty(Q) \underset{Q\to\infty}{\approx} \frac{1}{1+\delta}\Im^{\mathrm{peak}}(Q). \qquad (16.17)$$

In the special case where $\gamma = 1/2$ (see Chapter 10), we recover an asymptotically square-root impact corresponding to $\delta = 1/2$. This is essentially the argument of Farmer et al. for a square-root impact law. In this case, the permanent impact is such that

$$\mathfrak{I}_\infty(Q) \approx \frac{2}{3}\mathfrak{I}^{\text{peak}}(Q).$$

According to the fair-pricing condition, if impact is square-root, then two-thirds of the peak impact should remain permanent.

The argument that we have presented in this section appears to be quite general, and is in fact very similar to the one in Section 13.4.5, where we studied impact in the context of the propagator model when trade signs arrive in sequences whose lengths are distributed according to a power-law. However, this theory suffers from several drawbacks. First, the relation between the impact exponent δ and the power-law exponent $1 + \gamma$ for the distribution of metaorder sizes does not hold universally.[7]

Second, it makes the highly unrealistic assumptions that metaorders appear sequentially, in isolation, and that the market-maker can detect their start and end. Third, the model suggests that impact decays instantaneously to its asymptotic value after the metaorder terminates (see Section 13.4.5), whereas empirical data reveals that this behaviour is not routinely observed in real markets. Instead, impact undergoes an initially steep decay, then relaxes very slowly over a period that can span several days. Finally, it only holds in a large-Q regime, where the volume distribution does follow a power-law. For small volumes, its prediction deviates from the square-root law.

However, a strong point for the model is that empirical data suggests real market dynamics to be compatible with the prediction of a permanent impact equal to about two-thirds of peak impact. This two-thirds ratio appears to hold for metaorders that originate from some sort of information, rather than just for liquidity purposes (see Section 12.7 for recent papers on this point). Like the Glosten–Milgrom argument, according to which the bid–ask spread compensates for adverse selection, the fair-pricing argument is more general than the detailed set-up of the above model.

16.2 The MRR Model

Although inspiring, the Glosten–Milgrom model suffers from a problem: central to its formulation is the idea of a terminal time where the true price p_F is revealed, and at which people can transact as much as they want. The implicit idea (that also underlies Kyle's model; see Chapter 15) is that the market-maker's inventory can be fully transacted at the end of the day (say). The common lore is indeed to assume that one can trade large quantities at the market close without impacting the price, but this is totally unwarranted. Liquidating the market-maker's inventory incurs a potentially large impact cost and makes the Glosten–Milgrom break-even argument shaky.

Therefore, although the Glosten–Milgrom model is a useful starting point for thinking about how market-makers might choose to set the quotes that they

[7] See, e.g., Mastromatteo, I., Tóth, B., & Bouchaud, J.-P. (2014). Agent-based models for latent liquidity and concave price impact. *Physical Review E*, 89, 042805. In this paper, the LMF relation between the exponent of the order size distribution and the sign autocorrelation exponent γ was assumed. But in Donier, J., & Bonart, J. (2015). A million metaorder analysis of market impact on the Bitcoin. *Market Microstructure and Liquidity*, 1(02), 1550008, the metaorder size distribution is observed directly, and the relation $\delta = \gamma$ also fails.

offer to the rest of the market, its flaws provide motivation for considering other approaches where the notion of a terminal price is absent, while keeping the idea of a martingale price.

16.2.1 Martingale Evolution of the Traded Price

One such alternative is the **Madhavan–Richardson–Roomans** (MRR) model. In contrast to the Glosten–Milgrom framework, the MRR model does not rely on the existence of a terminal time or a terminal price. Instead, it considers an underlying fundamental price $p_{F,t}$, which evolves over time to reflect both the information content of trades and unanticipated news arriving in the market. The setting of the MRR model is very close to that of the propagator model that we discussed in Chapter 13.

In the MRR model, $p_{F,t}$ coincides with the traded price, and its evolution consists of two terms: one corresponding to the information content of trades, and the other to a noise term ξ_t that captures unanticipated news:

$$p_{F,t} - p_{F,t-1} = G^* \times (\varepsilon_t - \widehat{\varepsilon}_t) + \xi_t, \tag{16.18}$$

where the information content of trades G^* is assumed to be constant (i.e. independent of time). The first term on the right-hand side of Equation (16.18) is indeed proportional to the **sign surprise** $\varepsilon_t - \widehat{\varepsilon}_t$, where (as in Section 13.3) $\widehat{\varepsilon}_t$ is defined as:

$$\widehat{\varepsilon}_t = \mathbb{E}_{t-1}[\varepsilon_t].$$

By construction, the fundamental price (and the traded price) $p_{F,t}$ is a martingale. Therefore, the long-term expectation of the fundamental price is always such that

$$\mathbb{E}_t[p_{F,t+T}] = p_{F,t}.$$

In this market, how should Bob set his quotes so as to have no ex-post regrets? Similarly to the Glosten–Milgrom model, the key to finding the solution for the MRR model is to note that for both a buy order and a sell order, the expected realised transaction price at $t+1$ should be equal to the fundamental price:

$$\begin{aligned} a_{t+1} &= \mathbb{E}_t[p_{F,t+1}|\varepsilon_{t+1} = 1] = p_{F,t} + G^*(1 - \widehat{\varepsilon}_{t+1}); \\ b_{t+1} &= \mathbb{E}_t[p_{F,t+1}|\varepsilon_{t+1} = -1] = p_{F,t} - G^*(1 + \widehat{\varepsilon}_{t+1}). \end{aligned} \tag{16.19}$$

Another way to think about these expressions is that other traders only trade with Bob when their information allows them to break even. This is of course a somewhat unrealistic assumption that can at best be true on average.

In the following sections, we work out several directly testable predictions of the MRR models, in particular concerning the response function $\mathcal{R}(\ell)$ considered throughout the book.

16.2.2 Correlation and Response in the MRR model

By Equation (16.19), it immediately follows that[8]

$$s_{t+1} = a_{t+1} - b_{t+1} = 2G^*,$$

which is, by assumption, constant in time. Similarly, the mid-price just before the $(t+1)^{\text{th}}$ trade is given by

$$m_{t+1} = \frac{a_{t+1} + b_{t+1}}{2} = p_{\mathrm{F},t} - G^* \widehat{\varepsilon}_{t+1}.$$

Using Equation (16.18), the **mid-price dynamics** can be rewritten as

$$m_{t+1} - m_t = G^* \left(\varepsilon_t - \widehat{\varepsilon}_t\right) + G^* \left(\widehat{\varepsilon}_t - \widehat{\varepsilon}_{t+1}\right) + \xi_t. \tag{16.20}$$

This equation is somewhat similar to the propagator specification in Equation (13.22), but with an extra term on the right-hand side that is proportional to the change in the expected sign.

By considering a telescopic sum of expressions of the form in Equation (16.20), it follows that

$$m_{t+\ell} - m_t = G^* \sum_{n=t}^{t+\ell-1} \left(\varepsilon_n - \widehat{\varepsilon}_{n+1}\right) + \sum_{n=t}^{t+\ell-1} \xi_n. \tag{16.21}$$

By also using that $\mathbb{E}[\widehat{\varepsilon}_n \varepsilon_t] = \mathbb{E}[\varepsilon_n \varepsilon_t]$ when $n > t$, the **response function** $\mathcal{R}(\ell)$ (see Section 11.3.1) can then be expressed in terms of the sign correlation $C(\ell)$:

$$\mathcal{R}(\ell) = \mathbb{E}[(m_{t+\ell} - m_t) \cdot \varepsilon_t] = G^*(1 - C(\ell)) = \frac{s}{2}(1 - C(\ell)). \tag{16.22}$$

This remarkable relation can be tested empirically (as we will do in the next section).[9] When $\ell \to \infty$, the asymptotic value of $C(\ell)$ is 0. Therefore, the asymptotic value of the response function is

$$\mathcal{R}_\infty = G^* = s/2.$$

In the absence of any other costs, market-making ceases to be profitable when half the bid–ask spread only compensates the long-term impact of market orders. This is precisely the result we anticipated in Section 1.3.2.

Note that strictly speaking, the MRR model is a special case of the formalism that we have presented, with the additional assumption that the ε_t series is Markovian with $\widehat{\varepsilon}_t = \rho \varepsilon_{t-1}$, such that $C(\ell) = \rho^\ell$ with $\rho < 1$. (MRR did not consider the case where the ε_t series has long-range correlations).

[8] MRR consider the possibility that order processing costs c, or other costs incurred by the market-marker, can be added to the spread, to arrive at $s = 2(G^* + c)$. In this more general framework, the spread reflects both adverse impact and extra costs. We will neglect c throughout the rest of the chapter.

[9] In contrast with Equation (16.22), one finds that for the propagator model (for which the second term in Equation (16.20) is absent), the response function $\mathcal{R}(\ell)$ is a constant, equal to $1 - \mathbb{E}[\widehat{\varepsilon}^2]$. See Section 16.2.4.

16.2.3 Volatility versus Spread

In the MRR model, the long-term mid-price volatility per trade is given by

$$\widetilde{\sigma}_\infty^2 = \lim_{\ell\to\infty} \frac{\mathbb{V}[m_{t+\ell} - m_t]}{\ell},$$
$$= G^{*2}(1 - \mathbb{E}[\widehat{\varepsilon}^2]) + \Sigma^2,$$

where the tilde indicates that we work in trade time and Σ^2 is the variance of the noise term ξ_t in Equation (16.18). Approximating $\widehat{\varepsilon}_t$ by $C(1)\varepsilon_t$ and using the relation $s = 2G^*$, the above equation predicts an affine relation between the volatility per trade and the squared bid–ask spread:[10]

$$\widetilde{\sigma}_\infty^2 = \frac{1 - C(1)^2}{4} s^2 + \Sigma^2. \tag{16.23}$$

We will perform empirical tests of this relation in Section 16.3. Observe that if φ denotes the number of trades per unit (calendar) time, then the relationship between the volatility per trade $\widetilde{\sigma}$ and the (usual) volatility per unit time σ is simply

$$\sigma = \widetilde{\sigma} \times \sqrt{\varphi}. \tag{16.24}$$

In the Markovian MRR case $C(\ell) = \rho^\ell$, an explicit formula can be derived for $\widetilde{\sigma}^2(\ell)$:

$$\widetilde{\sigma}^2(\ell) = \frac{\mathbb{V}[m_{t+\ell} - m_t]}{\ell} = G^{*2}\left[1 - \rho^2 - \frac{2\rho}{\ell}(1 - \rho^\ell)\right] + \Sigma^2. \tag{16.25}$$

Note in particular that $\widetilde{\sigma}_\infty^2 \geq \widetilde{\sigma}^2(1)$ (i.e. short-term mid-price volatility does not exceed long-term volatility).

16.2.4 Interpolating Between the MRR and the Propagator Model

The MRR model surmises that sign surprises impact the *traded price*, whereas the propagator model assumes that sign surprises impact the *mid-price*. In order to illustrate the difference between these two models, it is conceptually useful to introduce a mixture model that interpolates between the propagator specification in Equation (13.19) and the MRR specification in Equation (16.20):

$$m_{t+1} - m_t = G^*(\varepsilon_t - \widehat{\varepsilon}_t) + G^{**}(\widehat{\varepsilon}_t - \widehat{\varepsilon}_{t+1}) + \xi_t, \tag{16.26}$$

where $G^{**} = G^*$ recovers the MRR model and $G^{**} = 0$ the propagator model. Alternatively, one can write this as an evolution for the traded price

$$p_t = m_t + G^*\varepsilon_t,$$

as

$$p_{t+1} - p_t = G^*(\varepsilon_{t+1} - \widehat{\varepsilon}_{t+1}) + (G^{**} - G^*)(\widehat{\varepsilon}_t - \widehat{\varepsilon}_{t+1}) + \xi_t, \tag{16.27}$$

where the second term vanishes in the MRR model, but leads to a "bid–ask bounce" (i.e. a short-term mean-reversion of the traded price) in the propagator model.

In this case, Equation (16.22) instead becomes:

$$\mathcal{R}(\ell) = G^*(1 - C(\ell)) + (G^{**} - G^*)(1 - \mathbb{E}[\widehat{\varepsilon}^2]), \tag{16.28}$$

which is an affine relation between $\mathcal{R}(\ell)$ and $1 - C(\ell)$.

[10] One can show that $\mathbb{E}[\widehat{\varepsilon}^2] = C(1)^2$ in fact holds with high numerical accuracy, with corrections smaller than 1%.

16.3 Empirical Analysis of the MRR Model

In this section, we perform empirical tests of two core predictions of the generalised MRR model: the relationship between the response function and the sign-correlation function (see Equation (16.22)) and the relationship between volatility and the bid–ask spread (see Equation (16.23)).

16.3.1 Response Function and Sign-Correlation Function

Figure 16.3 shows the values of $(1 - C(1)) \times \mathcal{R}(\ell)/\mathcal{R}(1)$ versus $(1 - C(\ell))$ for a large pool of stocks. The MRR model predicts the relationship between these quantities to be a straight line with slope 1 and intercept 0 (see Equation (16.22)). Overall, the MRR equation holds surprisingly well, but has some discrepancies:

- The MRR model appears to fare *better* for large-tick stocks than for small-tick stocks, in that the intercept of the regression is statistically different from zero (and weakly negative) and the slope is statistically larger than one for small-tick stocks. These biases can be explained by the mixture model; see Equation (16.28).
- A systematic positive (resp. negative) concavity appears for large ℓ for small-tick (resp. large-tick) stocks. This means that real price changes exhibit less (resp. more) resistance than what is apparent at small ℓ. As we discuss in the next chapter, this suggests that market-makers suffer more (resp. less) from adverse selection on medium time scales than on short time scales (see Equation (17.12)).

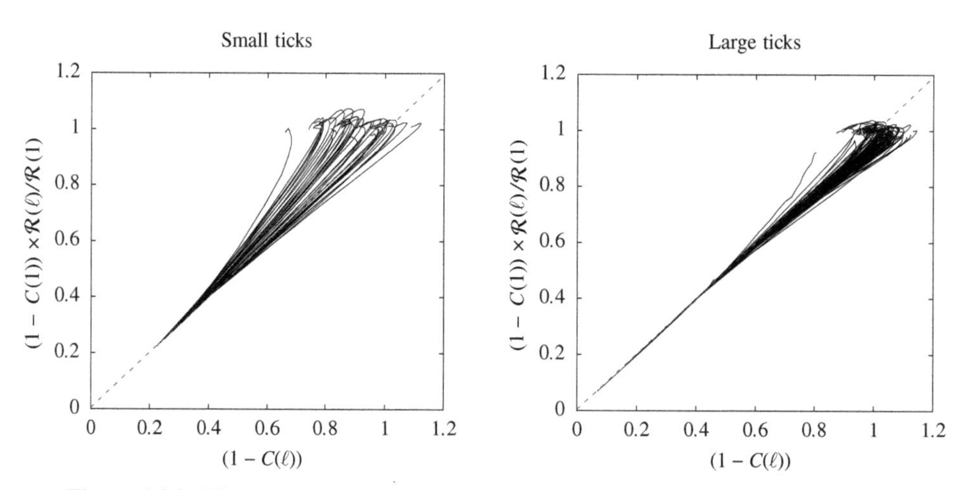

Figure 16.3. The quantity $(1 - C(1)) \times \mathcal{R}(\ell)/\mathcal{R}(1)$ versus $(1 - C(\ell))$ for a pool of 120 liquid stocks and $\ell \in \{1, 2, \cdots, 20\}$, divided into (left panel) small-tick stocks and (right panel) large-tick stocks. The dashed lines are regressions fitted to short lags only (such that $1 - C(\ell) \leq 0.6$), with slopes (left panel) 0.956 and (right panel) 1.01, and very small, statistically insignificant intercepts.

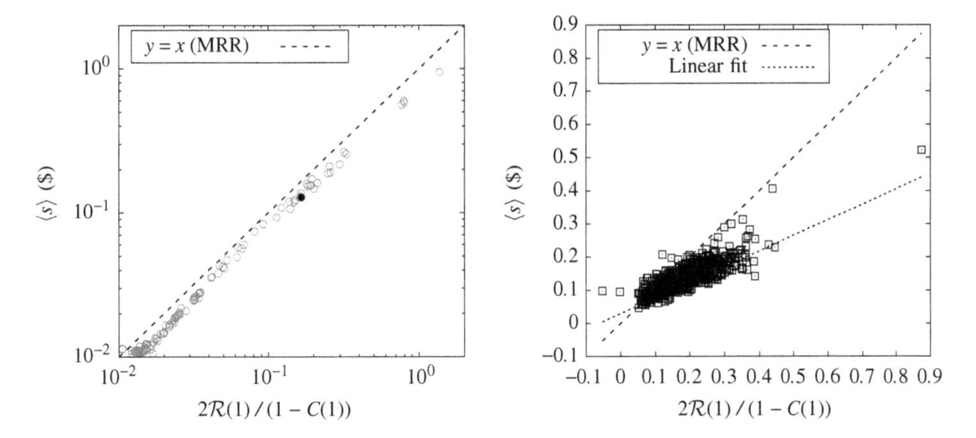

Figure 16.4. (Left panel) The average spread versus $2\mathcal{R}(1)/(1 - C(1))$ for a pool of 120 liquid stocks. Large-tick stocks are characterised by values of s close to \$0.01. Small-tick stocks are characterised by significantly larger values of the spread. The dashed line $y = x$ is the MRR prediction. (Right panel) The average spread $\langle s_t \rangle$ versus $2\mathcal{R}_t(1)/(1 - C_t(1))$, calculated for each 1-hour interval of TSLA trading during 2015. The dotted line is a linear fit $y = 0.47x + 0.03$. The dashed line is the MRR prediction $y = x$. The average point in the right plot corresponds to the black point in the left plot.

Note that while Figure 16.3 tests the functional relationship between $\mathcal{R}(\ell)$ and $1 - C(\ell)$, it does not question the MRR model's core prediction that $\mathcal{R}(1)$ should equal $s \times (1 - C(1))/2$. This is what we report in Figure 16.4, both cross-sectionally over our pool of stocks (left plot), and for one given small-tick stocks over time, for which the local spread varies appreciably (right plot). Overall, the MRR relation between spread, lag-1 impact and correlation is again remarkably well obeyed across different stocks (left plot). As we will argue in Chapter 17, this relationship is actually enforced by **competition between market-makers**, and states that the fastest market-makers break even on average.

The right panel in Figure 16.4 shows that although the MRR relation approximately holds *on average*, some significant deviations occur locally. On the one hand, market-makers do not increase spreads quite enough in situations where the impact of trades increases, or when the autocorrelation increases. On the other hand, the spread is systematically too wide in quieter situations, compensating the losses incurred in volatile situations.

16.3.2 Volatility and the Bid–Ask Spread

We now turn to the predicted affine relationship between long-term volatility and the bid–ask spread, Equation (16.23). We approximate the long-term volatility per trade $\widetilde{\sigma}_\infty$ by $\widetilde{\sigma}(\ell = 20)$, which is less noisy while still being close to $\widetilde{\sigma}_\infty$.

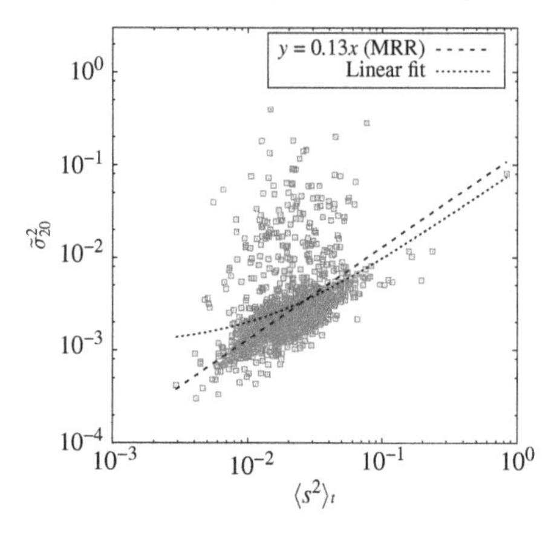

Figure 16.5. Scatter plot of the volatility $\widetilde{\sigma}_{20}^2$ versus average squared spread $\langle s^2 \rangle_t$ (measured in squared dollars) for one-hour intervals of TSLA, plotted in doubly logarithmic coordinates. The dash-dotted lines are linear regressions applied after smoothing the data with a kernel estimator, yielding $y = 0.087x + 0.0011$. The dotted line is the MRR prediction with $\Sigma^2 = 0$, leading to $y = 0.13x$.

Figure 16.5 shows a scatter plot of the squared volatility versus average squared spread $\langle s^2 \rangle_t$ for one-hour intervals during all trading days in 2015 for TSLA. The linear fit leads to $y = 0.087x + 0.0011$, which has a slope smaller than the slope 0.13 predicted by Equation (16.23) with $C(1) = 0.69$ (from Table 11.2). Note that with $\langle s^2 \rangle \approx 0.025$ for TSLA, one finds that the news component explains roughly a third of the price variance. This is similar to what Figure 13.2 (right) conveys in the context of the propagator model.

Interestingly, the plot suggests that there are two regimes: a relatively smooth cloud of points where the MRR relation is extremely well obeyed with a zero intercept (i.e. no contribution of the news component Σ^2) and more extreme volatility periods corresponding to scattered points, which are presumably dominated by exogenous events. These points outside the regular cloud are at the origin of the non-zero intercept of the linear fit, which agrees well with its intuitive interpretation.

Turning now to a cross-sectional test of Equation (16.23), we show in Figure 16.6 the average square volatility per trade $\widetilde{\sigma}_{20}^2$ as a function of $s^2 \times (1 - C(1)^2)$. The overall agreement with such a simple prediction is quite striking. In particular, the approximately linear relation between volatility per trade and spread is clearly vindicated.[11] The absence of any visible intercept suggests that the news contribution Σ^2 is itself proportional to the average squared spread. This means that

[11] This was first emphasised in Wyart, M., Bouchaud, J. P., Kockelkoren, J., Potters, M., & Vettorazzo, M. (2008). Relation between bid-ask spread, impact and volatility in order-driven markets. *Quantitative Finance*, 8(1), 41–57.

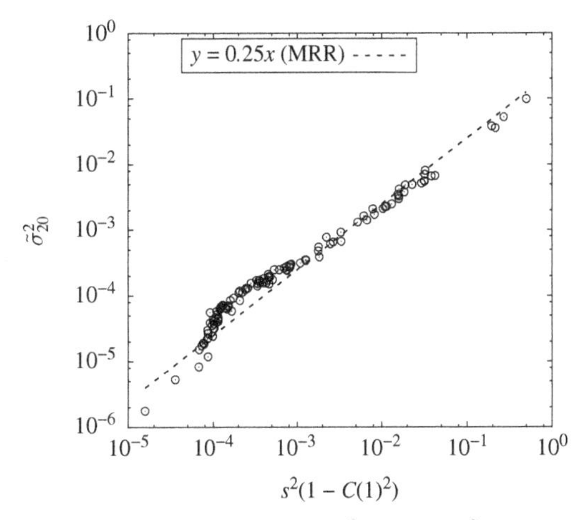

Figure 16.6. Cross-sectional scatter plot of $\widetilde{\sigma}^2_{20}$ versus $s^2 \times [1 - C(1)^2]$ (measured in squared dollars) for a pool of 120 liquid stocks traded on NASDAQ. Each point corresponds to one stock in 2015. The dashed line is the MRR prediction $y = x/4$.

all individual price movements (jumps included) are commensurate to the average spread.

16.4 Conclusion

We started this chapter with the traditional partitioning of traders into "market-makers" and "liquidity takers", a fraction of the latter possessing some information about the future value of the asset. In this context, the argument of Glosten and Milgrom states that the market-maker must open up a spread between the bid-price and the ask-price, in such a way that the losses incurred due to informed trades are compensated by spread gains (paid by the noise traders).

As we emphasised in the very first chapter of this book, markets can only function if liquidity providers are present. The Glosten–Milgrom model elicits a further constraint: liquidity providing is only viable if the fraction of informed trades is sufficiently small – or, equivalently, if the average amount of information in each trade is small. When the information gap is too large, market-makers retract, leading to a liquidity drought and a breakdown of the market. If the model is taken seriously, the fraction ϕ of trades that predict future price moves on a time horizon T should decrease at least as $1/\sqrt{T}$, with $\phi \approx 1\%$ for $T = 1$ day.

An important limitation of the model is that it assumes, as in the Kyle framework, the existence of a terminal time at which the fundamental price p_{F} (say the closing price of the market) is revealed and the game ends. In reality, some predictability exists over a wide spectrum of time scales, from high-frequency signals, which predict price moves over a few seconds, to intra-day signals and even signals on much longer horizons. The question of the profitability of market-making requires a framework where trading is open-ended, with no particular prediction horizon. We will discuss such a framework in the next chapter.

A simple setting where this question can be addressed is the Madhavan, Richardson and Roomans (MRR) model, which posits that the traded price p_t is a martingale, with returns linearly related to the sign of the order. This model is very close in spirit to the "surprise" version of the propagator model, which rather focuses on the mid-price m_t (see Section 13.3). The MRR makes falsifiable predictions about the impact response function $\mathcal{R}(\ell)$ and the volatility per trade $\widetilde{\sigma}$. In view of the simplicity of the model, the agreement with empirical data is remarkable.

In particular, spread and volatility appear as two sides of the same coin, with a causality that is hard to disentangle. Is spread merely compensating for volatility, as the Glosten–Milgrom arguments would have it, or is volatility also influenced by the spread, in the sense that the last transaction price shifts the expectations of the rest of the market? The success of the MRR model suggests that either a majority of trades in the markets are truly informed (which is doubtful), or that the market as a whole statistically adapts to the last traded price.

Take-Home Messages

(i) Marker makers are exposed to adverse selection because they offer to trade with other market participants, some of whom might be informed about the future value of the asset.

(ii) The Glosten–Milgrom framework is a stylised model for how market-makers should set their bid and ask quotes to exactly compensate for the costs of this adverse selection.

(iii) When liquidity takers are sufficiently uninformed (or misinformed), market-makers can solve the equations in the Glosten–Milgrom framework to find break-even quotes. In this framework, transactions always occur at the *a posteriori* fair price, and the price process is a martingale.

(iv) When liquidity takers are too informed, such a solution does not exist, so market-makers must withdraw from offering trades (or otherwise accept to make a loss on average).

(v) The MRR model is an example of a market where the Glosten–Milgrom conditions are met, and in which the transaction prices follow exactly a lag-1 propagator model.

(vi) The MRR model makes falsifiable predictions that agree very well with empirical data; in particular, the volatility per trade is found to be proportional to the spread.

16.5 Further Reading

The Determinants of the Bid–Ask Spread

Bagehot, W. (1971). The only game in town. *Financial Analysts Journal*, 27(2), 12–14.

Glosten, L. R., & Milgrom, P. R. (1985). Bid, ask and transaction prices in a specialist market with heterogeneously informed traders. *Journal of Financial Economics*, 14(1), 71–100.

Glosten, L. R., & Harris, L. E. (1988). Estimating the components of the bid/ask spread. *Journal of Financial Economics*, 21(1), 123–142.

Perold, A. F. (1988). The implementation shortfall: Paper versus reality. *The Journal of Portfolio Management*, 14(3), 4–9.

Subrahmanyam, A. (1991). Risk aversion, market liquidity, and price efficiency. *The Review of Financial Studies*, 4, 417–441

Krishnan, M. (1992). An equivalence between the Kyle (1985) and the Glosten-Milgrom (1985) models. *Economics Letters*, 40(3), 333–338.

Huang, R. D., & Stoll, H. R. (1997). The components of the bid–ask spread: A general approach. *Review of Financial Studies*, 10(4), 995–1034.

Handa, P., Schwartz, R., & Tiwari, A. (2003). Quote setting and price formation in an order driven market. *Journal of Financial Markets*, 6(4), 461–489.

Stoll, H. R. (2003). Market microstructure. In Constantinides, G. M., Harris, M., & Stulz, R. M. (Eds.), *Handbook of the economics of finance* (Vol. 1, pp. 553–604). Elsevier.

Amihud, Y., Mendelson, H., & Pedersen, L. H. (2006). Liquidity and asset prices. *Foundations and Trends' in Finance*, 1(4), 269–364.

Foucault, T., Pagano, M., & Röell, A. (2013). *Market liquidity: Theory, evidence, and policy*. Oxford University Press.

Tannous, G., Wang, J., & Wilson, C. (2013). The intra-day pattern of information asymmetry, spread, and depth: Evidence from the NYSE. *International Review of Finance*, 13(2), 215–240.

Break-Even Conditions and Metaorder Impact

Donier, J. (2012). Market impact with autocorrelated order flow under perfect competition. https://papers.ssrn.com/sol3/papers.cfm?abstract_id=2191660.

Farmer, J. D., Gerig, A., Lillo, F., & Waelbroeck, H. (2013). How efficiency shapes market impact. *Quantitative Finance*, 13(11), 1743–1758.

see also Rogers, K., https://mechanicalmarkets.wordpress.com/2016/08/15/price-impact-in-efficient-markets/.

The MRR Model and Extensions

Madhavan, A., Richardson, M., & Roomans, M. (1997). Why do security prices change? A transaction-level analysis of NYSE stocks. *Review of Financial Studies*, 10(4), 1035–1064.

Wyart, M., Bouchaud, J. P., Kockelkoren, J., Potters, M., & Vettorazzo, M. (2008). Relation between bid-ask spread, impact and volatility in order-driven markets. *Quantitative Finance*, 8(1), 41–57.

Bonart, J., & Lillo, F. (2016). A continuous and efficient fundamental price on the discrete order book grid. https://ssrn.com/abstract=2817279.

The Role of the Tick Size

Harris, L. E. (1994). Minimum price variations, discrete bid-ask spreads, and quotation sizes. *Review of Financial Studies*, 7(1), 149–178.

Bessembinder, H. (2000). Tick size, spreads, and liquidity: An analysis of NASDAQ securities trading near ten dollars. *Journal of Financial Intermediation*, 9(3), 213–239.

Goldstein, M. A., & Kavajecz, K. A. (2000). Eighths, sixteenths, and market depth: Changes in tick size and liquidity provision on the NYSE. *Journal of Financial Economics*, 56(1), 125–149.

Zhao, X., & Chung, K. H. (2006). Decimal pricing and information-based trading: Tick size and informational efficiency of asset price. *Journal of Business Finance & Accounting*, 33(5–6), 753–766.

Dayri, K., & Rosenbaum, M. (2015). Large tick assets: Implicit spread and optimal tick size. *Market Microstructure and Liquidity*, 1(01), 1550003.

Bonart, J. (2016). What is the optimal tick size? A cross-sectional analysis of execution costs on NASDAQ. https://ssrn.com/abstract=2869883.

17

The Profitability of Market-Making

A market-maker knows the price of everything and the value of nothing.

(After Oscar Wilde)

In Chapter 16, we discussed two models in which the transaction price is a martingale. This encodes in a strict manner that both the market-maker and the other trader have on average "no ex-post regrets" about their trades, because the price of each trade is also the best estimate of future prices. In this set-up, there is no arbitrage and no profits for market-makers. Although extremely convenient, one might ask whether this theoretical framework is too strict to account for the behaviour that occurs in real markets.

In this chapter, we revisit the question of whether simple market-making strategies can be profitable. Throughout our discussion, we employ as little theoretical prejudice as possible. This allows us to discuss issues such as inventory constraints and holding times, which are absent from the martingale approach.

Before embarking into this agnostic, model-free analysis, we first rephrase the **martingale** hypothesis for traded prices in a more illuminating way, which makes it obvious that market-making cannot, in this case, be profitable. The basic observation is as follows. Since the conditional expectation of *all* future traded prices is equal to the last traded price p_t, this is true in particular for the expectation of the next traded price at the bid and the next traded price at the ask. More precisely, let us denote by $t_a > t$ (respectively, $t_b > t$) the time of the next trade at the ask (respectively, bid), and $a_{\text{next}} := a_{t_a}$ the corresponding value of the ask-price (respectively, $b_{\text{next}} := b_{t_b}$ the corresponding value of the bid-price). Then:

$$\mathbb{E}[a_{\text{next}}] = \mathbb{E}[b_{\text{next}}] = p_t.$$

Therefore, although the next trade may happen either at the bid or at the ask, these two prices must both on average equal p_t, which is a surprising result given that $b_t < a_t$ at all times!

The intuition behind this apparent puzzle is the following: if the next market order is a buy, then the corresponding ask-price is most likely to be larger than p_t. If instead one or more sell market orders arrive first, they will impact the price downwards, such that the next buy order will hit a depressed ask-price, $a_{\text{next}} \leq a_t$. On average, the two effects cancel out in a competitive setting and $\mathbb{E}[a_{\text{next}}] = p_t = \mathbb{E}[b_{\text{next}}]$, as we now demonstrate.[1]

Let us assume that $\mathbb{E}[a_{\text{next}}] > \mathbb{E}[b_{\text{next}}]$. In this case, one could devise a market-making strategy that would profit from the difference: imagine observing the best bid and ask quotes posted by other traders, then simultaneously pegging infinitesimal (impact-less) buy and sell quantities at the bid- and ask-prices, respectively, and waiting for execution. The pegged order to buy (respectively, sell) will by definition be executed at b_{next} (respectively, a_{next}), so the expected profit of the round-trip is $\mathbb{E}[a_{\text{next}} - b_{\text{next}}]$. If this quantity was positive on average, it would mean that the market-makers are not competitive enough, because they could offer better quotes; if it was negative on average, market-makers would lose money on average and would have to provide less aggressive quotes. Hence, necessarily $\mathbb{E}[a_{\text{next}}] = \mathbb{E}[b_{\text{next}}]$.

Note that the above argument implicitly assumes that any limit order placed at the bid-price or at the ask-price is executed against the next market order. We will use this hypothesis repeatedly in the next section. The hypothesis is reasonable for small-tick stocks, where queue-priority effects are negligible. This can be understood in two different ways. The first is that for small-tick stocks, the queue lengths at the best quotes are usually short. Because the probability that an arriving market order consumes the whole best queue is large, traders who place limit orders at the best quotes are very likely to receive a matching, even if their limit orders are at the back of the queue. The second is that if the relative tick size is small, it costs a trader very little to place an order one tick ahead of the prevailing best bid- or ask-prices, thereby effectively jumping the queue (although this explanation neglects the possibility that other participants can undercut this new limit order by placing other limit orders at even better prices).

For large-tick stocks, on the other hand, queues are long and priority considerations become crucial. Both of the arguments about small-tick stocks fail: placing a limit order at the best quotes typically means being last in a long queue, with a very small probability of executing against the next market order, and it is usually impossible to jump the queue because there is typically no gap between the bid and the ask. For large-tick stocks, market-makers must therefore compete for priority, rather than competing for spreads.

[1] This argument is originally due to Jaisson, T. (2015). Liquidity and impacts in fair markets. *Market Microstructure and Liquidity*, 1(02), 1550010.

17.1 An Infinitesimal Market-Maker

In line with the argument of the previous section, imagine a market-maker that submits limit orders to maintain a very small volume υ_0 at both the bid- and ask-prices. For now, we assume that the market-maker has no inventory constraints, and can therefore conduct any number of buy or sell trades consecutively.

What is the **Profit & Loss** (P&L) balance of this market-maker? The market-maker's t^{th} transaction takes place either at a_t (when the sign of the market order is $\varepsilon_t = +1$) or at b_t (when the sign of the market order is $\varepsilon_t = -1$). One way to assign a value to each of the market-maker's transactions is to compare their price to the mid-price at some later time T. In this framework, the gain or loss from the tth individual trade, **marked-to-market** at time T, can be written as

$$g_t = \upsilon_0 \varepsilon_t \left(m_t + \varepsilon_t \frac{s_t}{2} - m_T \right). \tag{17.1}$$

To understand Equation (17.1), consider a buy market order ($\varepsilon_t = +1$), which hits the ask-queue at time t. When this trade occurs, the money received by the market-maker is $\upsilon_0 a_t = \upsilon_0 \times (m_t + s_t/2)$. The market-maker is then short υ_0 of the asset. At some later time T, the value of this position is marked-to-market at the mid-price m_T, at value $-\upsilon_0 m_T$.[2] The same arguments apply for a sell market order.

We will also assume that there is an additional payment ϖ associated with each trade. This payment could be a processing or transaction cost (for which $\varpi < 0$), a rebate (for which $\varpi > 0$), or could simply be zero. The effect of this additional payment is to create an extra gain or loss for the market-maker, equal to $\upsilon_0 \varpi$.

In this framework, the total P&L of the market-maker, marked to the mid-price at time T, is given by

$$\mathcal{G}_T = \upsilon_0 \left[\sum_{t=0}^{T-1} \theta_t \varepsilon_t \left(m_t + \varepsilon_t \frac{s_t}{2} - m_T \right) + \theta_t \varpi \right], \tag{17.2}$$

where θ_t is an **execution indicator**

$$\theta_t = \begin{cases} 1, & \text{if the limit order was matched at } t, \\ 0, & \text{otherwise.} \end{cases}$$

There are no partial fills when $\upsilon_0 = 1$ lot. Therefore, upon a market order arrival, the market-maker's limit order at the relevant best quote either fully matches (to the incoming market order) or does not match at all (due to its unfavourable position in the queue). For very small-tick stocks, the size of the queue at the best quotes is typically very small, so it is reasonable to assume that $\theta_t = 1$ most of the time. For large-tick stocks, it is much more difficult to obtain a position near the front of

[2] This assumption is not crucial in the following, as it would only change a boundary term to the total P&L of the market-maker, which is negligible when the market-maker inventory is bounded.

the queue, and large market order arrivals are also more rare. For these large-tick stocks, it is more reasonable to assume that $\theta_t = 0$ most of the time, except when one manages to be at the front of the queue.

By taking expectations of Equation (17.2), we arrive at the expected gain of our market-maker between times 0 and T:

$$\mathbb{E}[\mathcal{G}_T] = \upsilon_0 \left(\sum_{t=0}^{T-1} \mathbb{E}\left[\theta_t \varepsilon_t \left(m_t + \varepsilon_t \frac{s_t}{2} - m_T \right) + \theta_t \varpi \right] \right).$$

To proceed with our calculations, we now assume that θ_t is uncorrelated with the value of the spread and the price history.[3] Within this approximation,

$$\mathbb{E}[\mathcal{G}_T] = \upsilon_0 \mathbb{E}[\theta] \sum_{t=0}^{T-1} \left(\frac{\mathbb{E}[s]}{2} + \varpi - \mathcal{R}(T-t) \right),$$

where we have used the definition of the response function, Equation (16.22).

From this formula, it is clear that when $\varpi > 0$ (i.e. when the liquidity taker pays an additional fee per trade), the effect is tantamount to an upward shift of the ask-price and downward shift of the bid-price, both of size $\varpi > 0$, leading to an increased effective spread from s to $s + 2\varpi$.

Assuming that $\mathcal{R}_\infty$ exists and is finite, one has, for very large T:

$$\sum_{t=0}^{T-1} \mathcal{R}(T-t) = \sum_{u=1}^{T} \mathcal{R}(u) \approx T\mathcal{R}_\infty.$$

The expected long-term gain of the market-maker is thus given by a balance between spread, fees and long-term impact:

$$\mathbb{E}[\mathcal{G}_T] \approx T\upsilon_0 \mathbb{E}[\theta] \left(\frac{\mathbb{E}[s]}{2} + \varpi - \mathcal{R}_\infty \right). \tag{17.3}$$

Recalling the results of Section 13.2, we have seen that when $\beta > (1-\gamma)/2$, impact decays so quickly that the price is sub-diffusive (i.e. mean-reverting) and $\mathcal{R}(\ell)$ diverges to $-\infty$ in the limit $\ell \to \infty$. This situation would be a boon for market-makers, since Equation (17.3) predicts a strongly positive gain in that case. If, however, it instead holds that $\beta < (1-\gamma)/2$, then impact decay is too slow to prevent price trends. In this case, $\mathcal{R}(\ell)$ instead diverges to $+\infty$ when $\ell \to \infty$, rendering market-making extremely difficult. The important take-home message is that market-making is easy when prices mean-revert but difficult when prices trend. We will present a similar discussion when comparing the profitability of market orders to that of limit orders in Section 21.3.

[3] This assumption is reasonable for small-tick assets, for which $\theta_t \approx 1$, but is not justified for large-tick assets, for which the execution probability increases either when a large market order arrives or when the queue is very short. This leads to increased adverse selection, which we neglect here – but see Section 17.3.

17.2 Inventory Control for Small-Tick Stocks

An important problem with the result in Equation (17.3) is that it completely disregards the fact that the market-maker's inventory imbalance can become arbitrarily large. Even if market order signs were uncorrelated, the random imbalance between buy and sell volumes up to time T would grow like $\sqrt{T}$, which would lead to an unbounded market-maker inventory at large times. In real markets, market order signs are long-range autocorrelated, with an autocorrelation function $C(\ell) \sim l^{-\gamma}$, with $\gamma < 1$, which leads to an imbalance that grows even more quickly, as $T^{(1-\gamma)/2} \gg \sqrt{T}$. In any case, this imbalance diverges, positively or negatively, when $T \to \infty$. The risk of the market-maker therefore also diverges, which indicates that market-makers face the possibility of ruin due to an adverse market move. To account for this risk, it is necessary for market-makers to implement some form of **inventory control**. In this section, we consider the question of calculating a market-maker's P&L when applying such inventory-control limits.

17.2.1 The Problem

Deriving the optimal strategy of our market-maker Bob in the presence of inventory risk and long-range autocorrelations of market order signs is a very difficult problem. To gain some intuition, we will assume that the Bob's inventory-risk control amounts to him following a mean-reversion strategy that consists of offering less liquidity at the ask and more liquidity at the bid if he is already short, and vice-versa if he is already long.

Let ψ_{t-1} denote Bob's net position after trade $t-1$. Given ψ_{t-1}, we assume that the market-maker participates with a volume $\upsilon_t(\varepsilon)$ at the ask ($\varepsilon = +1$) and at the bid ($\varepsilon = -1$), with:

$$\upsilon_t(\varepsilon) = \upsilon_0 \max(1 + \kappa\psi_{t-1}\varepsilon, 0), \tag{17.4}$$

where $\kappa > 0$ is a parameter that describes how tightly Bob controls his inventory. To understand Equation (17.4), imagine that Bob's inventory is $\psi_{t-1} = 1$, (i.e. he is net long one unit). At time t, he will participate with volume proportional to $1 + \kappa$ at the ask and with volume proportional to $1 - \kappa$ at the bid, such that he is more likely to sell and reduce his inventory than to buy and increase his inventory. An alternative inventory control strategy is described in Appendix A.5.

Recall that for a small-tick stock, one can assume that $\theta_t = 1$ for most t. In this situation, the evolution equation for ψ_t becomes

$$\begin{aligned} \psi_t - \psi_{t-1} &= -\upsilon_t(\varepsilon_t)\theta_t\varepsilon_t, \\ &\approx -\upsilon_t(\varepsilon_t)\varepsilon_t. \end{aligned}$$

Mathematically, the difficulty with Equation (17.4) is the truncation at zero, which prevents the market-maker from offering negative volumes (which would have no meaning). To proceed with analytical calculations, we simply drop that constraint,[4] which amounts to neglecting the possibility that $\kappa|\psi|$ exceeds 1. This simplification allows one to write the evolution of the inventory ψ as a discrete **Ornstein–Uhlenbeck** (or autoregressive) process:

$$\psi_t = (1 - \kappa \upsilon_0)\psi_{t-1} - \upsilon_0 \varepsilon_t. \tag{17.5}$$

Introducing $\alpha := 1 - \kappa\upsilon_0$ yields the solution

$$\psi_t = -\upsilon_0 \sum_{\ell=0}^{\infty} \alpha^\ell \varepsilon_{t-\ell}. \tag{17.6}$$

Since we assume that the mean of the order signs is 0, the mean inventory is also 0. Its unconditional variance is given by

$$\mathbb{V}[\psi] = \upsilon_0^2 \frac{1 + 2\sum_{\ell=1}^{\infty} \alpha^\ell C(\ell)}{1 - \alpha^2}, \tag{17.7}$$

where $C(\ell)$ is the market order sign autocorrelation function. The expression (17.6) shows that in the presence of inventory control, the position of the market-maker remains finite, and only diverges when $\kappa \to 0$ (i.e. when $\alpha \to 1$).

Neglecting the fees for now (i.e. assuming that $\varpi = 0$) and assuming zero initial position (i.e. setting $\psi_0 = 0$), the total expected P&L of the market-maker can be expressed as

$$\begin{aligned}\mathbb{E}[\mathcal{G}] &= \psi_T m_T - \sum_{t=0}^{T-1} \mathbb{E}\left[(\psi_t - \psi_{t-1}) \cdot \left(m_t + \frac{\varepsilon_t s_t}{2}\right)\right] \\ &\approx T\mathbb{E}\left[\left(\upsilon_0 \varepsilon_t - \kappa \upsilon_0^2 \sum_{\ell=0}^{\infty} \alpha^\ell \varepsilon_{t-1-\ell}\right) \cdot \left(m_t + \frac{\varepsilon_t s_t}{2}\right)\right],\end{aligned}$$

where we have neglected in the second equation the sub-leading boundary term corresponding to the value of the inventory at $t = T$. This is justified because the inventory remains finite as long as $\kappa > 0$ (i.e. the inventory does not grow with T, see Equation (17.7)).

Let us first look at the term coming from the spread. Neglecting any possible correlations between the spread and market order signs, one finds

$$\frac{\mathbb{E}[s]}{2}\upsilon_0 \mathbb{E}\left[\left(1 - \kappa\upsilon_0 \sum_{\ell=0}^{\infty} \alpha^\ell \varepsilon_{t-1-\ell}\varepsilon_t\right)\right] = \frac{1-\alpha}{2\alpha}\mathbb{E}[s]\upsilon_0 \sum_{\ell=1}^{\infty} \alpha^\ell (1 - C(\ell)). \tag{17.8}$$

[4] We will return to our discussion of this approximation at the end of this section.

The impact term is slightly more subtle; it reads:

$$\upsilon_0 \mathbb{E}\left[\left(\varepsilon_t - \kappa\upsilon_0 \sum_{\ell=0}^{\infty} \alpha^\ell \varepsilon_{t-1-\ell}\right) \cdot m_t\right] = \upsilon_0 \mathbb{E}\left[\varepsilon_t \cdot m_t\right] - \upsilon_0 \frac{1-\alpha}{\alpha} \left\langle \sum_{\ell=1}^{\infty} \alpha^\ell \mathbb{E}\left[\varepsilon_{t-\ell} \cdot m_t\right] \right\rangle. \tag{17.9}$$

In the second term on the right-hand side, we write as an identity:

$$m_t \equiv m_t - m_{t-\ell} + m_{t-\ell},$$

and recall that by definition

$$\mathcal{R}(\ell) = \mathbb{E}\left[\varepsilon_{t-\ell} \cdot (m_t - m_{t-\ell})\right]. \tag{17.10}$$

Therefore, one can further transform the above expression as:

$$\upsilon_0 \mathbb{E}\left[\left(\varepsilon_t - \kappa\upsilon_0 \sum_{\ell=0}^{\infty} \alpha^\ell \varepsilon_{t-1-\ell}\right) \cdot m_t\right] = -\upsilon_0 \frac{1-\alpha}{\alpha} \sum_{\ell=1}^{\infty} \alpha^\ell \mathcal{R}(\ell), \tag{17.11}$$

where we have used the fact that $\mathbb{E}\left[\varepsilon_t \cdot m_t\right] = \mathbb{E}\left[\varepsilon_{t-\ell} \cdot m_{t-\ell}\right]$.

Gathering all the terms, we finally recover the profit per unit transaction and unit volume of an inventory-constrained market-maker:

$$\frac{\mathbb{E}[\mathcal{G}]}{\upsilon_0 T} \approx \frac{1-\alpha}{\alpha} \left(\frac{1}{2} \mathbb{E}[s] \sum_{\ell=1}^{\infty} \alpha^\ell (1 - C(\ell)) - \sum_{\ell=1}^{\infty} \alpha^\ell \mathcal{R}(\ell) \right). \tag{17.12}$$

This expression is only approximate, because we have neglected the non-linearity in Equation (17.4). Besides, the mean-reversion strategy (17.4) has no reason to be optimal, so more profitable strategies could possibly exist. Still, in spite of all our approximations, we recover that for the benchmark MRR case, for which $\mathcal{R}(\ell) = (1 - C(\ell))\mathbb{E}[s]/2$, the average profit of the market-maker is equal to 0 for any value of α. This makes sense in this case: since the transaction price is fair, any market-making strategy must break even.

We briefly return to the validity of our approximation that relies on the assumption that $|\kappa\psi| \ll 1$ with large probability, which we rephrase as $\kappa^2 \mathbb{V}(\psi) \ll 1$. In the limit $\kappa \to 0$, and assuming that $C(\ell) \sim c_\infty/\ell^\gamma$ for large ℓ, with $0 < \gamma < 1$, one finds

$$\kappa^2 \mathbb{V}[\psi] = (\kappa\upsilon_0)^2 \frac{1 + 2\sum_{k=1}^{\infty} \alpha^k C(k)}{1 - \alpha^2} \approx c_\infty \Gamma[1-\gamma](\kappa\upsilon_0)^\gamma \to 0.$$

Therefore, our approximation that $1 - \kappa\psi$ is positive is clearly supported for slow market-making.

In the other limit $\alpha = 1 - \kappa\upsilon_0 \to 0$, one finds:

$$\kappa^2 \mathbb{V}[\psi] = (\kappa\upsilon_0)^2 \frac{1 + 2\sum_{k=1}^{\infty} \alpha^k C(k)}{1 - \alpha^2} \approx 1,$$

so our approximation is questionable here. However, in this limit, Equation (17.5) simply reads

$$\psi_t = -\upsilon_0 \varepsilon_t,$$

so that

$$\kappa|\psi| = \upsilon_0\kappa \to 1, \text{ with } \upsilon_0\kappa < 1.$$

Therefore, the approximation remains true in this limit as well, and recovers the exact criterion in the MRR case. In the intermediate cases however, the hypothesis has no reason to hold precisely.

17.2.2 Break-Even Conditions and Empirical Results

We now study two extreme limits of the general formula in Equation (17.12). We consider both expressions with ϖ set to zero; for $\varpi \neq 0$ one simply needs to replace s by $s + 2\varpi$.

(i) **Slow market-making** corresponds to a weakly mean-reverting inventory (i.e. $\kappa \to 0$ and thus $\alpha \to 1$). In this limit, and assuming that $\mathcal{R}_\infty$ is finite, one recovers to leading order the result of the last section:

$$\frac{\mathbb{E}[\mathcal{G}]}{\upsilon_0 T} = \frac{1}{2}\mathbb{E}[s] - \mathcal{R}_\infty. \tag{17.13}$$

This expression is the negative of the average ex-post gain of a market order, which pays the half-spread but makes the long-term average impact $\mathcal{R}_\infty$. In the MRR model, $\mathcal{R}_\infty = s/2$ so the slow market-maker makes no profit.

(ii) **Fast market-making** corresponds to a strongly mean-reverting inventory (i.e. $\kappa\upsilon_0 \to 1$ and thus $\alpha \to 0$), where the inventory only depends on the last trade. In this limit

$$\frac{\mathbb{E}[\mathcal{G}]}{\upsilon_0 T} = \frac{1}{2}\mathbb{E}[s](1 - C(1)) - \mathcal{R}(1). \tag{17.14}$$

Within this framework, the least risky market-making strategy (which corresponds to the fastest possible mean-reversion to zero inventory) has zero average gain provided

$$\mathbb{E}[s] = \frac{2\mathcal{R}(1)}{1 - C(1)}, \tag{17.15}$$

which is precisely the MRR **break-even condition** (see Equation (16.22)) for $\ell = 1$. Hence, as noted in the last section, the MRR relation is such that market-makers break even on average on all time scales. For $\ell = 1$, the intuition here is that when trades tend to be strongly correlated ($C(1) \to 1$), Bob cannot easily unwind his position, and the adverse impact is much larger than the short-term impact $\mathcal{R}(1)$ (see also the discussion in Appendix A.5).

Figure 17.1 shows the empirically determined P&L of Bob's simple strategy normalised by the average spread, as a function of the inventory control parameter $\alpha = 1 - \kappa\upsilon_0$, where Equation (17.12) is computed using the empirically determined $C(\ell)$ and $\mathcal{R}(\ell)$. If one does not account for rebates, the average gain of Bob's simple strategy is always negative, with fast market-making strategies ($\alpha = 0$) faring better

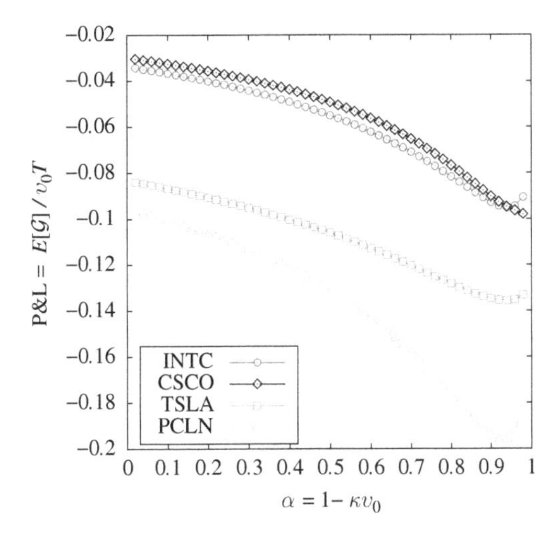

Figure 17.1. The normalised P&L from Equation (17.12), divided by the average spread, as a function of the inventory control parameter $\alpha = 1 - \kappa \upsilon_0$, and with the rebate fees ϖ set to zero, for INTC, CSCO, TSLA and PCLN. Note that in the last two cases, which correspond to large-tick stocks, Equation (17.12) is a priori not warranted (see Section 17.3).

than slow market-making strategies. The cost per trade is of the order of $0.10\langle s\rangle$ for the small-tick stocks, which is typically larger than the rebate fee $\varpi \approx \$0.0025$. This suggests that limit orders are in fact used as a cheaper alternative to market orders for small-tick assets, leading to spreads that are too small on average. For large-tick stocks, a more detailed discussion is needed (see Section 17.3).

Figure 17.2 shows the phase diagram of market-making in the plane $\mathcal{R}(1)$ vs. $\mathbb{E}[s]/2$, with two lines of slope $\mathcal{R}_\infty/\mathcal{R}(1)$ and $(1 - C(1))^{-1}$, corresponding to profitable slow market-making, and profitable fast market-making, respectively. In the MRR model, the two slopes are equal.

17.3 Large-Tick Stocks

For large-tick stocks, the spread is usually equal to its minimum value of one tick ϑ – which is, by definition, large. Naively, this should benefit market-makers, since the spread is bounded from below and is therefore artificially large. However, Bob's life is not that easy. His P&L (without inventory control) can be written as

$$\mathbb{E}[\mathcal{G}_T] = \upsilon_0 T \left[\mathbb{E}[\theta_t] \frac{\vartheta}{2} - \sum_{t=0}^{T-1} \mathbb{E}[\theta_t \varepsilon_t (m_T - m_t)] \right].$$

The position of Bob in the queue is now crucial. If he manages to always be at the front of the queue (for example if he is the only market-maker in town), he will

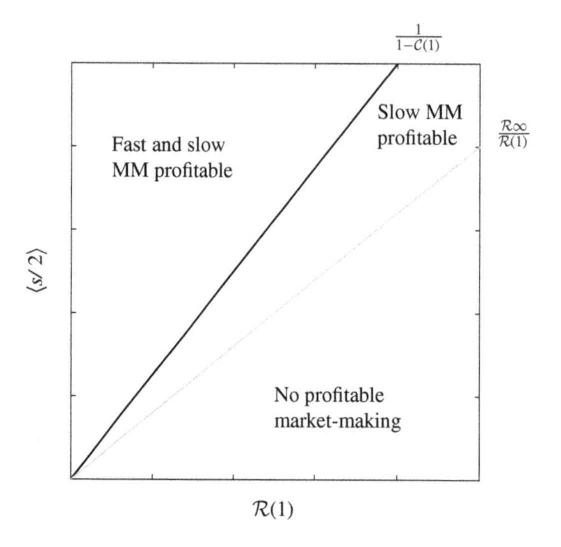

Figure 17.2. The theoretical bounds on the spread as a function of $\mathcal{R}(1)$. The grey line shows the slow market-making profitability bound $\mathcal{R}(\infty)/\mathcal{R}(1)$: slow market-making is profitable *above* this line. The black line shows the fast market-making bound slope $1/(1-C(1))$: fast market-making becomes profitable *above* this line. Note that some markets are such that the black line is below the grey line.

be executed against *all* incoming market orders, so $\theta_t = 1$. In this case, we recover exactly the same result as (17.3) for large T,

$$\mathbb{E}[\mathcal{G}_T^{\text{top}}] = \upsilon_0 T \left[\frac{\vartheta}{2} - \mathcal{R}_\infty\right],$$

for **top-priority limit orders**. In the presence of inventory control, Equation (17.12) also continues to hold if Bob is always at the front of the queue. Hence, the MRR relation $\mathcal{R}(\ell) = (1 - C(\ell))\vartheta/2$ ensures that even the smartest market-maker, always at the front of the queue, breaks even (in the absence of further rebate fees).

Now, consider that Bob has a random position in the queue. His limit order will only be executed at the next transaction if the incoming market order is large enough, but in these cases one expects the impact of these market orders to be larger than the unconditional impact across all market orders. More precisely, it is interesting to study the conditional response function

$$\mathcal{R}_\phi(\ell) = \langle \varepsilon_t \cdot (m_{t+\ell} - m_t) | \upsilon_t \geq \phi V_t \rangle, \tag{17.16}$$

where υ_t is the volume of the incoming market order and V_t is the total volume at the best quote. In words, this response function computes the impact of a market order that consumes more than a fraction ϕ of the available volume. In particular, one recovers the usual unconditional response function when $\phi = 0$ and the response function of price-changing market orders when $\phi = 1$:

$$\mathcal{R}_{\phi=0}(\ell) \equiv \mathcal{R}(\ell), \qquad \mathcal{R}_{\phi=1}(\ell) \equiv \mathcal{R}^1(\ell).$$

In Figure 17.3, we show $\mathcal{R}_\phi(\ell)$ as a function of ϕ for $\ell = 1$ (corresponding to $\alpha = 0$) and $\ell = 20$ (corresponding to $\alpha = 0.95$), for our two large-tick stocks. One clearly observes that $\mathcal{R}_\phi(\ell)$ is an increasing function of ϕ, which confirms that impact is larger for more aggressive market orders. Now, consider a limit order with a relative queue position equal to ϕ. The corresponding execution indicator θ_t is equal to 1 when $v_t \geq \phi V_t$ and equal to 0 otherwise. Since $\mathcal{R}_\phi(\ell)$ is an increasing function of ϕ, the following inequality holds true:

$$\mathbb{E}[\theta_t]\mathbb{E}\left[\varepsilon_t(m_T - m_t)\right]] \leq \mathbb{E}\left[\theta_t\varepsilon_t(m_T - m_t)\right]],$$

where the equality holds only for top-priority limit orders. The adverse selection suffered by a randomly placed limit order is stronger than the adverse selection of a top priority limit order, because there is a positive correlation between the probability to be executed and a large subsequent adverse price move. Hence:

$$\mathbb{E}[\mathcal{G}_T^{\text{random}}] < \mathbb{E}[\theta] \times \mathbb{E}[\mathcal{G}_T^{\text{top}}].$$

Looking again at Figure 17.1, one observes that for large-tick stocks, and for a fast market-maker ($\alpha \to 0$) with top priority,

$$\frac{\mathbb{E}[\mathcal{G}_T^{\text{top}}]}{v_0 T} \approx -0.05\vartheta.$$

In the NASDAQ markets,[5] typical rebate fees are of the order of 0.25ϑ, leaving an average total gain per trade equal to about 0.2ϑ for top priority limit orders on large-tick stocks.[6] As the priority of the limit order degrades, one expects adverse selection costs to increase until the gain per trade (including rebate fees) vanishes. From Figure 17.3, one estimates a difference of about 0.2ϑ between the average of $\mathcal{R}_\phi(20)$ over all values of ϕ (corresponding to a randomly placed limit order) and the unconditional response $\mathcal{R}(20) = \mathcal{R}_0(20)$ (corresponding to a top priority limit order). Hence, randomly placed limit orders roughly break even on average.

17.4 Conclusion

Market-makers earn the spread but suffer from price impact. By formalising the basic Glosten–Milgrom intuition in a model-free framework, we have shown how

[5] Brogaard, J., Hendershott, T., & Riordan, R. (2014). High-frequency trading and price discovery. *Review of Financial Studies*, 27(8), 2267–2306.

[6] This order of magnitude will be confirmed by an independent calculation in Section 21.4.

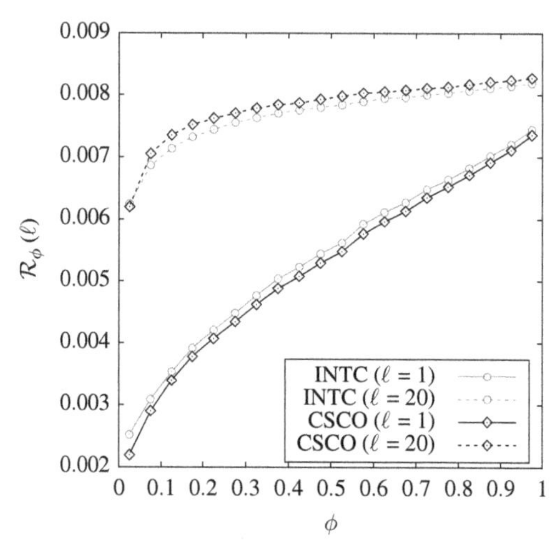

Figure 17.3. The response function $\mathcal{R}_\phi(\ell) = \langle \varepsilon_t \cdot (m_{t+\ell} - m_t) | \upsilon_t > \phi V_t \rangle$ as a function of ϕ for (circles) INTC and (diamonds) CSCO. The solid curves show the results for $\ell = 1$ and the dotted curves show the results for $\ell = 20$.

to compute the long-term P&L of a market-maker who is always at the best quotes and manages inventory risk. The market-maker's gains depend on the relative size of the spread and on the lag-dependent response function. For loosely controlled inventory risk, corresponding to slow market-making, the balance is between the half-spread and the long-term impact. For tightly controlled inventories, the balance is between the half-spread and the short-term impact, corrected by the autocorrelation of the trades. Within the MRR model, spread and impact turn out to balance each other exactly, for any market-making frequency. Reality is more complex, however.

Empirical data confirms that market-making in modern electronic markets is highly competitive. Simple market-making strategies with inventory control are found to yield negative profits (see Figure 17.1). Smarter market-making strategies, which include high-frequency signals that help market-makers to decide more precisely when and where to submit or cancel a limit order, can presumably be made to eke out a small profit. For large-tick stocks, the main challenge is to gain time priority in the queue: top-priority limit orders are found to be profitable on average, after including rebate fees.

In any case, the main practical conclusion is that, like in the MRR model, market orders and limit orders incur on average roughly equivalent costs – at least in the absence of short-term information about the order flow and price changes (see Section 21.3). This equivalence can be seen as a consequence of competition between market-makers, which compresses the spread to its minimum value.

Take-Home Messages

(i) If prices are martingales, the expected price of the next trade that will take place at the ask must equal the expected price of the next trade that will take price at the bid.
(ii) This illustrates that market-making is a more subtle task than it may first appear. The main difficulty of market-making stems from the problem that market-makers cannot choose the execution time and price at which they trade – and thus earn the instantaneous bid–ask spread.
(iii) Price impact and fees both offset the potential profits that market-makers could earn.
(iv) For large-tick stocks, trade prices are not a martingale, due to the frequent occurrence of bid–ask bounce. However, obtaining limit order executions for these stocks is difficult, because it typically requires waiting in a long queue. Therefore, the profitability of market-making for such stocks is not obvious.

17.5 Further Reading

Trading Costs and the Profitability of Market-Making

Madhavan, A., Richardson, M., & Roomans, M. (1997). Why do security prices change? A transaction-level analysis of NYSE stocks. *Review of Financial Studies*, 10(4), 1035–1064.

Handa, P., Schwartz, R. A., & Tiwari, A. (1998). The ecology of an order-driven market. *The Journal of Portfolio Management*, 24(2), 47–55.

Madhavan, A., & Sofianos, G. (1998). An empirical analysis of NYSE specialist trading. *Journal of Financial Economics*, 48(2), 189–210.

Jones, C. M. (2002). A century of stock market liquidity and trading costs. https://ssrn.com/abstract=313681.

Handa, P., Schwartz, R., & Tiwari, A. (2003). Quote setting and price formation in an order driven market. *Journal of Financial Markets*, 6(4), 461–489.

Stoll, H. R. (2003). Market microstructure. In Constantinides, G. M., Harris, M., & Stulz, R. M. (Eds.), *Handbook of the economics of finance* (Vol. 1, pp. 553–604). Elsevier.

Wyart, M., Bouchaud, J. P., Kockelkoren, J., Potters, M., & Vettorazzo, M. (2008). Relation between bid-ask spread, impact and volatility in order-driven markets. *Quantitative Finance*, 8(1), 41–57.

Dayri, K., & Rosenbaum, M. (2015). Large tick assets: Implicit spread and optimal tick size. *Market Microstructure and Liquidity*, 1(01), 1550003.

Mastromatteo, I. (2015). Apparent impact: The hidden cost of one-shot trades. *Journal of Statistical Mechanics: Theory and Experiment*, 2015(6), P06022.

Bonart, J., & Lillo, F. (2016). A continuous and efficient fundamental price on the discrete order book grid. https://ssrn.com/abstract=2817279.

High-Frequency Trading and Market-Making

Avellaneda, M., & Stoikov, S. (2008). High-frequency trading in a limit order book. *Quantitative Finance*, 8(3), 217–224.

Jones, C. M. (2013). What do we know about high-frequency trading? https://ssrn.com/abstract=2236201.

Menkveld, A. J. (2013). High frequency trading and the new market-makers. *Journal of Financial Markets*, 16(4), 712–740.

Biais, B., & Foucault, T. (2014). HFT and market quality. *Bankers, Markets & Investors*, 128, 5–19.

Brogaard, J., Hendershott, T., & Riordan, R. (2014). High-frequency trading and price discovery. *Review of Financial Studies*, 27(8), 2267–2306.

Menkveld, A. J. (2016). The economics of high-frequency trading: Taking stock. *Annual Review of Financial Economics*, 8, 1–24.

Guéant, O. (2016). Optimal market-making. arXiv:1605.01862.

Jovanovic, B., & Menkveld, A. J. (2016). Middlemen in limit order markets. https://ssrn.com/abstract=1624329.

Menkveld, A. J., & Zoican, M. A. (2016). Need for speed? Exchange latency and liquidity. https://papers.ssrn.com/sol3/Papers.cfm?abstract_id=2442690.

PART VIII

Market Dynamics at the Meso-scale

Introduction

At this point in our journey into financial markets, we have discussed a large number of major empirical facts, developed microscopic models of the LOB, and proposed several different paths for explaining impact. We have also seen how statistical models of order flow can capture some intricacies of the price-formation process, and how economic models can shed light on some of the effects induced by strategic behaviour. However, none of the tools or approaches that we have considered so far can properly account for a crucial empirical fact: the square-root impact of traded volumes on prices. This suggests that a piece of the puzzle is still missing.

The square-root impact law has been found in many different markets, with completely different microstructural organisations. This suggests that the explanation should not be sought within the strict framework of an LOB, but relies on a more general – perhaps more fundamental – underlying mechanism.

In this part, we will introduce a very general framework designed for understanding the behaviour of supply and demand in a complex and dynamic environment. At the core of this approach lies the following observation, which has arisen at many points throughout the book: the liquidity shortage common to all markets suggests that the visible LOB is only the tip of the iceberg, beneath which lies a large amount of *latent liquidity*. In the coming chapters, we will argue that understanding and modelling directly this latent liquidity is key for understanding how markets behave – and how best to act in them.

The starting point of this discussion will be the free evolution of supply and demand: how do they behave, irrespective of any transactions that might occur, and thus also of the clearing mechanism in place? To address this question, we will introduce a model that provides insight into the vanishing nature of liquidity around the market price. In contrast to the models that we have studied so far, this model successfully reproduces the square-root law of impact, while still resembling the propagator models that we have discussed in the previous parts.

Of the puzzles that we evoked at the beginning of this book, many answers may now be found. The flat signature plot can be explained by the actions of market participants seeking arbitrage opportunities, irrespective of the question of market efficiency. Excess volatility can be explained by the sheer impact of noise trading and self-referential feedback loops, without creating any obvious arbitrage opportunities. Price moves can be explained by the interplay between information and statistical reaction, in a never-ending tit-for-tat game between liquidity providers and liquidity takers.

Of course, this is not the end of the story. Although the latent liquidity framework is compatible with economic approaches (under some conditions), fully unifying the approaches still requires a huge amount of work. Throughout this part, we will highlight the challenges associated with the latent liquidity idea, and describe some open questions that remain the topic of active research today.

18

Latent Liquidity and Walrasian Auctions

The world we live in is vastly different from the world we think we live in.

(Nassim N. Taleb)

18.1 More than Meets the Eye

LOBs are designed to aggregate all the buy and sell limit orders from patient traders, so that they can be matched with the market orders of more hurried investors. It would therefore be easy to confuse the content of an LOB with the total supply and demand of a given asset at a given time. However, there are many reasons to believe that an LOB only reflects a tiny fraction of the true total supply and demand – which, for the most part, actually remain latent.

One argument for the existence of latent liquidity comes from examining LOB volumes and order flow. Recalling our discussion in Section 10.5, it is not uncommon for large players to seek to buy or sell a substantial fraction (say, a few tenths of a percent up to a few percent) of the total market cap of a given asset. Since the total outstanding volume in the LOB of a typical small-tick stock is roughly 10^{-3} of the market cap, and since the average daily volume (ADV) of trade is rarely above 0.5% of the market cap, such traders' execution horizons must be on the scale of days to weeks. Therefore, at any given time, a large component of the supply or demand does not appear in the LOB, but instead reveals itself gradually.

Why do such traders not simply submit very large limit orders to the LOB? By doing so, they could clearly gain substantial queue priority. However, these limit orders would also signal to the rest of the market that a very large trader is present, and would immediately create price impact (remember again the dialogue at the beginning of Chapter 1). For small-tick stocks, this information leakage clearly outweighs the gain in queue priority. For larger-tick stocks, the total volume in the LOB is perhaps ten times larger, but is still just a small fraction of the ADV, so the drawbacks still far outweigh the benefits.

Another argument for the existence of latent liquidity concerns the shape of the LOB. As we saw in Chapter 4, real LOBs exhibit a hump-shape that first increases then decreases with increasing distance from the best quotes, and that eventually decays to zero at large distances. There is no reason why the true underlying supply/demand for an asset should be so sharply localised around the market price.

These arguments suggest the existence of large reserves of liquidity that are invisible in the LOB but that are present in the minds of market participants. Most of the orders that would exist if slow traders submitted all their intended limit orders immediately are simply not revealed in the LOB.

When seeking to understand the impact of large metaorders (including the square-root law in Section 12.3), this latent liquidity is more important than the liquidity visible in the LOB. This is because large metaorders are executed on a time scale much longer than the typical renewal time of the visible LOB, and they represent volumes that exceed the prevailing volumes at the best quotes. We therefore need a modelling strategy that is able to describe the dynamics of the (often unobservable) latent liquidity, including how latent liquidity is perturbed by a metaorder. This is what we pursue in this chapter.

18.2 A Dynamic Theory for Supply and Demand Curves

The dynamics of the liquidity results from two distinct mechanisms: (i) order matching and market clearing when transactions take place, and (ii) evolution of the intentions of traders between transactions. Let us first focus on the market clearing mechanism.

18.2.1 Walrasian Auctions and Market Clearing

Consider a set of agents in a market that operates in continuous time. At any given time t, some agents will be considering buying some quantity of the asset, while others will be considering selling. Each agent has a certain **reserve price** p, which is the minimum price at which he or she is willing to sell (respectively, maximum price at which he or she is willing to buy), and a corresponding volume. The classical supply curve $\mathscr{S}(p,t)$ (respectively, demand curve $\mathscr{D}(p,t)$) represents the aggregate quantities that the agents are willing to sell below price p (respectively, buy above price p) at time t (see Chapter 1). However, as we discussed in the previous section, only a small fraction of the total supply and demand is revealed in the LOB. Instead, most such supply and demand remains as unexpressed trading intentions. This is what we define as **latent liquidity**.

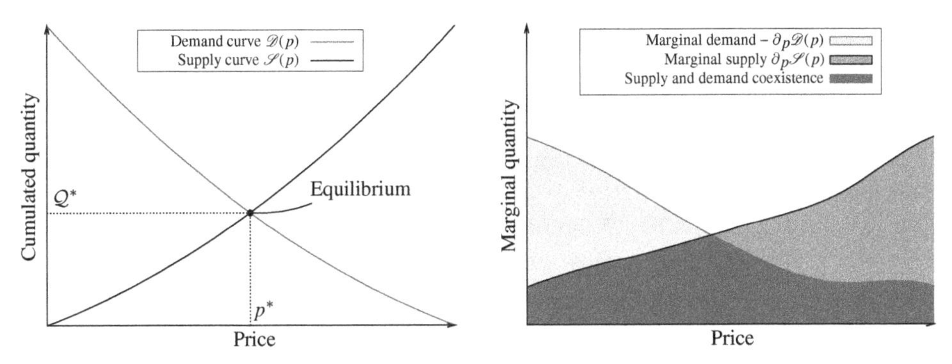

Figure 18.1. (Left panel) Illustration of the supply and demand curves (SD), and the resulting clearing price p^* according to Equation (18.1). (Right panel) Corresponding marginal supply and demand curves (MSD).

To model the dynamics of the total liquidity (both expressed and latent), we introduce the **marginal supply and demand curves** (MSD curves)

$$\rho^-(p,t) = \partial_p \mathscr{S}(p,t) \geq 0;$$
$$\rho^+(p,t) = -\partial_p \mathscr{D}(p,t) \geq 0.$$

As illustrated by the right panel of Figure 18.1, the MSD curves can be regarded as the density of supply and demand intentions in the vicinity of a given price. It is the analogue of the LOB in the present theoretical framework.

Let us assume the market is cleared by a succession of auctions, such that the kth auction takes place at time t_k. In the classical Walrasian framework (see Chapter 1), the transaction or clearing price p_k^* is then set to the value that matches supply and demand, such that (see Equation (1.5)):

$$\mathscr{D}(p_k^*, t_k) = \mathscr{S}(p_k^*, t_k). \tag{18.1}$$

This price is unique provided that the curves are strictly monotonic. The left panel of Figure 18.1 shows some example of supply and demand curves and the resulting transaction price p^*.

At time t_k, market clearing occurs and instantly removes all matched orders. Assuming that all the supply and demand intentions near the transaction price were matched during the auction, the state of the MSD curves immediately after the auction is simple to describe (see Figures 18.1 and 18.5):

$$\rho^-(p, t \downarrow t_k) = \begin{cases} \rho^-(p, t_k), & \text{for } p > p_k^*, \\ 0, & \text{for } p \leq p_k^*; \end{cases}$$

$$\rho^+(p, t \downarrow t_k) = \begin{cases} \rho^+(p, t_k), & \text{for } p < p_k^*, \\ 0, & \text{for } p \geq p_k^*. \end{cases}$$

In other words, the two MSD curves are *càglàd* (see Chapter 3): they jump immediately after the auction.

In the next section, we seek to set up a plausible framework for the dynamics of the supply and demand curves after the auction has been settled, until the next auction at time t_{k+1}. This will allow us to describe, among other things, how the supply and demand curves evolve from the truncated shape given by Equation (18.2) just after time t_k until t_{k+1}. This framework will rely on only two relevant parameters (the price volatility and the volume traded per unit time), and in certain limits will give rise to a universal evolution of the MSD curves.

18.2.2 Evolution of Intentions

We now introduce some assumptions about how agents behave. The framework that we present assumes that market participants only have partial knowledge of the fundamental price process $p_F(t)$, which they try to estimate in their own way, with some idiosyncratic error around $p_F(t)$. More precisely, we assume that between two auctions, the MSD curves evolve according to three distinct mechanisms:

(i) **New intentions**: New intentions to buy/sell, not present in the MSD before time t, can appear. The probability for new buy/sell intentions to appear between times t and $t + \mathrm{d}t$ and between prices p and $p + \mathrm{d}p$ is given by $\lambda^{\pm}(p - p_F)\mathrm{d}p\mathrm{d}t$, where $\lambda^{\pm}(x)$ are two arbitrary positive functions of the relative price $x := p - p_F$. This is close in spirit to the Santa Fe model, although the incoming flow is now centred around the current fundamental price $p_F(t)$, rather than the current mid-price.

(ii) **Cancellations**: Existing intentions to buy/sell at price p can be cancelled and disappear from the supply and demand curves. The probability for an existing buy/sell intention at price p to disappear between t and $t + \mathrm{d}t$ is given by $\nu^{\pm}(p - p_F)\mathrm{d}t$.

(iii) **Price revisions**: Existing intentions to buy/sell at price p can be revised (e.g. due to new information or idiosyncratic factors). Between times t and $t + \mathrm{d}t$, each agent i revises his/her reservation price p_t^i as

$$p_t^i \longrightarrow p_t^i + \underbrace{\beta_t^i}_{\text{sensitivity}} \times \underbrace{\mathrm{d}\xi}_{\text{public news}} + \underbrace{\mathrm{d}W_i}_{\text{idiosyncratic noise}}. \tag{18.2}$$

The term $\mathrm{d}\xi_t$ is a noise term that is common to all i, representing some public fundamental information on the price, such that $\mathrm{d}p_F = \mathrm{d}\xi$. The coefficient β_t^i is the sensitivity of agent i to the particular piece of news. Some agents may over-react ($\beta_t^i > 1$), others may under-react ($\beta_t^i < 1$), or even anti-react ($\beta_t^i < 0$). The term $\mathrm{d}W_i$ models all other idiosyncratic noise contributions (i.e. all other reasons for which an agent might change his/her valuation).

It is convenient to assume that the news term $\mathrm{d}\xi$ is a Wiener noise with variance $\sigma^2\mathrm{d}t$, and that the idiosyncratic terms $\mathrm{d}W_i$ are Wiener noises, independent across different agents, with mean zero and variance $\Sigma^2\mathrm{d}t$.

In the model, the variation of β_t^i across agents and the noise contribution $\mathrm{d}W_i$ both reflect the **heterogeneity of agents**' beliefs. To simplify our calculations, we will assume that the idiosyncratic terms $\mathrm{d}W_i$ average out in the limit of many agents, in the sense that no single participant accounts for a non-zero fraction of the total supply or demand. While not strictly necessary, this assumption leads to a deterministic aggregate behaviour for the supply and demand curves and allows us to sidestep some difficult mathematics.

We pause to make two remarks about the dynamical rules detailed above. First, this model shares several similarities with the Sante Fe model for the LOB, which we discussed in Chapter 8. In the Santa Fe model, orders appear and are cancelled with Poisson rates, much as postulated above for the latent supply and demand curves. However, the price revision process, with the common news driver $\mathrm{d}\xi$, is absent from the Santa Fe model. In other words, while the Santa Fe model fully explains the dynamics of the price in terms of order flow, the present specification allows us to include some **exogenous** factors as well. The present model also permits agents to continuously change their view about the future price without leaving the market, whereas the Santa Fe model only allows existing orders to be cancelled outright. This will turn out to be a crucial ingredient for explaining the empirical shape of price impact.

Second, one can regard the dynamical equations of the present framework not as a set of rules imposed in the formulation of the model, but rather as the consequences of actions undertaken by utility-optimising, heterogeneous agents. Suppose each agent i has a certain utility $\mathcal{U}_i(p,\varphi|\widehat{p}_t^i,\mathcal{F}_t)$ for buying or selling a unit quantity at price p, given his/her estimate of the fundamental price $\widehat{p}_t^i$ and all the other information available at time t, as encoded in $\mathcal{F}_t$. At each time t, each agent computes his/her optimal action (p_t^i,φ_t^i) from the choices of buying ($\varphi = +1$), selling ($\varphi = -1$) or remaining inactive ($\varphi = 0$), as the result of the following optimisation program:

$$(p_t^i,\varphi_t^i) = \underset{p,\varphi}{\operatorname{argmax}}\ \mathcal{U}_i(p,\varphi|\widehat{p}^i(t),\mathcal{F}_t). \tag{18.3}$$

Because of the random evolution of the outside world, which is summarised by $(\widehat{p}^i(t),\mathcal{F}_t)$, the values of $\varphi_t^i \in \{-1,0,+1\}$ and p_t^i can change between t and $t+\mathrm{d}t$, leading to the above rules. In this way, the agent's reserve price evolves as the solution of Equation (18.3) changes. Writing $p_t^i = f_i(\widehat{p}^i(t),t)$, where f_i is a regular function if $\mathcal{U}_i$ is sufficiently regular, and applying Itô's lemma we arrive at

$$\begin{aligned}\mathrm{d}p^i &= \frac{\partial f_i}{\partial t}\mathrm{d}t + \frac{\partial f_i}{\partial p}\mathrm{d}\widehat{p}^i + \frac{\sigma_i^2}{2}\frac{\partial^2 f_i}{\partial p^2}\mathrm{d}t,\\ &= \alpha_t^i\mathrm{d}\widehat{p}^i + \gamma_t^i\mathrm{d}t,\end{aligned}$$

i.e. the stochastic evolution of the world is naturally reflected in the stochastic evolution of the agent's reserve price. The drift term γ_t^i will have little influence in the following, so we will neglect it henceforth. To recover the specification above, we further decompose the price revision $\mathrm{d}p^i = \alpha_t^i\mathrm{d}\widehat{p}^i$ into a common component $\beta_t^i\mathrm{d}\xi$, corresponding to public information, and an idiosyncratic component $\mathrm{d}W_i$, as above.

18.2.3 The Free Evolution of the MSD Curves

Endowed with the above hypotheses, one can obtain stochastic partial differential equations for the evolution of the marginal supply ($\rho^-(p,t) := \partial_p \mathscr{S}(p,t)$) and the marginal demand ($\rho^+(p,t) := -\partial_p \mathscr{D}(p,t)$) in the absence of transactions (i.e. between two auctions). We will supplement these equations by market clearing conditions when an auction takes place.

The full derivation of these equations is beyond the scope of this book.[1] However, their final form is quite intuitive and easy to interpret. It turns out that these equations take a simple form in the moving frame of the reference price p_F. Introducing the relative price $x = p - p_F$,

$$\begin{aligned}\partial_t \rho^+(x,t) &= D\partial^2_{xx}\rho^+(x,t) - \nu^+(x)\rho^+(x,t) + \lambda^+(x);\\ \partial_t \rho^-(x,t) &= \underbrace{D\partial^2_{xx}\rho^-(x,t)}_{\text{revisions}} - \underbrace{\nu^-(x)\rho^-(x,t)}_{\text{cancellations}} + \underbrace{\lambda^-(x).}_{\text{new orders}}\end{aligned} \tag{18.4}$$

The diffusion coefficient is given by

$$D = \frac{1}{2}\left(\Sigma^2 + \sigma^2 \mathbb{V}[\beta^i_t]\right).$$

Part of the diffusion (Σ^2) comes from the purely idiosyncratic, noisy updates of agents; another part ($\sigma^2 \mathbb{V}[\beta^i_t]$) comes from the inhomogeneity of their reaction to news and indeed vanishes if all the β^i_t are equal to 1. One could argue that extra drift terms should be added to these equations, describing the agents' propensity to revise their price estimates towards the reference price ($x = 0$). However, these drift terms play a relatively minor role for the following discussion, so we do not consider them further.

The expressions in Equation (18.4) describe the structural evolution of supply and demand around the reference price p_F, which in fact does not appear explicitly in these equations. Interestingly, the dynamics of the MSD curves can be treated independently of the dynamics of the reference price, provided that the MSD are expressed in a moving frame centred around p_F.

The linear Equations (18.4) for $\rho^+(x,t)$ and $\rho^-(x,t)$ can be formally solved in the general case, by starting from an arbitrary initial condition such as Equation (18.2). However, this general solution is not very illuminating, so we instead focus on the special case where the cancellation rate $\nu^\pm(x) \equiv \nu$ depends neither on x nor on the side of the latent LOB. The evolution of the MSD curves can then be written in a

[1] For a full derivation, see Donier, J., Bonart, J., Mastromatteo, I., & Bouchaud, J. P. (2015). A fully consistent, minimal model for non-linear market impact. *Quantitative Finance*, 15(7), 1109–1121. See also Lasry, J. M., & Lions, P. L. (2007). Mean field games. *Japanese Journal of Mathematics*, 2(1), 229–260; and Lachapelle, A., Lasry, J. M., Lehalle, C. A., & Lions, P. L. (2016). Efficiency of the price formation process in presence of high-frequency participants: A mean field game analysis. *Mathematics and Financial Economics*, 10(3), 223–262, for related ideas in the context of mean-field games.

transparent way, as

$$\rho^{\pm}(x,t) = \int_{-\infty}^{+\infty} \frac{\mathrm{d}x'}{\sqrt{4\pi Dt}} \rho^{\pm}(x',t=0^{+}) e^{-\frac{(x'-x)^2}{4Dt} - \nu t} + \int_0^t \mathrm{d}t' \int_{-\infty}^{+\infty} \frac{\mathrm{d}x'}{\sqrt{4\pi D(t-t')}} \lambda^{\pm}(x') e^{-\frac{(x'-x)^2}{4D(t-t')} - \nu(t-t')}, \tag{18.5}$$

where $\rho^{\pm}(x,t=0^{+})$ is the initial condition from just after the kth auction, for which we choose $t_k = 0$ for simplicity. The first term describes the free evolution of this initial condition, whereas the second term describes new intentions that appear and evolve between times 0 and t.

We now explore the properties of the above solution at time $t = t_{k+1} := t_k + \tau$ (i.e. just before the next auction), in the two asymptotic limits: $\tau \to \infty$, which corresponds to very infrequent auctions, and $\tau \to 0$, which corresponds to continuous-time auctions (as used in most modern financial markets).

18.3 Infrequent Auctions

The aim of this section is to show that in the limit of very infrequent auctions (such as Walrasian auctions), the shape of the marginal supply and demand curves can be computed explicitly. This allows one to characterise Kyle's lambda (i.e. the impact parameter; see Chapter 15) in this framework.

In Equation (18.5), if we let $t = \tau \to \infty$, then the first term disappears exponentially fast, with rate ν. This means that the system reaches a **stationary state** $\rho^{\pm}_{\mathrm{st.}}(x)$, independent of the initial conditions. In this case, the integral over t' in second term in Equation (18.5) can be computed explicitly, to give the following general solution:

$$\rho^{\pm}_{\mathrm{st.}}(x) = \frac{1}{2\sqrt{\nu D}} \int_{-\infty}^{+\infty} \mathrm{d}x' \lambda^{\pm}(x') e^{-\sqrt{\frac{\nu}{D}}|x'-x|}. \tag{18.6}$$

A particularly simple case occurs when $\lambda^{\pm}(x) = \Omega_{\pm} e^{\mp x/\Delta}$. This case corresponds to the situation in which the probability that buyers (respectively, sellers) are interested in a transaction decays exponentially with increasing (respectively, decreasing) relative price. The parameter Δ can be regarded as the price interval over which buy and sell intentions coexist.

In this toy example, a stationary state only exists when the cancellation rate is strong enough to prevent the accumulation of orders far away from $y = 0$. When $\nu\Delta^2 > D$:

$$\rho^{\pm}_{\mathrm{st.}}(x) = \frac{\Omega_{\pm}\Delta^2}{\nu\Delta^2 - D} e^{\mp x/\Delta}.$$

The right panel of Figure 18.1 shows the generic shape of $\rho_{\mathrm{st.}}(x)$, with an overlapping region where buy and sell orders coexist. The auction price $p^* :=$

$p_F + x^*$ is determined by Equation (18.1):

$$V^* := \int_{x^*}^{\infty} dy \rho_{st.}^+(x) = \int_{-\infty}^{x^*} dy \rho_{st.}^-(x),$$

where V^* is, by definition, the volume exchanged during the auction. This equality is the analogue of Equation (1.6) in the present context.

For the exponential case, this equation can be solved as:

$$x^* = \frac{\Delta}{2} \ln\left(\frac{\Omega_+}{\Omega_-}\right). \tag{18.7}$$

This solution has a clear interpretation: if the new buy-order intentions that have accumulated since the last auction outsize the new sell-order intentions during the same period, then the auction price will exceed the reference price, and vice-versa. In this case, one expects the imbalance to revert in the next period (by definition of the reference price), leading to short-time mean-reversion around p_F. If $\Omega_+ \approx \Omega_- :\approx \Omega_0$, then

$$x^* \approx \frac{\Omega_+ - \Omega_-}{2\Omega_0} \Delta,$$

and

$$V^* = \frac{\Omega_0 \Delta^3}{\nu \Delta^2 - D}.$$

After the auction, the MSD curves start again from $\rho_{\mp}^{st.}(x)$, truncated below (respectively, above) x^*, as in Equation (18.2).

We now turn to price impact in this model, by imagining that a buy metaorder of volume Q and with a very high reserve price is introduced in an otherwise balanced market (i.e. $\Omega_+ = \Omega_-$). For sufficiently small Q, the market clearing price p_Q can be computed by writing $\mathscr{S}(p_Q) = \mathscr{D}(p_Q) + Q$ and Taylor-expanding $\mathscr{S}(p)$ and $\mathscr{D}(p)$ around p_F to first order in Q. One readily gets a **linear impact law**:

$$\mathscr{S}(p_Q) = \mathscr{D}(p_Q) + Q,$$
$$\mathscr{S}(p_F) + (p_Q - p_F)\partial_p \mathscr{S}(p_F) = \mathscr{D}(p_F) + (p_Q - p_F)\partial_p \mathscr{D}(p_F) + Q + O(Q^2),$$
$$\Rightarrow \quad \mathfrak{I}(Q) := p_Q - p_F = \Lambda Q + O(Q^2), \tag{18.8}$$

where we have used that $\mathscr{S}(p_F) = \mathscr{D}(p_F)$. **Kyle's lambda**, defined as the slope of the linear impact law, is given by the relationship

$$\Lambda^{-1} = \partial_p \mathscr{S}(p_F) - \partial_p \mathscr{D}(p_F) = \rho_{st.}^+(0) + \rho_{st.}^-(0). \tag{18.9}$$

Whenever the $\rho_{st.}^+(x)$ and $\rho_{st.}^-(x)$ do not simultaneously vanish at $x = 0$, the price response to a perturbation must be linear, as in the Kyle model (see Chapter 15). For the exponential case, one finds

$$\Lambda = \frac{\Delta}{2V^*}.$$

This has an immediate interpretation: the price impact of a metaorder of volume Q is proportional to the typical disagreement range Δ times the ratio of Q over the typical transacted volume at each auction V^*. Note that for the exponential model, the impact law can be computed beyond the linear regime, and reads

$$\mathfrak{I}(Q) = \Delta \sinh^{-1}\left(\frac{Q}{2V^*}\right), \tag{18.10}$$

which is linear for small Q and logarithmic at large Q.

The main take-home message is that when the inter-auction time is sufficiently long, each auction clears an equilibrium supply with an equilibrium demand, both given by the long-time solution of Equation (18.6). This corresponds to the standard representation of market dynamics in the Walrasian context, since in this case only the long-term properties of supply and demand matter, and the transient behaviour is discarded. In the next section, we will depart from this limiting case by introducing a finite inter-auction time, such that the transient dynamics of supply and demand become a central feature.

18.4 Frequent Auctions

We now turn to the other limit, where the inter-auction time $\tau = t_{k+1} - t_k$ tends to zero. Since all the supply curve left of (respectively, demand curve right of) the auction price is wiped out by the auction process, one intuitively expects that after a very small time τ, the density of buy/sell orders in the immediate vicinity of the transaction price will always be small. This is indeed the case, and one can specify exactly the shape of the stationary MSD after many auctions have taken place, as follows.

Consider again Equation (18.5) just before the $(k+1)^{\text{th}}$ auction at time t_{k+1}, in the case where the flow of new orders is symmetric (i.e. $\lambda^+(x) = \lambda^-(-x)$), such that the transaction price is always at the equilibrium price ($x^* = 0$). Focusing on the supply side, one can postulate that when $\tau \to 0$, then in the vicinity of $x = 0$, the term $\rho^-(x, t = k\tau)$ can be written as

$$\rho^-(x, t = k\tau) = \sqrt{\tau}\,\phi_k\left(\frac{y}{\sqrt{D\tau}}\right) + O(\tau) \tag{18.11}$$

(and symmetrically for the demand side). Now, consider Equation (18.5), with the change of variables

$$\begin{aligned} x &\to \sqrt{D\tau}u, \\ x' &\to \sqrt{D\tau}v, \\ t' &\to \tau(1 - z). \end{aligned}$$

This leads to:

$$\begin{aligned} \rho^-(u\sqrt{D\tau}, (k+1)\tau) = &\int_{-\infty}^{+\infty} \frac{\mathrm{d}v}{\sqrt{4\pi}} \rho^-(\sqrt{D\tau}v, t \downarrow k\tau) e^{-\frac{(v-u)^2}{4} - \nu\tau} + \\ &+ \tau \int_0^1 \mathrm{d}z \int_{-\infty}^{+\infty} \frac{\mathrm{d}v}{\sqrt{4\pi z}} \lambda(v\sqrt{D\tau}) e^{-\frac{(u-v)^2}{4z} - \nu z\tau}. \end{aligned} \tag{18.12}$$

Just after the auction, ρ^- is strictly zero for $w < 0$ (see Equation 18.2), so the first integral over v indeed only runs from 0 to ∞.

Using the ansatz Equation (18.11) and taking the limit $\tau \to 0$ leads to the following iteration equation, exact up to order $\sqrt{\tau}$:

$$\phi_{k+1}(u) = \int_0^{+\infty} \frac{\mathrm{d}v}{\sqrt{4\pi}} \phi_k(v) e^{-(u-v)^2/4} + \sqrt{\tau}\,\lambda(0) + O(\tau).$$

Note that ν has disappeared from the equation (but will appear in the boundary condition; see below), and only the value of λ close to the transaction price is relevant at this order.

After a very large number of auctions, one therefore finds that the stationary shape of the demand curve close to the price and in the limit $\tau \to 0$ is given by the non-trivial solution of the following fixed-point equation:

$$\phi_\infty(u) = \int_0^{+\infty} \frac{\mathrm{d}w}{\sqrt{4\pi}} \phi_\infty(w) e^{-(u-w)^2/4}, \tag{18.13}$$

supplemented by the boundary condition $\phi_\infty(u) \approx \mathcal{L}\sqrt{D}u$ for $u \gg 1$, where $\mathcal{L}$ can be fully determined by matching with the explicit solution for $\tau = 0$ (see Equation (19.4) in the next chapter). The final expression reads

$$\mathcal{L} = \frac{1}{D} \int_0^{\infty} \mathrm{d}x' \left[\lambda(-x') - \lambda(x')\right] e^{-\sqrt{\nu/D}x'}.$$

In the exponential toy model, it follows that

$$\mathcal{L} = \frac{2\Omega_0 \Delta}{\Delta\nu - D}, \tag{18.14}$$

which is a measure of liquidity (see Equation (18.17) below) that increases with the flux of incoming intentions Ω_0 and decreases with the cancellation rate ν. Note that $\mathcal{L}$ has dimensions of $1/[\text{price}]^2$.

Equation (18.13) is of the so-called Wiener–Hopf type, and its analytical solution can be found in the literature.[2] We plot this solution numerically in Figure 18.3. As the figure illustrates, the solution is very close to an affine function for $u > 0$. Hence, for $\tau \to 0$

$$\rho^-_{\text{st.}}(x \geq 0) \approx \mathcal{L}(x + x_0); \quad \rho^+_{\text{st.}}(x \leq 0) \approx \mathcal{L}(-x + x_0); \quad x_0 \approx 0.824\sqrt{D\tau}, \tag{18.15}$$

as illustrated in Figure 18.2. Note in particular that for $x = 0$:

$$\rho^-_{\text{st.}}(x = 0) = \rho^+_{\text{st.}}(x = 0) \approx 0.824\mathcal{L}\sqrt{D\tau}; \qquad (\tau \to 0). \tag{18.16}$$

[2] See: Atkinson, C. (1974). A Wiener-Hopf integral equation arising in some inference and queueing problems. *Biometrika*, 61, 277–283.

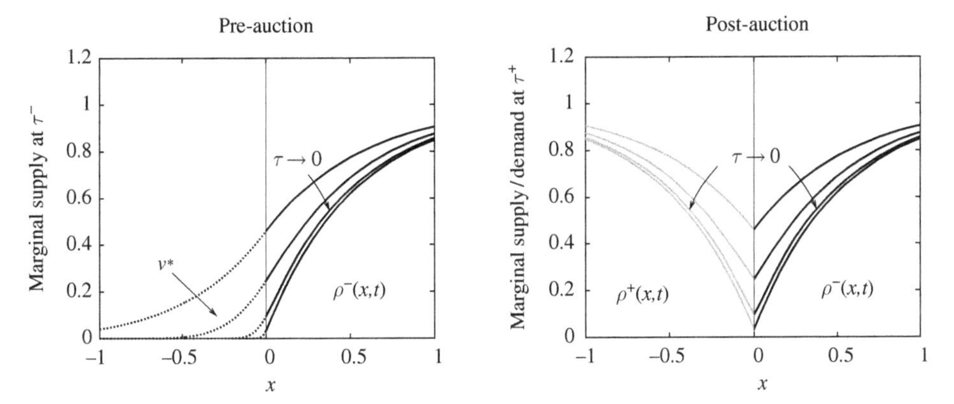

Figure 18.2. (Left panel) Shape of the marginal supply curve immediately before the auctions, for different inter-auction times τ. The integral of the dashed line represents the volume V^* that will be cleared at the auction. (Right panel) Shape of the MSD immediately after the auctions, again for different inter-auction times τ. As $\tau \to 0$, the MSD acquires a characteristic V-shape.

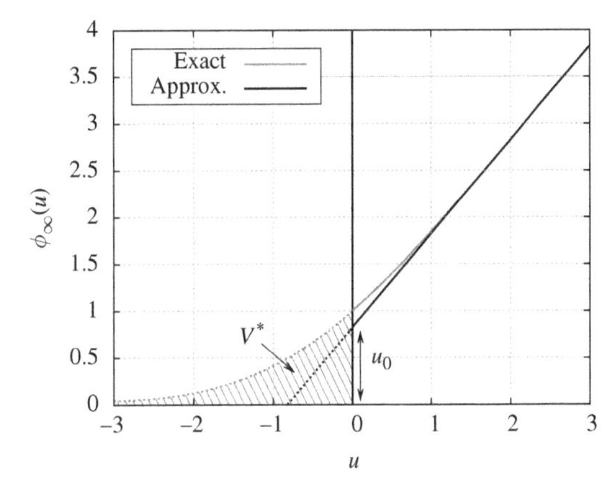

Figure 18.3. Numerical solution to Equation (18.13). The grey line represents the density of sell orders prior to an auction. This density increases with increasing price. The hatched region for negative u, with total volume V^*, represents the sell orders that will be executed at the next auction (the buy curve is not represented here). Above the auction price, the order density is well approximated by (black line) an affine function.

Together with Equation (18.9), this leads to

$$\Lambda^{-1} \approx 1.65\mathcal{L}\sqrt{D\tau}. \tag{18.17}$$

In summary, in the limit of frequent auctions ($\tau \to 0$), the stationary shape $\rho_{\text{st.}}(x)$ of the MSD curves close to the transaction price $x = O(\sqrt{D\tau})$ is universal for a wide class of model specifications (i.e. the functions $\nu^{\pm}(x)$ and $\lambda^{\pm}(x)$). This MSD curve is given by $\sqrt{\tau}\phi_\infty(x/\sqrt{D\tau})$, where ϕ_∞ can itself be approximated by a simple affine function.

We now turn to the interpretation of this result in terms of market liquidity and price impact. In the $\tau \to 0$ limit, one can show that the volume V^* cleared at each auction is given by:

$$V^*(\tau) = \mathcal{L}D\tau. \tag{18.18}$$

The total transacted volume V_T within a finite time interval of length T is given by $V_T = V^*(\tau)T/\tau$. When $\tau \to 0$, this volume remains finite, and is given by

$$V_T = \mathcal{L}DT.$$

This observation should be put in perspective with the recent evolution of financial markets, where the time between transactions τ has become very small and the volume of each transaction has decreased, in such a way that the daily volume has remained roughly constant.[3]

Looking back at Equation (18.17), we note that Kyle's lambda diverges as $\Lambda \propto 1/\sqrt{\tau}$ when $\tau \to 0$. This is the pivotal result of the present section. It means that as the auction frequency increases, the marginal supply and demand becomes very small around the transaction price (see Figure 18.4). Intuitively, this is due to the fact that close to the transaction price, liquidity has no time to rebuild between two auctions. From the point of view of price impact, the divergence of Kyle's lambda as $1/\sqrt{\tau}$ means that the auction price becomes more and more susceptible to any imbalance between supply and demand. In other words, the market becomes *fragile* in the high-frequency limit.

18.5 From Linear to Square-Root Impact

When the inter-auction time τ tends to 0, one can compute the supply and demand curves just before an auction from the shape of the MSD curves close to the transaction price, given by Equation (18.15). Using the results of the previous section, one then can work out the corresponding impact $\mathfrak{I}(Q) = p_Q - p_{\mathrm{F}}$ for $\tau \to 0$, with the tantalising result:

$$\mathfrak{I}(Q) = \sqrt{D\tau} \times \mathcal{Y}\left(\frac{Q}{V^*}\right); \qquad V^* = \mathcal{L}D\tau, \tag{18.19}$$

where $\mathcal{Y}(u)$ is a function with the asymptotic behaviour:

$$\mathcal{Y}(u) \approx_{u \ll 1} 0.555u;$$
$$\mathcal{Y}(u) \approx_{u \gg 1} \sqrt{2u}.$$

[3] Between 1995 and 2015, the total daily transacted volume (as a fraction of the market capitalisation) on the US stock market has roughly doubled, when the daily number of transactions has been multiplied by a factor 10 to 100 depending on stocks.

The fact that $\mathcal{Y}(1)$ is a number of order 1 means that trading the typical auction volume typically moves the price by $\sqrt{D\tau}$ (i.e. its own volatility on time τ).

By Equation (18.19), the impact $\Im(Q)$ is linear in a region where the volume Q is much smaller than $V^* \sim \mathcal{L}D\tau$ (i.e. when the extra volume Q is small compared to the typical volume exchanged during auctions). In the other limit ($Q \gg V^*$), however, one recovers the **square-root impact law**:

$$\Im(Q \gg V^*) \approx \sqrt{\frac{2Q}{\mathcal{L}}}. \tag{18.20}$$

Note that Equation (18.20) can be rewritten exactly as the empirical result (Equation (12.6)). Using the relation $V_T = \mathcal{L}DT$ and $\sigma_T = \sqrt{DT}$, one indeed finds

$$\sqrt{\frac{2Q}{\mathcal{L}}} = \sqrt{2}\sigma_T \sqrt{\frac{Q}{V_T}}.$$

If Q is small enough for the linear approximation of the MSD to hold (see Equation (18.15)), then the impact is universally given by Equation (18.19). For very large Q, however, this linear approximation breaks down and one instead enters a different, presumably non-universal regime that is beyond the scope of the present discussion (see Section 12.3.3). As Figure 18.4 illustrates, impact is Kyle-like for $Q < V^*$, and crosses over to a square-root regime when Q becomes greater than V^*. For $\tau \to 0$, as in the case of continuous-time double auctions, the volume per auction $V^* = \mathcal{L}D\tau$ also tends to zero, so that the region where impact

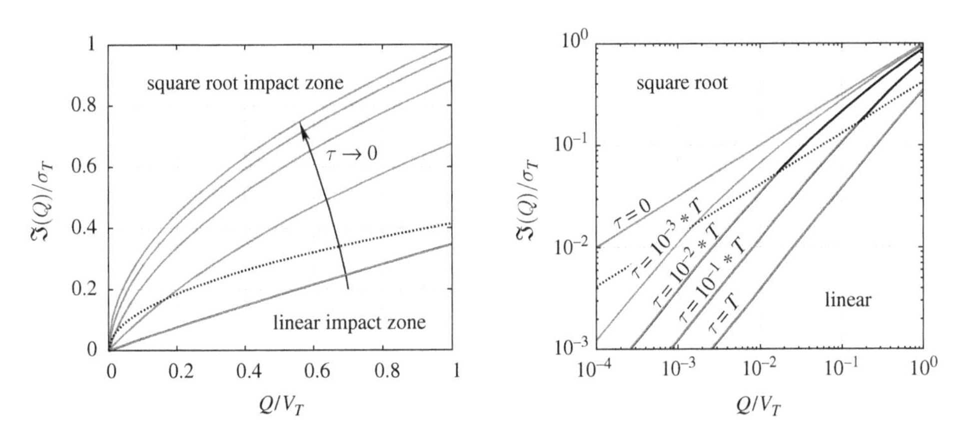

Figure 18.4. The impact of a traded volume Q for a given inter-auction time τ. The impact is linear for $Q \ll V^* = \mathcal{L}D\tau$ and square-root for $Q \gg V^* = \mathcal{L}D\tau$. The linear impact zone shrinks to zero when $\tau \to 0$, where one recovers a pure square-root impact (i.e. a diverging Kyle's Λ).

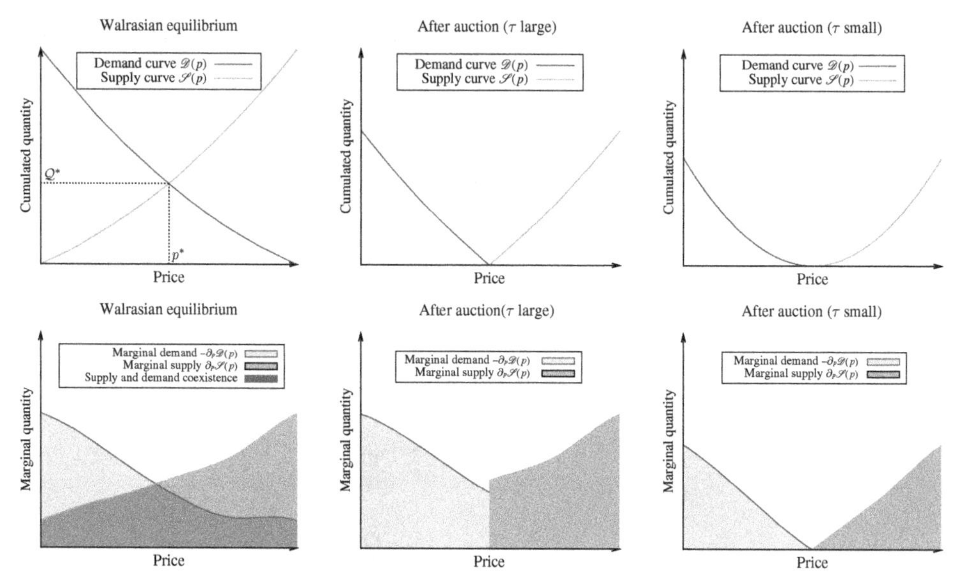

Figure 18.5. (Top row) Supply and demand curves in (left) Walrasian auctions, (centre) immediately after infrequent auctions and (right) immediately after frequent auctions. (Bottom row) Corresponding MSD curves. When transactions occur, supply and demand cannot cross (centre and right). When the market is cleared frequently, marginal supply and demand curves are depleted close to the price and exhibit a characteristic V-shape (right).

is linear shrinks to zero. In other words, when the market clearing time becomes infinitely small, the impact of small trades is never linear.[4]

18.6 Conclusion

The big-picture message of this chapter is quite simple, and is well summarised by the graphs in Figure 18.5. In this figure, we show:

- the standard Walrasian supply and demand curves just before the auction (from which the clearing price p^* can be deduced);
- the stationary supply and demand curves just after an auction, when the inter-auction time τ is large enough (in which case, the marginal supply and demand are both finite at p^*);
- the supply and demand curves in the continuous-time limit $\tau \to 0$, for which the marginal supply and demand curves vanish linearly around the current price.

Interestingly, it is possible to check the above prediction on the shape of the MSD curves using Bitcoin data. Due to its relatively recent invention, the Bitcoin

[4] However, in practice, many discretisation effects (minimum lot size, tick size, etc.), as well as high-frequency liquidity, can restore the linear regime when $Q \to 0$. See also the discussion in Benzaquen, M. & Bouchaud, J.P. (2017). Impact with multi-timescale liquidity, arXiv:1710.03734.

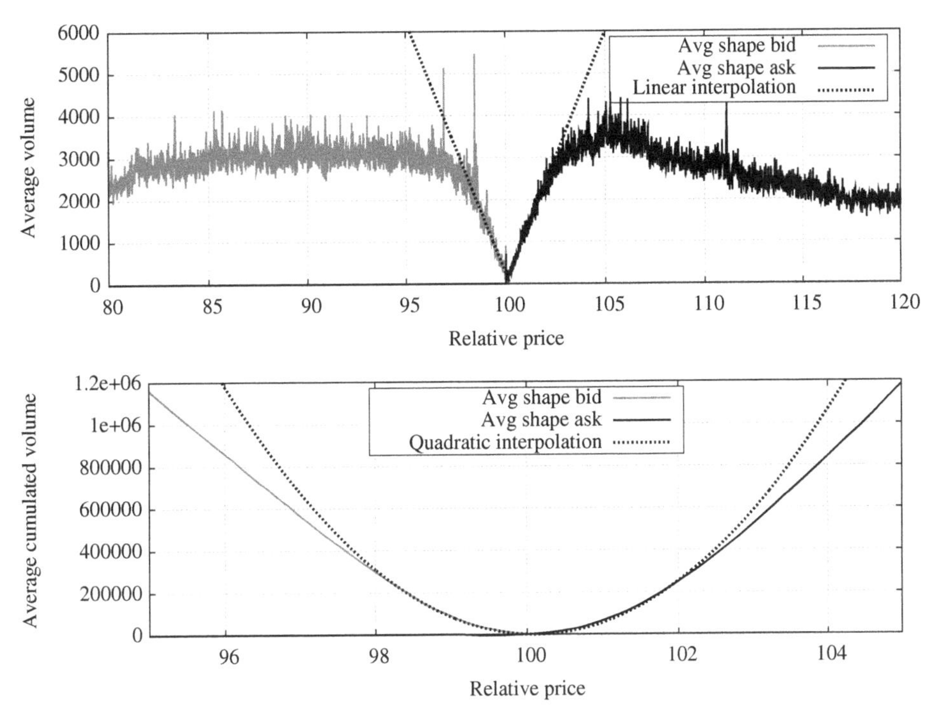

Figure 18.6. (Top panel) The mean relative volume profile of a Bitcoin LOB, which we argue provides a representative sample of the MSD curves, at least close to the mid-price. The data comes from successive snapshots of a full LOB in the Bitcoin market, taken every 15 minutes from May 2013 to September 2013, centred around the current mid-price. (Bottom panel) Integrated shape of the visible LOB. The LOB/MSD curves grow linearly with increasing distance from the mid-price, resulting in a quadratic shape for supply and demand.

market is much less mature than equities or futures markets. Therefore, the activity generated by market participants in this market is typically less strategic and less sophisticated than that of traditional financial markets (this was certainly the case in 2013, which is the period described by our data). One example of such activity is that many Bitcoin traders display their orders in the visible LOB, even far from the mid-price. Therefore, a much larger fraction of the total liquidity is visible (and a much smaller fraction is undisclosed) than is the case in other markets.

Figure 18.6 shows the mean relative volume profile for a Bitcoin LOB, which we argue is a good proxy for the MSD curves, at least not too far from the mid-price. From the Bitcoin data, one can see that the MSD curves are indeed linear in the vicinity of the mid-price. This behaviour persists over a price range of about 5%, in agreement with the dynamical theory of supply and demand in the limit of frequent auctions. Correspondingly, one expects that the impact of metaorders should be well described by a square-root law in this region. This is indeed also found empirically.[5]

In this chapter, we have reviewed how simple ideas regarding the time-dependent preferences of agents can be implemented in terms of a partial differential equation, Equation (18.4), governing the evolution of *latent liquidity*. This equation describes the dynamics of liquidity between auctions, whose role is to clear the market by matching buy and sell orders that overlap at the end of each period of free evolution of the supply and demand curves. Note that the modelling strategy here does not rely on any equilibrium or fair-pricing conditions, but rather relies on purely statistical considerations.

When auctions are allowed to happen at high frequency, the **liquidity vanishes linearly** around the price, which in turn leads to an anomalous, square-root impact of small orders that increases with market volatility. This result highlights an apparent paradox: the more frequently transactions occur (thereby theoretically increasing market efficiency), the more fragile the resulting price is! In continuous-time double-auction markets, the price can be seen as the point at which a *vanishing supply meets a vanishing demand.*

In the next chapter, we will focus on the dynamics of impact in this continuous limit. In particular, we will consider the impact of a metaorder executed using an arbitrary trading schedule, and we will see that directly describing the evolution of agents' intentions offers a much deeper level of understanding than does postulating ad-hoc stochastic models for prices.

Take-Home Messages

(i) In modern financial markets, most liquidity remains latent because agents try to avoid revealing their true intentions, for fear of adverse impact. Therefore, there is more to model than visible liquidity.

(ii) In a market with many small, heterogeneous, stochastic agents, the dynamics of latent liquidity can be modelled using forward PDEs. This framework is mostly relevant for zero-intelligence agents, but can also account for some models with heterogeneous, optimising agents.

(iii) The dynamics of liquidity can in this case be decomposed according to two distinct mechanisms: the free evolution of liquidity (trading intentions) and the clearing mechanism. Whereas the second depends on the particular choice of market organisation (e.g. auctions or continuous-time LOB trading), the first is in principle general.

[5] See: Donier, J., & Bonart, J. (2015). A million metaorder analysis of market impact on the Bitcoin. *Market Microstructure and Liquidity*, 1(02), 1550008.

(vi) In a set-up where agents can submit, cancel and revise their orders, the free evolution of liquidity can be described by a diffusion-type equation.

(v) The more frequent the auctions in a market, the less liquidity in the vicinity of the price, and thus the higher the impact of an arriving market order. In the limit of a continuous-time double-auction mechanism, liquidity near the price decays linearly, which results in square-root impact.

18.7 Further Reading

The material of this chapter is heavily based on the following publications:

Donier, J., Bonart, J., Mastromatteo, I., & Bouchaud, J. P. (2015). A fully consistent, minimal model for non-linear market impact. *Quantitative Finance*, 15(7), 1109–1121.

Donier, J., & Bonart, J. (2015). A million metaorder analysis of market impact on the Bitcoin. Market *Microstructure and Liquidity*, 1(02), 1550008.

Donier, J., & Bouchaud, J. P. (2016). From Walras' auctioneer to continuous time double auctions: A general dynamic theory of supply and demand. *Journal of Statistical Mechanics: Theory and Experiment*, 2016(12), 123406.

For related ideas, see:

Lasry, J. M., & Lions, P. L. (2007). Mean field games. *Japanese Journal of Mathematics*, 2(1), 229–260.

Gao, X., Dai, J. G., Dieker, T., & Deng, S. (2016). Hydrodynamic limit of order book dynamics. *Probability in the Engineering and Informational Sciences*, 1–30.

Lachapelle, A., Lasry, J. M., Lehalle, C. A., & Lions, P. L. (2016). Efficiency of the price formation process in presence of high frequency participants: A mean field game analysis. *Mathematics and Financial Economics*, 10(3), 223–262.

19

Impact Dynamics in a Continuous-Time Double Auction

Details that could throw doubt on your interpretation must be given, if you know them. You must do the best you can – if you know anything at all wrong, or possibly wrong – to explain it. If you make a theory, for example, and advertise it, or put it out, then you must also put down all the facts that disagree with it, as well as those that agree with it.

(Richard P. Feynman)

In the previous chapter, we discussed a simple, theoretical framework that seeks to describe the evolution of latent liquidity between two auctions. In this chapter, we dive more deeply into the dynamics of prices under the pressures created by market participants seeking to trade. To achieve this, we consider from the outset the limiting case $\tau \to 0$, which corresponds to the continuous-time double-auction setting of modern financial markets, in which transactions occur whenever market orders meet outstanding limit orders. Within this framework, we amend the dynamical model encapsulated in Equation (18.4), which allows us to study the impact of a metaorder executed with an arbitrary trading schedule over a specific time interval T.

19.1 A Reaction-Diffusion Model

The trick that we will employ is to add to Equation (18.4) a term that mimics auctions in a continuous setting, by removing orders as soon as they can be matched. As an analogy with chemical reactions, when a buy order (B) meets a sell order (S) of the same size at the same price, they react and are removed from the supply and demand curves, such that $B+S \to \emptyset$. The corresponding dynamical equations then read

$$\partial_t \rho^+(x,t) = D\partial^2_{xx}\rho^+(x,t) - \nu^+(x)\rho^+(x,t) + \lambda^+(x) - R(x,t),$$
$$\partial_t \rho^-(x,t) = \underbrace{D\partial^2_{xx}\rho^-(x,t)}_{\text{revisions}} - \underbrace{\nu^-(x)\rho^-(x,t)}_{\text{cancellations}} + \underbrace{\lambda^-(x)}_{\text{new orders}} - \underbrace{R(x,t)}_{\text{clearing}}, \quad (19.1)$$

where $D\partial^2_{xx}\rho$ is the **diffusion term** and $R(x,t)$ is the **reaction term** that removes any buy and sell orders that meet at a given relative price $x = p - p_F(t)$ from the corresponding densities $\rho^{\pm}(x,t)$.[1] Crucially, the same term $R(x,t)$ appears in both equations, to reflect the fact that the disappearance of any buy volume is accompanied by the disappearance of the same sell volume.

Instantaneous market clearing corresponds to continuous-time auctions, such that the supports of $\rho^+(x,t)$ and $\rho^-(x,t)$ are mutually exclusive at all times: buy orders live on $(-\infty, x^*(t))$ while sell orders live on $(x^*(t), +\infty)$, where $x^*(t)$ is the relative clearing price $p^* - p_F$. This is the case that we will consider henceforth.

Although it may look innocuous, the $R(x,t)$ term makes these equations highly non-trivial. However, we can sidestep this difficulty by introducing an auxiliary field

$$\chi(x,t) := \rho^+(x,t) - \rho^-(x,t) = \begin{cases} \rho^+(x,t), & \text{for } x < x^*(t); \\ \rho^-(x,t), & \text{for } x > x^*(t); \\ 0, & \text{for } x = x^*(t). \end{cases}$$

Recall that only buy orders are present for $x < x^*(t)$, and that only sell orders are present for $x > x^*(t)$, so $\chi(x,t)$ in fact captures all information about buy and sell orders (see Figure 3.1). The equation

$$\chi(x^*, t) = 0$$

can be regarded as the definition of the (relative) clearing price x^*.

For simplicity, we henceforth assume that the cancellation rates $\nu^{\pm}(x)$ are constant (i.e. independent of x). By taking the difference of the two equations in (19.1), one finds that the dynamical evolution of $\chi(x,t)$ is linear and does not depend at all on the reaction term $R(x,t)$:

$$\partial_t \chi(x,t) = D\partial^2_{xx}\chi(x,t) - \nu\chi(x,t) + \lambda^+(x) - \lambda^-(x). \tag{19.2}$$

By noting the parallel to Equation (18.6), we arrive at the stationary solution

$$\chi_{\text{st.}}(x) = \frac{1}{2\sqrt{\nu D}} \int_{-\infty}^{+\infty} dx' \, [\lambda^+(x') - \lambda^-(x')] \, e^{-\sqrt{\frac{\nu}{D}}|x'-x|}. \tag{19.3}$$

In a balanced market, buy/sell intentions are symmetric about the reference price p_F, so we can describe them both with a single function

$$\lambda(x) := \lambda^+(x) = \lambda^-(-x).$$

By inspection of the integral in Equation (19.3), this leads to $\chi_{\text{st.}}(0) = 0$, so that (by construction) the long-term clearing price in the absence of a perturbing metaorder is $x^* = 0$.

[1] We recall that $p_F(t)$ is the time-dependent reference price, around which the order flow is organised (see Section 18.2.2).

In the balanced case, $\chi_{\text{st.}}(x)$ is an odd function of x and goes to zero at the origin with a finite slope, given by:

$$\left.\frac{\mathrm{d}\chi_{\text{st.}}(x)}{\mathrm{d}x}\right|_{x=0} = \frac{1}{D}\int_0^{+\infty} \mathrm{d}x' \left[\lambda(x') - \lambda(-x')\right] e^{-\sqrt{\frac{\nu}{D}}x'} = -\mathcal{L}, \tag{19.4}$$

where $\mathcal{L}$ is the liquidity parameter introduced in Chapter 18. Therefore, we recover the V-shaped liquidity curves discussed in Figure 18.2 (see Section 18.4).[2]

This result shows (as is well known) that a diffusion equation has a locally linear stationary distribution close to an absorbing boundary at $x = 0$. The local gradient $\mathcal{L}$ is related to the current J (i.e. the total volume of transactions per unit time) via the equality

$$D\mathcal{L} = J := \lim_{\tau \to 0} \frac{V^*(\tau)}{\tau},$$

where $V^*(\tau)$ is the volume exchanged during each time interval τ (see Equation (18.18)).

The width w of the linear region (beyond which the detailed shape of $\lambda(x)$ influences the equilibrium profile) is of the order

$$w \sim \sqrt{\frac{D}{\nu}}.$$

It is reasonable to expect that the latent liquidity has a memory time ν^{-1}, whose order of magnitude is several hours to several days – remember that we are speaking here of slow actors, not of market-makers contributing to the high-frequency dynamics of the revealed LOB. Taking D to be of the order of the price volatility (i.e. typically a few percent per day), one finds that w is itself a few percent of the price.

In the remaining sections of this chapter, we consider price changes that are small compared to w (i.e. price changes of less than a few percent). To perform our calculations, we make use of the locally linear shape of the MSD curves (see Section 18.2) which is only valid when price changes are of less than a few percent. For larger price changes, the following considerations would need to be amended, with the addition of other modelling ingredients.

19.2 A Metaorder in an Equilibrated Market

We now extend the above framework by introducing a **metaorder arrival** and calculating its impact on the price. We model this metaorder as an extra current of buy or sell orders in an otherwise equilibrated market. We assume that the total size Q of the arriving metaorder is small enough not to influence the behaviour

[2] We in fact already obtained this result via a different approach in Section 8.7.

of the rest of the market (i.e. that its arrival does not modify the values of D, ν or λ).[3] We also assume that each of the metaorder's child orders arrive at exactly the transaction price $p^*(t)$ at their time of arrival, and are therefore executed immediately as market orders.

Consider a metaorder that arrives at some random time t, with no conditioning on the particular state of the MSD. For mathematical convenience, we set $t = 0$. Let T denote the lifetime of the metaorder. For $0 \leq t \leq T$, let $j(t)$ denote the signed trading flux due to the metaorder at time t, with $j(t) > 0$ for a buy metaorder and $j(t) < 0$ for a sell metaorder (and $j(t) = 0$ for $t < 0$ and $t > T$).[4] For example, a buy metaorder of size Q with constant execution rate is such that $j(t) = Q/T$. Introducing $x^*(t) := p^*(t) - p_{\rm F}$, Equation (19.2) becomes

$$\partial_t \chi(x,t) = D\partial^2_{xx}\chi(x,t) + \underbrace{j(t)\delta(x - x^*(t))}_{\text{metaorder arrival}} - \nu\chi(x,t) + \lambda(x) - \lambda(-x). \tag{19.5}$$

Because we assume that the metaorder arrives in a stationary state with no conditioning on the state of the MSD curves, it holds that at $t = 0$, the MSD can be described by $\chi_{\rm st.}(x) \approx -\mathcal{L}x$ for sufficiently small x. Hence, remarkably, liquidity on average decreases linearly close to the current price; this is at the heart of the anomalous square-root impact law. For $t > 0$, but small enough that both new cancellations and new depositions remain negligible, only the first two terms in Equation (19.5) need to be considered, and the evolution of $\chi(x,t)$ simplifies to:[5]

$$\partial_t \chi(x,t) \approx D\partial^2_{xx}\chi(x,t) + j(t)\delta(x - x^*(t)).$$

This is the standard heat equation with a source term localised at $x = x^*(t)$. Applying the heat kernel to the source term readily gives the MSD profile in the presence of the metaorder:

$$\chi(x,t) = \chi_{\rm st.}(x) + \int_0^t \frac{du\, j(u)}{\sqrt{4\pi D(t-u)}} \exp\left[-\frac{(x - x^*(u))^2}{4D(t-u)}\right], \tag{19.6}$$

where $x^*(u)$ is the relative transaction price at time u and $\chi_{\rm st.}(x) = -\mathcal{L}x$. Using that $\chi(x^*(t),t) \equiv 0$ provides a **self-consistent integral equation** for the price at time $t > 0$:

$$\mathcal{L}x^*(t) = \int_0^t \frac{du\, j(u)}{\sqrt{4\pi D(t-u)}} \exp\left[-\frac{(x^*(t) - x^*(u))^2}{4D(t-u)}\right]. \tag{19.7}$$

Equation (19.7) is the central equation of this chapter. It is formally justified in the limit $\nu, \lambda \to 0$, holding $\mathcal{L}$ fixed. As we noted above, taking the limit $\nu, \lambda \to 0$ allows

[3] Large metaorders might lead to a sudden increase of the cancellation rate ν and a corresponding drop of the liquidity $\mathcal{L}$, which may in turn result in a flash crash; see Chapter 22.

[4] In fact, the present formalism can accommodate more complex trading profiles, such that $j(t)$ changes sign (see Section 19.4).

[5] In this limit, the precise price around which the deposition takes place is irrelevant. In Equation (19.5), we have assumed that $x = 0$ remains the reference price for deposition for all t, but we could have chosen instead the transaction price as the reference price, as in the Santa Fe model specification (i.e. $\lambda(x) \to \lambda(x - x^*(t))$).

one to discard the corresponding terms in Equation (19.5). The modifications brought in by a finite value of ν will be discussed at the end of the next section.

Provided impact is small, in the sense that $|x^*(u) - x^*(t)| \ll \sqrt{D(t-u)}$ for all t and u, the exponential term in Equation (19.7) is approximately equal to 1. Equation (19.7) then boils down to the continuous-time linear propagator model of Chapter 13, with a square-root impact decay (i.e. the decay exponent is $\beta = 1/2$):

$$x^*(t) = \frac{1}{\mathcal{L}} \int_0^t \frac{\mathrm{d}u \, j(u)}{\sqrt{4\pi D(t-u)}}. \tag{19.8}$$

One can show that this linear approximation is valid for sufficiently small trading rates $j(u)$, but breaks down for more aggressive execution (for which a more precise analysis is needed, as we explore in the next section). Equation (19.7) can therefore be regarded as a non-linear generalisation of the propagator model from Chapter 13, based on a realistic description of the stochastic dynamics of the underlying supply and demand curves. The existence of a sound micro-foundation ensures that many desirable properties hold at the mesoscopic scale (such as the absence of arbitrage, as we discuss in Section 19.5).

19.3 Square-Root Impact of Metaorders

In this section, we will show how the square-root impact law is recovered within the present formalism. The simplest case in which a fully non-linear analysis is possible is that of a metaorder executed at a constant rate $j_0(t) = Q/T$ for $t \in [0,T]$. In this case, one can check by substitution that $x^*(u) = A\sqrt{Du}$ is an exact solution of Equation (19.7), where $A(j_0)$ is the solution of the equation

$$A(j_0) = \frac{j_0}{J} \int_0^1 \frac{\mathrm{d}u}{\sqrt{4\pi(1-u)}} e^{-\frac{A^2(1-\sqrt{u})}{4(1+\sqrt{u})}}.$$

One can obtain the asymptotic behaviour of $A(j_0)$ in the two limits $j_0 \ll J$ and $j_0 \gg J$. In the first case, $A \approx j_0/J\sqrt{\pi}$; in the second case, $A \approx \sqrt{2j_0/J}$.

Assuming for now that the metaorder is uninformed (in the sense that $j(t)$ is uncorrelated with the future evolution of the reference price p_{F}), the peak impact of the metaorder $\mathfrak{I}^{\mathrm{peak}}(Q)$ is simply given by the price at the end of the metaorder $x^*(T)$. This leads to an **exact** square-root impact law:

$$\mathfrak{I}^{\mathrm{peak}}(Q) \approx \sqrt{\frac{Q}{\mathcal{L}}} \times \begin{cases} \sqrt{\frac{j_0}{\pi J}} & \text{for } j_0 \ll J, \text{ (i.e. the slow limit);} \\ \sqrt{2} & \text{for } j_0 \gg J, \text{ (i.e. the fast limit).} \end{cases} \tag{19.9}$$

Recall from Section 12.3 that the empirical result can be written as

$$\mathfrak{I}^{\mathrm{peak}}(Q) = Y\sigma_T\sqrt{Q/V_T},$$

where σ_T is the volatility and V_T is the traded volume (both on scale T), and Y is a constant of order 1. In the present context, $\sigma_T^2 = DT$ and $V_T = JT$ with $J = D\mathcal{L}$. Equation (19.9) is thus identical to Equation (12.6) with $\mathcal{L} = V_T/\sigma_T^2$ and a prefactor Y obtained as:

$$Y \approx \begin{cases} \sqrt{\frac{j_0}{\pi J}} & \text{for } j_0 \ll J, \text{ (i.e. the slow limit);} \\ \sqrt{2} & \text{for } j_0 \gg J, \text{ (i.e. the fast limit).} \end{cases} \tag{19.10}$$

More generally, Equation (19.7) is amenable to an exact treatment in the large-$j(t)$ limit, provided $j(t)$ does not change sign and is a sufficiently regular function of time.[6] The leading term yields the average impact trajectory:

$$x^*(t) \approx \sqrt{\frac{2q(t)}{\mathcal{L}}},$$

where

$$q(t) := \int_0^t \mathrm{d}u\, j(u)$$

is the volume executed up to time t. Note that this price impact depends only on the executed volume $q(t)$, not on the execution schedule. This generalises the above result for a uniform execution schedule, for which we found impact to be independent of the trading intensity j_0 for large j_0. Such a weak dependence on the execution schedule is indeed in qualitative agreement with empirical results.

How do these results change if ν is not infinitesimal? Calculations show that the square-root regime only holds for metaorders whose duration T is small compared to ν^{-1}. For longer metaorders, most intentions are cancelled (i.e. agents change their mind on the valuation), erasing all the memory of the supply and demand curves. One can show that in that regime, impact becomes linear in the size of the metaorder Q, as we would expect on general grounds (see Section 12.3).[7]

19.4 Impact Decay

An interesting topic that was difficult to address within the discrete-auction formalism of Chapter 18 is that of **impact decay**. Specifically, how does the price behave for times $t > T$, after the metaorder has been fully executed? In the case of a uniform execution schedule, the impact decay is given by the solution of

$$x^*(t) = \frac{Dj_0}{J}\int_0^T \frac{\mathrm{d}u}{\sqrt{4\pi D(t-u)}} \exp\left[-\frac{(x^*(t) - A\sqrt{Du})^2}{4D(t-u)}\right]. \tag{19.11}$$

[6] Details can be found in Donier, J., Bonart, J., Mastromatteo, I., & Bouchaud, J. P. (2015). A fully consistent, minimal model for non-linear market impact. *Quantitative Finance*, 15(7), 1109–1121.

[7] See Benzaquen, M. & Bouchaud, J.P. (2017). Impact with multi-timescale liquidity, arXiv:1710.03734.

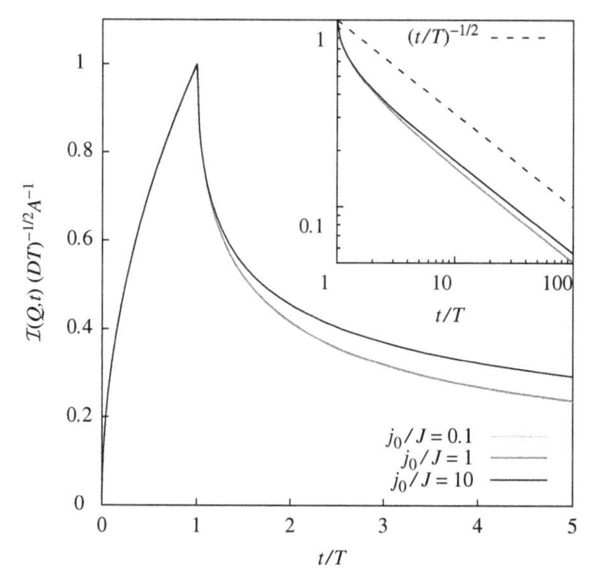

Figure 19.1. Impact $\Im(Q,t)$ as a function of rescaled time t/T for various trading rate parameters j_0/J. The initial growth of the impact follows exactly a square-root law, and is followed by a sudden regime shift after the end of the metaorder. For $t \downarrow T$, the slope of the impact function becomes infinite, and at large times one observes an inverse-square relaxation proportional to $\sqrt{T/t}$ with a pre-factor that depends on j_0/J. Note that the curves for $j_0/J = 0.1$ and $j_0/J = 1$ are nearly indistinguishable.

In the limit of small j_0, the term in the exponential is small, and it is appropriate to use the linear propagator model. In this case, the model predicts the following normalised impact relaxation:

$$\frac{\Im(Q,t>T)}{\Im^{\text{peak}}(Q)} = \frac{\sqrt{t} - \sqrt{t-T}}{\sqrt{T}}. \tag{19.12}$$

This quantity starts at 1 for $t = T$ (by definition), behaves as $1 - \sqrt{(t-T)/T}$ very shortly after time T, and behaves as $(T/t)^{1/2}$ for larger t.

For large j_0, the analysis of Equation (19.11) is quite complicated and beyond the scope of the book. The full analysis reveals that the rescaled initial decay of impact is in fact exactly given by $1 - \sqrt{(t-T)/T}$, independently of j_0/J. For large times, the relaxation also behaves as $(T/t)^{1/2}$, but now with a different prefactor (see Figure 19.1). It is noteworthy that the time scale over which impact decays, say by a factor 2, is proportional to T itself (i.e. the time over which the metaorder was executed). Therefore, the longer an arriving metaorder pushes the market, the longer it takes to come back. This is another manifestation of the long-range resilience of markets (see Chapter 13).

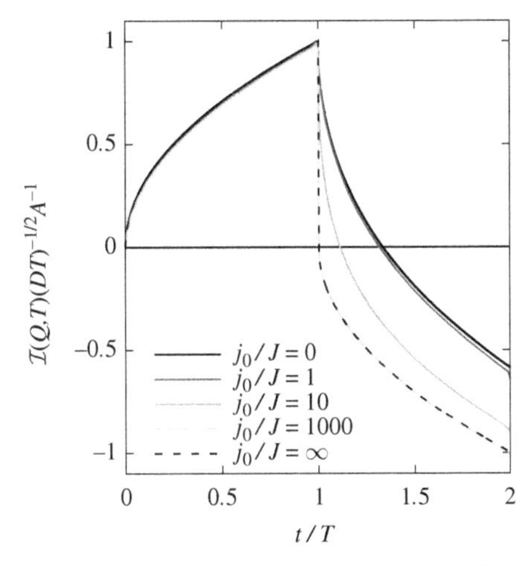

Figure 19.2. Trajectory of the average price before and after a sudden switch of the sign of a metaorder. Here, $j(t) = j_0$ for $t < T$ and $j(t) = -j_0$ for $t > T$. The solid curves show the results for different choices of j_0. The dotted curve shows the theoretical benchmark $j_0 \to 0$, which corresponds to the propagator model. The dash-dotted curve shows the $j_0 = \infty$ limit. In the large-j_0 regime, non-linear effects render the propagator model invalid, and considerably increase the impact of the reversal trade.

The main features of Equation (19.12) are broadly consistent with empirical studies (see Section 12.4 and Section 19.7): impact is a function of t/T, the initial slope of decay is infinite and the final decay is slow.

Another interesting case to consider is when trading is completely reverted after time T, in the sense that $j(t) = j_0$ for $t \in [0,T]$ and $j(t) = -j_0$ for $t \in [T,2T]$. Within the linear propagator approximation, the time needed for the price to revert to its initial value (before continuing to be pushed down by the sell metaorder) is $T/4$. In the non-linear regime $j_0 \gg J$, the price goes down much faster. One can show that it reaches its initial value after a time of order $(J/j_0)T \ll T/4$ (see Figure 19.2). This is because a fast buy metaorder leaves behind an almost empty demand curve, which cannot resist efficiently against the subsequent sell trade.

19.5 Absence of Price Manipulation

One consequence of the asymmetry of the directional trajectory (i.e. on the way up and on the way down) is that the impact of a sell trade after a buy trade is much larger than that of a buy trade after a buy trade. Such an asymmetry means that conducting a simple round-trip trade is costly, because the average sell price is below the average buy price. Hence, one cannot manipulate the price to one's own advantage, first pushing it up by buying aggressively, in the hope of selling back at a higher price.

In fact, this **absence of price manipulation** holds in full generality within the non-linear propagator Equation (19.7). Any viable model of price impact should indeed be such that mechanical price manipulation (i.e. leading to a positive profit after a closed trading loop with no information) is impossible. Using an impact model with profitable closed trading trajectories (e.g. in optimal execution algorithms) would lead to instabilities. This is one of the many desirable properties of latent liquidity models. This benefit makes such models superior to effective models based on prices only, for which ensuring the absence of price manipulation requires extra constraints – see Section 19.7 for references.

To establish the absence of price manipulation, we start by noting that the average cost of a closed trajectory is given by

$$\mathscr{C} = \int_0^T \mathrm{d}u\, j(u)x^*(u), \quad \text{with} \quad \int_0^T \mathrm{d}u\, j(u) = 0,$$

where $x^*(u)$ given by Equation (19.7). This formula simply means that the executed quantity $j(u)\mathrm{d}u$ between time u and $u+\mathrm{d}u$ has price $x^*(u)$. Without loss of generality, we can assume that the initial and final positions are both zero, so there is no additional marked-to-market boundary term.

Using Equation (19.7), it is straightforward to show that $\mathscr{C}$ can be expressed as a quadratic form

$$\mathscr{C} = \frac{1}{2}\int_0^T \int_0^T \mathrm{d}u\mathrm{d}u'\, j(u)\mathbb{M}(u,u')\, j(u'),$$

where $\mathbb{M}$ is a non-negative operator. More precisely, introducing

$$K_z(u,v) := \Theta(u-v)e^{-Dz^2(u-v)+izx^*(u)},$$

one can rewrite the matrix $\mathbb{M}$ as a "square", i.e.:

$$\mathbb{M}(u,u') = D\int_{-\infty}^{\infty} \mathrm{d}z z^2 \int_{-\infty}^{+\infty} \mathrm{d}v\, K_z(u,v)\, K_z^*(u',v).$$

Therefore, $\mathscr{C} \geq 0$ for any execution schedule, so price manipulation is impossible within the present framework. This generalises a result first obtained for the linear propagator model (see Section 19.7): any trading incurs positive expected impact costs. Hence, trading is only justified if some reliable prediction signal is available. Still, many investors trade without a clear notion of the impact cost they incur.

19.6 Conclusion and Open Problems

In this chapter, we have shown that the model for the dynamics of intentions, as developed in the previous chapter, naturally leads to a non-linear propagator model in the limit of continuous-time double-auction markets. This framework allowed us to compute the average price trajectory in the presence of a metaorder, and therefore to generalise the results of Section 13.4.4.

The central result of this chapter is Equation (19.7), which reproduces the observed square-root impact law but also predicts non-trivial trajectories when trading is interrupted or reversed. We also showed that the framework is free of price manipulation, which we argued makes it useful for applications. Again, describing directly the evolution of agents' intentions leads to a richer and more consistent framework than postulating ad-hoc stochastic models for price changes.

Should we be fully satisfied by this model, or is there room for further improvements? In fact, several problems remain, including:

(i) As we emphasised in Section 19.2, the slow-metaorder limit recovers exactly the propagator model, but with a propagator that decays as the inverse square-root of time, corresponding to an exponent $\beta = 1/2$. This value of β is larger than the critical value $\beta_c = (1 - \gamma)/2$ that allows one to balance the long memory in the order flow, described by the exponent γ, and the mean-reversion of the impact, described by the exponent β (see Section 13.2). In other words, a propagator decaying as the inverse square-root of time would lead to a strongly mean-reverting signature plot (see Equation (2.9)).[8] This effect is not observed in real markets. In the model presented in this chapter and the previous chapter, the dynamics of liquidity is itself a diffusion, which in turn imposes $\beta = 1/2$. In order to tune the exponent β to a value less than $1/2$, one should generalise the dynamics of the latent liquidity.[9]

(ii) When the participation rate $f = j_0/J$ is small, the model predicts the prefactor Y of the square-root law to depend on the trading speed j_0 (see Equation (19.10)). However, most of the available empirical data is in the small-f regime, and suggests little dependence on f.

One possible resolution is to assume that a large fraction of the market trading speed J is due to high-frequency trading. Writing $J = J_{\text{HFT}} + J_{\text{slow}}$, with $J_{\text{HFT}} \gg J_{\text{slow}}$, one can show that the criterion for Y to be trading-rate independent is to be in the regime where $j_0 \gg J_{\text{slow}}$, which is much easier to satisfy.[10]

(iii) The model assumes that the marginal supply and demand profile is composed of a very large number of infinitesimal orders. Fluctuations induced by the granularity of these orders, or the presence of a very broad distribution of order sizes, are completely neglected. These fluctuations might play an important role in reality.

(iv) The model neglects strategic behaviour and therefore lacks realism, particularly at the micro-scale. For example, the model has no reason to satisfy the

[8] This assumes that the fundamental price $p_F(t)$ itself follows an independent random walk. One could consider coupling the dynamics of $p_F(t)$ to the dynamics of the order flow such that the traded price becomes exactly diffusive, but this would be ad-hoc.

[9] Work in this direction can be found in Benzaquen, M., & Bouchaud, J. P. (2017). A fractional dynamics description of supply and demand. arXiv:1704.02638.

[10] On this point, see: Benzaquen, M., & Bouchaud, J. P. Market impact with multi-timescale liquidity. arXiv:1710.03734.

Glosten–Milgrom condition for the bid–ask spread without including sophisticated market-making strategies. Unifying this latent liquidity framework with both economic intuitions and micro-level behaviours still requires some work.

(v) There are potential feedback loops between price moves, order flow and the shape of the latent/visible LOB. In particular, the latent liquidity theory assumes that latent orders instantaneously materialise in the LOB when the distance to the price becomes small. Delay in this conversion mechanism may contribute to liquidity droughts and lead to unstable feedback loops, especially when price movements accelerate. This mechanism could be at the heart of the universal power-law distribution of returns, which appears to be related to self-induced liquidity crises but unrelated to exogenous news. Understanding how this happens is an important open problem with much work still to be done!

Take-Home Messages

(i) In a market model with many heterogeneous agents, an extra market order flux induces a distortion of the otherwise V-shaped latent order book, and its impact can be fully expressed as a self-consistent integral equation.

(ii) In the regime of fast executions, one recovers the square-root law for impact. This framework thus gives a micro-foundation to price impact, which appears as a universal phenomenon in real markets.

(iii) In the regime of slow executions, one recovers the linear propagator model for impact, with an inverse square-root kernel.

(iv) Within this framework, all closed-loop trading trajectories result in positive costs. This means that prices cannot be manipulated, and therefore makes the framework suitable for trading strategy optimisation.

(v) Financial markets show complex feedback loops that are not accounted for by this zero-intelligence version of the model. Incorporating strategic behaviour within the framework remains a challenge.

19.7 Further Reading

The material of this chapter is heavily based on the following publication:

Donier, J., Bonart, J., Mastromatteo, I., & Bouchaud, J. P. (2015). A fully consistent, minimal model for non-linear market impact. *Quantitative Finance*, 15(7), 1109–1121.

For related ideas, see:

Lasry, J. M., & Lions, P. L. (2007). Mean field games. *Japanese Journal of Mathematics*, 2(1), 229–260.
Obizhaeva, A. A., & Wang, J. (2013). Optimal trading strategy and supply/demand dynamics. *Journal of Financial Markets*, 16(1), 1–32.
Gao, X., Dai, J. G., Dieker, T., & Deng, S. (2016). Hydrodynamic limit of order book dynamics. *Probability in the Engineering and Informational Sciences*, 1–30.
Lachapelle, A., Lasry, J. M., Lehalle, C. A., & Lions, P. L. (2016). Efficiency of the price formation process in presence of high frequency participants: A mean field game analysis. *Mathematics and Financial Economics*, 10(3), 223–262.

Reaction-Diffusion Models

Barkema, G. T., Howard, M. J., & Cardy, J. L. (1996). Reaction-diffusion front for $A + B \to \emptyset$ in one dimension. *Physical Review E*, 53(3), R2017.
Bak, P., Paczuski, M., & Shubik, M. (1997). Price variations in a stock market with many agents. *Physica A: Statistical Mechanics and Its Applications*, 246(3–4), 430–453.
Tang, L. H., & Tian, G. S. (1999). Reaction-diffusion-branching models of stock price fluctuations. *Physica A: Statistical Mechanics and Its Applications*, 264(3), 543–550.

Scaling of Impact Decay

Moro, E., Vicente, J., Moyano, L. G., Gerig, A., Farmer, J. D., Vaglica, G., & Mantegna, R. N. (2009). Market impact and trading profile of hidden orders in stock markets. *Physical Review E*, 80(6), 066102.
Bershova, N., & Rakhlin, D. (2013). The non-linear market impact of large trades: Evidence from buy-side order flow. *Quantitative Finance*, 13(11), 1759–1778.
Brokmann, X., Serie, E., Kockelkoren, J., & Bouchaud, J. P. (2015). Slow decay of impact in equity markets. *Market Microstructure and Liquidity*, 1(02), 1550007.
Gomes, C., & Waelbroeck, H. (2015). Is market impact a measure of the information value of trades? Market response to liquidity vs. informed metaorders. *Quantitative Finance*, 15(5), 773–793.

Absence of Price Manipulation

Huberman, G., & Stanzl, W. (2004). Price manipulation and quasi-arbitrage. *Econometrica*, 72(4), 1247–1275.
Alfonsi, A., & Schied, A. (2010). Optimal trade execution and absence of price manipulations in limit order book models. *SIAM Journal on Financial Mathematics*, 1(1), 490–522.
Gatheral, J. (2010). No-dynamic-arbitrage and market impact. *Quantitative Finance*, 10(7), 749–759.
Gatheral, J., Schied, A., & Slynko, A. (2012). Transient linear price impact and Fredholm integral equations. *Mathematical Finance*, 22(3), 445–474.

20

The Information Content of Prices

Most, probably, of our decisions to do something positive, the full consequences of which will be drawn out over many days to come, can only be taken as the result of animal spirits – a spontaneous urge to action rather than inaction, and not as the outcome of a weighted average of quantitative benefits multiplied by quantitative probabilities.

(John Maynard Keynes)

After our deep-dive into the microstructural foundations of price dynamics, the time is ripe to return to one of the most important (and one of the most contentious!) questions in financial economics: what information is contained in prices and price moves? This question has surfaced in various shapes and forms throughout the book, and we feel that it is important to devote a full chapter to summarise and clarify the issues at stake. We briefly touched on some of these points in Section 2.3. Now that we have a better handle on how markets really work at the micro-scale, we return to address this topic in detail.

20.1 The Efficient-Market View

Traditionally, market prices are regarded to reflect the fundamental value (of a stock, currency, commodity, etc.), up to small and short-lived mispricings. In this way, a financial market is regarded as a measurement apparatus that aggregates all private estimates of an asset's true (but hidden) value and, after a quick and efficient digestion process, provides an output price. In this view, private estimates should only evolve because of the release of a new piece of information that objectively changes the value of the asset. Prices are then martingales because (by definition) new information cannot be anticipated or predicted. In this context, neither microstructural effects nor the process of trading itself can affect prices, except perhaps on very short time scales, due to discretisation effects like the tick size.

20.1.1 Major Puzzles

This Platonian view of markets is fraught with a wide range of difficulties that have been the subject of thousands of academic papers in the last 30 years (including those with renewed insights from the perspective of market microstructure). The most well known of these puzzles are:

- The **excess-trading puzzle**: If prices really reflect value and are unpredictable, why are there still so many people obstinately trying to eke out profits from trading? Although some amount of excess trading might be justified by risk management or cash management, it is difficult to explain the sheer volume of trading activity in many markets.
- The **excess-volatility puzzle**: Prices (of stocks, currencies, etc.) move around too much to be explained by fluctuations of the fundamental value. This is Shiller's volatility puzzle. In particular, many large price jumps seem to occur without any substantial news. Conversely, prices do not move enough if news is released while markets are closed.[1]
- The **trend-following puzzle**: Price returns tend to be positively autocorrelated on long time scales, such as weeks to months. In other words, some information about future price moves seems to be contained in the past price changes themselves. Trend following is present in all asset classes (including stock indices, commodities, FX, bonds, and so on) and has persisted for at least two centuries. This is in stark contradiction with the Efficient Market story, under which such autocorrelations should not be present. Given that the CTA/trend-following industry managed an estimated 325 billion dollars at the end of 2013, it is hard to argue that these strategies are economically insignificant.

20.1.2 Noise Traders

Faced with the excess-volatility and excess-trading puzzles, research in the 1980s proposed to break away from the strict orthodoxy of rational market participants, and to instead introduce a new category of uninformed agents (or **noise traders**) – see Section 20.5. Including such noise traders allows one to eliminate the **no-trade theorem**[2] and to account for excess trading. As illustrated by the Kyle model and Glosten–Milgrom model (see Chapters 15 and 16), the existence of noise traders is actually crucial for liquidity providers – and thus for the very existence of markets. However, this line of thought refrains from having the price deviate

[1] On this last point, see, e.g., French, K. R., & Roll R., (1986). Stock return variances: The arrival of information and the reaction of traders. *Journal of Financial Economics*, 17, 5–26. On excess volatility in currency markets, see Lyons, R. (2001). *The microstructure approach to foreign exchange rates*. MIT Press.

[2] If I see you want to buy, I infer you know something I do not know and therefore I should not sell. Hence, in a perfectly informed, rational world, no trading should ever take place.

from the fundamental value, in the sense that there is no long-term impact of noise trading on prices. This is quite clear in the continuous Kyle model (see Section 15.4), where noise traders participate in the price-formation process, but the impact of their trades does not contribute to price volatility. As we emphasised in the introductory section of Chapter 11, price impact in this story merely means that trades *forecast* fundamental prices – but noise trading by itself cannot increase long-term volatility, or create long-lived mispricings.

20.2 Order-Driven Prices

After accepting the presence of non-rational agents, the next conceptual step is to accept that all trades (informed or random, large or small) possibly contribute to long-term volatility. This corresponds to a paradigm change: instead of fundamental value driving prices, the main driver of price changes is the order flow itself – whether informed or random. This is the **order-driven view of markets**, which we review in this section.

20.2.1 Trades Impact Prices

At least naively, the order-driven theory of price changes offers a solution to the excess-volatility puzzle: if trades *by themselves* move prices, then excess trading could create excess volatility. This hypothesis is enticing for several reasons:

- As we have argued in Chapters 12 and 13, impact is surprisingly large (a metaorder representing 1% of the average daily volume moves the price by about 10% of its average daily volatility) and long-lived (the impact of an order decays slowly, as a power-law of the lag). Even small, random trades impact prices significantly.
- If order flows become by themselves a dominant cause for price changes, market participants should attempt to predict future order flows themselves, rather than attempting to predict fundamentals (which may play a role, but only on much longer time scales). This ties up neatly with Keynes' famous beauty contest:

  > Investment based on genuine long-term expectation is so difficult [...] as to be scarcely practicable. He who attempts it must surely [...] run greater risks than he who tries to guess better than the crowd how the crowd will behave

  which is much closer to the way market participants operate. As reformulated by A. Shleifer and L. Summers:[3]

  > The key to investment success is not just predicting future fundamentals, but also *predicting the movement of other active investors*. Market professionals spend considerable resources tracking price trends, volume, short interest, odd-lot volume, investor

[3] Shleifer, A. & Summers, L. H. (1990). The noise trader approach to finance. *The Journal of Economic Perspectives*, 4(2), 19–33. The emphasis is ours.

sentiment indexes and numerous other gauges of demand for equities. Tracking these possible indicators of demand makes no sense if prices responded only to fundamental news and not to investor demand.

This is likely to be the mechanism for the universal trend-following effect mentioned as the third major puzzle in Section 20.1. In fact, artificial market experiments show that even when the fundamental value is known to all agents, they are tempted to forecast the behaviour of their fellow subjects and end up creating trends, bubbles and crashes. The temptation to outsmart our peers seems to be too strong to resist!

- The order-flow-driven price scenario allows one to understand why self-exciting feedback effects are so prevalent in financial markets. As we have seen on various occasions, volatility and activity are clustered in time, which strongly suggests that the activity of the market itself leads to more trading. This, in turn, impacts prices and generates more volatility (see Chapters 2 and 9). In fact, calibrating any of these self-exciting models on data leads to the conclusion that a very large fraction (~80%) of market activity and volatility is *self-generated*, rather than exogenous. This fits nicely with Shiller's excess-volatility puzzle. Feedback is also a natural mechanism for the universal nature of return distributions, including their power-law tails, which appear to be asset-independent and unrelated to news (see Section 2.2.1).

20.2.2 The Statistically Efficient View

At this point, we recall that prices are approximately martingales (or diffusive) over time scales spanning from seconds to weeks. This is the content of the flat signature plot, repeatedly discussed throughout the book, which tells us that volatility is approximately independent of time lag (see Section 2.1.4).

In the Efficient Market picture, this phenomenon is a consequence of market prices only reacting to unpredictable news, and almost immediately digesting the corresponding information content. In practice, this explanation is hard to believe, because the impact of news on an asset's fundamental value is so difficult to assess that at least some amount of short-term mispricing should be present, even in very liquid markets. But this should induce **excess short-term volatility** and mean-reversion. For example, to make the signature plot for the S&P 500 futures contract approximately flat, mispricings must be less than 0.05% of the asset's price and have a reversion time of only 10 minutes. How can prices be so precise when there is so much uncertainty? Even under the Efficient Market Hypothesis, there must therefore be some other mechanism in place to explain why signature plots are universally flat.

In fact, the long-range autocorrelation in order flows (see Chapter 10) is clear proof of the presence of **long-lived imbalances** between supply and demand, which markets cannot quickly digest and equilibrate (as assumed in the Efficient Market picture). Naively, these long-lived imbalances should create trends and mispricings. In Chapter 13, we saw that these effects are mitigated by liquidity providers who (in normal market conditions) compete to remove any exploitable price pattern and thereby buffer these imbalances. This is essentially the content of the propagator model, in which impact decay is fine-tuned to compensate the long memory of order flow, and causes the price to be close to a martingale. As we discussed in Section 2.3, this makes prices **statistically efficient** without necessarily being **fundamentally efficient**. In other words, competition at high frequencies is enough to whiten the time series of returns, but not necessarily to ensure that prices reflect fundamental values.

One could thus imagine a hypothetical market where traders only seek to exploit short-term statistical arbitrage opportunities, without any long-term changes in fundamental value. By doing so, such traders' activity makes prices unpredictable (i.e. flattens the signature plot) and simply propagates the high-frequency value of volatility to long time scales. The resulting volatility would in this case have no reason whatsoever to match the fundamental volatility (which is zero in our hypothetical market). Hence, one plausible explanation for the excess-volatility puzzle is that the trading-induced volatility is larger than the fundamental volatility.

Some economists wince at this scenario, because it would imply that prices decouple from fundamental values. For example, in the words of French and Roll:[4]

> Under the trading noise hypothesis, stock returns should be serially correlated [...]: *unless market prices are unrelated to the objective economic value of the stock*, pricing errors must be corrected in the long run. These corrections would generate negative autocorrelations.

In other words, noise trading should imply negative autocorrelations and a decaying signature plot; otherwise, prices would err away from the economic value of the asset. In the context of classical economics, however, this cannot occur and the only possible conclusion is that a flat signature plot is proof that prices must always be close to their fundamental values.

However, another conclusion is warranted if we pay heed to Black:

> We might define an efficient market as one in which price is within a factor of 2 of value, i.e. the price is more than half of value and less than twice value. The factor of 2 is arbitrary, of course. Intuitively, though, it seems reasonable to me, in the light of sources of uncertainty about value and the strength of the forces tending to cause price to return to value. By this definition, I think almost all markets are efficient almost all of the time. Almost all means at least 90%.[5]

[4] French, K. R., & Roll, R. (1986). Stock return variances: The arrival of information and the reaction of traders. *Journal of Financial Economics*, 17(1), 5–26. The emphasis is ours.

[5] Black, F. (1986). Noise. *The Journal of Finance*, 41(3), 528–543. In his introduction, Black also writes: "I recognize that most researchers [...] will regard many of my conclusions as wrong, or untestable, or unsupported by existing evidence. [...] In the end, my response to the skepticism of others is to make a prediction: someday, these conclusions will be widely accepted. The influence of noise traders will become apparent."

Black's statement is that the actions of traders who seek to exploit statistical arbitrage lead to martingale prices that do indeed err away from their fundamental values by large amounts, and for long times. In Section 2.3.2, we estimated that the order of magnitude for mean-reversion around the fundamental value can be as long as a few years. Such a long-term mean-reversion hardly leaves any trace that econometricians or traders can exploit (see Equation (20.3)).

How can such substantial mispricings remain for so long? In other words, why does the aggregation process that financial markets are supposed to realise fail so badly? One commonly cited reason is that the fundamental value is determined so vaguely that the usual arbitrage mechanism (i.e. pushing the price back to its fundamental value) is not strong enough. In contrast to Kyle's world, where the fundamental price is revealed at some time in the future, there is not any terminal time where the true level of, say, the S&P500 is announced. While this **anchoring** does happen at maturity for bond prices or option prices, this idea is clearly not relevant for stock prices or currency levels. Therefore, the spring driving prices back to fundamental values is either non-existent, or otherwise very loose. Another possible reason for persistent mispricings is that **self-referential effects** can actually prevent the market as a whole from converging to an unbiased best guess of the fundamental value (see Section 20.3 for an explicit scenario).

20.2.3 Permanent Impact and Information

If order flow is the dominant cause of price changes, "information" is chiefly about correctly anticipating the behaviour of others, as Keynes envisioned, and not about fundamental value. The notion of information should then be replaced by the notion of **correlation with future returns**, induced by future flows. For example, when all market participants interpret a positive piece of news as negative and sell accordingly, the correct move for an arbitrageur is to interpret the news as negative, even if doing so does not make economic sense. Of course, if all market participants are rational and make trading decisions based on their best guess of the fundamental value, order flow will just reflect deviations from fundamentals and the Efficient Market picture is recovered.

In a market where price formation is dominated by order flow, the difference between a noise trader and an informed trader is merely the strength of correlation between the trader's trades and the future order flow. This allows informed traders to forecast future price changes. For noise traders, there is no such correlation. The short-term reaction part of impact is however similar in both cases.[6] In the

[6] See "Impact Decay" in Section 19.7 for empirical studies on how informed and noise trades impact prices similarly on the short run and differently on the long run; see in particular Tóth, B., Eisler, Z., & Bouchaud, J.-P. (2017). The short-term price impact of trades is universal. https://ssrn.com/abstract=2924029.

long run, the price slowly reverts to its initial value for noise trades, but retains a permanent impact component for informed trades, due to the continued pressure of future order flow. The propagator model (see Chapter 13) suggests an even more complex picture, where the impact of all individual trades tends to zero, but the long-range autocorrelation of trades causes order flow to create a non-zero contribution to volatility.

It could well be the case that even random trades have a small but non-zero permanent impact. This is what happens in the Santa Fe model (see Chapter 8), where all trades are (by definition) random, but the rest of the market acts as if they contain some information. As the model illustrates, random trades are enough to generate volatility. This is also the case in the Kyle model (see Chapter 15), if we also assume that market-makers think there is information (i.e. $\sigma_F > 0$) when actually there is none. In the absence of strong anchoring to a fundamental value, the price ends up wandering around because of the order flow itself, with a volatility generated by the price-formation mechanism at the micro-scale (as we argued in the previous section).

20.3 A Self-Referential Model for Prices

In the order-driven theory of asset prices, fundamental values matter less than market participants' trades. Within this framework, an important question arises: why does some kind of law of large numbers not remove mispricings, as the classical arbitrage argument would predict? After all, in the Efficient Market view, even if individuals have a vague idea about the fundamental value, the aggregation of all opinions should make the price converge to its true value p_F. In a nutshell, the toy model below illustrates two important situations where such convergence to p_F may fail: strong correlations (all agents observe the same news and the same price) and/or extreme financial disparities (the financial weight of some agents remains dominant even for large population sizes). This toy model is of course not meant to be taken literally, but it provides some useful insights on why large crowds may fail to behave wisely.

We now discuss this point more formally. For a given agent $i \in \{1,\ldots,N\}$, let $\widehat{p}_t^i$ denote the agent's estimate of the asset's value at time t, where this estimate is based on his/her preferences, information set and information-processing capabilities. To keep things simple, we do not consider the detailed price-formation process, but instead assume that the market price at any given time t is simply given by the weighted average price estimate across all agents.[7] In this framework,

$$p_{t+1} - p_t = \sum_{i=1}^{N} w_i \times \left(\widehat{p}_{t+1}^i - \widehat{p}_t^i\right),$$

where w_i is the weight of agent i, with $\sum_{i=1}^{N} w_i = 1$. Let

$$\delta_t^i = \widehat{p}_t^i - p_{F,t}$$

denote the mispricing of agent i at time t, where $p_{F,t}$ is the true fundamental price at time t (which the agents strive to discover), and let

$$\bar{\delta}_t := \sum_{i=1}^{N} w_i \delta_t^i = p_t - p_{F,t}$$

denote the average mispricing across all agents.

[7] This is actually in line, when all $w_i \to 0$, with a Walrasian auction mechanism as detailed in Chapter 18.

In this model, we assume that agents change their forecasts based on their own assessment of fundamental information, and also on the actions of other agents, which reflect the **wisdom of the crowd** on the price. In other words, we account for the fact that in uncertain conditions, people tend to rely on the judgement of others as well as on true information.

A simple dynamical equation incorporating these effects is:

$$\delta^i_{t+1} = \underbrace{\rho\delta^i_t}_{\text{error correction}} + \underbrace{\gamma\bar{\delta}_t}_{\text{self-referential}} + \underbrace{\eta^i_t}_{\text{noise}} + \underbrace{(\beta^i_t - 1)\xi_t}_{\text{news}}, \tag{20.1}$$

where $\rho \leq 1$, $\gamma \geq 0$, and $\xi_t := p_{F,t+1} - p_{F,t}$, which we assume to have mean zero. (This equation generalises the model of Chapter 18: compare in particular with Equation (18.2), which corresponds to the case $\rho = \gamma = 0$.)

The terms of Equation (20.1) can be interpreted as follows:

- The first term indicates that due to **information processing**, $\rho \leq 1$ and agents tend to correct their idiosyncratic pricing errors.
- The second term captures the **self-referential effect**. It indicates that agents tend to react to the last mispricing error of the market, for fear they are missing some information that may be revealed later, but that is already available to some other investors. One should expect the value of γ to be dependent on the amplitude of the mispricing and to decrease for large mispricings. In other words, people pay less and less attention to large mispricings (both positive and negative) that look more and more unreasonable. To model these effects, we assume that γ is an even function of $\bar{\delta}$ and choose:[8]

$$\gamma(\bar{\delta}) = \gamma_0 - \gamma_2\bar{\delta}^2 + O(\bar{\delta}^4); \qquad \gamma_0, \gamma_2 \geq 0. \tag{20.2}$$

- The third term is **idiosyncratic noise**, which we assume to have mean zero and be independent both in time and across agents.
- The last term describes how news affects the **fundamental value**, while having different impact on different agents, some of whom over-react ($\beta^i_t > 1$); others of whom under-react ($\beta^i_t < 1$).

Multiplying Equation (20.1) by w_i and summing over i, we arrive at the following evolution equation for $\bar{\delta}_t$:

$$\bar{\delta}_{t+1} = (\rho + \gamma_0)\bar{\delta}_t - \gamma_2\bar{\delta}^3_t + \sum_{i=1}^{N} w_i\eta^i_t + \xi_t\left(\sum_{i=1}^{N} w_i\beta^i_t\right).$$

We first consider the case of a very large number of agents N, with weights w_i all going to zero in the $N \to \infty$ limit. In this case, the idiosyncratic term averages to zero (as in Chapter 19), while the last term can be expressed in terms of the average bias at time t,

$$\bar{\beta}_t = \sum_{i=1}^{N} w_i\beta^i_t.$$

The temporal evolution of the average mispricing then becomes:

$$\bar{\delta}_{t+1} = (\rho + \gamma_0)\bar{\delta}_t - \gamma_2\bar{\delta}^3_t + (\bar{\beta}_t - 1)\xi_t.$$

This is a well-known equation in the context of phase transitions in physics or bifurcations in dynamical systems. The equation has several different regimes:

[8] Adding a term linear in $\bar{\delta}$ would lead to a systematic positive or negative bias in the mispricing, but would not dramatically change the following discussion.

(i) In the case $\gamma_0 = \gamma_2 = 0$ and $\bar{\beta}_t = 1$ (which corresponds to the last mispricing having no influence on the agents' present forecasts, and there being no average bias in the interpretation of news), the temporal evolution of the average mispricing is given by

$$\bar{\delta}_{t+1} = \rho\bar{\delta}_t.$$

This leads to $\bar{\delta}_t = \rho^t\bar{\delta}_0$, which is an exponential relaxation towards zero. In this case, agents individually correct their mistakes and all idiosyncratic errors disappear after a time $\approx |\ln\rho|^{-1}$. This is the Efficient Market story.

(ii) In the case $\gamma_0 = \gamma_2 = 0$ but $\bar{\beta} \neq 1$ (which corresponds to the agents being collectively biased in their interpretation of news), the temporal evolution of the average mispricing is given by[9]

$$\bar{\delta}_{t+1} = \rho\bar{\delta}_t + (\bar{\beta}_t - 1)\xi_t. \tag{20.3}$$

This is a standard AR process, with mean zero, and mean-reverting pricing errors of variance

$$\mathbb{V}[\bar{\delta}] = \frac{\mathbb{E}_t[(\bar{\beta} - 1)^2]\sigma_{\mathrm{F}}^2}{1 - \rho^2},$$

and a mean-reversion time given by $|\ln\rho|^{-1}$. When $\rho \to 1$ (i.e. when the relaxation time is very large, say of the order of years), the pricing errors are themselves close to a random walk, leading to some **excess volatility** in the transaction price. Interestingly, in the same limit, pricing errors can become very large even if $\bar{\beta}$ is close to 1. The opposite limit ($\rho \to 0$), where mispricings quickly revert to zero, corresponds exactly to the high-frequency noise model introduced in Section 2.1.3, with a strongly decreasing signature plot. In this regime, the near-absence of excess high-frequency volatility in data can only be explained if mispricings disappear in a few seconds, which is quite unrealistic.

(iii) In the case $\gamma_0 > 0$ but $\rho + \gamma_0 < 1$, which corresponds to weak feedback across agents, the situation is similar to the one just described (with an effective value $\rho_{\mathrm{eff}} = \rho + \gamma_0 > \rho$). Interestingly, the influence of past mispricings on agents' perceptions slows down the mean-reversion time and increases the magnitude of mispricings and of excess volatility.

(iv) When $\rho_{\mathrm{eff}} = \rho + \gamma_0 \geq 1$, a **bifurcation** takes place. The Efficient Market fixed point $\delta_{\mathrm{F}} = 0$ becomes unstable and is replaced by two fixed points

$$\delta^* = \pm\sqrt{\frac{\rho_{\mathrm{eff}} - 1}{\gamma_2}},$$

around which mispricings fluctuate due to the noise term. In other words, when the feedback parameter γ_0 becomes sufficiently strong, two **self-fulfilling states** can appear: one where the market collectively overvalues the asset ($\delta^* > 0$); the other where the market collectively undervalues the asset ($\delta^* < 0$). The mispricing sets in because agents are over-influenced by the behaviour of others. These periods of over-valuation or under-valuation can last for a very long time. In the language of physics, this is similar to a point particle in a "Mexican-hat" potential. The transition from positive to negative mispricing requires the particle to hop over the energy barrier separating the two minima – a process that can take a very long time when δ^* is large and/or when the noise amplitude is small.

The above toy model provides a concrete example of a situation where a market with an infinite number of agents fails to produce the right price, because of self-referential effects. Even in the regime where feedback effects are not strong enough to destabilise the Efficient Market fixed point, one sees that price errors might

[9] This situation is precisely what is envisaged in Summers, L. (1986). Does the stock market rationally reflect fundamental values? *The Journal of Finance*, 41(3), 591–601.

persist for long times because of the weakness of the mean-reverting (arbitrage) forces when $\rho \uparrow 1$.

Within this modelling framework, one can also discuss the case of strongly heterogeneous agents, where the weights w_i are so broadly distributed that the law of large numbers ceases to apply. Concretely, this means that if one agent represents a finite fraction of the total market, the idiosyncratic noise η_t^i does not average to zero. In this case, the market price remains dominated by the pricing errors of a handful of individuals, even when $N \to \infty$. This situation may well occur in less-liquid financial markets.

20.4 Conclusion

In summary, there are two fully consistent yet hardly reconcilable views about the dynamics of financial markets.

(i) One story posits that the fundamental value evolves as a random walk, and that the price follows suit, almost immediately and without errors. High-frequency dynamics and microstructural effects then define the so-called *price-discovery* process – again referring to a Platonian word where the price exists *in abstracto*, and the market must "discover" it. Markets digest real economic information and are fundamentally efficient. Beating the market is difficult and requires having a good model of the fundamental value.

(ii) The other story posits that prices are impacted by order flows, irrespective of whether or not these order flows are informed. High-frequency dynamics and microstructural effects then define the so-called *price-formation* process. Importantly, in this view, the price would not exist in the absence of trading. The fine balance between liquidity providers and liquidity takers leads to the removal of any short-term predictability, such that prices are statistically efficient regardless of whether they reflect any fundamental value. Beating the market is difficult and requires having a good model of the future order flow – i.e. how the crowd will behave.

Deciding which is of these stories is closest to reality should not be a matter of belief, but rather of scientific investigation. Although no smoking-gun proof is yet available, research in recent decades has uncovered a considerable accumulation of puzzles and anomalies that are unexplained by the fundamental-value story but quite naturally accounted for in the order-flow story. In our opinion, this strongly argues in favour of Keynes, Shiller and Black (among others) and against the Efficient Market school. The recent accumulation of microstructural stylised facts, which allow ever-greater focus on the price formation mechanism, all but confirm that fundamental information plays a relatively minor role in the dynamics of financial markets, at least on short to medium time scales.

Perhaps the most relevant contribution of the field of market microstructure to this debate is an understanding of the order of magnitude of price impact, and its slow decay, both of which reveal that even small trades impact prices. The order-flow theory of price changes opens the door to a quantitative, principled understanding of the various self-exciting and self-referential effects in financial markets. These effects are probably responsible for the most salient stylised facts reviewed in Chapter 2, including both volatility clustering and the heavy tails of the return distribution.

Take-Home Messages

(i) In the standard economic view, markets are supposed to reflect the true value of assets. By definition, prices are unpredictable and diffusive.

(ii) This view is faced with several puzzles, including excess trading, excess volatility and trend-following. Noise traders are often added to models to account for the first puzzle (and to make trading happen at all), but the other two puzzles remain. In particular, fundamental value often plays a major role in price modelling.

(iii) A competing view is that prices are driven by order flow – regardless of their informational content. In this view, the force ensuring that prices remain close to their fundamental value is too weak to play a decisive role, and markets are dominated by other effects.

(iv) In order-driven markets, traders should predict the behaviours of others, rather than fundamentals. The notion of information is replaced by the notion of correlation with other agents' behaviours. This might explain why markets are observed to be so endogenous and volatile, and can result in long-lasting mispricings.

(v) Because of price impact, arbitrage strategies iron out the very predictability that they exploit. The proliferation of such strategies thus contributes to making prices unpredictable. As a consequence, markets can be statistically efficient without being fundamentally efficient – i.e. unpredictable while erring away from fundamentals.

(vi) The question of whether markets are driven by fundamental values or by human behaviours is far from being settled. However, accumulating empirical evidence brings the Efficient Markets picture further and further into question.

20.5 Further Reading

Critique of the Efficient Market Hypothesis

Grossman, S. J., & Stiglitz, J. E. (1980). On the impossibility of informationally efficient markets. *The American Economic Review*, 70(3), 393–408.

Summers, L. H. (1986). Does the stock market rationally reflect fundamental values? *The Journal of Finance*, 41(3), 591–601.
Shiller, R. J. (1990). Speculative prices and popular models. *The Journal of Economic Perspectives*, 4(2), 55–65.
Shleifer, A., & Vishny, R. W. (1997). The limits of arbitrage. *The Journal of Finance*, 52(1), 35–55.
Brunnermeier, M. K. (2001). *Asset pricing under asymmetric information: Bubbles, crashes, technical analysis, and herding*. Oxford University Press on Demand.
Schwert, G. W. (2003). Anomalies and market efficiency. In G. M. Constantinides, M. Harris, & R. M. Stulz (Eds.), *Handbook of the Economics of Finance* (Vol. 1, pp. 939–974). Elsevier.
Farmer, J. D., & Geanakoplos, J. (2009). The virtues and vices of equilibrium and the future of financial economics. *Complexity*, 14(3), 11–38.
Kirman, A. (2010). *Complex economics: Individual and collective rationality*. Routledge.
Ang, A., Goetzmann, W. N., & Schaefer, S. M. (2011). The efficient market theory and evidence: Implications for active investment management. *Foundations and Trends in Finance*, 5(3), 157–242.
Alajbeg, D., Bubas, Z., & Sonje, V. (2012). The efficient market hypothesis: Problems with interpretations of empirical tests. *Financial Theory and Practice*, 36(1), 53–72.
Lo, A. W. (2017). *Adaptive markets*. Princeton University Press.

Excess Volatility and Volatility without News

Shiller, R. J. (1980). Do stock prices move too much to be justified by subsequent changes in dividends? *American Economic Review*, 71, 421–436.
LeRoy, S. F., & Porter, R. D. (1981). The present-value relation: Tests based on implied variance bounds. *Econometrica: Journal of the Econometric Society*, 49, 555–574.
French, K. R., & Roll, R. (1986). Stock return variances: The arrival of information and the reaction of traders. *Journal of Financial Economics*, 17(1), 5–26.
Cutler, D. M., Poterba, J. M., & Summers, L. H. (1989). What moves stock prices? *The Journal of Portfolio Management*, 15(3), 4–12.
Fair, R. C. (2002). Events that shook the market. *The Journal of Business*, 75(4), 713–731.
Joulin, A., Lefevre, A., Grunberg, D., & Bouchaud, J. P. (2008). Stock price jumps: News and volume play a minor role. *Wilmott Magazine*, September/October, 1–7.
Dichev, I. D., Huang, K., & Zhou, D. (2014). The dark side of trading. *Journal of Accounting, Auditing & Finance*, 29(4), 492–518.

Noise Traders

Black, F. (1986). Noise. *The Journal of Finance*, 41(3), 528–543.
Shleifer, A., & Summers, L. H. (1990). The noise trader approach to finance. *The Journal of Economic Perspectives*, 4(2), 19–33.
Barber, B. M., & Odean, T. (1999). Do investors trade too much? *American Economic Review*, 89(5), 1279–1298.
Barber, B. M., & Odean, T. (2000). Trading is hazardous to your wealth: The common stock investment performance of individual investors. *The Journal of Finance*, 55(2), 773–806.
Bouchaud, J.-P., Ciliberti, C., Majewski, A., Seager, P., & Sin-Ronia, K. (2017). Black was right: price is within a factor 2 of value, arXiv:1711.04717.

Order-Flow-Driven Price Changes

Hasbrouck, J. (1991). The summary informativeness of stock trades: An econometric analysis. *Review of Financial Studies*, 4(3), 571–595.

Lyons, R. (2001). *The microstructure approach to foreign exchange rates*. MIT Press.

Evans, M. D., & Lyons, R. K. (2002). Order flow and exchange rate dynamics. *Journal of Political Economy*, 110(1), 170–180.

Chordia, T., & Subrahmanyam, A. (2004). Order imbalance and individual stock returns: Theory and evidence. *Journal of Financial Economics*, 72(3), 485–518.

Hopman, C. (2007). Do supply and demand drive stock prices? *Quantitative Finance*, 7, 37–53.

Bouchaud, J. P., Farmer, J. D., & Lillo, F. (2009). How markets slowly digest changes in supply and demand. In Hens, T. & Schenk-Hoppe, K. R. (Eds.), *Handbook of financial markets: Dynamics and evolution*. North-Holland, Elsevier.

Self-Referential, Endogenous Models

Challet, D., Marsili, M., & Zhang, Y. C. (2013). *Minority games: Interacting agents in financial markets*. OUP Catalogue.

Sornette, D., Malevergne, Y., & Muzy, J. F. (2004). Volatility fingerprints of large shocks: Endogenous versus exogenous. In Takayasu, H. (Ed.), *The application of econophysics* (pp. 91–102). Springer Japan.

Sornette, D. (2006). Endogenous versus exogenous origins of crises. In Albeverio, S., Jentsch, V., & Kantz, H. (Eds.), *Extreme events in nature and society* (pp. 95–119). Springer, Berlin-Heidelberg.

Wyart, M., & Bouchaud, J. P. (2007). Self-referential behaviour, overreaction and conventions in financial markets. *Journal of Economic Behavior & Organisation*, 63(1), 1–24.

Bouchaud, J. P. (2011). The endogenous dynamics of markets: Price impact, feedback loops and instabilities. In Berd, A. M. (Ed.), *Lessons from the Credit Crisis*. Risk Publications.

Harras, G., Tessone, C. J., & Sornette, D. (2012). Noise-induced volatility of collective dynamics. *Physical Review E*, 85(1), 011150.

Thurner, S., Farmer, J. D., & Geanakoplos, J. (2012). Leverage causes fat tails and clustered volatility. *Quantitative Finance*, 12(5), 695–707.

Bouchaud, J. P. (2013). Crises and collective socio-economic phenomena: Simple models and challenges. *Journal of Statistical Physics*, 151(3–4), 567–606.

Galla, T., & Farmer, J. D. (2013). Complex dynamics in learning complicated games. *Proceedings of the National Academy of Sciences*, 110(4), 1232–1236.

Caccioli, F., Shrestha, M., Moore, C., & Farmer, J. D. (2014). Stability analysis of financial contagion due to overlapping portfolios. *Journal of Banking & Finance*, 46, 233–245.

Self-Reflexive Models and Calibration

Bollerslev, T., Engle, R. F., & Nelson, D. B. (1994). ARCH models. In Engle, R. & McFadden, D. (Eds.), *Handbook of econometrics* (Vol. 4, pp. 2959–3038). North-Holland.

Filimonov, V., & Sornette, D. (2012). Quantifying reflexivity in financial markets: Toward a prediction of flash crashes. *Physical Review E*, 85(5), 056108.

Hardiman, S., Bercot, N., & Bouchaud, J. P. (2013). Critical reflexivity in financial markets: A Hawkes process analysis. *The European Physical Journal B*, 86, 442–447.

Chicheportiche, R., & Bouchaud, J. P. (2014). The fine-structure of volatility feedback I: Multi-scale self-reflexivity. *Physica A: Statistical Mechanics and Its Applications*, 410, 174–195.

Bacry, E., Mastromatteo, I., & Muzy, J. F. (2015). Hawkes processes in finance. *Market Microstructure and Liquidity*, 1(01), 1550005.
Blanc, P., Donier, J., & Bouchaud, J. P. (2016). Quadratic Hawkes processes for financial prices. *Quantitative Finance*, 17, 1–18.

Trend Following

De Bondt, W. P. (1993). Betting on trends: Intuitive forecasts of financial risk and return. *International Journal of Forecasting*, 9(3), 355–371.
Jegadeesh, N., & Titman, S. (2011). Momentum. *Annual Review of Financial Economics*, 3(1), 493–509.
Moskowitz, T. J., Ooi, Y. H., & Pedersen, L. H. (2012). Time series momentum. *Journal of Financial Economics*, 104(2), 228–250.
Lempérière, Y., Deremble, C., Seager, P., Potters, M., & Bouchaud, J. P. (2014). Two centuries of trend following. *Journal of Investing Strategies*, 3, 41–61.
Barberis, N., Greenwood, R., Jin, L., & Shleifer, A. (2015). X-CAPM: An extrapolative capital asset pricing model. *Journal of Financial Economics*, 115(1), 1–24.
Geczy, C., & Samonov, M. (2015). 215 years of global multi-asset momentum: 1800–2014 (equities, sectors, currencies, bonds, commodities and stocks). https://papers.ssrn.com/sol3/papers.cfm?abstract_id=2607730.
Covel, M. (2017). *Trend following: How to make a fortune in bull, bear and black swan markets*. Wiley.

Experimental Markets

Smith, V. L., Suchanek, G. L., & Williams, A. W. (1988). Bubbles, crashes, and endogenous expectations in experimental spot asset markets. *Econometrica: Journal of the Econometric Society*, 56, 1119–1151.
Kagel, J. H., & Roth, A. E. (1995). *The handbook of experimental economics* (pp. 111–194). Princeton University Press.
Nagel, R. (1995). Unraveling in guessing games: An experimental study. *The American Economic Review*, 85(5), 1313–1326.
Hommes, C. (2013). *Behavioral rationality and heterogeneous expectations in complex economic systems*. Cambridge University Press.
Batista, J. D. G., Massaro, D., Bouchaud, J. P., Challet, D., & Hommes, C. (2017). Do investors trade too much? A laboratory experiment. *Journal of Economic Behavior and Organization*, 140, 18–34.

PART IX
Practical Consequences

Introduction

We are now approaching the end of a long empirical, theoretical and conceptual journey. In this last part, we will take the important step of accumulating all this knowledge about markets to illuminate some important decisions about how to best act in them. In the following chapters, we will address two practical topics of utmost importance for market participants and regulators: optimal trade execution and market fairness and stability.

For both academics and practitioners, the question of how to trade and execute optimally is central to the study of financial markets. Given an investment decision with a given direction, volume and time horizon, how should a market participant execute it in practice? Addressing this question is extremely difficult, and requires a detailed understanding of market dynamics that ranges from the microscopic scale of order flow in the LOB (e.g. for optimising the positioning of orders within a queue or for choosing between a limit order and a market order) to the mesoscopic scale of slow liquidity (e.g. for scheduling the execution of large investment decisions over several hours, days or months). Many important questions on these topics have been active areas of research in the last two decades.

The final chapter of this book will address the critical question of market fairness and stability. These topics have been at the core of many debates in recent years – especially with the rise of competition between exchanges and high-frequency trading. In this last chapter, we will discuss how some market instabilities can arise from the interplay between liquidity takers and liquidity providers, and can even be created by the very mechanisms upon which markets are built.

In this jungle, regulators bear the important responsibility of designing the rules that define the ecology in financial markets. As we will discuss, because the resulting system is so complex, sometimes these rules can have unintended consequences on both the way that individual traders behave and the resulting price-formation process. Therefore, sometimes the most obvious solutions are not necessarily the best ones!

We conclude with a discussion of an interesting hypothesis: that markets intrinsically contain – and have always contained – some element of instability. From long-range correlations to price impact, this book has evidenced a collection of phenomena that are likely inherent to financial markets, and that we argue market observers, actors and regulators need to understand and accept. Therefore, understanding how to make sense of the many mysterious aspects of the market price may well be the most important question in financial markets today.

21

Optimal Execution

I will remember that I didn't make the world, and it doesn't satisfy my equations.
(Emanuel Derman and Paul Wilmott "The Modeler's Oath")

Optimal execution is a major issue for financial institutions, in which asset managers, hedge funds, derivative desks, and many more seek to minimise the costs of trading. Such costs can consume a substantial fraction of the expected profit of a trading idea, which perhaps explains why most active managers fail to beat passive index investing in the long run. **Market friction** is especially relevant for high-turnover strategies. Given the importance of these considerations, it is unsurprising that a whole new branch of the financial industry has emerged to address the notion of best execution. Nowadays, many brokerage firms propose trading costs analysis (TCA) and optimised execution solutions for their buy-side clients.

Trading costs are usually classified into two groups:

- **Direct trading costs** are fees that must be paid to access a given market. These include brokerage fees, direct-access fees, transaction taxes, regulatory fees and liquidity fees.[1] These fees are all relatively straightforward to understand and to measure. As a general rule, direct trading costs are of the order of 0.1–1 basis points and SEC regulatory fees are 0.01 basis points.
- **Indirect trading costs** arise due to market microstructure effects and due to the dynamics of the supply and demand. In contrast to direct trading costs, indirect trading costs are quite subtle. These costs include the bid–ask spread and **impact costs**. As we discussed in Chapter 16, the bid–ask spread is a consequence of the information asymmetry between different market participants and is determined endogenously by market dynamics. Spread costs are typically a few basis points in liquid markets, but can vary substantially over time, according to market

This chapter is not essential to the main story of the book.

[1] Liquidity fees are typically charged per order and are generally different for market and limit orders. With these so-called "maker-taker fees", exchanges usually aim to incentivise liquidity provision.

conditions. Note that the tick size can have a considerable effect on the value of the spread, especially for large-tick stocks.

Impact costs are also a consequence of the bounded availability of liquidity. These costs also arise endogenously, but are deceptive because they are of a statistical nature. Impact is essentially invisible to the naked eye (after a buy trade, the price actually goes down about half the time, and vice-versa), and only appears clearly after careful averaging (see Chapter 12). As such, impact costs have historically been disregarded by the industry, although they actually represent the lion's share of costs for moderate to large metaorders. The square-root law in Equation (12.6) can be used to estimate the unitary impact cost to be $2/3 \times \sigma_T \sqrt{Q/V_T}$. For example, when trading 1% of the daily traded volume V_T on a stock with a 2% daily volatility σ_T, the square-root law estimates the impact to be about 15 basis points. This is an order of magnitude larger than direct costs and spread costs (which are of the order of 1 basis point). Therefore, modelling impact is critical to devising good execution strategies.

In this chapter, we discuss the so-called **optimal execution** problem: given that a financial institution has taken the decision to buy or sell a quantity Q of an asset, how is this decision implemented in the market such that the desired quantity gets executed at the best possible price? As we have emphasised throughout this book, liquidity rationing is a fundamental property of markets (see in particular Sections 10.5 and 18.1). Large orders cannot be executed instantaneously, so trading must instead occur over an investment horizon T, which can sometimes be quite long. What is the optimal horizon T? What is the corresponding trading schedule? Should one trade at a uniform rate during the time interval $[0,T]$, or should one front- or back-load the execution? Should one use market orders or limit orders? Is it wise to join a long bid- or ask-queue? In the coming sections, we address these important questions and give a short account of the enormous literature on optimal execution.

21.1 The Many Facets of Optimal Execution

The optimal execution problem first requires setting a suitable loss function to minimise. One possible example is the average **execution shortfall**, or slippage, which compares the average execution price and the decision price (see Equation (21.2) below). Other objectives can be considered, such as minimising the tracking error between the execution price and the VWAP or TWAP prices (see Section 10.5).

The optimisation problem can then be subdivided into several approximately independent decisions:

(i) **Execution schedule**: One should determine how much volume to execute during each time interval. This will depend on the liquidity available in the market and the model selected for describing price impact.
(ii) **Market venue**: In modern financial markets, it is common for assets to be traded on several different exchanges simultaneously. As well as their geographical differences, many such exchanges differ by their fee structure and the details of their trading mechanism (e.g. whether trading is in a lit exchange or a dark pool).
(iii) **Type of order**: Several orders can be used to execute a trade, such as a limit order (iceberg or visible) or a market order (immediate or cancel, fill or kill, etc.) – see Section 3.2.2.
(iv) **Timing of order placement**: Both the choice of order type and the timing of order submissions depend on many factors, including the instantaneous state of the LOB and the availability of high-frequency prediction signals.

21.2 The Optimal Scheduling Problem

21.2.1 Linear Impact Models

As we have discussed many times throughout the book, market participants who wish to trade large quantities of an asset must split their intended volume into many child orders, which they then submit incrementally. When traders do so, they must decide how to schedule these child orders. Since each child order impacts the price, the execution price of subsequent child orders depends on the time at which previous child orders were submitted (and possibly on their volume). Traders attempt to minimise their total impact costs by optimally scheduling the arrival time of child orders.

This problem can be formulated precisely in the context of propagator models in discrete time (see Chapter 13). Consider a trader[2] who wants to buy a given quantity Q of an asset over a fixed, specified time horizon T. We assume that the trader chooses to execute the metaorder via a sequence of child market orders with identical size $\upsilon_t = \upsilon_0$, interspersed with periods of inactivity (during which $\upsilon_t = 0$), such that

$$\sum_{t=1}^{T} \upsilon_t = Q. \tag{21.1}$$

For such traders, the primary performance metric is the execution shortfall (or slippage)

$$\Delta = \sum_{t=1}^{T} \upsilon_t (p_t - m_0), \tag{21.2}$$

[2] By symmetry, the same arguments hold for sellers.

which measures the total price paid for executing the metaorder, relative to the initial mid-price m_0, including the bid–ask spread. Using a linear propagator specification for the mid-price dynamics (see Equation (13.7)), the average execution shortfall is given by[3]

$$\mathbb{E}[\Delta] = \sum_{t=1}^{T} \upsilon_t \left(\sum_{1 \leq t' < t} G(t - t')\upsilon_{t'} \right) + \sum_{t=1}^{T} \frac{\upsilon_t s_t}{2}, \tag{21.3}$$

where we have used $p_t = m_t + s_t/2$. We assume that the sign of the metaorder is uncorrelated with the order flow from the rest of the market, and that in the absence of our investor, the price is a martingale, so its evolution does not contribute to the average execution shortfall. These assumptions are far from innocuous, since the information used to make a trading decision could well be shared with a number of other investors, who may trade in the same direction during the same time interval. Note that in treating $\mathbb{E}[\Delta]$ as a cost, we make the implicit assumption that the impact will eventually decay to zero (i.e. $G(t \to \infty) = 0$), so that none of the impact can be captured by the investor when unwinding the position.[4]

Equation (21.3) shows that the expected execution shortfall is partly due to the spread and partly due to the impact of past trades. For simplicity, we assume that the spread is constant, such that $s_t = s$ for all t. If the spread was time-varying, the optimisation problem would be more complicated, since traders would try to take advantage of moments where the spread is small, and the two terms in Equation (21.3) would interact in a non-trivial way. Assuming s to be constant simplifies the problem considerably, because doing so allows us to disregard the last term in Equation (21.3) in the optimisation program.

By permuting the dummy variables t and t', we can express the impact term in a symmetric form:

$$\sum_{t=1}^{T} \upsilon_t \left(\sum_{t'=1}^{t} G(t - t')\upsilon_{t'} \right) = \frac{1}{2} \sum_{\substack{t,t'=1 \\ t \neq t'}}^{T} \upsilon_t G(|t - t'|)\upsilon_{t'}. \tag{21.4}$$

This is a quadratic form in υ_t that must be minimised under the condition in Equation (21.1).

The difficulty in performing this minimisation comes from the discrete nature of the υ_t terms. A simplification occurs if the propagator $G(t)$ varies sufficiently slowly that the sum in Equation (21.4) can be replaced by integrals. Defining the

[3] Strictly speaking, the propagator $G(t)$ considered here differs from that of Equation (13.7) by a factor υ_0, equal to the size of each individual trade.

[4] If impact was fully permanent, the average cost of a round-trip would just be equal to the bid–ask spread.

trading intensity $j(t)$ such that $v_t = j(t)\mathrm{d}t$, Equation (21.4) becomes[5]

$$\frac{1}{2}\sum_{t,t'=1;(t\neq t')}^{T} v_t G(|t-t'|) v_{t'} \longrightarrow \frac{1}{2}\iint_0^T \mathrm{d}t\mathrm{d}t'\, j(t) G(|t-t'|) j(t'), \tag{21.5}$$

which must be minimised under the constraint $\int_0^T \mathrm{d}t\, j(t) = Q$. We are now in the realm of standard **variational calculus**. Introducing a Lagrange multiplier ζ that enforces the constraint Equation (21.1), and setting the derivative with respect to $j(t)$ to zero, it follows that $j(t)$ must satisfy the following condition for all t:

$$\int_0^T \mathrm{d}t'\, G(|t-t'|)\, j(t') = \zeta. \tag{21.6}$$

Equation (21.6) is difficult to solve exactly. However, in the particular case of an exponentially decaying propagator $G(t) = G_0 e^{-\omega t}$, one can check that the following (so-called "bucket") solution solves the above equation:[6]

$$j^*(t) = \frac{Q}{1+\omega T/2}\left[\delta(t) + \delta(T-t) + \frac{\omega}{2}\right]. \tag{21.7}$$

Under this optimal schedule, a fraction $1/(2+\omega T)$ of the total volume should be executed at the open and at the close, while the rest should be executed at a constant speed throughout the interval $(0, T)$. Interestingly, in the limit where $\omega T \gg 1$ (i.e. for a quasi-instantaneous impact decay), one finds that a TWAP (constant trading speed) execution

$$j^*(t) \to \frac{Q}{T}; \qquad (\omega T \to \infty) \tag{21.8}$$

is optimal. The minimal cost associated with this trading schedule is given by

$$\mathbb{E}[\Delta^*] = G_0\left(\frac{Q}{2+\omega T}\right)^2 (1+\omega T) + \frac{Qs}{2}. \tag{21.9}$$

The last term comes from the spread and is independent of the trading schedule, whereas the first term (which represents the impact cost) is a decreasing function of the horizon T. As expected within a quadratic cost model, the slower the execution, the smaller the impact costs. Observe that in practice the execution cannot be infinitely slow; otherwise, the expected gain that motivates trading would be totally erased. Therefore, the execution horizon T must remain smaller than the prediction horizon. We will discuss this issue in more detail in Section 21.2.3.

Another interesting case is the power-law propagator $G(t) \propto (\omega t)^{-\beta}$, for which the optimal solution has a U-shaped profile that diverges as $t^{(\beta-1)/2}$ (or as

[5] Note that $j(t)$ should be interpreted as the probability that a given market order belongs to the metaorder times the total rate of market orders per unit time.

[6] Note that the δ-functions in Equation (21.7) are such that $\int_0^T \mathrm{d}t\,\delta(t) = 1/2$, since both are localised right at the edges of the interval $[0, T]$.

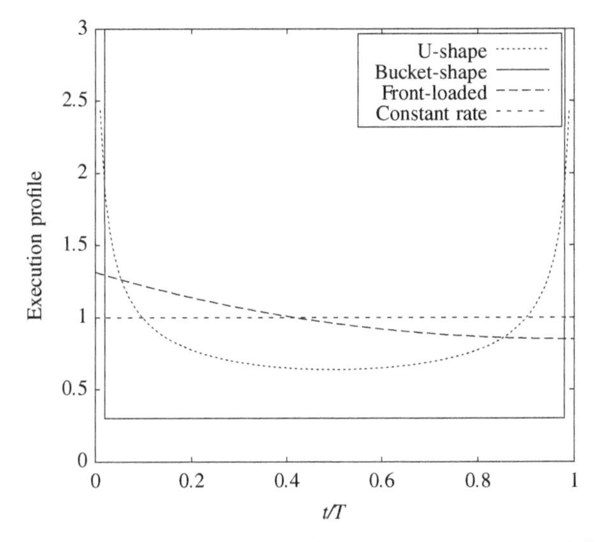

Figure 21.1. Optimal schedules for (solid curve) an exponential propagator, which leads to a bucket-shaped execution profile, (dotted curve) a power-law shaped propagator, which leads to a U-shaped execution profile, and (dashed curve) an Almgren–Chriss schedule penalising the variance of the shortfall, which leads to a front-loaded execution profile. The black dashed line shows a TWAP schedule.

$(T - t)^{(\beta-1)/2}$) close to the edges (see Figure 21.1). The solution is given by the function[7]

$$j^*(t) = A(\beta)QT^{-\beta}\sqrt{t^{\beta-1}(T-t)^{\beta-1}},$$

where $A(\beta) = \beta\Gamma[\beta]/\Gamma^2[(\beta+1)/2]$, such that Equation (21.1) is satisfied.

More generally, without any specific assumptions on the shape of $G(t)$, one can prove that the optimal solution must be symmetric about $T/2$:

$$j^*(t) = j^*(T-t), \text{ for all } t. \tag{21.10}$$

To see why this is the case, substitute t with $T - t$ in Equation (21.6) and then change variables from t' to $T - t'$. Performing these transformations illustrates that $j^*(t)$ and $j^*(T-t)$ obey the same equation, so j^* is symmetric about $T/2$.

In reality, $G(t)$ cannot be completely arbitrary for the solution to be reasonable, in the sense that for all t, $j(t) \geq 0$ for buy metaorders and $j(t) \leq 0$ for sell metaorders. This is a desirable property, because we would not want the solution to oscillate between buy and sell orders in an attempt to arbitrage one's own impact. This would be the counterpart, in the context of execution, of a price manipulation strategy (see Section 19.5). A sufficient condition for this not to occur is that the

[7] See: Gatheral, J., Schied, A., & Slynko, A. (2012). Transient linear price impact and Fredholm integral equations. *Mathematical Finance*, 22(3), 445–474.

propagator is absolutely monotonic:

$$(-1)^n \frac{d^n G(t)}{dt^n} \geq 0, \text{ for all } t \geq 0, \text{ for all } n \geq 1.$$

The exponential function $G(t) = G_0 e^{-\omega t}$ and the power-law function $G(t) \propto (\omega t)^{-\beta}$ are both examples of absolutely monotonic functions.

21.2.2 The Almgren–Chriss Problem

So far, we have assumed that the investor's only concern is the *expected* execution shortfall. This assumes that the investor seeks to execute a large number of metaorders, so that shortfall fluctuations average out in his/her P&L curve. However, the price in a given instance can wander during the execution period, and can accidentally lead to a very high implementation shortfall (or, in good cases, to a very negative implementation shortfall). For traders who submit metaorders infrequently, this is uncomfortable.

To deal with these fluctuations, the natural idea (first considered in a seminal paper by Almgren and Chriss) is to add a penalty term proportional to the *variance* of the shortfall. Neglecting the impact contribution to the variance, which is indeed small in practice, the **shortfall variance** can be computed from Equation (21.2) in the continuum limit and after integration by parts:

$$\mathbb{V}[\Delta] = \mathbb{V}\left[\int_0^T \mathrm{d}t\, j(t)\,(p(t) - p_0)\right] = \mathbb{V}\left[\int_0^T \mathrm{d}p(t)\,(Q - q(t))\right],$$

where $q(t)$ is the executed volume up to time t. Assuming that the price is a Brownian motion with volatility σ and that the execution schedule is independent of the price path leads to

$$\mathbb{V}[\Delta] = \sigma^2 \int_0^T \mathrm{d}t\,(Q - q(t))^2.$$

The new objective function then reads:

$$\mathbb{E}[\Delta] \longrightarrow \mathbb{E}[\Delta] + \Gamma\sigma^2 \int_0^T \mathrm{d}t\,(Q - q(t))^2,$$

where Γ is a measure of **risk aversion** and the last term penalises any quantity that remains unexecuted. For a propagator that decays very quickly ($\omega T \gg 1$), one can again solve this optimisation problem, yielding the celebrated Almgren–Chriss optimal schedule:[8]

$$q^*(t) := \int_0^t \mathrm{d}t'\, j^*(t') = Q\left[1 - \frac{\sinh(\Omega(T-t))}{\sinh \Omega T}\right], \qquad \Omega^2 = \frac{\Gamma\sigma^2}{\bar{G}_0}.$$

[8] We have rescaled $G_0 \to \bar{G}_0 = G_0/\omega$ such that the role of impact remains non-trivial in the limit $\omega \to \infty$.

The initial trading speed is given by

$$j^*(0) = Q\frac{\Omega}{\tanh \Omega T}. \tag{21.11}$$

If $\Omega \to 0$, then $j^*(0) \to Q/T$. In this case, the shortfall risk is irrelevant and one recovers a TWAP profile. For very large risk aversion $\Gamma \to \infty$, by contrast, $j^*(0) = Q\Omega$, which tends to infinity. In this case, trading is very intense early on, so that execution is quickly completed and the trader is immune to the fluctuations of the underlying asset. In other words, risk aversion leads to a **front-loaded execution profile**.

21.2.3 Trading with a Signal

Until now, we have assumed that in the absence of the trader, the price is a martingale. In this framework, we saw that impact costs are minimised when the execution horizon is infinite. This conclusion is of course modified if we assume that our investor buys because he/she anticipates a price increase over some time horizon T_α. In this case, a delayed execution is tantamount to losing part of the profitability of the trade. Intuitively, the optimal execution schedule should be such that the horizon T is somewhat shorter than T_α. What is the optimal trading profile in this case?

We will assume that the predicted price trajectory is given by

$$\widehat{p}(t) = p_0 + \alpha(1 - e^{-t/T_\alpha}).$$

The average execution shortfall $\mathbb{E}[\Delta]$ for the trader (including impact costs and opportunity costs – see Section 21.3) can be written as

$$\mathbb{E}[\Delta] = \underbrace{-\int_0^T \mathrm{d}t \alpha e^{-t/T_\alpha} j(t)}_{\text{opportunity costs}} + \underbrace{\frac{1}{2}\iint_0^T \mathrm{d}t\mathrm{d}t'\, j(t)G(|t-t'|)j(t')}_{\text{impact costs}} + \underbrace{\frac{Qs}{2}}_{\text{spread costs}},$$

The optimisation problem can be fully solved in the limit of instantaneous impact $\omega \to \infty$, where $G(\cdot)$ becomes a δ-function. Introducing again a Lagrange multiplier to impose that the executed quantity is Q, the solution reads

$$\bar{G}_0 j^*(t) = \zeta + \alpha e^{-t/T_\alpha}.$$

To keep the algebra simple, let us first assume that $T \ll T_\alpha$ (i.e. the trading horizon is much smaller than the prediction horizon), and check at the end whether the assumption is self-consistent. In this case, the optimisation condition becomes:

$$\bar{G}_0 j^*(t) = \zeta + \alpha(1 - t/T_\alpha),$$

which, using the constraint that $\int_0^T \mathrm{d}t\, j^*(t) = Q$, finally leads to

$$j^*(t) = \frac{Q}{T} + \frac{\alpha}{\bar{G}_0}\left(\frac{T-2t}{2T_\alpha}\right). \tag{21.12}$$

In this solution, trading is initially fast, to benefit as much as possible from the trading signal. The total cost corresponding to the optimal schedule $j^*(t)$ is readily computed to be

$$\mathscr{C}(Q,T) = \frac{\alpha QT}{2T_\alpha} + \frac{\bar{G}_0 Q^2}{T} + \frac{\alpha^2 T^3}{6T_\alpha^2 \bar{G}_0} + \frac{Qs}{2}. \tag{21.13}$$

For a given Q, maximising this function with respect to T leads to

$$\frac{\alpha Q}{2T_\alpha} - \frac{\bar{G}_0 Q^2}{T^2} + \frac{S^2 T^2}{2T_\alpha^2 \bar{G}_0} = 0,$$

which in turn leads to a second-degree equation for $X = T^2$:

$$\alpha^2 X^2 + \alpha \bar{G}_0 T_\alpha Q X - 2T_\alpha^2 \bar{G}_0^2 Q^2 = 0 \Rightarrow X^* = \frac{T_\alpha \bar{G}_0 Q}{\alpha}.$$

Therefore, provided $T^* \ll T_\alpha$,

$$T^* = \sqrt{\frac{T_\alpha \bar{G}_0 Q}{\alpha}}. \tag{21.14}$$

Interestingly, the optimal execution horizon T^* grows as the square-root of the target volume Q. During this horizon, the price has increased on average by an amount $\alpha T^*/T_\alpha$, which is proportional to $\sqrt{Q}$ because of the realisation of the trader's signal. This is essentially the argument put forth by Gabaix et al. to explain the square-root impact law:[9] in their picture, the square-root law is not an intrinsic property of the latent liquidity, as we argued in Chapters 18 and 19, but rather emerges from the fact that trades are initiated by informed investors who use knowledge of how much the price is going to move in the future to optimise their trading schedule. We are again back to our discussion on the origin of impact from Section 11.1! As discussed in Section 12.5, this scenario would not explain why the square-root impact law also holds with Q replaced by $q(t)$ for all intermediate times $t \in [0,T]$.

Plugging Equation (21.14) into Equation (21.13) yields the optimal cost corresponding to $T = T^*$:

$$\mathscr{C}(Q,T^*) = \frac{5}{3}\sqrt{\frac{\bar{G}_0 \alpha}{T_\alpha}}\, Q^{3/2}.$$

[9] Gabaix, X., Gopikrishnan, P., Plerou, V., & Stanley, H. E. (2003). A theory of power-law distributions in financial market fluctuations. *Nature*, 423, 267–270.

Throughout this analysis, we have assumed Q to be fixed at the outset of trading. The optimal metaorder size Q^* that results from a global cost-benefit optimisation is obtained as

$$\text{P\&L} \approx (\widehat{p}_\infty - p_0)Q - \mathscr{C}(Q, T^*) \Rightarrow Q^* = \frac{4\alpha T_\alpha}{\bar{G}_0}.$$

Hence, the optimal quantity depends linearly on the information α, as in the Kyle model (see Chapter 15). Plugging this Q^* back into the result for T^* finally leads to

$$T^* = \frac{2}{5} T_\alpha,$$

which shows that our calculation is only approximately correct in that case (since it assumes that $T^* \ll T_\alpha$) but the main message is captured correctly.

21.2.4 Optimal Scheduling with a Non-linear Propagator

Until now, we have used the linear propagator model to determine the optimal execution schedule. This model is adequate when the trading rate is small. For larger trading rates, however, non-linear effects play a role. In Chapter 19, we discussed how the linear propagator framework can be extended to describe the dynamics of the underlying (latent) liquidity. In the presence of an additional metaorder with trading intensity $j(t)$, the self-consistent equation of motion of the price is given by Equation (19.7), which boils down to the propagator model with $\beta = 1/2$ when $j(t)$ is small.

In the opposite limit, where large $j(t)$ is large, one can derive the following differential equation for the price:[10]

$$\begin{aligned} x(t)|\dot{x}(t)| &\approx \frac{j(t)}{\mathcal{L}} + \frac{D}{\mathcal{L}}\left[\frac{3\ddot{x}(t)j(t)}{\dot{y}^3(t)} - \frac{2\dot{j}(t)}{\dot{y}^2(t)}\right], \\ &\approx \frac{j(t)}{\mathcal{L}} + D\left[\frac{2\dot{j}(t)Q(t)}{j^2(t)} - 3\right], \end{aligned}$$

where $x(t) = p(t) - p_F$ (see Chapter 18). When $j(t)$ has a constant sign, this equation can be integrated, and leads to the following final result for the execution shortfall:

$$\mathbb{E}[\Delta] = \mathbb{E}\left[\int_0^t dt'\, j(t')x(t')\right] = \frac{1}{3}\sqrt{2\mathcal{L}Q^3}\left[1 - \frac{3D\mathcal{L}T}{2Q}\right] + O(T^2).$$

The remarkable result here is that $j(t)$ has disappeared: the execution shortfall is, in the large-intensity limit, completely independent of the execution schedule! In this limit, the optimal trading problem is dominated by sub-leading effects such as the finite trading speed and finite deposition and cancellation rates. This is a somewhat counter-intuitive result of the non-linear propagator framework, Equation (19.7), which suggests that the dependence of impact costs on the execution schedule could be weaker than naively anticipated.

[10] For details, see Donier, J., Bonart, J., Mastromatteo, I., & Bouchaud, J. P. (2015). A fully consistent, minimal model for non-linear market impact. *Quantitative Finance*, 15(7), 1109–1121.

21.3 Market Orders or Limit Orders?

21.3.1 Adverse Selection and Opportunity Costs

It is one thing to decide on a theoretical execution schedule, but it is quite another to actually send orders to the market. When doing so, another major conundrum appears: whether to use market orders or limit orders.

Sending market orders ensures immediate execution, but incurs a cost equal to half the instantaneous bid–ask spread. It also signals to the rest of the market that there is an impatient buyer or seller out there. This leakage of information will most likely degrade the condition of future orders.

Sending limit orders avoids paying half the spread, but suffers from two predicaments.

The first predicament is what we will call (slightly abusively) **opportunity cost**. Consider a buyer sending a limit order at the best bid. If the buyer's limit order is executed very soon after its arrival, then he/she has saved paying half the spread. However, if the price starts drifting upwards before the buyer's limit order is executed, then he/she has lost the opportunity to buy the asset at the old price. If the buyer still intends to buy at the end of the day (say), he/she will have to pay a higher price. There is therefore an *opportunity cost* associated with not having traded.

The second predicament is **adverse selection** (see Chapter 16). Consider again the same buyer, who sends a limit order at the best bid. Imagine now that the buy limit order is matched by an informed sell market order that correctly anticipates a short-term drop in price. Had the buyer also been informed, he/she would have cancelled the limit order and resubmitted it later (maybe even as a market order) to benefit from the lower price.

The arguments on the profitability of market-making strategies, as we put forth in Chapters 16 and 17, already suggest that market orders and limit orders are roughly equivalent from the point of view of execution. This might not be true under certain market conditions or for certain configurations of the LOB, but must be true on average. Otherwise, if (say) market orders were more favourable than limit orders, their arrival rate would increase and the arrival rate of limit orders would decrease, opening the spread s_t and making market orders less favourable (and vice-versa for limit orders).

21.3.2 A Martingale Argument

It is interesting to detail the argument for why limit orders do not allow traders to buy cheap, as one might naively think. While Chapters 16 and 17 have detailed the argument from a market-maker's point of view, we will now adopt the investor's point of view, and therefore include the possibility of submitting limit orders at any price, and also impose the condition that the order *must* be executed, one way or another, before the end of the time horizon T.[11]

Suppose that a trader has no particular information about the future price path between now ($t = 0$) and the time by which the trade must be executed ($t = T$). Here, we assume that T is a short time scale, say a day or less, such that any long-term drift or interest rate can be neglected. Mathematically, given the information at hand $\mathcal{F}_0$, the expectation of the price change $p_T - p_0$ is zero, i.e. the price is a **martingale**

$$\mathbb{E}[p_T - p_0|\mathcal{F}_0] = 0. \tag{21.15}$$

Assume now that the trader decides to submit a buy limit order at a distance d *below* the initial mid-price. For now, assume also that the volume of the limit order is sufficiently small that its impact on the price trajectory is negligible (we address the case with larger order sizes in Section 21.3.4). One of the following two situations must occur:

- At some intermediate time $t \in (0, T)$ the traded price p_t is such that $p_t \leq p_0 - d$. When this happens, we deem the order to have been executed, irrespective of its place in the order queue (this assumption clearly fails for large-tick assets; see Section 21.4).
- The traded price never satisfies $p_t \leq p_0 - d$ for any $t \in (0, T)$, so the limit order is still unexecuted at time T. In this case, we assume that the trader must then use a market order that is executed at price p_T.

If the price process p_t is a martingale, then the expectation of p_T conditional on execution at price $p_0 - d$ is given by

$$\mathbb{E}[p_T|\text{exec.}] = \mathbb{E}[p_T|p(\text{exec.}) = p_0 - d] = p_0 - d. \tag{21.16}$$

This execution occurs with probability $\phi_{\text{exec.}}$. With probability $1 - \phi_{\text{exec.}}$, the order is not executed, and the mid-price terminates on average at $\mathbb{E}[p_T|\overline{\text{exec.}}]$. Decomposing Equation (21.15) into the two scenarios yields:

$$\begin{aligned} &(1 - \phi_{\text{exec.}})\mathbb{E}[p_T|\overline{\text{exec.}}] + \phi_{\text{exec.}}\mathbb{E}[p_T|\text{exec.}] = p_0, \\ &\quad \Rightarrow (1 - \phi_{\text{exec.}})\mathbb{E}[p_T - p_0|\overline{\text{exec.}}] = \phi_{\text{exec.}} d, \end{aligned} \tag{21.17}$$

where we have used Equation (21.16).

The average execution shortfall (or slippage cost) $\mathbb{E}[\Delta](d)$ of a buy limit order placed at distance d below the initial mid-price is given by

$$\mathbb{E}[\Delta](d) = -\phi_{\text{exec.}} d + \underbrace{(1 - \phi_{\text{exec.}})\mathbb{E}[p_T - p_0|\overline{\text{exec.}}]}_{\text{opportunity cost}},$$

[11] As will become apparent, this is actually just another way of looking at the same problem as we did for market-makers.

where the second term is the cost of not having traded until T. Using Equation (21.17), one then concludes that:[12]

$$\mathbb{E}[\Delta](d) = -\phi_{\text{exec.}}d + \phi_{\text{exec.}}d = 0.$$

Importantly, due to the martingale property of the price, the potential gain due to an execution at a favourable price is exactly compensated by the adverse price move in the case where the order was not executed, independently of the distance d where the limit order is placed. The attractiveness of limit orders, which may appear to allow one to execute at a cheaper price, is therefore an optical illusion – unless one has some information about future price moves.

21.3.3 Trends and Mean-Reversion

This discussion can also be extended to the case where the mid-price is trending or mean-reverting. Assume that the mid-price started at p_0 and has reached the level $p_0 - d$. If the price is trending, then it is likely to continue going down; if the price is mean-reverting, then it is likely to go back up. Therefore:

$$\mathbb{E}\left[p_T|\text{exec.}\right] = p_0 - d - d \times \rho,$$

where ρ is a measure of the returns autocorrelation, with $\rho > 0$ for trending returns and $\rho < 0$ for mean-reverting returns. Extending the calculation in the previous section, the average cost of a limit order is given by

$$\mathbb{E}[\Delta](d,\rho) = \phi_{\text{exec.}}d\rho. \tag{21.18}$$

Equation (21.18) reveals an interesting rule-of-thumb: when considering the profitability of limit orders, trending is detrimental and mean-reversion is favourable. We have already reached this conclusion in the context of market-making (see Section 17.1). This makes a lot of sense: market-making should be easy when the price mean reverts, but is difficult when trends are present. From the point of view of a market-maker, trending implies adverse selection.

21.3.4 Larger Order Sizes

So far in this section, we have considered only limit orders whose size is so small that impact effects can be neglected. In reality, however, the presence of an extra buy limit order at distance d from the initial price will change the trajectories of

[12] This is exactly the content of the so-called optional stopping theorem for martingales: if we define the stopping time $t^* = \min(\tau, T)$, where $t^* = \tau < T$ corresponds to the limit order being executed before T and $t^* = T < \tau$ corresponds to a market order execution at time T, then the theorem stipulates that $\mathbb{E}[p^*_t] = p_0$, meaning that the expected execution price is exactly p_0. In the absence of a signal on the price, this is in fact a property shared by all (bounded) investment strategies.

the future price and, all else being equal, shift the expected price move to positive territories. Hence, we generalise Equation (21.15) as

$$\mathbb{E}[p_T - p_0|\mathcal{F}_0] := \mathcal{I}_{\text{LO}}(d), \tag{21.19}$$

where $\mathcal{I}_{\text{LO}}(d) > 0$ is (by definition) the impact at time T of the limit order.

For simplicity, we assume that once executed, the price process is a martingale, such that Equation (21.16) holds. The whole argument of Section 21.3.2 can be repeated, finally leading to

$$\mathbb{E}[\Delta](d) = \mathcal{I}_{\text{LO}}(d).$$

The average cost of a limit order, including opportunity costs, is now positive and equal to its own impact. Therefore, we again reach the conclusion that limit orders can only be profitable when some predictive signal on future price moves is available. In reality, even non-executed limit orders do indeed impact prices (and, by the same token, change their probability of execution).

21.4 Should I Stay or Should I Go?

So far in this chapter, we have completely disregarded the effects of price discretisation and queue priority. This is clearly inadequate for large-tick assets, for which the limit order queues at the best quotes are typically very long. For such assets, one cannot assume that a given limit order is always executed as soon as the (transaction) price p_t coincides with its price. For the reason explained in Equation (21.18), limit orders with high priority should benefit from short-term mean-reversion (or "bid–ask bounce"), while limit orders with low priority will suffer from the adverse selection of sweeping market orders MO^1 (see Section 10.2). One therefore expects that a limit order's **queue position** determines whether it is likely to be profitable or should be cancelled.

In this section, we introduce and study a simplified framework to discuss this question. We consider the problem from the perspective of a trader who seeks to maximise the expected gain $\mathcal{G}(h, V)$ earned by a limit order with unit size $\upsilon_0 = 1$ and with position h in a limit order queue of size V (h is the "height" in the queue). We assume that the trade is for a large-tick asset, such that the spread is almost always equal to $\vartheta = 1$ tick. We assume that the trader has an outstanding limit order at the bid, and that he/she leaves this limit order in the queue until either it is matched by an incoming market order or the ask-queue depletes to 0, in which case we assume that the trader cancels the limit order and immediately resubmits a new limit order in the incipient bid-queue one tick higher. We assume that the trader will not cancel the limit order at any time before one of these two events happens, such that there is no possible optimisation over more complex cancel-and-replace strategies.

To keep the modelling framework as simple as possible, we will make the following assumptions:

- All limit orders have a fixed volume, which we normalise to $\upsilon_0 = 1$, and arrive with a rate $\lambda(V)$ that depends on the total volume of limit orders in the queue.
- Each existing limit order is cancelled at a rate $\nu(h, V)$, where $h \in [0, V)$ is the total volume of limit orders *ahead* of the given order. In this way, the cancellation rate of a given order depends on both the total queue length and the order's position in the queue.
- Market orders with size υ arrive with rate $\mu(\upsilon|V)$, which is conditional on the queue size V.
- All event arrival rates are independent of the opposite queue volume.

Consider observing this system at some time step t. The set of possible events that could occur at the next time step are:[13]

(i) A new limit order arrives in the same queue as the trader's limit order. This new arrival will increase the size of the queue, but will not change the value of h for the trader's limit order.
(ii) A cancellation occurs ahead of the trader's limit order, thereby improving its position by 1 and decreasing the queue size by 1.
(iii) A cancellation occurs behind the trader's limit order, leaving its position unchanged but decreasing the queue size by 1.
(iv) A market order with volume $\upsilon < h$ arrives. This improves the queue position of the trader's order and decreases the total size of the queue.
(v) A market order with volume $\upsilon \geq h$ arrives. This executes the limit order. In this case, the gain or loss experienced by the limit order owner comes from the balance between adverse selection and the value of $s/2 + \varpi$, where ϖ denotes any fee associated with the matching and s is the bid–ask spread, which we assume in the present case to be equal to 1 tick (because we are considering a large-tick asset).
(vi) The queue at the opposite-size best quote depletes to zero, leading to an adverse price move of one tick. In this case, the trader experiences a loss (due to the opportunity cost) equal to $\mathscr{C}$, which is on the order of a tick.

The **expected ex-post gain** $\mathcal{G}(h, V)$ of the trader is given by the weighted sum of all these possibilities, as a recursion relation:

[13] We do not include events such as a new limit order arriving at the opposite-side best queue, which are irrelevant for our problem in the present formulation.

$$\mathcal{G}(h,V) = \lambda(V)\mathcal{G}(h,V+1) + \sum_{y=1}^{h-1} \nu(y,V)\mathcal{G}(h-1,V-1)$$

$$+ \sum_{y=h+1}^{V} \nu(y,V)\mathcal{G}(h,V-1) + \sum_{q=1}^{h-1} \mu(\upsilon|V)\mathcal{G}(h-\upsilon,V-\upsilon)$$

$$+ \sum_{\upsilon=h}^{\infty} \mu(\upsilon|V)\left[\frac{\vartheta}{2} + \varpi - \mathcal{R}_\infty(\upsilon,V)\right] - \gamma\mathscr{C}, \tag{21.20}$$

where γ denotes the probability that the opposite-side queue depletes[14] and $\mathcal{R}_\infty(\upsilon,V)$ is the permanent impact of a market order of size υ impinging on a queue of size V. Note that these events exhaust all possibilities, such that the rates obey the following normalisation condition:

$$\lambda(V) + \sum_{y \neq h} \nu(y,V) + \sum_{\upsilon=1}^{\infty} \mu(\upsilon|V) + \gamma = 1. \tag{21.21}$$

We skip the term $y = h$ in the first summation because we have assumed that the trader will not cancel his/her order.

Studying this model in full generality is difficult. To make analytical progress, we therefore introduce the following additional assumptions:

- The rate of limit order arrivals is constant and independent of the queue size V, such that $\lambda(V) = \lambda$.
- Cancellations occur only at the tail of the queue, with a rate ν that is independent of the queue size V, such that $\nu(h,V) = \nu\delta_{h,V}$.
- Market orders arrive at a constant rate μ, with a size υ that is either equal to $\upsilon_0 = 1$ (with probability $1-\phi$) or $\upsilon = V$ (with probability ϕ), such that

$$\mu(\upsilon|V) = \mu\left[(1-\phi)\delta_{\upsilon,1} + \phi\delta_{\upsilon,V}\right].$$

  In other words, there is a fixed probability ϕ for a price-changing market order MO^1.
- The impact of a market order is independent of V and depends only on whether or not it changes the price, such that

$$\mathcal{R}_\infty(1,V) = \mathcal{R}_\infty^0; \qquad \mathcal{R}_\infty(V,V) = \mathcal{R}_\infty^1.$$

Under these simplifications, the normalisation condition becomes

$$\lambda + \nu + \mu + \gamma = 1.$$

[14] In reality, this probability depends on both queue volumes, but we discard these complications here.

Assuming that $v_0 = 1 \ll h, V$ means that relative changes in queue lengths are small and that we can expand Equation (21.20) in much the same way as we did to derive the Fokker–Planck equation in Section 5.3.4. To first-order, the continuum limit reads:

$$(\mu\phi + \gamma)G(h,V) = (\lambda - \nu)\,\partial_V G(h,V) - \mu(1-\phi)\left[\partial_V G(h,V) + \partial_h G(h,V)\right] \\ + \mu\phi\left[\frac{\vartheta}{2} + \varpi - \mathcal{R}^1_\infty\right] - \gamma\mathscr{C}. \tag{21.22}$$

The boundary condition at $h \downarrow 0$ expresses the expected gain of a limit order with highest priority:

$$G(0,V) = G_0 := (1-\gamma)\left[\frac{\vartheta}{2} + \varpi - (1-\phi)\mathcal{R}^0_\infty - \phi\mathcal{R}^1_\infty\right] - \gamma\mathscr{C}.$$

For realistic empirical values, G_0 is generally positive. The first-order PDE in Equation (21.22) can be solved by the **method of characteristics.**[15] In the present case, the characteristics are simply straight trajectories in the plane (h, V), and the general method boils down to setting $G(h,V) := \mathscr{G}(h,u)$ with:

$$u := V - \varkappa h; \qquad \varkappa := 1 - \frac{\lambda - \nu}{\mu(1-\phi)}. \tag{21.23}$$

For a given, fixed value of u, Equation (21.22) becomes an ordinary differential equation for $\mathscr{G}$:

$$\mathscr{G}(h,u) + h^*\frac{d\mathscr{G}(h,u)}{dh} = \chi\left[\frac{\vartheta}{2} + \varpi - \mathcal{R}^1_\infty\right] - (1-\chi)\mathscr{C},$$

where we have introduced $\chi := \mu\phi/(\mu\phi + \gamma)$ and $h^* := \chi(1-\phi)/\phi$.

The solution for $\mathscr{G}(h,u)$ then reads:

$$\mathscr{G}(h,u) = \mathscr{G}(0,u)e^{-h/h^*} + \left[\chi\left(\frac{\vartheta}{2} + \varpi - \mathcal{R}^1_\infty\right) - (1-\chi)\mathscr{C}\right](1 - e^{-h/h^*}).$$

Using the boundary condition $\mathscr{G}(0,u) = G_0$, we finally obtain the full expected-gain profile:

$$G(h,V) = G_0 e^{-h/h^*} + \left[\chi\left(\frac{\vartheta}{2} + \varpi - \mathcal{R}^1_\infty\right) - (1-\chi)\mathscr{C}\right](1 - e^{-h/h^*}).$$

Despite our strong assumptions, which cause G to no longer depend on V, this result is interesting because the first term on the right-hand side is generally

[15] See, e.g., Courant, R., & Hilbert, D. (1962). *Methods of mathematical physics* (Vol. 2: *Partial differential equations*). Interscience.

positive (the impact of small market orders $\mathcal{R}^0_\infty$ is smaller than half a tick) while the second term may be negative (the impact of large market orders $\mathcal{R}^1_\infty$ is larger than half a tick). Hence there is a critical queue position h_c below which limit orders are on average profitable and beyond which they are costly. One finds:

$$\mathcal{G}(h_c, V) = 0 \Rightarrow h_c \approx h^* \ln\left[1 + \frac{\mathcal{G}_0}{(1-\chi)\mathscr{C} - \chi\left(\frac{\vartheta}{2} + \varpi - \mathcal{R}^1_\infty\right)}\right]. \tag{21.24}$$

Empirical data suggests that $\mathcal{R}^1_\infty \cong 0.8\vartheta$ and $\mathcal{R}^0_\infty \cong 0.5\vartheta$ (see the top right panel of Figure 14.2). Table 6.1 gives $\chi \approx 0.5$, $\phi \approx 0.25$ and $\gamma < 0.01$. On NASDAQ, the rebate fee for executed limit orders is $\varpi \approx 0.25\vartheta$.[16] Injecting these numbers into Equation (21.24) leads to $\mathcal{G}_0 \approx 0.175\vartheta$, close to the number obtained in Section 17.3 for the gain of top-priority limit orders.

Since the opportunity cost $\mathscr{C}$ can be estimated as one tick minus the gain of top-priority limit orders (since we have assumed that a limit order is immediately placed one tick above the previous bid if the ask moves up), one finally obtains $h_c \approx 0.5\upsilon_0$. In other words, limit orders must have a very high priority to be profitable, since queues on large-tick stocks are typically much longer than υ_0. Forgetting the rebate fee, Equation (21.24) would lead to $h_c < 0$, so even top-priority limit orders would be costly (see Figure 17.1). Of course, if the rebate fee was really 0, impact and order flow would adapt in such a way that top-priority limit orders would remain (barely) profitable.

The framework we have detailed in this section is admittedly an over-simplification of a real market. For example, both V_a and V_b should be taken into account when designing real optimal-execution strategies. The strategy that we have studied is also sub-optimal because we have not considered the situation in which the given trader can cancel the limit order when signals suggest that the price is about to move. However, our main aim was to understand the value of queue priority for large-tick stocks, and Equation (21.24) provides an important first step in this direction. The results that we have presented illustrate why HFT firms invest huge sums of money to reduce their latency, as doing so helps them both to achieve the best possible queue position after deciding to submit a limit order and to monitor (and react to) a variety of high-frequency indicators that may protect them against adverse selection. In any case, the results of our calculations in this section illustrate that just a few years after the heydays of HFT, it has become very difficult to make substantial profits using simple market-making strategies.

[16] See, e.g., Brogaard, J., Hendershott, T., & Riordan, R. (2014). High-frequency trading and price discovery. *Review of Financial Studies*, 27(8), 2267–2306.

21.5 Conclusion

Making savvy trading decisions is notoriously hard. As we have discussed in this chapter, making good executions is no less difficult. Traders seeking to execute large metaorders must consider a wide variety of important decisions, including how to slice-and-dice their metaorder, which venue or platform to trade on, whether to use limit orders or market orders, and (in the case of large-tick assets) how to assess the value of queue position.

Addressing these issues requires the study of mathematical problems that are abundant in recent literature. These problems are complex, and their solutions often rely heavily on the quality of the modelling assumptions used to formulate them. In some cases, solutions are only available via simulation, where one numerically replays the real order flow with some extra orders that must be executed. However, one should keep in mind that those extra orders were not in the historical flow. If they were, they would have generated some reaction from the market, leading to some impact that is absent from the data, making such back-testing a bit dodgy. The impact of these phantom trades is often comparable in magnitude to the effects one attempts to study, for example the performance of an execution strategy. One therefore needs some adequate modelling, using for example accurate models of queue dynamics (as in Chapter 7) or a multi-event propagator model (as in Chapter 14) where the impact of past events is described explicitly.

As a result of competition, markets are very close to being an even playing field for liquidity providers and liquidity takers. Although they have different risk profiles, market orders and typical limit orders are roughly equivalent in terms of their costs. This ties up with the conclusions of Chapter 17: attempting to earn the spread using limit orders is very difficult in today's highly competitive markets. For large-tick stocks, limit orders are only profitable if they have high priority. Obtaining this high priority is a difficult task, and usually requires, among other things, low-latency technology.

Take-Home Messages

(i) Traders experience direct and indirect trading costs. Direct costs include market access fees, brokerage fees, transaction taxes, and liquidity fees. Indirect costs are costs that are determined endogenously by the state of supply and demand, such as the bid–ask spread and impact costs.

(ii) Because of indirect costs, a good approach to execution can make the difference between a profitable and a losing strategy.

(iii) Investors typically break their execution pipeline into three stages: the macroscopic level (deciding on a time horizon and a volume to execute), the mesoscopic level (deciding on the scheduling of these orders within the defined horizon) and the microscopic level (choosing which order to send to the market, on which venue and whether as a limit order or a market order).
(iv) Optimising at the macro- and meso-scales requires a good understanding of price impact. Propagator models provide a convenient framework for tractable optimal execution design.
(v) Optimising at the micro-scale requires a good model of LOB dynamics, and especially of the best queue dynamics. When it comes to choosing between limit orders or market orders, investors face the same problem as market-makers. Sending market orders ensures immediate execution, but incurs a cost equal to half the instantaneous bid–ask spread and creates information leakage. Sending limit orders incurs uncertainty, opportunity costs and exposure to adverse selection. In terms of cost, the two types of orders are thus roughly equivalent.
(vi) Optimal execution has remained an active research topic for several decades. This sustained research activity has led to the growth of a vast literature on this topic.

21.6 Further Reading

Implementation Shortfall and Trading Costs

Perold, A. F. (1988). The implementation shortfall: Paper versus reality. *The Journal of Portfolio Management*, 14(3), 4–9.
Kissell, R. (2006). The expanded implementation shortfall: Understanding transaction cost components. *The Journal of Trading*, 1(3), 6–16.
Hendershott, T., Jones, C., & Menkveld, A. J. (2013). Implementation shortfall with transitory price effects. In Easley, D., Lopez de Prado, M, & O'Hara, M. (Eds.), *High frequency trading: New realities for trades, markets and regulators*. Risk Books.

Optimal Execution: General

Hasbrouck, J. (2007). *Empirical market microstructure: The institutions, economics, and econometrics of securities trading*. Oxford University Press.
Aldridge, I. (2009). *High-frequency trading: A practical guide to algorithmic strategies and trading systems* (Vol. 459). John Wiley and Sons.
Cartea, A., Jaimungal, S., & Penalva, J. (2015). *Algorithmic and high-frequency trading*. Cambridge University Press.
Guéant, O. (2016). *The financial mathematics of market liquidity: From optimal execution to market- making* (Vol. 33). CRC Press.

Optimal Execution: Specialised

Bertsimas, D., & Lo, A. W. (1998). Optimal control of execution costs. *Journal of Financial Markets*, 1(1), 1–50.
Almgren, R., & Chriss, N. (2001). Optimal execution of portfolio transactions. *Journal of Risk*, 3, 5–40.
Huberman, G., & Stanzl, W. (2005). Optimal liquidity trading. *Review of Finance*, 9(2), 165–200.
Kissell, R., & Malamut, R. (2006). Algorithmic decision-making framework. *The Journal of Trading*, 1(1), 12–21.
Avellaneda, M., & Stoikov, S. (2008). High-frequency trading in a limit order book. *Quantitative Finance*, 8(3), 217–224.
Schied, A., & Schöneborn, T. (2009). Risk aversion and the dynamics of optimal liquidation strategies in illiquid markets. *Finance and Stochastics*, 13(2), 181–204.
Alfonsi, A., Fruth, A., & Schied, A. (2010). Optimal execution strategies in limit order books with general shape functions. *Quantitative Finance*, 10(2), 143–157.
Kharroubi, I., & Pham, H. (2010). Optimal portfolio liquidation with execution cost and risk. *SIAM Journal on Financial Mathematics*, 1(1), 897–931.
Gatheral, J., & Schied, A. (2011). Optimal trade execution under geometric Brownian motion in the Almgren and Chriss framework. *International Journal of Theoretical and Applied Finance*, 14(03), 353–368.
Forsyth, P. A., Kennedy, J. S., Tse, S. T., & Windcliff, H. (2012). Optimal trade execution: A mean quadratic variation approach. *Journal of Economic Dynamics and Control*, 36(12), 1971–1991.
Guéant, O., Lehalle, C. A., & Fernandez-Tapia, J. (2012). Optimal portfolio liquidation with limit orders. *SIAM Journal on Financial Mathematics*, 3(1), 740–764.
Gârleanu, N., & Pedersen, L. H. (2013). Dynamic trading with predictable returns and transaction costs. *The Journal of Finance*, 68(6), 2309–2340.
Obizhaeva, A. A., & Wang, J. (2013). Optimal trading strategy and supply/demand dynamics. *Journal of Financial Markets*, 16(1), 1–32.
Schied, A. (2013). Robust strategies for optimal order execution in the Almgren-Chriss framework. *Applied Mathematical Finance*, 20(3), 264–286.
Guéant, O., & Royer, G. (2014). VWAP execution and guaranteed VWAP. *SIAM Journal on Financial Mathematics*, 5(1), 445–471.
Cartea, A., & Jaimungal, S. (2015). Optimal execution with limit and market orders. *Quantitative Finance*, 15(8), 1279–1291.
Guéant, O., & Lehalle, C. A. (2015). General intensity shapes in optimal liquidation. *Mathematical Finance*, 25(3), 457–495.
Alfonsi, A., & Blanc, P. (2016). Dynamic optimal execution in a mixed-market-impact Hawkes price model. *Finance and Stochastics*, 20(1), 183–218.
Cartea, A., & Jaimungal, S. (2016). Incorporating order-flow into optimal execution. *Mathematics and Financial Economics*, 10(3), 339–364.
Guéant, O. (2016). Optimal market-making. arXiv:1605.01862.
Curato, G., Gatheral, J., & Lillo, F. (2017). Optimal execution with non-linear transient market impact. *Quantitative Finance*, 17(1), 41–54.
Moreau, L., Muhle-Karbe, J., & Soner, H. M. (2017). Trading with small price impact. *Mathematical Finance*, 27(2), 350–400.

Absence of Price Manipulation

Gatheral, J. (2010). No-dynamic-arbitrage and market impact. *Quantitative Finance*, 10(7), 749–759.
Alfonsi, A., Schied, A., & Slynko, A. (2012). Order book resilience, price manipulation, and the positive portfolio problem. *SIAM Journal on Financial Mathematics*, 3(1), 511–533.

Gatheral, J., Schied, A., & Slynko, A. (2012). Transient linear price impact and Fredholm integral equations. *Mathematical Finance*, 22(3), 445–474.
Donier, J., Bonart, J., Mastromatteo, I., & Bouchaud, J. P. (2015). A fully consistent, minimal model for non-linear market impact. *Quantitative Finance*, 15(7), 1109–1121.

Market Orders versus Limit Orders

Handa, P., & Schwartz, R. A. (1996). Limit order trading. *The Journal of Finance*, 51(5), 1835–1861.
Harris, L., & Hasbrouck, J. (1996). Market vs. limit orders: The SuperDOT evidence on order submission strategy. *Journal of Financial and Quantitative Analysis*, 31(02), 213–231.
Handa, P., Schwartz, R. A., & Tiwari, A. (1998). The ecology of an order-driven market. *The Journal of Portfolio Management*, 24(2), 47–55.
Kaniel, R., & Liu, H. (2006). So what orders do informed traders use? *The Journal of Business*, 79(4), 1867–1913.
Wyart, M., Bouchaud, J. P., Kockelkoren, J., Potters, M., & Vettorazzo, M. (2008). Relation between bid-ask spread, impact and volatility in order-driven markets. *Quantitative Finance*, 8(1), 41–57.

Cost of Latency

Glosten, L. R. (1994). Is the electronic open limit order book inevitable? *The Journal of Finance*, 49(4), 1127–1161.
Stoikov, S. (2012). http://market-microstructure.institutlouisbachelier.org/uploads/91_3%20STOIKOV%20Microstructure_talk.pdf.
Farmer, J. D., & Skouras, S. (2013). An ecological perspective on the future of computer trading. *Quantitative Finance*, 13(3), 325–346.
Moallemi, C., & Yuan, K. (2014). The value of queue position in a limit order book. In *Market microstructure: Confronting many viewpoints*. Louis Bachelier Institute.

22
Market Fairness and Stability

When men are in close touch with each other, they no longer decide randomly and independently of each other, they each react to the others. Multiple causes come into play which trouble them and pull them from side to side, but there is one thing that these influences cannot destroy and that is their tendency to behave like Panurges sheep.

(Henri Poincaré, Comments on Bachelier's thesis)

As is stated in any book on market microstructure (including the present one!), markets must be organised such that trading is fair and orderly. This means that markets should make the best efforts to be an even playing field for all market participants, and should operate such that prices are as stable as possible. As we emphasised in Chapter 1, market stability relies heavily on the existence of liquidity providers, who efficiently buffer instantaneous fluctuations in supply and demand, and smooth out the trading process. It seems reasonable that these liquidity providers should receive some reward for stabilising markets, since by doing so they expose themselves to the risk of extreme adverse price moves. However, rewarding these liquidity providers too heavily is in direct conflict with the requirement that markets are fair. In summary, if bid–ask spreads are too small, then liquidity providers are not sufficiently incentivised, so liquidity becomes *fragile*; if bid–ask spreads are too wide, then the costs of trading become unacceptable for other investors.

The rise of electronic markets with competing venues and platforms is an elegant way to solve this dual problem, through the usual argument of competition. In a situation where all market participants can act either as liquidity providers or as liquidity takers (depending on their preferences and on market conditions), the burden of providing liquidity is shared, and bid–ask spreads should settle around fair values, as dictated by the market. As we saw in Chapter 17, this indeed seems to be the case in most modern liquid markets, such as major stock markets, futures and FX markets, in which the average bid–ask spread and the costs associated with adverse selection offset each other almost exactly. Note that this was not the case in the past: bid–ask spreads have decreased significantly during the last 30 years, from an average of 60 basis points for US stocks throughout the period 1900–1980, down to only a few basis points today (at least when the tick size is not too large).[1]

From this point of view, modern continuous-time double-auction markets have greatly decreased the economic rents available to liquidity providers. Trading costs – at least when measured using the bid–ask spread – have gone down. But how has the corresponding "collectivisation" of adverse selection influenced market stability? Has the advent of electronic markets also made markets more stable – or has the decrease in economic rents available to liquidity providers made them more fragile?

Perhaps surprisingly, empirical evidence suggests that markets are neither significantly more nor significantly less stable than they have ever been, at least seen with a long view. For example, despite major evolutions in market microstructure and the rise of algorithmic, high-frequency trading, the distribution of daily returns (including the tail region; see Section 2.2.1) has barely changed in the last century. To illustrate this point, Figure 22.1 shows the probability that any given stock in the S&P 500 has an absolute daily return $|r|$ larger than ten times the average daily volatility in the given year. It is hard to detect any significant trend in the number of market jumps between 1992 and the 2013 – i.e. between a pre-HFT era and the present HFT era. Other measures of market (in-)stability, such as the Hawkes feedback parameter g (see Chapter 9), also show remarkable stability over that period.[2]

This finding clearly runs contrary to widespread anecdotal evidence. In recent years, major financial markets have experienced several "flash crashes", which have received a great deal of attention from the media, who typically point the finger at HFT. The S&P500 flash crash of 6 May 2010 is still fresh in our memory – but who remembers that a nearly identical event with a similar speed and amplitude occurred on 28 May 1962, long before algorithms and HFT existed? While it is true that technical glitches and algorithmic bugs are at the origin of some recent hiccups, the more relevant question is whether the new ecology of markets has *structurally* created more instability.

In this discussion, one should bear in mind that the S&P volatility has fluctuated significantly in the period 1992–2013 (and, more generally, throughout the last century), with annual volatility ranging from less than 10% in calm periods such as 1965 or 2005, to more than 60% in crisis periods such as 2008. These volatility

[1] See: Jones, C. M. (2002). A century of stock market liquidity and trading costs. https://ssrn.com/abstract=313681.

[2] See: Hardiman, S., Bercot, N., & Bouchaud, J. P. (2013). Critical reflexivity in financial markets: A Hawkes process analysis. *The European Physical Journal B*, 86: 442–447.

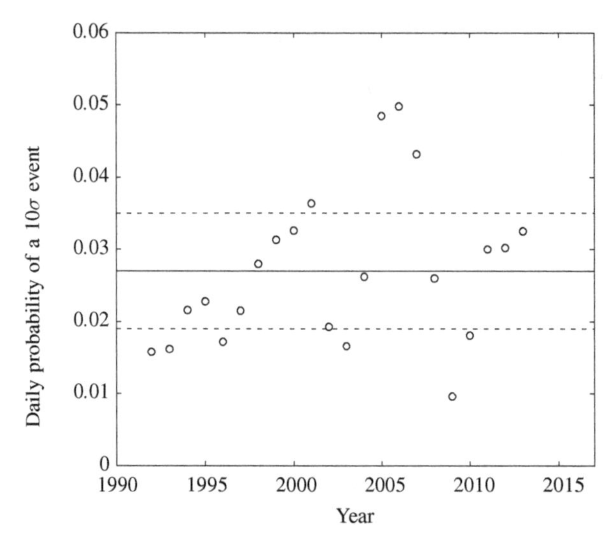

Figure 22.1. Empirical probability of a 10-σ price jump on a given day for any of the stocks in the S&P 500, from 1992 to 2013. We estimate the daily volatility σ for each stock and each year separately. The horizontal line shows the full-sample mean 0.027, and the dashed lines at ± 0.008 correspond to one standard deviation (neglecting correlations, which are in fact present and would lead to wider error-bars). There is no significant change between the pre-HFT era (before about 2000) and the explosion of HFT (after about 2008). Note that the average jump probability is compatible with a Student's-t distribution with a tail exponent μ between 3 and 4 (see Section 2.2.1).

swings are certainly very large, but it is hard to ascribe them to microstructural effects or the presence of HFT. It would be absurd to blame HFT for triggering the 2008 Great Recession (not to mention the 1929 Great Depression, which predated HFT by several decades). The worst one-day drop of the US market in 2008 was actually less than 10%, compared to more than 20% in 1929 or 1987. Ultimately, the cause of chronic volatility fluctuations and extreme tail events is still uncertain.

Why do markets have such a persistent propensity to crash, in spite of many changes in market microstructure and regulation? One possible explanation is simply that markets will always be fragile, because such fragility is an unavoidable consequence of trading itself. An alternative explanation is that we (collectively) have not yet understood the basic mechanisms that lead to instabilities. If this is the case, then more efforts in this direction may lead to the design or discovery of efficient regulation and optimal microstructure design. Our hope is that this book opens up new ways of thinking about these issues, and highlights the necessity to develop a multi-scale picture that ties together the high-frequency and low-frequency dynamics of financial markets.

22.1 Volatility, Feedback Loops and Instabilities

As we have discussed in Chapters 2 and 20, there is strong evidence to suggest that many major price changes are self-induced by markets' amplifying dynamics rather than due to large shocks affecting fundamental value. In our view, identifying the precise **feedback loop**s at the origin of extreme events and power-law tails is one of the major scientific challenges for future research. This task is of utmost importance not only from a theoretical point of view (why is the distribution of extreme events so universal?) but also from the point of view of regulation. Can one change the micro-organisation of markets so as to curb excess volatility and reduce the number of spurious price jumps, as has been argued by the proponents of a financial transaction tax (FTT)?

Market efficiency is intimately associated with the idea that markets are stable (see Section 20.3). In this view, whenever prices deviate from fundamental values, arbitrageurs are supposed to intervene and correct mispricings. Irrational behaviour can amplify pricing errors, but "big money" is assumed to be smart and not prone to irrational bouts. Because the price is supposed to reflect all news available at a given time, only exogeneous news can trigger large price moves. This point of view is very common in financial economics and implies there is little we can do to alleviate large price changes, which merely reflect fundamental changes in the valuation of assets.

A more pragmatic view is to be agnostic about the rationality of prices, and to think instead about how and why markets can go awry. After all, as with any engineering project (even if run by the brightest minds), unpleasant surprises can occur.[3] While it is undeniable that stabilising forces are at play in financial markets (e.g. more buyers enter the market when prices go down, and more sellers when prices are high), some impact-driven, destabilising feedback loops are also clearly present. Whether or not stabilising forces are always sufficient to keep Black Swans at bay should not remain a question of axioms, but should rather be answered using empirical evidence. If engineering better market stability is possible at all, then understanding the nature of these feedback loops is a prerequisite. In this section, we list some of the impact-driven feedback loops that are present in financial markets. In a nutshell, these feedback loops all lead to an acceleration of the market, as buy (respectively, sell) pressure begets buy (respectively, sell) pressure, while liquidity fails to replenish fast enough.

[3] On this point, recall the economists' 2009 "Letter to the Queen", when asked why nobody had anticipated the 2008 crisis: "In summary, Your Majesty, the failure to foresee the ... crisis and to head it off, while it had many causes, was principally a failure of the collective imagination of many bright people ... to understand the risks to the system as a whole."

22.1.1 Trend-Following

Trend-following is a severe problem for the Efficient Market Hypothesis, since its mechanism is the exact opposite of what should make markets efficient (i.e. market forces should drive prices back to their fundamental values, not amplifying mispricings).[4] However, provided trades impact prices, it is not difficult to understand how trends can emerge endogenously from behavioural biases.[5] Imagine that a fraction of the market believes that some useful information is contained in the past behaviour of prices. These investors buy when the price has gone up and sell when the price has gone down. The impact of these investors' trades will then mechanically sustain the trend. In this scenario, the only way to prevent trends from developing is for a sufficiently large fraction of mean-reverting investors to act against the trend.

While slow trend-following (on time scales of months) generates relatively few trades, so is unlikely to generate short-term price swings, fast trend-following (on time scales of minutes to hours) can potentially enter an unstable mode where trends beget trends. In fact, the way to make profit from a trend is to join in as early as possible. Market participants who do so cause the trend to accelerate. But for market-makers, trends are akin to **adverse selection** (see Section 21.3). Therefore, local trends put liquidity providers under strain and can destabilise markets, as we now discuss.

22.1.2 Market-Maker Panic

In Chapter 16, we saw that the Glosten–Milgrom framework predicts that a market breakdown can occur: when liquidity providers believe that the quantity of information revealed by trades exceeds some threshold, there is no longer any value of the bid–ask spread that allows them to break even. Whether real or perceived, the risk of adverse selection is detrimental to liquidity. This creates a clear amplification channel that can lead to liquidity crises. As the saying goes: "liquidity is a coward – it's never there when you need it".

To illustrate this point, imagine that the price has recently had a "lucky streak" of many changes in the same direction – say upwards. This creates anxiety for liquidity providers, who fear that some information about the future price, unknown to them, is the underlying reason for the recent trend. The consequence is an increased reluctance to provide liquidity on the ask side: such liquidity providers become more likely to cancel their existing limit orders and less likely to refill the LOB with new limit orders. Statistically, the ask-price is more likely

[4] One could argue that trends actually *slowly* drive prices towards fundamental values. However, this would mean that information is not quickly reflected in prices, which is problematic for the claim that markets are informationally efficient.

[5] For an explicit model, see, e.g., Wyart, M., & Bouchaud, J. P. (2007). Self-referential behaviour, overreaction and conventions in financial markets. *Journal of Economic Behavior & Organisation*, 63(1), 1–24.

to move up, reinforcing and accelerating the trend. Buyers observe the ask-price move, and are tempted to execute their orders more rapidly. They tend to cancel their limit orders on the bid side and send market orders on the ask side instead, which results in less liquidity on both sides. This self-reinforcing feedback loop can enter an unstable mode and lead to a crash. In some form or another, this scenario is probably responsible for the appearance of many price jumps that occur in the absence of any news – including the infamous flash crash of 6 May 2010 and the lesser-known flash crash of 28 May 1962.

How might this runaway trajectory be prevented? One possible answer is that some investors, who were previously waiting on the side-lines, might decide that the large price swing makes it a bargain to sell, and might thereby provide liquidity to buffer the price movement. As we argued in Chapter 18, most liquidity is latent and not present in the LOB. When the speed of the market is such that this latent liquidity cannot be transmuted into real liquidity sufficiently quickly, a crash becomes unavoidable. A possible solution is to put in place a **circuit breaker** (see Section 22.2), which allows the market to "take a breath", so that some (slow) latent liquidity can replenish the LOB and stabilise the price.

22.1.3 Deleveraging Spirals

Stop-loss strategies can also create major destabilising feedback spirals. Suppose that a trader has decided to sell his/her position if the asset price falls below a certain price level. This sale could be a voluntary decision by the trader, but could also be caused by a mandatory margin requirement imposed by a trading counterparty. If the price hits this threshold, the trader will start selling. The impact of these sales will push the price down further, possibly triggering more stop-loss orders from other investors holding similar positions. This is again a self-reinforcing mechanism that can only be stopped by a sufficient number of buyers stepping in quickly enough. If prices decline too steeply, these buyers might actually be deterred from doing so, in which case a crash is all but inevitable.

Stop-loss strategies are also implicit in option **hedging**. Consider a put option, which is an insurance contract that protects an investor against falling prices. Having sold a put option, the hedging strategy of the insurer amounts to selling when the price return is negative, potentially leading to the very same feedback loops described above. This is precisely what happened in October 1987, when $80 billion of US stocks were insured in this manner (compared to a daily turnover of just $5 billion). In other words, a mere 5% change in hedge position required the market to digest roughly 80% of the average daily liquidity, which wrought havoc for market stability.

22.1.4 Contagion

Another destabilising mechanism is **inter-market contagion**. A price-drop for asset A can create anxiety for liquidity providers on a correlated asset B, which then itself becomes prone to a large price move, even if the event concerning asset A is idiosyncratic and only weakly affects the fundamental value of B. Inter-market contagion again creates potentially unstable feedback loops, because the initial price-drop for asset A appears to be vindicated by the subsequent price-drop for asset B. This may reinforce the sell pressure on asset A, and spill over on other correlated assets as well, leading to a crash of the whole market. Empirical evidence suggests that this effect was fully at play during the May 2010 flash crash.

Correlated portfolios can also produce a more subtle type of impact-mediated contagion. If, for whatever reason, an asset manager has to liquidate his/her portfolio, the impact of the resulting trades will lower the marked-to-market valuation not only of his/her own portfolio (which might amplify the urge to liquidate), but also of all similar portfolios held by other asset managers. This cross-asset deleveraging spiral took place during the "quant crunch" of August 2007. ETF index-trading also mechanically increases inter-asset correlations.[6] The last two cases are example of so-called **crowded trades** (or "overlapping portfolios").

22.2 A Short Review of Micro-Regulatory Tools

Exchanges, trading platforms and regulators all have access to a range of tools to improve the way markets operate, in terms of both fairness and stability. Some of these tools concern the rules of trading, others concern fee structures and taxes, and yet others concern emergency measures that can be taken in crisis situations to bring markets back on an even keel. We now provide a brief discussion of some tools and topics that have been the focus of recent attention in the academic literature, and consider whether these micro-prudential solutions might alleviate some of the unstable feedback loops discussed in the previous section. As we will see, the possible consequences of applying many of these mechanisms can be surprising, such that easy solutions often have detrimental consequences.

22.2.1 Microstructural Rules

Tick Size

As we saw in Chapter 4, larger tick sizes stimulate larger volumes at the best quotes, lower cancellation rates and have a more stable LOB (small-tick assets

[6] Several studies suggest that the rise of algorithmic trading and high-frequency market-making has substantially increased the cross-correlation between assets, particularly during crisis periods. See, e.g., Gerig, A. (2015). High-frequency trading synchronizes prices in financial markets. https://ssrn.com/abstract=2173247.

typically lead to "flickering" liquidity that is much harder for both market participants and regulators to monitor). However, by making time priority more important, large tick sizes encourage liquidity providers to post orders early, thereby increasing their exposure to adverse selection. Furthermore, large spreads deter uninformed trades, so that the large tick size equilibrium is likely to be skewed towards informed traders, which can lead to a **market breakdown**, as shown in Section 16.1.4. This would go against the initial goal of improving liquidity!

From the viewpoint of market stability, even stimulating larger quantities at the best quotes is not necessarily without its drawbacks. As queue priority becomes an increasingly important consideration, liquidity providers might end up investing important resources in high-frequency technologies, so as to mitigate the risk of increased adverse selection by being able to cancel their limit orders quickly when the situation appears to deteriorate. In this case, apparently large volumes at the best quotes can be deceptive, because they are likely to evaporate when they are needed most.

Interestingly, a *decrease* in the tick size of US stocks in 2000 is generally believed to have improved liquidity and decreased investors' trading costs (although the evidence is not always compelling). In terms of market liquidity and fairness, the question of whether there exists an optimal tick size remains open.

Priority Rules

As we discussed in Section 3.2.1, different priority mechanisms encourage traders to behave in different ways. Price–time priority rewards traders for posting limit orders early, and discourages cancellations (since priority is lost), but also favours a race to low-latency trading, which many market commentators argue provides an unfair advantage to wealthy financial institutions with large sums to invest in new technology. Price–size priority favours large investors who gain priority by placing large orders, often much larger than their desired trading volume. This leads to enhanced volumes at the best quotes and possibly to a more resilient liquidity. But precisely as for large tick sizes, this increases liquidity providers' exposure to sweeping market orders and adverse selection. Qualitatively, this leads to a similar situation where informed traders benefit from high liquidity, but where displayed volumes can quickly vanish, potentially leading to price jumps.

Order Types

Most platforms offer traders the ability to submit a wide range of different order types, including iceberg, immediate-or-cancel and fill-or-kill orders (see Section 3.2.2). Although such orders are typically popular with their users, recent evidence suggests that some order types might have been at the root of some market instabilities. For example, so-called *inter-market sweep orders (ISOs)*[7] have been identified as the major source of ultra-fast flash crashes. Such events may be regarded more as technical glitches than genuine market instabilities, due to the additional complexity or even the flawed design of these instruments.

Minimum Resting Time

Some researchers have argued that limit orders should be required to remain active for some specified minimum time before they can be cancelled. Supporters of this proposal argue that it would make liquidity more stable. However, one should again remember that imposing a minimum resting time also increases the risk that a limit order is picked off by a trader with access to private information about the likely future value of the asset. Therefore, the introduction of a minimum resting time would cause the bid–ask spread to widen, to compensate for this increased level of adverse selection, and would likely discourage market participants from providing liquidity. The resulting overall impact on liquidity remains unclear.

Speed Bumps

One possible way to reduce the speed advantage of HFT is to introduce artificial speed bumps within trading platforms. For example, the trading platform IEX placed a 38-mile coil of optical fibre in front of its trading engine, to induce a 350-microsecond delay for all orders. Although IEX's market share has increased since its inception, it is still rather small (a few percent). This suggests that investors do not really feel ripped off by HFT, contrary to the wild and somewhat uninformed claim in the general press that markets are "rigged".[8]

Discrete Batch Auctions

Instead of allowing trading to occur at any moment in continuous time, some market commentators have suggested that trades should occur at random auction times – perhaps one every 100 ms on average. The aim of this approach is to mitigate the opportunities for cross-market arbitrage by HFT. From the point of view of market stability, we concluded in Chapter 18 that the amount of liquidity in the immediate vicinity of the traded price should increase when auctions occur in discrete time. However, this improvement only concerns very small volumes, less than the average volume traded in the market during the inter-auction time. To

[7] See Golub, A., Keane, J., & Poon, S. H. (2012). High-frequency trading and mini flash crashes. arXiv:1211.6667, where ISO orders, specific to the US market, are described as follows: "An ISO is a limit order designated for quick and automatic execution in a specific market center and can be executed at the target market even when another market center is publishing a better quotation. Without the need to search for the best quotation during their execution, such orders can possibly achieve faster execution than regular market orders, even though both types of orders demand liquidity and move [the] price."

[8] For two opposite viewpoints, see Lewis, M. (2014). *Flash boys: Cracking the money code*. Allen Lane; and Asness, C. S. (2014). My top 10 peeves. *Financial Analysts Journal*, 70(1), 22–30. As far as we are concerned, we believe that modern electronic markets with HFT are much less rigged now than they have been in the past.

have a substantial effect and reduce the impact of – say – buying 1% of the average daily volume, the inter-auction time would need to be of the order of minutes, not milliseconds. Therefore, trying to improve market stability through frequent batch auctions is only really useful if the frequency of market clearing is very low (perhaps once every hour or so). Clearing markets so infrequently entails many other problems, such as hampering the diffusion of information across markets and the need to synchronise these auctions across different markets – not to mention the possible development of secondary markets in which transactions would take place between these batch auctions, rendering the whole exercise of enforcing infrequent batch auctions to be redundant.

Circuit Breakers

Some platforms implement mechanisms designed to halt trading if prices change too quickly. In this way, circuit breakers aim to provide latent liquidity with the opportunity to become active. The NYSE circuit breaker has several levels, which are triggered if the price of an asset falls a given percentage below the average closing price during the previous months. If a 10% fall occurs, then the circuit breaker triggers a temporary halt. This halt lasts for one hour if it occurs before 14:00 or for 30 minutes if it occurs between 14:00 and 14:30 (no halt occurs if the circuit breaker triggers after 14:30). If a 30% fall occurs, then the circuit breaker halts trading for the rest of the day. These rules raise several questions: it is not clear whether they are in any sense optimal, nor is it clear that they do not create unintended consequences. For example, if traders fear that the circuit breaker is about to trigger, and that they will therefore be denied access to liquidity, they may create a sell pressure in an attempt to trade before the price falls further, which only accelerates the process. Obviously, circuit breakers can only work if there is some slow liquidity about to materialise (or if some rogue algorithm is unplugged). One way or the other, this proved to be true for the flash crash of 6 May 2010, since the price indeed bounced back up as soon as the trading halt was over.

22.2.2 Fees and Taxes

Maker/Taker Fees

To encourage liquidity provision and slow down liquidity consumption, many platforms charge a fee for market orders and/or offer a rebate for executed limit orders. On some platforms, these fees are inverted, such that liquidity providers pay the fee and liquidity takers receive the rebate. In either case, maker/taker fees are tantamount to changing the size of the bid–ask spread, as they effectively allow trades to occur at fractional tick values (because limit orders are effectively closer or farther away from the mid-price, depending on the fee structure). It is therefore likely that such maker/taker fees have an effect that is qualitatively similar to modifying the tick size, in addition to introducing small asymmetries between exchange platforms that so-called "smart order routers" (SOR) try to arbitrage. In practice, the effect of static maker/taker fees on liquidity must be rather small, except if behavioural biases play a large role.

Dynamical Maker/Taker Fees

Another idea worth investigating is to make maker/taker fees explicitly depend on market conditions, such as the current size of the bid–ask spread or the amplitude of recent price moves, in such a way as to promote liquidity provision more strongly when it is needed most (e.g. by strongly increasing the maker fees when the spread is large). These dynamical fees would not only concern new market and limit orders, but also pre-existing limit orders, in an attempt to reduce the cancellation rate. These rules would be endogenised by HFT/execution algorithms, and perhaps prevent self-induced short-term liquidity droughts. However, the way market participants adapt to such new rules is crucial for the success of a dynamical control of markets.

Financial Transaction Taxes (FTT)

Many market commentators have advocated the introduction of an FTT (or **Tobin Tax**[9]) as a way to "throw sand" in the wheels of financial markets (as Tobin once said) and deter short-term trading. In most proposals, the magnitude of the FTT is envisaged to be substantial (perhaps $\approx$ 10–20 basis points), and certainly much larger than the maker/taker fee. Supporters of this proposal claim that such a tax would reduce trading volumes, and hence reduce any volatility directly induced by trading.

However, this idea faces an important problem: if this tax is applied to market-makers (who by design trade extremely frequently), they might change the way that they provide liquidity. Therefore, the bid–ask spread will most likely increase to compensate for the tax. Since liquidity takers would also pay the tax, only highly informed agents would consider trading, potentially leading to a market breakdown, as suggested by the Glosten–Milgrom model (see Section 16.1.4). Clearly, this would fail to reduce volatility by any significant margin.

Several pieces of empirical evidence do not fit well with the seemingly compelling "sand in the wheels" argument. By this logic, reducing transaction costs (or reducing the bid–ask spread) should lead to an increase in volatility. Empirical evidence suggests that if anything, the opposite happens.

In recent years, there has been a huge decrease in bid–ask spreads for many stocks. In the period 1900–1980, it was common for stocks to have an average

[9] J. M. Keynes, and much later J. Tobin, were the first to propose a proportional tax, paid on all transactions.

bid–ask spread of around 60 basis points; today, these same stocks typically have a bid–ask spread of just a few basis points. This enormous change has not led to an explosion in volatility, as arguments in favour of an FTT would suggest. Moreover, several natural experiments, such as the introduction of an FTT in Sweden or the abrupt change of tick size that occurs after a stock split or when the price of some stocks crosses a specified threshold (like 10, 50 and 100 euros on the Paris Bourse) all suggest that volatility tends to increase when trading costs or the bid–ask spread increase. A cross-sectional analysis, comparing volatility to relative tick size, also points in the same direction.

The mechanism behind this effect is not well understood. This puzzle illustrates the difficulty in devising adequate regulation that avoids unintended consequences. From what we have learned throughout this book, we may put forward the following hypothesis. We have seen that the asymptotic response function for price-changing events is roughly equal to the bid–ask spread, particularly for large-tick stocks (see Figures 13.1 and 16.4). This suggests that the market interprets – rightly or wrongly – any price-changing trade as containing an amount of information proportional to the bid–ask spread itself.[10] By increasing the value of the elementary price movement, a larger bid–ask spread could bias investors' interpretations of microscopic price changes, which could then propagate to longer time horizons and contribute to a larger volatility. Of course, the relationship between spread and volatility is more complex than this simple picture (see Section 16.3 and Figure 16.6), so the above hypothesis clearly needs more scrutiny, but it provides a simple illustration of the important effects.

In summary, FTT is a prime example of a seemingly good idea that could actually lead to major detrimental effects.

22.2.3 Discussion

In view of the universality of market dynamics across different time periods, instruments and microstructures, it is not obvious that any of the above proposals can prevent markets from undergoing their frequent "liquidity tantrums". However, short of a deep understanding of the mechanisms that lead to such instabilities, one cannot exclude the possibility of some smart micro-regulation that would improve the way markets operate.

As this section has made clear, most of the effects of new rules are more subtle than what might be initially anticipated, because markets always adapt to any new, necessarily imperfect, regulation. For any new rule, one effect is certain:

[10] As we have mentioned in previous chapters, this seems to be vindicated by the fact that all trades, independently of their motivation, have roughly the same average short-term impact on prices; see Tóth, B., Eisler, Z., & Bouchaud, J. P. (2017). The short-term price impact of trades is universal. https://ssrn.com/abstract=2924029.

to make some old profit opportunities disappear in favour of new ones. Because competition is likely to have eroded the profitability of old opportunities, it might even occur that the new opportunities are even more profitable – at least during an initial transition period. It may take time for traders to fully understand new regulations, and for markets to reach competitive equilibrium.

22.3 Conclusion: In Prices We Trust?

One of the most important messages of this book is that impact transforms trades into price changes. This mechanism is necessary for prices to reflect information, but it is also a transmission belt for feedback loops.

Under normal market conditions, these feedback loops are buffered by liquidity providers, who must constantly balance the rewards provided by the bid–ask spread with the risk of being adversely selected by informed traders. Electronic markets have made it possible to spread this risk of adverse selection over a large number of agents, who compete to provide liquidity (instead of all the burden falling on a single market-maker, as happened in the past). However, when liquidity providers come to believe – rightly or wrongly – that the order flow is toxic, in the sense that the adverse selection risk is large, they tend to withdraw from the market. This causes liquidity to vanish (see Section 16.1.4).

Empirical evidence suggests that the "new" market-makers in modern financial markets all react in a similar way to risk indicators, such as short-term volatility, short-term activity bursts and short-term trends. This implies that the apparent diversification benefit of having many different liquidity providers is absent at the times when it is needed most. The reason for this relative failure is that fierce competition has driven bid–ask spreads down to levels where liquidity provision is only marginally profitable, so liquidity providers in modern markets are also extremely sensitive to risk.

Optimal situations are often fragile situations. How should liquidity provision be rewarded such that markets are maximally resilient while ensuring an even playing field? In our view, this remains an open question. The fundamental paradox is that a market is like a pyramid standing on its tip: the current price is precisely where supply and demand are at their minimum. In the stylised model of Chapters 18 and 19, the marginal supply and demand curves actually *vanish* at the traded price, so the "price" only makes sense for infinitesimally small transactions. As soon as the traded quantity becomes substantial, impact is of paramount importance.

A direct consequence of this observation is that instantaneous market prices are *not* a suitable benchmark against which to assess the value of large portfolios, as is assumed by **mark-to-market** accounting rules. At the very least, a haircut should be applied to account for the impact of liquidating a position. Using the square-root law of Chapter 12, the impact-corrected value of a 1% holding of the market capitalisation of a stock with 2% daily volatility should be ~1.5% below the market price! Therefore, using the market price to evaluate the value of a large portfolio is at best fallacious, and at worst dangerous, because doing so can contribute to deleveraging spirals and market instabilities.[11] The use of instantaneous, fragile prices as a reference for important transactions also makes markets vulnerable to a wide variety of manipulation strategies, such as pushing the daily closing price above some level to trigger a contingent claim.

Prices are fleeting quantities, hyper-sensitive to fluctuations in order flow and prone to endogenous crashes. So why do we put such blind faith in them? The belief that prices are faithful estimates of the value of assets, portfolios and firms has dominated the study of financial markets for decades. Despite widespread adoption, this approach has many unfortunate consequences, including stifling both macro-prudential research and the microstructural issues that we have addressed throughout this book.

Perhaps it is time for a better paradigm.

22.4 Further Reading

HFT, Algorithmic Trading and Market Quality

Zhang, F. (2010). High-frequency trading, stock volatility, and price discovery. https://ssrn.com/abstract=1691679.

Chordia, T., Roll, R., & Subrahmanyam, A. (2011). Recent trends in trading activity and market quality. *Journal of Financial Economics*, 101(2), 243–263.

Hendershott, T., Jones, C. M., & Menkveld, A. J. (2011). Does algorithmic trading improve liquidity? *The Journal of Finance*, 66(1), 1–33.

Amihud, Y., Mendelson, H., & Pedersen, L. H. (2012). *Market liquidity: Asset pricing, risk, and crises*. Cambridge University Press.

Golub, A., Keane, J., & Poon, S. H. (2012). High frequency trading and mini flash crashes. arXiv preprint arXiv:1211.6667.

Farmer, J. D., & Skouras, S. (2013). An ecological perspective on the future of computer trading. *Quantitative Finance*, 13(3), 325–346.

Jones, C. M. (2013). What do we know about high-frequency trading? https://ssrn.com/abstract=2236201.

Mackintosh, P. (2013). Want to hurt HFT? Careful what you wish for. Credit Suisse Market Commentary (September 2013).

Menkveld, A. J. (2013). High frequency trading and the new market-makers. *Journal of Financial Markets*, 16(4), 712–740.

Subrahmanyam, A. (2013). Algorithmic trading, the flash crash, and coordinated circuit breakers. *Borsa Istanbul Review*, 13(3), 4–9.

Biais, B., & Foucault, T. (2014). HFT and market quality. *Bankers, Markets & Investors*, 128, 5–19.

Brogaard, J., Hendershott, T., & Riordan, R. (2014). High-frequency trading and price discovery. *Review of Financial Studies*, 27(8), 2267–2306.

[11] On this point, see, e.g., Bouchaud, J. P., Caccioli, F., & Farmer, D. (2012). Impact-adjusted valuation and the criticality of leverage. *Risk*, 25(12), 74.

Easley, D., Prado, M. L. D., & O'Hara, M. (2014). *High-frequency trading: New realities for traders, markets and regulators*. Risk Books.
Boehmer, E., Fong, K. Y., & Wu, J. J. (2015). International evidence on algorithmic trading. https://ssrn.com/abstract=2022034.
Gao, C., & Mizrach, B. (2016). Market quality breakdowns in equities. *Journal of Financial Markets*, 28, 1–23.
Menkveld, A. J. (2016). The economics of high-frequency trading: Taking stock. *Annual Review of Financial Economics*, 8, 1–24.

The Flash Crash

Menkveld, A. J., & Yueshen, B. Z. (2013). Anatomy of the flash crash. Available at SSRN, 2243520.
Kirilenko, A. A., Kyle, A. S., Samadi, M., & Tuzun, T. (2016). The flash crash: High frequency trading in an electronic market. https://ssrn.com/abstract=1686004

Feedback Loops

Treynor, J. L. (1988). Portfolio insurance and market volatility. *Financial Analysts Journal*, 44(6), 71–73.
Chiarella, C., He, X. Z., & Hommes, C. (2006). Moving average rules as a source of market instability. *Physica A: Statistical Mechanics and Its Applications*, 370(1), 12–17.
Wyart, M., & Bouchaud, J. P. (2007). Self-referential behaviour, overreaction and conventions in financial markets. *Journal of Economic Behavior & Organisation*, 63(1), 1–24.
Brunnermeier, M. K., & Pedersen, L. H. (2009). Market liquidity and funding liquidity. *Review of Financial Studies*, 22(6), 2201–2238.
Marsili, M., Raffaelli, G., & Ponsot, B. (2009). Dynamic instability in generic model of multi-assets markets. *Journal of Economic Dynamics and Control*, 33(5), 1170–1181.
Bouchaud, J. P. (2011). The endogenous dynamics of markets: Price impact, feedback loops and instabilities. In Berd, A. (Ed.), *Lessons from the credit crisis*. Risk Publications.
Khandani, A. E., & Lo, A. W. (2011). What happened to the quants in August 2007? Evidence from factors and transactions data. *Journal of Financial Markets*, 14(1), 1–46.
Thurner, S., Farmer, J. D., & Geanakoplos, J. (2012). Leverage causes fat tails and clustered volatility. *Quantitative Finance*, 12(5), 695–707.
Caccioli, F., Shrestha, M., Moore, C., & Farmer, J. D. (2014). Stability analysis of financial contagion due to overlapping portfolios. *Journal of Banking & Finance*, 46, 233–245.
Corradi, F., Zaccaria, A., & Pietronero, L. (2015). Liquidity crises on different time scales. *Physical Review E*, 92(6), 062802.
Gerig, A. (2015). High-frequency trading synchronizes prices in financial markets. https://ssrn.com/abstract=2173247.
Kyle, A. S., & Obizhaeva, A. A. (2016). Large bets and stock market crashes. https://ssrn.com/abstract=2023776.
Cont, R., & Wagalath, L. (2016). Fire sales forensics: measuring endogenous risk. *Mathematical Finance*, 26(4), 835–866.
See also Section 20.5.

Regulatory Policies, FTT

Umlauf, S. R. (1993). Transaction taxes and the behavior of the Swedish stock market. *Journal of Financial Economics*, 33, 227–240.

Aliber, R. Z., Chowdhury, B., & Yan, S. (2003). Some evidence that a Tobin tax on foreign exchange transactions may increase volatility. *European Finance Review*, 7, 481–510

Harris, L. (2003). *Trading and exchanges: Market microstructure for practitioners*. Oxford University Press.

Biais, B., Glosten, L., & Spatt, C. (2005). Market microstructure: A survey of microfoundations, empirical results, and policy implications. *Journal of Financial Markets*, 8(2), 217–264.

Hau, H. (2006). The role of transaction costs for financial volatility: Evidence from the Paris Bourse. *Journal of the European Economic Association*, 4(4), 862–890.

Bouchaud, J. P., Caccioli, F., & Farmer, D. (2012). Impact-adjusted valuation and the criticality of leverage. *Risk*, 25(12), 74.

Pomeranets, A. (2012). Financial transaction taxes: International experiences, issues and feasibility. *Bank of Canada Review*, 2, 3–13.

Wang, G. H., & Yau, J. (2012). Would a financial transaction tax affect financial market activity? Insights from future markets. *Policy Analysis*, 702, 1–23.

Colliard, J. E., & Hoffmann, P. (2013). *Sand in the chips: Evidence on taxing transactions in an electronic market*. Mimeo.

Brewer, P., Cvitanic, J., & Plott, C. R. (2013). Market microstructure design and flash crashes: A simulation approach. *Journal of Applied Economics*, 16(2), 223–250.

Foucault, T., Pagano, M., & Röell, A. (2013). *Market liquidity: Theory, evidence, and policy*. Oxford University Press.

Asness, C. S. (2014). My top 10 peeves. *Financial Analysts Journal*, 70(1), 22–30.

Budish, E., Cramton, P., & Shim, J. (2015). The high-frequency trading arms race: Frequent batch auctions as a market design response. *The Quarterly Journal of Economics*, 130(4), 1547–1621.

Bonart, J. (2016). What is the optimal tick size? A cross-sectional analysis of execution costs on NASDAQ. https://ssrn.com/abstract=2869883.

Jacob Leal, S., & Napoletano, M. (2016). Market stability vs. market resilience: Regulatory policies experiments in an agent-based model with low- and high-frequency trading. https://ssrn.com/abstract=2760996.

Appendix

A.1 Description of the NASDAQ Data

Throughout the book, our empirical calculations are based on historical data that describes the LOB dynamics during the whole year 2015 for 120 liquid stocks traded on NASDAQ. On the NASDAQ platform, each stock is traded in a separate LOB with a tick size of $\vartheta = \$0.01$. The platform enables traders to submit both visible and hidden limit orders. Visible limit orders obey standard price–time priority, while hidden limit orders have lower priority than all visible limit orders at the same price. The platform also allows traders to submit mid-price-pegged limit orders. These orders are hidden, but when executed, they appear with a price equal to the national mid-price (i.e. the mid-price calculated from the national best bid and offer) at their time of execution. Therefore, although the tick size is \$0.01, some orders are executed at a price ending with \$0.005.

The data that we study originates from the LOBSTER database (see below), which lists every market order arrival, limit order arrival, and cancellation that occurs on the NASDAQ platform during 09:30–16:00 each trading day. Trading does not occur on weekends or public holidays, so we exclude these days from our analysis. When we calculate daily traded volumes or intra-day patterns, we include all activity during the full trading day. In all other cases, we exclude market activity during the first and last hour of each trading day, to remove any abnormal trading behaviour that can occur shortly after the opening auction or shortly before the closing auction.

For each stock and each trading day, the LOBSTER data consists of two different files:

- The *message* file lists every market order arrival, limit order arrival and cancellation that occurs. The LOBSTER data does not describe hidden limit order arrivals, but it does provide some details whenever a market order matches to a hidden limit order (see discussion below).

- The *orderbook* file lists the state of the LOB (i.e. the total volume of buy or sell orders at each price). The file contains a new entry for each line in the message file, to describe the market state immediately after the corresponding event occurs.

Both files are time-stamped to the nanosecond. Whenever a market order matches to several different limit orders, the LOBSTER data lists each separate matching as a new line in the message and orderbook files, but with the same time-stamp. Thanks to this extremely high time resolution, it is straightforward to infer which LOB executions correspond to which market orders. Whenever this happens, we batch these events together and sum the individual executed volumes to recover the original market order.

Although the LOBSTER data reports the arrival of market orders that match to hidden limit orders, it does not provide information about whether such orders were to buy or to sell. Therefore, whenever we calculate statistics involving the trade sign, we disregard market orders that only match to hidden limit orders, because we are unable to allocate such orders a sign.

When analysing the LOBSTER data, we perform several checks for internal consistency. For example, we confirm that the ask-price is always larger than the bid-price, and that all changes to the LOB, as listed in the orderbook file, correspond to the order flow described in the message file. In our pool of stocks, we find fewer than ten stocks with any such inconsistencies during the whole of 2015. In each case, the number of inconsistencies never exceeds one per month. Therefore, the LOBSTER database is of extremely high quality, easily rivalling the databases used by the top financial institutions in the world. Nevertheless, to ensure that these inconsistencies do not impact our results, we discard any inconsistent data points from our analysis.

A.2 Laplace Transforms and CLT

Convolutions and Laplace transforms

Let us recall Equation (5.10), valid for $V \geq 1$:

$$\Phi(\tau, V) = \int_0^\tau \mathrm{d}\tau_1 \lambda e^{-\lambda\tau_1} \Phi(\tau - \tau_1, V+1) e^{-(\mu+\nu)\tau_1} + \int_0^\tau \mathrm{d}\tau_2 (\mu+\nu) e^{-(\mu+\nu)\tau_2} \Phi(\tau - \tau_2, V-1) e^{-\lambda\tau_2}. \tag{A.1}$$

Such equations are readily solved in Laplace space (with respect to τ). Recall that the Laplace transform of $\Phi(\tau, V)$ is given by

$$\widehat{\Phi}(z, V) := \int_0^\infty \mathrm{d}\tau \Phi(\tau, V) e^{-z\tau}.$$

Table A.1. *The 120 small-, medium- and large-tick stocks that we include in our sample, along with their mean price and mean bid–ask spread during 2015. The stocks are ordered in ascending mean price.*

ticker	mean price [$]	mean spread [$]	ticker	mean price [$]	mean spread [$]
AMD	2.33	0.0110	SNDK	67.33	0.0287
SIRI	3.87	0.0108	CHRW	68.09	0.0291
BAC	16.58	0.0124	CTXS	68.75	0.0304
AMAT	19.11	0.0116	NTRS	71.82	0.0292
FITB	19.81	0.0112	VRSK	72.91	0.0408
MU	20.71	0.0124	DLTR	74.40	0.0257
SYMC	22.57	0.0115	MAR	75.55	0.0275
NVDA	23.04	0.0119	LRCX	76.06	0.0402
MAT	24.98	0.0120	TROW	77.30	0.0280
ATVI	26.47	0.0121	ADBE	80.33	0.0243
GE	26.60	0.0123	CHKP	81.92	0.0405
CSCO	27.81	0.0113	XOM	82.49	0.0139
TERP	28.23	0.0480	WDC	83.63	0.0428
JD	29.34	0.0147	FISV	83.70	0.0323
CA	29.72	0.0120	ADP	84.22	0.0230
DISCA	30.51	0.0127	WBA	85.01	0.0300
CSX	31.08	0.0123	ESRX	85.93	0.0232
FOX	31.21	0.0117	TSCO	87.30	0.0679
INTC	31.96	0.0116	FB	87.53	0.0147
PFE	33.70	0.0119	SWKS	88.38	0.0543
T	33.81	0.0116	NXPI	90.60	0.0558
VOD	34.55	0.0117	CME	93.65	0.0428
AMTD	35.48	0.0142	INTU	96.73	0.0422
TMUS	35.96	0.0165	INCY	100.40	0.1877
IBKR	37.11	0.0421	GILD	106.16	0.0258
LMCA	37.94	0.0153	EXPE	106.48	0.0746
YHOO	38.20	0.0122	QQQ	108.21	0.0124
EBAY	38.33	0.0121	SBAC	114.99	0.0811
WFM	39.34	0.0133	BMRN	115.06	0.201
MDLZ	40.48	0.0119	CELG	118.02	0.0549
ORCL	40.64	0.0119	AAPL	119.72	0.0138
FAST	40.90	0.0140	VRTX	123.14	0.166
LVNTA	41.00	0.0428	AVGO	124.10	0.085
XLNX	43.63	0.0130	NTES	127.10	0.295
LLTC	44.34	0.0135	MNST	135.37	0.118
AAL	44.74	0.0156	HSIC	143.01	0.134
MSFT	46.26	0.0118	COST	147.31	0.052
LBTYK	47.06	0.0126	ULTA	156.02	0.178
PAYX	48.91	0.0142	AMGN	157.73	0.080
LBTYA	49.34	0.0154	JAZZ	165.30	0.407

Table A.1. (*cont.*)

ticker	mean price [$]	mean spread [$]	ticker	mean price [$]	mean spread [$]
NDAQ	51.66	0.0213	ALXN	176.51	0.239
C	53.28	0.0118	PNRA	179.54	0.276
TXN	53.67	0.0124	CHTR	179.60	0.198
CINF	53.97	0.0240	BIDU	187.66	0.144
NCLH	54.32	0.0326	NFLX	188.93	0.256
ADSK	55.53	0.0201	ILMN	189.26	0.254
VIAB	55.93	0.0175	LMT	202.74	0.0986
MYL	56.10	0.0232	TSLA	227.40	0.214
KLAC	58.64	0.0234	ORLY	229.37	0.239
CMCSA	58.90	0.0120	SHPG	230.84	0.315
ADI	59.35	0.0194	EQIX	259.21	0.333
PCAR	59.45	0.0177	SHW	271.55	0.268
QCOM	61.89	0.0125	AGN	282.51	0.278
SBUX	61.93	0.0151	BIIB	344.48	0.394
CTSH	62.05	0.0166	AMZN	452.08	0.326
CTRP	62.50	0.0687	REGN	491.74	0.975
JPM	63.58	0.0124	ISRG	505.92	0.989
ROST	64.88	0.0281	GOOG	591.23	0.392
CERN	66.15	0.0259	CMG	659.40	1.054
DISH	66.57	0.0421	PCLN	1211.77	1.567

The Laplace transform of the first integral on the right-hand side of Equation (A.1) reads

$$\mathcal{L}_1 = \int_0^\infty d\tau \int_0^\tau d\tau_1 \lambda e^{-\lambda\tau_1} \Phi(\tau - \tau_1, V+1) e^{-(\mu+\nu)\tau_1} e^{-z\tau}. \tag{A.2}$$

Applying the change of variables $\tau_1 \to \tau - \tau_1$, this becomes

$$\mathcal{L}_1 = \int_0^\infty d\tau \int_0^\tau d\tau_1 \lambda e^{(\lambda+\mu+\nu)\tau_1} \Phi(\tau_1, V+1) e^{-(z+\lambda+\mu+\nu)\tau}. \tag{A.3}$$

Switching the two integrals, one obtains

$$\mathcal{L}_1 = \int_0^\infty d\tau_1 \int_{\tau_1}^\infty d\tau \lambda e^{(\lambda+\mu+\nu)\tau_1} \Phi(\tau_1, V+1) e^{-(z+\lambda+\mu+\nu)\tau}. \tag{A.4}$$

The inner integral can now be easily computed to yield

$$\mathcal{L}_1 = \frac{\lambda}{(z+\lambda+\mu+\nu)} \int_0^\infty d\tau_1 \Phi(\tau_1, V+1) e^{-z\tau_1} = \frac{\lambda}{(z+\lambda+\mu+\nu)} \widehat{\Phi}(z, V+1). \tag{A.5}$$

The second integral of Equation (A.1) can be computed in a similar way, by replacing λ with $\mu+\nu$ and $V+1$ with $V-1$. Ultimately, we obtain the following recursion equation for $V \geq 1$:

$$\widehat{\Phi}(z, V) = \frac{\lambda}{(z+\lambda+\mu+\nu)}\widehat{\Phi}(z, V+1) + \frac{\mu+\nu}{(z+\lambda+\mu+\nu)}\widehat{\Phi}(z, V-1), \quad \text{(A.6)}$$

which is exactly Equation (5.13). We may also consider the generating function in V, defined as:

$$\widetilde{\Phi}(z, y) := \sum_{V=1}^{\infty} y^V \widehat{\Phi}(z, V). \quad \text{(A.7)}$$

Multiplying both sides of Equation (A.6) and summing over V leads to:

$$y(z+\lambda+\mu+\nu)\widetilde{\Phi}(z, y) = \lambda\left[\widetilde{\Phi}(z, y) - \widehat{\Phi}(z, V=1)\right] + y^2(\mu+\nu)\widetilde{\Phi}(z, y), \quad \text{(A.8)}$$

and hence a closed-form solution for $\widetilde{\Phi}(z, y)$ as a function of y,

$$\widetilde{\Phi}(z, y) = \frac{\lambda\widehat{\Phi}(z, V=1)}{y^2(\mu+\nu) - y(z+\lambda+\mu+\nu) + \lambda}. \quad \text{(A.9)}$$

Small-z Expansion of Laplace Transforms

Let $\widehat{\Phi}(z) := \int_0^{\infty} \mathrm{d}\tau \Phi(\tau) e^{-z\tau}$ be the Laplace transform of the probability density Φ of some random variable τ. The k-th derivative of $\widehat{\Phi}(z)$ is

$$\frac{\partial^k \widehat{\Phi}}{\partial z^k}(z) = \int_0^{\infty} \mathrm{d}\tau \Phi(\tau) \frac{\partial^k}{\partial z^k} e^{-z\tau} = (-1)^k \int_0^{\infty} \mathrm{d}\tau \tau^k \Phi(\tau) e^{-z\tau}. \quad \text{(A.10)}$$

Therefore, its value at $z = 0$ is related to the kth moment of the distribution:

$$\left.\frac{\partial^k \widehat{\Phi}}{\partial z^k}\right|_{z=0} = (-1)^k \int_0^{\infty} \mathrm{d}\tau \tau^k \Phi(\tau) = (-1)^k \mathbb{E}[\tau^k]. \quad \text{(A.11)}$$

In particular, $\widehat{\Phi}(0) = 1$. The small-z expansion of $\ln\widehat{\Phi}(z)$ defines the *cumulants* of the distribution Φ. In particular:

$$\ln\widehat{\Phi}(z) \underset{z\to 0}{=} \frac{\widehat{\Phi}'(0)}{\widehat{\Phi}(0)} z + \frac{1}{2}\frac{\widehat{\Phi}''(0)\Phi(0) - \widehat{\Phi}'(0)^2}{\widehat{\Phi}(0)^2} z^2 + O(z^3). \quad \text{(A.12)}$$

By using the moment expansion obtained above, one finally obtains

$$\ln\widehat{\Phi}(z) \underset{z\to 0}{=} -\mathbb{E}[\tau]z + \frac{1}{2}\mathbb{V}[\tau]z^2 + O(z^3). \quad \text{(A.13)}$$

Linearly Growing Cumulants and the Central Limit Theorem

Let $\Phi(\tau, V)$ be the probability distribution of some random variable τ for some parameter V (here, one can think of the distribution of first-hitting times given an initial queue volume V). Assume that $\Phi(z,V)$ can be written as $\phi(z)^V$. By using $\ln\widehat{\Phi}(z,V) = V\ln\phi(z)$, one finds that the nth cumulant $c_n(V)$ of $\Phi(\tau,V)$ is equal to $V \times c_n(1)$.

Now assume that all cumulants are finite, and consider the rescaled random variable $\tilde{\tau} := \frac{\tau-\mathbb{E}[\tau]}{\sqrt{V}}$, whose cumulants are $\tilde{c}_n(V) = V^{1-n/2}c_n(1)$. One can then see that all cumulants of order $n \geq 3$ tend to 0 when $V \to \infty$, which shows that $\tilde{\tau}$ tends to a Gaussian random variable with zero mean and variance $c_2(1)$.

Laplace Transform of Power-Law Distributions

Let us consider a probability distribution with a power-law tail for large arguments:

$$\Phi(\tau) = \frac{A}{\tau^{1+\mu}} + o(\tau^{-2}), \qquad (\tau \to \infty), \tag{A.14}$$

with $0 < \mu < 1$ (i.e. in the case where the mean value of τ is infinite). What is the small-z behaviour of the Laplace transform $\widehat{\Phi}(z)$ in that case?

We first introduce an auxiliary function

$$F(\tau) := \Theta(\tau - \tau^*)\frac{A}{\tau^{1+\mu}}, \tag{A.15}$$

where τ^* is an arbitrary cut-off, below which $F(\tau) = 0$. Then, the Laplace transform $\widehat{\Phi}(z)$ can be rewritten as

$$\begin{aligned}
\widehat{\Phi}(z) &:= \int_0^\infty \mathrm{d}\tau e^{-z\tau}\Phi(\tau), \\
&= 1 + \int_0^\infty \mathrm{d}\tau [e^{-z\tau} - 1]\,\Phi(\tau), \\
&= 1 + \int_0^\infty \mathrm{d}\tau [e^{-z\tau} - 1]\,[F(\tau) + (\Phi(\tau) - F(\tau))],
\end{aligned}$$

where in the second line we have used the fact that $\int_0^\infty \mathrm{d}\tau\Phi(\tau) = 1$. Let us now analyse these different terms when $z \downarrow 0$.

The first term I_1 in the integral of the last line can be analysed at small z. Changing variable from τ to $u = z\tau$, one has

$$\int_0^\infty \mathrm{d}\tau [e^{-z\tau} - 1]\,F(\tau) = Az^\mu \int_{z\tau^*}^\infty \mathrm{d}u \frac{e^{-u} - 1}{u^{1+\mu}}. \tag{A.16}$$

Whenever $0 < \mu < 1$, the integral over u is convergent both for $u \to \infty$ and for $u \to 0$ (note that $e^{-u} - 1 \approx -u$ for $u \to 0$). Hence

$$I_1 = A\Gamma[-\mu]z^\mu + \frac{A}{1-\mu}\tau^{*1-\mu}z + o(z). \tag{A.17}$$

The second term I_2 of the integral can be estimated by noting that, by construction, $\Phi(\tau) - F(\tau)$ decays faster than τ^{-2}, such that

$$\int_0^\infty \mathrm{d}\tau\, \tau\, (\Phi(\tau) - F(\tau)) = b \tag{A.18}$$

is finite. Therefore, for $z \to 0$,

$$\int_0^\infty \mathrm{d}\tau \left[e^{-z\tau} - 1\right] (\Phi(\tau) - F(\tau)) = -bz + o(z). \tag{A.19}$$

The small-z behaviour of $\widehat{\Phi}(z)$ is thus given by

$$\widehat{\Phi}(z) = 1 + A\Gamma[-\mu] z^\mu + \left(\frac{A}{1-\mu}\tau^{*1-\mu} - b\right) z + o(z). \tag{A.20}$$

Conversely, when the Laplace transform of a distribution behaves as $1 - Cz^\mu$ for small z and $0 < \mu < 1$, this means that the distribution itself decays as a power law with exponent $1 + \mu$.

The same analysis can be redone for $n < \mu < n + 1$, where n is an integer. The result is

$$\widehat{\Phi}(z) = 1 + \sum_{k=1}^{n} \frac{(-1)^k}{k!} \mathbb{E}[\tau^k] z^k + A\Gamma[-\mu] z^\mu + O(z^{n+1}). \tag{A.21}$$

The special case where μ is an integer leads to extra terms containing $\ln z$. For example, when $\mu = 1$,

$$\widehat{\Phi}(z) = 1 - Az \ln z + O(z). \tag{A.22}$$

A.3 A Propagator Model with Volume Fluctuations

How can volume fluctuations be included in the formalism developed in Chapter 13? Since the volume of trades υ is broadly distributed, the impact of trades could itself be a highly fluctuating quantity. We have seen in Section 11.3.2 that this is in fact not so, because large trade volumes mostly occur when a comparable volume is available at the opposite-best quote, in such a way that the impact of large trades is in fact quite similar to that of small trades. Mathematically, we have seen that the average impact is a power-law function υ^ζ, or perhaps a logarithm $\log \upsilon$.

To keep calculations simple, let us postulate a logarithmic impact and a broad, log-normal distribution of υ. More precisely, we assume that the volume $\upsilon \geq \upsilon_0$ is such that:

$$P(u) = \frac{1}{\sqrt{2\pi\Sigma^2}} e^{-\frac{u^2}{2\Sigma^2}}, \tag{A.23}$$

where $u := \ln(\upsilon/\upsilon_0)$. Assume also that the impact of a trade of sign ε and volume υ is given by $z = \varepsilon \ln(\upsilon/\upsilon_0)$. The resulting impact is then a zero-mean Gaussian random variable, which inherits long-range autocorrelations from the sign process.

In line with Section 13.3, suppose now that only the surprise in u moves the price. By construction, this ensures that price returns are uncorrelated. An elegant way to write this mathematically is to express the (correlated) Gaussian variables u_t in terms of a set of auxiliary uncorrelated Gaussian variables $\widehat{u}_m$, through:

$$u_t = \sum_{m \leq t} K(t-m)\widehat{u}_m, \qquad \mathbb{E}[\widehat{u}_m \widehat{u}_{m+\ell}] = \delta_{\ell,0}, \tag{A.24}$$

where $K(\cdot)$ is a kernel solving the Yule–Walker equation such that the u_t have the required correlations:[1]

$$C(\ell) = \mathbb{E}[u_t u_{t+\ell}] = \sum_{m \geq 0} K(m+n)K(m). \tag{A.25}$$

In the case where C decays as $c_\infty \ell^{-\gamma}$ with $0 < \gamma < 1$, it is easy to show that the asymptotic decay of $K(n)$ should also be a power law $k_\infty n^{-\delta}$ with $2\delta - 1 = \gamma$ and $k_\infty^2 = c_\infty \Gamma[\delta]/\Gamma[\gamma]\Gamma[1-\delta]$. Note that $1/2 < \delta < 1$.

Inverting Equation (A.24) then leads to

$$\widehat{u}_n = \sum_{m \leq n} Q(n-m) u_m, \tag{A.26}$$

where Q is the matrix inverse of K, such that $\sum_{m=0}^{\ell} K(\ell - m)Q(m) = \delta_{\ell,0}$ for all ℓ. For a power-law kernel $K(n) \sim k_\infty n^{-\delta}$, one obtains for large n

$$Q(n) \sim (\delta - 1)\sin \pi\delta / (\pi k_\infty) n^{\delta - 2} < 0. \tag{A.27}$$

Note that whenever $\delta < 1$, one can show that $\sum_{m=0}^{\infty} Q(m) \equiv 1$.

When the u_m are known for all $m \leq t-1$, the corresponding $\widehat{u}_m$ can also be computed. The predicted value of the yet unobserved u_t is then given by

$$\mathbb{E}_{t-1}[u_t] = \sum_{m<t} K(t-m)\widehat{u}_m, \tag{A.28}$$

and the surprise in u_t is simply

$$u_t - E_{t-1}[u_t] = K(0)\widehat{u}_n. \tag{A.29}$$

The generalisation of the price equation of motion (see Equation (13.22)) is therefore

$$m_{t+1} - m_t = G(1)K(0)\widehat{u}_t + \xi_t, \tag{A.30}$$

[1] The following equation can be solved to extract $K(\ell)$ from $C(\ell)$, using the so-called Levinson–Durbin algorithm for solving Toepliz systems; see, e.g., Percival, D. B. (1992). Simulating Gaussian random processes with specified spectra. *Computer Science and Statistics*, 24, 534.

which, again by construction, removes any predictability in the price returns. From this equation of motion, one can derive $G(\ell)$ and $\mathcal{R}(\ell)$.

From the expression of the $\widehat{u}$s in terms of the us, one finds:

$$G(\ell) = G(1)K(0)\sum_{m=0}^{\ell-1} Q(m) = -G(1)K(0)\sum_{m=\ell}^{\infty} Q(m). \tag{A.31}$$

Using the asymptotic estimate of $Q(\cdot)$ given in Equation (A.27), we finally obtain

$$G(\ell) \sim_{\ell\gg 1} G(1)\frac{\sin(\pi\delta)K(0)}{\pi k_\infty}\ell^{\delta-1} := \Gamma_\infty \ell^{-\beta}. \tag{A.32}$$

Identifying the exponents leads to $\beta = 1-\delta = (1-\gamma)/2$, recovering Equation (13.17) derived in Section 13.2.1. The quantity $G(1)$, relating surprise in order flow to price changes, measures the so-called "information content" of the trades. It can be measured from empirical data using the above relation between prefactors.

Finally, from Equation (A.30), one computes the full impact function for all ℓ:

$$\mathcal{R}(\ell) = E[(m_{n+\ell} - m_n)\cdot u_n] = G(1)K(0)^2. \tag{A.33}$$

This impact function is completely flat, independent of ℓ, as in the simplified MRR model described in Section 16.2.4. However, adopting the MRR framework but assuming that the transaction price, rather than the mid-price, is impacted by the surprise in v_t, we find that the full impact function is again given by Equation (16.22), $\mathcal{R}(\ell) = G(1)[1 - C(\ell)]$, which now increases with ℓ.

In summary, all the results obtained within the context of a volume-independent impact and the DAR model can be extended to a model where log-volumes are correlated Gaussian variables.

A.4 TIM and HDIM

TIM: Variogram

We give here the (rather ugly) explicit expressions for the variogram $\mathcal{V}(\ell)$ for the TIM:

$$\begin{aligned}
\mathcal{V}(\ell) = \Sigma^2\ell &+ \sum_{0\le n<\ell}\sum_{\pi} G_\pi(\ell-n)^2\mathbb{P}(\pi) + \sum_{n>0}\sum_{\pi}[G_\pi(\ell+n) - G_\pi(n)]^2\,\mathbb{P}(\pi)\\
&+2\sum_{0\le n<n'<\ell}\sum_{\pi,\pi'}\mathbb{P}(\pi)\mathbb{P}(\pi')G_\pi(\ell-n)G_{\pi'}(\ell-n')C_{\pi,\pi'}(n'-n)\\
&+2\sum_{0<n<n'<\ell}\sum_{\pi,\pi'}\mathbb{P}(\pi)\mathbb{P}(\pi')[G_\pi(\ell+n)-G_\pi(n)][G_{\pi'}(\ell+n')-G_{\pi'}(n')]C_{\pi,\pi'}(n-n')\\
&+2\sum_{0\le n<\ell}\sum_{n'>0}\sum_{\pi,\pi'}\mathbb{P}(\pi)\mathbb{P}(\pi')G_\pi(\ell-n)[G_{\pi'}(\ell+n')-G_{\pi'}(n')]C_{\pi',\pi}(n'+n),
\end{aligned} \tag{A.34}$$

where Σ^2 is the variance of the public news term ξ_t, which is absent only when *all* price-changing events are accounted for.

HDIM: Fitting the Influence Kernel

We give here an approximate method for fitting the influence matrix kernel κ. Consider again the defining equation:

$$\Delta(\pi^*, \varepsilon_t, t) = \Delta_{\pi^*} + \sum_{t_1<t} \sum_{\pi_1} I(\pi_{t_1} = \pi_1) \kappa_{\pi_1,\pi^*}(t - t_1) \varepsilon_{t_1} \varepsilon_t + \xi'_t. \tag{A.35}$$

Multiplying this equation by $I(\pi_t = \pi^*)$ and averaging over t yields

$$\langle \Delta(\pi_t, \varepsilon_t, t) I(\pi_t = \pi^*) \rangle_t = \Delta_{\pi^*} \mathbb{P}(\pi^*) + \sum_{t_1<t} \sum_{\pi_1} \kappa_{\pi_1,\pi^*}(t - t_1) C_{\pi_1,\pi^*}(t - t_1). \tag{A.36}$$

Multiplying Equation (A.35) by $I(\pi_t = \pi^*) I(\pi_{t'} = \pi) \varepsilon_{t'} \varepsilon_t$, with $t' < t$, and again averaging over t, yields a second equation:

$$\begin{aligned} \langle \Delta(\pi_t, \varepsilon_t, t) I(\pi_t = \pi^*) I(\pi_{t'} = \pi) \varepsilon_{t'} \varepsilon_t \rangle_t &= \Delta_{\pi^*} \mathbb{P}(\pi^*) \mathbb{P}(\pi) C_{\pi,\pi^*}(t - t') \\ &+ \sum_{t_1<t} \sum_{\pi_1} \kappa_{\pi_1,\pi^*}(t - t_1) \langle I(\pi_t = \pi^*) I(\pi_{t'} = \pi) I(\pi_{t_1} = \pi_1) \varepsilon_{t_1} \varepsilon_{t'} \rangle. \end{aligned} \tag{A.37}$$

Note, however, that the last term includes a three-body correlation function, which is difficult to estimate (except when $t_1 = t'$). At this stage, we need to make some approximation to estimate these higher-order correlations.[2] We assume that all three-body correlation functions can be factorised in terms of two-body correlation functions. More precisely, for $t_1 \neq t'$:

$$\langle I(\pi_t = \pi^*) I(\pi_{t'} = \pi) I(\pi_{t_1} = \pi_1) \varepsilon_{t_1} \varepsilon_{t'} \rangle \approx \mathbb{P}(\pi^*) \mathbb{P}(\pi_1) \mathbb{P}(\pi) C_{\pi_1,\pi}(t' - t_1). \tag{A.38}$$

This allows us to extract κ_{π,π^*} from a numerically convenient but approximate expression:

$$\begin{aligned} \frac{\langle \Delta(\pi_t, \varepsilon_t, t) I(\pi_t = \pi^*) I(\pi_{t-\ell} = \pi) \varepsilon_{t-\ell} \varepsilon_t \rangle_t}{\mathbb{P}(\pi^*) \mathbb{P}(\pi)} &= \Delta_{\pi^*} C_{\pi,\pi^*}(\ell) + \\ + \sum_{n>-\ell}{}' \sum_{\pi_1} \mathbb{P}(\pi_1) \kappa_{\pi_1,\pi^*}(n + \ell) C_{\pi_1,\pi}(n) &+ \kappa_{\pi,\pi^*}(\ell) \frac{\langle I(\pi_t = \pi^*) I(\pi_{t-\ell} = \pi) \rangle}{\mathbb{P}(\pi^*) \mathbb{P}(\pi)}, \end{aligned} \tag{A.39}$$

where $\sum'$ indicates that the term $n = 0$ is skipped. Since the left-hand side of this equation can be readily estimated from the data, one can use the last equation to infer (approximately) the influence kernels.

[2] Very recently, F. Patzelt has shown how to estimate these three-point correlation functions from data. See: Patzelt, F., & Bouchaud, J. P. (2017). Nonlinear price impact from linear models. *Journal of Statistical Mechanics*, 12, 123404.

A.5 An Alternative Market-Making Strategy

The calculation given in Section 17.2 is based on the assumption that the market-maker's (MM) inventory follows an Ornstein–Uhlenbeck type of dynamics driven by the order flow. This approach leads to interesting predictions, but it is based on an uncontrolled approximation. To overcome this limitation, we present an alternative argument. Let us consider the following simple MM strategy, where the inventory ψ_t *after* trade t can only be $\psi_t = 0$ or $\psi_t = \pm 1$ (i.e. the MM can only sell when he/she has bought, and vice-versa).

Recall that the sign of the tth trade is ε_t. In this simple MM strategy, this leads to the following dynamical equation for ψ_t:

$$\psi_t = -(1 - \psi_{t-1}^2)\varepsilon_t + \psi_{t-1}\frac{1 - \varepsilon_t\psi_{t-1}}{2}. \tag{A.40}$$

The first term of Equation (A.40) means that if the previous inventory was flat, the next inventory becomes the opposite of the sign of the trade (i.e. selling to buyers and vice-versa). The second term means that if the MM is, say, long one stock, the inventory stays long one stock if there is a sell trade is coming, but becomes flat if a buy trade is coming (in which case $\varepsilon_t\psi_{t-1} = 1$).

The trick now is to see that ψ_t can be rewritten as $\psi_t = -\theta_t\varepsilon_t$ with $\theta_t = 0$ or 1. Equation (A.40) then reads:

$$\theta_t = \left(1 - \frac{\theta_{t-1}}{2}\right) + \frac{\theta_{t-1}}{2}\varepsilon_t\varepsilon_{t-1}, \tag{A.41}$$

which can be rewritten as

$$\psi_t = \eta_t(1 - \psi_{t-1}); \qquad \text{with} \quad \psi_t = 1 - \theta_t; \quad \eta_t = \frac{1}{2}(1 - \varepsilon_t\varepsilon_{t-1}), \tag{A.42}$$

The explicit solution is thus simply

$$\psi_t = \eta_t - \eta_t\eta_{t-1} + \eta_t\eta_{t-1}\eta_{t-2} + \cdots, \tag{A.43}$$

from which one can compute the fraction of the time f the MM is either long or short (i.e. $f = \langle\theta\rangle$), since:

$$1 - f = \langle\psi_t\rangle = \langle\eta_t\rangle - \langle\eta_t\eta_{t-1}\rangle + \langle\eta_t\eta_{t-1}\eta_{t-2}\rangle + \cdots. \tag{A.44}$$

The interpretation is quite clear: $\langle\eta_t\rangle$ is the probability that η is equal to 1, given by $(1 - C(1))/2$. The term $\langle\eta_t\eta_{t-1}\ldots\eta_{t-k+1}\rangle$ is the probability that ε_t, ε_{t-1}, ... ε_{t-k} are all of alternating sign. For the Markovian MRR model, this occurs with probability $((1 - C(1))/2)^{k-1}$, so one gets

$$1 - f = \frac{1}{2}(1 - C(1))\frac{1}{1 + (1 - C(1))/2} = \frac{1 - C(1)}{3 - C(1)} \Rightarrow f = \frac{2}{3 - C(1)}. \tag{A.45}$$

This approximation is actually numerically very good in practical cases even when the sign evolution is not Markovian.

The total P&L of the MM is (minus) the sum over all trades of the price paid times the variation of the position, i.e.:

$$\mathcal{G} = -\sum_{t=1}^{T}(\psi_t - \psi_{t-1}) \cdot (m_t + \frac{\varepsilon_t s_t}{2}) \approx -T\left\langle (\psi_t - \psi_{t-1}) \cdot \left(m_t + \frac{\varepsilon_t s_t}{2}\right)\right\rangle, \tag{A.46}$$

where s_t is the spread before the t-th trade. Using the evolution equation above, one finds:

$$\psi_t - \psi_{t-1} = -\varepsilon_t(1 - \frac{\psi_t^2}{2}) - \frac{\psi_{t-1}}{2}, \tag{A.47}$$

and thus, neglecting spread fluctuations and rearranging terms:

$$\frac{1}{T}\mathcal{G} = \langle \psi_t \cdot (m_{t+1} - m_t)\rangle + \frac{s}{2}\left\langle\left(1 - \frac{1}{2}\psi_t^2 + \frac{\psi_{t-1}\varepsilon_t}{2}\right)\right\rangle. \tag{A.48}$$

The second term is clearly positive, and is the gain coming from the spread. The first term is negative because if $\psi_t = -1$, then ε_t must have been $+1$, and the price has on average gone up. Using the explicit solution for ψ_t, one finds:

$$\langle \psi_t \cdot (m_{t+1} - m_t)\rangle = -\langle(\eta_t - \eta_t\eta_{t-1} + \eta_t\eta_{t-1}\eta_{t-2} + \ldots.) \cdot \varepsilon_t \cdot (m_{t+1} - m_t)\rangle, \tag{A.49}$$

which can be expanded as

$$\begin{aligned}\langle \psi_t \cdot (m_{t+1} - m_t)\rangle = &-(1 - \langle\eta_t\rangle)\langle\varepsilon_t \cdot (m_{t+1} - m_t)\rangle_{\varepsilon_t\varepsilon_{t-1}=1}\\ &-\langle\eta_t\eta_{t-1}\rangle\langle\varepsilon_t \cdot (m_{t+1} - m_t)\rangle_{\varepsilon_t\varepsilon_{t-1}=-1\,\&\,\varepsilon_{t-1}\varepsilon_{t-2}=-1}\\ &+\langle\eta_t\eta_{t-1}\eta_{t-2}\rangle\langle\varepsilon_t \cdot (m_{t+1} - m_t)\rangle_{\varepsilon_t\varepsilon_{t-1}=-1\,\&\,\varepsilon_{t-1}\varepsilon_{t-2}=-1\,\&\,\varepsilon_{t-2}\varepsilon_{t-3}=-1} + \cdots.\end{aligned} \tag{A.50}$$

In general, this expression must be computed numerically. However, since the probability of observing a strictly alternating series decreases quickly, keeping only the first three or four terms is enough to retain good accuracy.

Alternatively, one can find a Markovian approximation for this term as well. Indeed, for the Markovian MRR model, conditioning on past signs is irrelevant, so all the quantities $\langle\varepsilon_t \cdot (m_{t+1} - m_t)\rangle_{\ldots}$ are equal to $\mathcal{R}(1) := \langle\varepsilon_t \cdot (m_{t+1} - m_t)\rangle$, hence the above equation simplifies to

$$\langle \psi_t \cdot (m_{t+1} - m_t)\rangle = -(1 - \langle\psi_t\rangle)\mathcal{R}(1) = -f\mathcal{R}(1). \tag{A.51}$$

This leads to the following P&L of the MM:

$$\frac{1}{T}\mathcal{G} = -f\mathcal{R}(1) + \frac{s}{2}\left(1 - \frac{f}{2} + \left\langle\frac{\psi_{t-1}\varepsilon_t}{2}\right\rangle\right). \tag{A.52}$$

One can also obtain the following equalities:

$$\langle \varepsilon_t \psi_t \rangle = -f; \tag{A.53}$$

$$\langle \psi_t \varepsilon_{t+1} \rangle = 2 - 3f; \tag{A.54}$$

$$\langle \psi_{t+1} \varepsilon_t \rangle = \frac{1}{2}(C(1) - 1)f - C(1); \tag{A.55}$$

$$\langle \psi_{t+1} \psi_t \rangle = 2f - 1. \tag{A.56}$$

Note the second equality imposes $f \geq 1/3$ and allows one to compute the average gain of the MM:

$$\frac{1}{T}\mathcal{G} = -f\mathcal{R}(1) + \frac{s}{2}\left(1 - \frac{f}{2} + \frac{2-3f}{2}\right) = -f\mathcal{R}(1) + s(1-f). \tag{A.57}$$

This result is quite transparent: since f is the fraction of the time the MM is either long or short, he/she can only benefit from the bid–ask spread a fraction $1 - f$ of the time but suffers from adverse impact a fraction f of the time.

In the MRR model, one furthermore has: $\mathcal{R}(1) = (1 - C(1))s/2$ and $f = 2/(3 - C(1))$, so one finally finds that the gain of the MM is in fact exactly 0, as it should be.

A.6 Acronyms, Conventions and Symbols

Acronyms

ARCH: Auto-Regressive Conditional Heteroskedasticity
ACF: Auto-Correlation Function
CIR: Cox–Ingersoll–Ross
CLT: Central Limit Theorem
DAR: Discrete Auto-Regressive
ECDF: Empirical Cumulative Density Function
FGLW: Farmer–Gerig–Lillo–Waelbroeck
FTT: Financial Transaction Tax
HDIM: History-Dependent Impact Models
HFT: High-Frequency Traders (or Trading)
IID: Independent Identically Distributed
LMF: Lillo–Mike–Farmer
LOB: Limit Order Book
LSE: London Stock Exchange
MLE: Maximum Likelihood Estimator (or Estimation)
MM: Market-Maker
MRR: Madhavan–Richardson–Roomans
MSD: Marginal Supply and Demand

NYSE: New York Stock Exchange
ODE: Ordinary Differential Equation
P&L: Profits and Losses
PDE: Partial Differential Equation
PP: Point Process
PPP: Poisson Point Process
SD: Supply and Demand
TIM: Temporary Impact Model
TWAP: Time-Weighted Average Price
VAR: Vector Auto-Regressive
VWAP: Volume-Weighted Average Price
ZI: Zero-Intelligence

Conventions

Alice: an informed trader
Bob: a market-maker
$\sim$: scales as, of the order of
$\approx$: approximately equal to (mathematically or numerically)
$\cong$: approximately equal to (empirically)
$\propto$: proportional to
$:=$: equal by definition to
$\equiv$: identically equal to
$t' \uparrow t$: limit as t' tends to t from below
$t' \downarrow t$: limit as t' tends to t from above
$\overline{f}(t) = \lim_{t' \downarrow t} f(t')$
$\underline{f}(t) = \lim_{t' \uparrow t} f(t')$
$\mathbb{E}[.]$: mathematical expectation
$\langle . \rangle$: empirical average
$[x]$: dimension of x
$\widehat{\Phi}$: Laplace transform of Φ
$\widetilde{\Phi}$: generating function of Φ

Roman Symbols

A: generic constant
$\mathcal{A}$: ask side of the LOB
a: ask-price
$a(\cdot)$: generic function
B: generic constant

$\mathcal{B}$: bid side of the LOB
b: bid-price
$C(\cdot)$: autocorrelation function (of returns, of the sign of the trades, etc.)
$\mathscr{C}$: execution cost (or shortfall)
CA^0: non-price-changing cancellation
CA^1: price-changing cancellation
$c(\cdot)$: rescaled covariance
c_∞: constant describing the asymptotic decay of the sign autocorrelation
D: diffusion coefficient
$D(\cdot)$: volume-dependent diffusion coefficient (queue models)
$\mathcal{D}(\cdot)$: lag-dependent diffusion coefficient
$\mathscr{D}(p,t)$: demand curve
d: distance from the same-side best quote
$d^\dagger$: distance from the opposite-side best quote
F: drift
$F(\cdot)$: volume-dependent drift (queue models)
$\mathcal{F}_t$: filtration at time t
$\mathscr{F}$: scaling function
f: generic fraction, or participation rate
$f(\cdot)$: generic function, or generic distribution density
$G(\cdot)$: lag-dependent propagator
$G_1 := G(1)$: lag-1 propagator
G_0, G^*: impact parameters
$\mathcal{G}$: gain (of a strategy)
g: feedback parameter in the Hawkes model
$\mathcal{H}(\cdot)$: kernels in Hasbrouck's model
h: position of an order in a queue ("height" of the order)
I: volume imbalance
$\mathcal{I}$: impact of an order
$\mathfrak{I}(Q,T)$: impact of a metaorder of volume Q executed over time T
$\mathbb{I}$: indicator function
i: generic integer index
J: exit flux (or current)
$\mathcal{J}$: total reinjection current in an extended Fokker–Planck equation
$j_0 := Q/T$: trading intensity
$j(\cdot)$: volume per unit time sent to the market (or trading rate)
$K(\cdot)$: discrete derivative of the propagator $G(\cdot)$
$\mathbb{K}(\cdot)$: DAR kernel
k: generic integer
$\mathcal{L}$: liquidity parameter in the square-root law/latent liquidity model
$\mathscr{L}(t)$: the state of the limit order book at time t

$\mathbb{L}$: likelihood function
LO^0: non-price-change limit order
LO^1: price-changing limit order
ℓ: discrete time lag
ℓ_i: size of the ith gap in the LOB
MO^0: non-price-changing market order
MO^1: price-changing market order
$\mathcal{M}$: total market capitalisation
m: mid-price
N: generic integer (number of agents, number of trades, number of events, etc.)
n: generic integer
$P(\cdot)$: probability distribution
$P_{\text{st.}}(\cdot)$: stationary probability distribution
$\mathbb{P}[.]$: probability of the event
p: generic price
p_{F}: fundamental price
$\widehat{p}$: predicted price
p^*: Walrasian auction price
p_+: probability that the ask-queue empties first, or probability that the next trade is a buy
p_-: probability that the bid-queue empties first, or probability that the next trade is a sell
Q: total volume of a metaorder
q: partial volume of a metaorder executed so far
$R(x,t)$: "reaction term", between buy and sell orders
$\mathcal{R}$: response function
$\mathcal{R}^0$: response function of a non-price-changing market order
$\mathcal{R}^1$: response function of a price-changing market order
$\mathcal{R}_\phi$: response function of a market order consuming more than a fraction ϕ of the volume at best
$\mathbb{R}$: aggregate impact
r: generic ratio
r_t: price change between t and $t+1$
S: size of a family in a branching process
$S(\cdot)$: return response function
$\mathcal{S}(p,t)$: supply curve
s: bid–ask spread
T: time horizon (execution time, etc.)
T_m: memory time
T_1: first-hitting time
t: time

$\mathcal{U}$: utility function
$U(\cdot)$: generic function
$u := V/\bar{V}$: queue volume rescaled by its average value
u, v, x, y: generic variables
$V_+(p)$: volume in the buy order book at price p
$V_-(p)$: volume in the sell order book at price p
V: total volume in a queue
$V_{a,b}$: volume at ask (respectively, bid)
V^*: most probable volume in a queue/volume cleared in an auction
$\bar{V}$: average volume in a queue
V_T: total exchanged volume within a time interval T
$\mathcal{V}$: price variogram
$\mathbb{V}[.]$: variance operator
υ: volume of a single order
υ_0: elementary volume (or lot size)
υ_t: volume of a single transaction at time t
$v_t := \varepsilon_t \upsilon_t$: signed volume of an order at time t
$W(t)$: Wiener noise
$W_\pm$: hopping rate in queue models
$\mathcal{W}(\cdot)$: effective "potential"
w: generic width
w_i: financial weight of agent i
$x := (\varepsilon_x, p_x, \upsilon_x, t_x)$: order specification
Y: numerical prefactor of the square-root law
$\mathcal{Y}$: scaling function
Z: normalisation
z: Laplace variable

Greek Symbols

α: parameter often associated to a decay, or to the strength of a predictor
β: exponent describing the decay of the propagator, or of a kernel
β^i: response of agent i to news
$\Gamma[.]$: Gamma function
Γ: risk-aversion parameter
Γ_∞: constant describing the asymptotic decay of the propagator
γ: exponent describing the decay of the autocorrelation of signs
δ: exponent describing the dependence of impact on volume; or difference between the price and the fundamental price; or inter-arrival times
$\delta\cdot$: difference
$\delta(\cdot)$: Dirac delta-function

Δ: step size (gap in the LOB), or implementation shortfall
$\Delta\cdot$: difference
ϵ: small quantity
ε: sign of a transaction
$\widehat{\varepsilon}$: predicted sign of a transaction
$\varepsilon^{\dagger}$: sign of the "side" of an event (bid or ask)
ζ: generic exponent or generic Lagrange parameter
η: Ornstein–Uhlenbeck noise
θ: polar angle; or binary variable: $\theta = 0, 1$
$\boldsymbol{\theta}$: set of parameters of a model
Θ: Heaviside function, $\Theta(x) := x^{+}$
ϑ: tick size (absolute)
ϑ_r: relative tick size
κ: decay rate, or inverse relaxation time
$\kappa_{\pi,\pi'}(\cdot)$: kernels in HDIM models
Λ: Kyle's impact parameter
λ: deposition rate
μ: market order rate or exponent describing the power-law tail of return distributions
Ξ: filter kernel
ξ: IID random noise
Π: probability of a sweeping market order
π: type of event
ϖ: transaction cost or rebate
ρ: correlation coefficient, or autoregression parameter
$\rho^{\pm}(p,t)$: marginal supply and demand curves
$\varrho(\cdot)$: distribution of initial volumes
Σ: generic root-mean-square (news term, volume of noise traders, etc.)
σ: volatility, or root-mean-square
σ_{τ} or $\sigma(\tau)$: volatility on scale T
σ_{F}: volatility of the fundamental price
ς: skewness
τ: time lag
τ_{liq}: time for liquidity to refill
$\Phi(\tau, V)$: probability of first-hitting time for a queue of size V
$\Phi(u)$: Hawkes kernel
ϕ: probability, fraction, or rate
ϕ_0: refill probability after depletion
$\phi_k(\cdot)$: auxiliary function
φ: rate per unit time (e.g. of Poisson events)

χ: generic exponent or constant
$\chi(\cdot)$: auxiliary function
Ψ: kernel of the generalised Hawkes model; or correlation-induced contribution to $\mathcal{V}$ in the propagator model
ψ: angle, or position of the market-maker
Ω, ω: inverse time scales (rates)

Index

Note

Words with an asterix are too frequent to be exhaustively referenced in the Index. Only the first page where they appear is given.

CPSIA information can be obtained
at www.ICGtesting.com
Printed in the USA
LVHW061014180922
728646LV00009B/716